STUDY GUIDE AND SELECTED SOLUTIONS MANUAL

Karen Timberlake
Los Angeles Valley College

Mark Quirie
Algonquin College

CHEMISTRY

An Introduction to General, Organic, and Biological Chemistry

Thirteenth Edition

330 Hudson Street, NY NY 10013

Courseware Portfolio Manager: Scott Dustan
Director, Courseware Portfolio Management: Jeanne Zalesky
Content Producer: Melanie Field
Managing Producer: Kristen Flathman
Courseware Analyst: Coleen Morrison
Courseware Director, Content Development: Jennifer Hart
Courseware Editorial Assistant: Fran Falk
Full Service Vendor: SPi Global
Manufacturing Buyer: Stacey Weinberger
Cover Image Credit: © Ralph Clevenger/Corbis Documentary/Getty Images

Copyright © 2018, 2017, 2011, 2008, 2005 Pearson Education, Inc. All rights reserved. Manufactured in the United States of America. This publication is protected by Copyright and permission should be obtained from the publisher prior to any prohibited reproduction, storage in a retrieval system, or transmission in any form or by any means, electronic, mechanical, photocopying, recording, or otherwise. For information regarding permissions, request forms and the appropriate contacts within the Pearson Education Global Rights & Permissions department, please visit www.pearsoned.com/permissions/.

This work is solely for the use of instructors and administrators in teaching courses and assessing student learning. Unauthorized dissemination, publication or sale of the work, in whole or in part (including posting on the internet) will destroy the integrity of the work and is strictly prohibited.

PEARSON, ALWAYS LEARNING and MasteringChemistry are exclusive trademarks in the U.S. and/or other countries owned by Pearson Education, Inc. or its affiliates.

Unless otherwise indicated herein, any third-party trademarks that may appear in this work are the property of their respective owners and any references to third-part trademarks, logos or other trade dress are for demonstrative or descriptive purposes only. Such references are not intended to imply any sponsorship, endorsement, authorization or promotion of Pearson's products by the owners of such marks, or any relationship between the owner and Pearson Education, Inc. or its affiliates, authors, licensees or distributors.

ISBN 10: 0-134-55398-5
ISBN 13: 978-0-134-55398-6

www.pearsonhighered.com

10 2023

Contents

Preface		v
To the Student		vi
Using This Study Guide and Selected Solutions Manual		vii
Instructor and Student Supplements		viii

1	**Chemistry in Our Lives**	1
	Study Guide	
	Selected Answers and Solutions to Text Problems	12
2	**Chemistry and Measurements**	15
	Study Guide	
	Selected Answers and Solutions to Text Problems	33
3	**Matter and Energy**	45
	Study Guide	
	Selected Answers and Solutions to Text Problems	60
	Selected Answers to Combining Ideas from Chapters 1 to 3	72
4	**Atoms and Elements**	73
	Study Guide	
	Selected Answers and Solutions to Text Problems	88
5	**Nuclear Chemistry**	96
	Study Guide	
	Selected Answers and Solutions to Text Problems	107
6	**Ionic and Molecular Compounds**	113
	Study Guide	
	Selected Answers and Solutions to Text Problems	137
	Selected Answers to Combining Ideas from Chapters 4 to 6	148
7	**Chemical Quantities and Reactions**	149
	Study Guide	
	Selected Answers and Solutions to Text Problems	169
8	**Gases**	182
	Study Guide	
	Selected Answers and Solutions to Text Problems	194
9	**Solutions**	201
	Study Guide	
	Selected Answers and Solutions to Text Problems	219
	Selected Answers to Combining Ideas from Chapters 7 to 9	227
10	**Acids and Bases and Equilibrium**	229
	Study Guide	
	Selected Answers and Solutions to Text Problems	246
11	**Introduction to Organic Chemistry: Hydrocarbons**	254
	Study Guide	
	Selected Answers and Solutions to Text Problems	2

Copyright © 2018 Pearson Education, Inc.

Contents

12 Alcohols, Thiols, Ethers, Aldehydes, and Ketones 279
  Study Guide
  Selected Answers and Solutions to Text Problems 292
  Selected Answers to Combining Ideas from Chapters 10 to 12 297

13 Carbohydrates 298
  Study Guide
  Selected Answers and Solutions to Text Problems 311

14 Carboxylic Acids, Esters, Amines, and Amides 316
  Study Guide
  Selected Answers and Solutions to Text Problems 331

15 Lipids 338
  Study Guide
  Selected Answers and Solutions to Text Problems 353
  Selected Answers to Combining Ideas from Chapters 13 to 15 359

16 Amino Acids, Proteins, and Enzymes 360
  Study Guide
  Selected Answers and Solutions to Text Problems 372

17 Nucleic Acids and Protein Synthesis 377
  Study Guide
  Selected Answers and Solutions to Text Problems 389

18 Metabolic Pathways and ATP Production 394
  Study Guide
  Selected Answers and Solutions to Text Problems 410
  Selected Answers to Combining Ideas from Chapters 16 to 18 415

Preface

This *Study Guide and Selected Solutions Manual* is intended to accompany *Chemistry: An Introduction to General, Organic, and Biological Chemistry,* thirteenth edition. Our purpose is to provide you with additional learning resources to increase your understanding of the key concepts in the text. Each chapter correlates with a chapter in the text. For every Section in each chapter, there are Learning Exercises and Answers that focus on problem solving, which in turn promote an understanding of the chemical principles. A Checklist of chapter concepts and a multiple-choice Practice Test provide a review of the chapter content. In the Selected Solutions, we have included the answers and solutions to all the odd-numbered problems in the text.

We hope that this Study Guide and Selected Solutions Manual will help in the learning of chemistry. If you wish to make comments or corrections, or ask questions, you can send us an e-mail message at *khemist@aol.com.*

Karen C. Timberlake
Los Angeles Valley College

Mark Quirie
Algonquin College

To the Student

> One must learn by doing the thing;
> though you think you know it, you
> have no certainty until you try.
> —*Sophocles*

Here you are in a chemistry class with your textbook in front of you. Perhaps you have already been assigned some reading or some problems to do in the book. Looking through the chapter, you may see words, terms, and pictures that are new to you. This may very well be your first experience with a science class like chemistry. At this point you may have some questions about what you can do to learn chemistry. This *Study Guide and Selected Solutions Manual* is written with those considerations in mind.

Learning chemistry is similar to learning something new such as tennis or skiing or driving. If I asked you how you learn to play tennis or ski or drive a car, you would probably tell me that you would need to practice often. It is the same with learning chemistry; understanding the chemical ideas and successfully solving the problems depends on the time and effort you invest in it. If you practice every day, you will find that learning chemistry is an exciting experience and a way to understand the current issues of the environment, health, and medicine.

Manage Your Study Time

I often recommend a study system to students in which you read one section of the text and immediately practice the questions and problems that are at the end of that section. In this way, you concentrate on a small amount of information and use what you learned to answer questions. This helps you to organize and review the information without being overwhelmed by the entire chapter. It is important to understand each section, because they build like steps. Information presented in each chapter proceeds from the basic to the more complex. Perhaps you can only study three or four sections of the chapter. As long as you also practice doing some problems at the same time, the information will stay with you.

Some Strategies for Learning Chemistry

Learning chemistry requires us to place new information in our long-term memory, which allows us to remember those ideas for an exam, a process called retrieval. We can develop new study habits that help us successfully recall new information by connecting it with our prior knowledge. This can be accomplished by doing a lot of practice testing. We can check how much we have learned by going back a few days later and retesting. Another useful learning strategy is to study different ideas at the same time, which allows us to connect those ideas and to differentiate between them. Although these study habits may take more time and seem more difficult, they help us find the gaps in our knowledge and connect new information with what we already know.

Form a Study Group

I highly recommend that you form a study group in the first week of your chemistry class. Working with your peers will help you use the language of chemistry. Scheduling a time to meet each week will help you study and prepare to discuss problems. You will be able to teach some things to other students in the group, and sometimes they will help you understand a topic that puzzles you. You won't always understand a concept right away. Your group will help you see your way through it. Most of all, a study group creates a strong support system when students help each other successfully complete the class.

Go to Office Hours

Try to go to your tutor's and/or professor's office hours. Your professor wants you to understand and enjoy learning this material. Often a tutor is assigned to a class or there are tutors available at your college. Don't be intimidated. Going to see a tutor or your professor is one of the best ways to clarify what you need to learn in chemistry.

Using This Study Guide and Selected Solutions Manual

Now you are ready to sit down and study chemistry. Let's go over some methods that can help you learn chemistry. This *Study Guide* is written specifically to help you understand and practice the chemical concepts that are presented in your class and in your text. The following features are part of this *Study Guide and Selected Solutions Manual:*

1. Learning Goals
The Learning Goals give you an overview of what each section in the chapter is about and what you can expect to accomplish when you complete your study and learning of that section.

2. Key Terms
Key Terms appear throughout each chapter in the Study Guide. As you review the description of the Key Terms, you will have an overview of the topics you have studied in that chapter. Because many of the Key Terms may be new to you, this is an opportunity to review their meaning.

3. Chapter Sections
Each chapter section begins with a list of the important ideas to guide you through each of the learning activities. When you are ready to begin your study, read the matching section in the textbook and review the *Sample Problems* in the text. Included are the *Key Math Skills* and *Core Chemistry Skills.*

4. Learning Exercises
The Learning Exercises give you an opportunity to practice problem solving related to the chemical principles in the chapter. Each set of Learning Exercises reviews one chemical principle. The answers are found immediately following each Learning Exercise. Check your answers right away. If they don't match the answer in the Study Guide, review that section of the text again. It is important to make corrections before you go on. Chemistry involves a layering of skills such that each one must be understood before the next one can be learned.

5. Checklist
Use the Checklist to check your understanding of the Learning Goals. This Checklist gives you an overview of the major topics in each section. If something does not sound familiar, go back and review it. One aspect of being a strong problem-solver is the ability to evaluate your knowledge and understanding as you study.

6. Practice Test
A Practice Test is found at the end of each chapter. When you feel you have learned the material in a chapter, you can check your understanding by taking the Practice Test. The Practice Test questions are keyed to the Learning Goals, which allows you to identify the sections you may need to review. Answers are found at the end of the Practice Test. If the results of the Practice Test indicate that you know the material, you are ready to proceed to the next chapter.

7. Selected Answers and Solutions
The Selected Answers and Solutions to the odd-numbered problems for each chapter in the text follow each Practice Test for that chapter. For certain chapters, Selected Answers to the Combining Ideas for groups of chapters are included.

Instructor and Student Supplements

Chemistry: An Introduction to General, Organic, and Biological Chemistry, thirteenth edition, provides an integrated teaching and learning package of support material for both students and professors.

Name of Supplement	Available in Print	Available Online	Instructor or Student Supplement	Description
Study Guide and Selected Solutions Manual (9780134553986)	✓		Supplement for Students	The *Study Guide and Selected Solutions Manual*, by Karen Timberlake and Mark Quirie, promotes active learning through a variety of exercises with answers as well as practice tests that are connected directly to the learning goals of the textbook. Complete solutions to odd-numbered problems are included.
MasteringChemistry™ (www.masteringchemistry.com) (9780134551272)		✓	Supplement for Students and Instructors	This product includes all of the resources of MasteringChemistry™ plus the now fully mobile eText 2.0. eText 2.0 mobile app offers offline access and can be downloaded for most iOS and Android phones/tablets from the Apple App Store or Google Play. Added integration brings videos and other rich media to the student's reading experience. MasteringChemistry™ from Pearson is the leading online homework, tutorial, and assessment system, designed to improve results by engaging students with powerful content. Instructors ensure students arrive ready to learn by assigning educationally effective content and encourage critical thinking and retention with in-class resources such as Learning Catalytics™. Students can further master concepts through traditional and adaptive homework assignments that provide hints and answer specific feedback. The Mastering™ gradebook records scores for all automatically-graded assignments in one place, while diagnostic tools give instructors access to rich data to assess student understanding and misconceptions. http://www.masteringchemistry.com.
Pearson eText enhanced with media (stand-alone: ISBN 9780134545684; within MasteringChemistry™: 9780134552170)		✓	Supplement for Students	The thirteenth edition of *Chemistry: An Introduction to General, Organic, and Biological Chemistry* features a Pearson eText enhanced with media within Mastering. In conjunction with Mastering assessment capabilities, new **Interactive Videos and 3D animations** will improve student engagement and knowledge retention. Each chapter contains a balance of interactive animations, videos, sample calculations, and self-assessments / quizzes embedded directly in the eText. Additionally, the Pearson eText offers students the power to create notes, highlight text in different colors, create bookmarks, zoom, and view single or multiple pages. Icons in the margins throughout the text signify that there is a new **Interactive Video** or animation located within MasteringChemistry™ for *Chemistry: An Introduction to General, Organic, and Biological Chemistry*, thirteenth edition.
Laboratory Manual by Karen Timberlake (9780321811851)	✓		Supplement for Students	This best-selling lab manual coordinates 35 experiments with the topics in *Chemistry: An Introduction to General, Organic, and Biological Chemistry*, thirteenth edition, uses laboratory investigations to explore chemical concepts, develop skills of manipulating equipment, reporting data, solving problems, making calculations, and drawing conclusions.
Instructor's Solutions Manual (9780134564661)		✓	Supplement for Instructors	Prepared by Mark Quirie, the Instructor's solutions manual highlights chapter topics, and includes answers and solutions for all Practice Problems in the text.

Instructor and Student Supplements

Name of Supplement	Available in Print	Available Online	Instructor or Student Supplement	Description
Instructor Resource Materials–Download Only (9780134552262)		✓	Supplement for Instructors	Includes all the art, photos, and tables from the book in JPEG format for use in classroom projection or when creating study materials and tests. In addition, the instructors can access modifiable PowerPoint™ lecture outlines. Also available are downloadable files of the Instructor's Solutions Manual and a set of "clicker questions" designed for use with classroom-response systems. Also visit the Pearson Education catalog page for Timberlake's *Chemistry: An Introduction to General, Organic, Biological Chemistry,* thirteenth edition, at www.pearsonhighered.com to download available instructor supplements.
TestGen Test Bank-Download Only (9780134564678)		✓	Supplement for Instructors	Prepared by William Timberlake, this resource includes more than 1600 questions in multiple-choice, matching, true / false, and short-answer format.
Online Instructor Manual for Laboratory Manual (9780321812858)		✓	Supplement for Instructors	This manual contains answers to report sheet pages for the *Laboratory Manual* and a list of the materials needed for each experiment with amounts given for 20 students working in pairs, available for download at www.pearsonhighered.com.

1
Chemistry in Our Lives

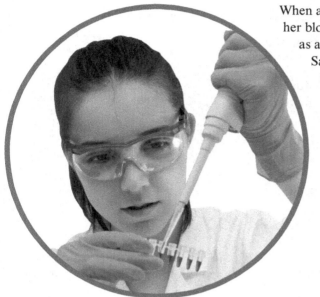

When a female victim was found dead in her home, samples of her blood and stomach contents were sent to Sarah, who works as a forensic scientist. After using a variety of chemical tests, Sarah concluded that the victim was poisoned when she ingested ethylene glycol. Since the initial symptoms of ethylene glycol poisoning are similar to those of alcohol intoxication, the victim was unaware of the poisoning. When ethylene glycol is oxidized, the products can cause kidney failure and may be toxic to the body. Two weeks later, police arrested the victim's husband after finding a bottle of antifreeze containing ethylene glycol in the laundry room of the couple's home. How does this result show the use of the scientific method?

Credit: nyaivanova / Shutterstock

LOOKING AHEAD

1.1 Chemistry and Chemicals
1.2 Scientific Method: Thinking Like a Scientist
1.3 Studying and Learning Chemistry
1.4 Key Math Skills for Chemistry
1.5 Writing Numbers in Scientific Notation

 The Health icon indicates a question that is related to health and medicine.

1.1 Chemistry and Chemicals

Learning Goal: Define the term chemistry and identify substances as chemicals.

- Chemistry is the study of the composition, structure, properties, and reactions of matter.
- A chemical is any substance that always has the same composition and properties wherever it is found.

♦ **Learning Exercise 1.1**

Is each of the following a chemical?

a. _____ ascorbic acid (vitamin C)
b. _____ time for a radioisotope to lose half its activity
c. _____ sodium fluoride in toothpaste
d. _____ amoxicillin, an antibiotic

Chapter 1

e. _____ distance of 2.5 mi walked on a treadmill during exercise

f. _____ carbon nanotubes used to transport medication to cancer cells

Answers a. yes b. no c. yes d. yes e. no f. yes

1.2 Scientific Method: Thinking Like a Scientist

Learning Goal: Describe the activities that are part of the scientific method.

- The scientific method is a process of explaining natural phenomena beginning with making observations, forming a hypothesis, and performing experiments.
- When the results of the experiments are analyzed, a conclusion is made as to whether the hypothesis is *true* or *false*.
- After repeated successful experiments, a hypothesis may become a scientific theory.

The scientific method develops a conclusion or theory using observations, hypotheses, and experiments.

♦ **Learning Exercise 1.2A**

Identify each of the following as an observation, a hypothesis, an experiment, or a conclusion:

a. _____ Sunlight is necessary for the growth of plants.

b. _____ Plants in the shade are shorter than plants in the sun.

c. _____ Plant leaves are covered with aluminum foil and their growth is measured.

d. _____ Fertilizer is added to plants.

e. _____ Ozone slows plant growth by interfering with photosynthesis.

f. _____ Ozone causes brown spots on plant leaves.

Answers a. hypothesis b. observation c. experiment
          d. experiment e. conclusion f. observation

♦ **Learning Exercise 1.2B**

Identify each of the following as an observation, a hypothesis, an experiment, or a conclusion:

a. _____ One half-hour after eating a wheat roll, Cynthia experiences stomach cramps.

b. _____ The next morning, Cynthia eats a piece of toast and one half-hour later has an upset stomach.

c. _____ Cynthia thinks she may be gluten intolerant.

d. _____ Cynthia tries a gluten-free roll and does not have any stomach cramps.

e. _____ Because Cynthia does not have an upset stomach after eating a gluten-free roll, she believes that she is gluten intolerant.

Answers a. observation b. observation c. hypothesis
          d. experiment   e. conclusion

1.3 Studying and Learning Chemistry

Learning Goal: Identify strategies that are effective for learning. Develop a study plan for learning chemistry.

- Components of the text that promote learning include: *Looking Ahead, Learning Goals, Review, Key Math Skills, Core Chemistry Skills, Try It First, Connect, Engage, Test, Analyze the Problem, Sample Problems, Solution Guides, Study Checks with Answers, Chemistry Link to Health, Chemistry Link to the Environment, Practice Problems, Clinical Applications, Interactive Videos, Clinical Updates, Concept Maps, Chapter Reviews, Key Terms, Review of Key Math Skills, Review of Core Chemistry Skills, Understanding the Concepts, Additional Problems, Challenge Problems, Answers to Practice Problems, Combining Ideas,* and *Glossary/Index.*
- A plan for learning chemistry utilizes the features in the text that help develop a successful approach to learning chemistry.

♦ **Learning Exercise 1.3**

Which of the following activities would be included in a successful study plan for learning chemistry?

a. ____ attending class occasionally
b. ____ working problems with friends from class
c. ____ attending review sessions
d. ____ planning a regular study time
e. ____ not doing the assigned problems
f. ____ visiting the instructor during office hours
g. ____ attempting to work a Sample Problem before checking the Solution
h. ____ asking yourself questions as you read
i. ____ testing yourself often by working Practice Problems

Studying in a group can be beneficial to learning.
Credit: Chris Schmidt/Schmidt/E+/Getty Images

Answers   **a.** no   **b.** yes   **c.** yes   **d.** yes   **e.** no
            f. yes   **g.** yes   **h.** yes   **i.** yes

1.4 Key Math Skills for Chemistry

Learning Goal: Review math concepts used in chemistry: place values, positive and negative numbers, percentages, solving equations, and interpreting graphs.

- Basic math skills are important in learning chemistry.
- In a number, we identify the place value of each digit.
- A positive number is greater than zero and has a positive sign (+); a negative number is less than zero and has a negative sign (−).
- A percentage is calculated as the parts divided by the whole, then multiplied by 100%.
- An equation is solved by rearranging it to place the unknown value on one side.
- A graph represents the relationship between two variables, which are plotted on perpendicular axes.

Chapter 1

> **Study Note**
>
> On your calculator, there are four keys that are used for basic mathematical operations. The change sign $\boxed{+/-}$ key is used to change the sign of a number.
>
> To practice these basic calculations on the calculator, work through the problem going from left to right doing the operations in the order they occur. If your calculator has a change sign $\boxed{+/-}$ key, a negative number is entered by entering the number and then pressing the change sign $\boxed{+/-}$ key. At the end, press the equals $\boxed{=}$ key or ANS or ENTER.
>
> **Addition and Subtraction**
>
> **Example 1:** $15 - 8 + 2 =$
> **Solution:** $15 \boxed{-} 8 \boxed{+} 2 \boxed{=} 9$
> **Example 2:** $4 + (-10) - 5 =$
> **Solution:** $4 \boxed{+} 10 \boxed{+/-} \boxed{-} 5 \boxed{=} -11$
>
> **Multiplication and Division**
>
> **Example 3:** $2 \times (-3) =$
> **Solution:** $2 \boxed{\times} 3 \boxed{+/-} \boxed{=} -6$
> **Example 4:** $\dfrac{8 \times 3}{4} =$
> **Solution:** $8 \boxed{\times} 3 \boxed{\div} 4 \boxed{=} 6$

♦ **Learning Exercise 1.4A**

KEY MATH SKILL
Identifying Place Values

a. Identify the place value for each of the digits in the number 825.10.

Digit	Place Value
8	
2	
5	
1	
0	

Answer

Digit	Place Value
8	hundreds
2	tens
5	ones
1	tenths
0	hundredths

b. A tablet contains 0.325 g of aspirin. Identify the place value for each of the digits in the number 0.325.

Digit	Place Value
3	
2	
5	

Answer

Digit	Place Value
3	tenths
2	hundredths
5	thousandths

♦ **Learning Exercise 1.4B**

KEY MATH SKILL
Using Positive and Negative Numbers in Calculations

Evaluate each of the following:

a. $\dfrac{-14 + 22}{-4}$

b. $\dfrac{-3 + (-15)}{3}$

c. $\dfrac{-2 \times 10}{-5}$

Answers   a. -2 b. -6 c. 4

Chemistry in Our Lives

♦ **Learning Exercise 1.4C**

KEY MATH SKILL
Calculating Percentages

a. There are 20 patients on the 3rd floor of a hospital. If 4 patients were discharged today, what percentage of patients were discharged?

b. In a clinical trial with 120 participants, 36 responded to the new drug. What percentage of the participants responded?

c. Of the 12 pairs of socks in a drawer, 3 pairs are white. What percentage of the socks are white?

Answers   a. 20%   b. 30%   c. 25%

Study Note

1. Solve an equation for a particular variable by performing the same mathematical operation on *both* sides of the equation.
2. If you eliminate a symbol or number by subtracting, you need to subtract that same symbol or number on the opposite side.
3. If you eliminate a symbol or number by adding, you need to add that same symbol or number on the opposite side.
4. If you cancel a symbol or number by dividing, you need to divide both sides by that same symbol or number.
5. If you cancel a symbol or number by multiplying, you need to multiply both sides by that same symbol or number.

♦ **Learning Exercise 1.4D**

KEY MATH SKILL
Solving Equations

Solve each of the following equations for y:

a. $2y + 14 = -2$ b. $3y - 8 = 22$ c. $-3y + 12 = -6$

Answers   a. $y = -8$ b. $y = 10$ c. $y = 6$

♦ Learning Exercise 1.4E

The dilution equation shows the relationship between the concentration (C) and volume (V) of a solution.

$$C_1V_1 = C_2V_2$$

Solve the dilution equation for each of the following:

a. C_1 **b.** V_2 **c.** V_1

Answers **a.** $C_1 = \dfrac{C_2V_2}{V_1}$ **b.** $V_2 = \dfrac{C_1V_1}{C_2}$ **c.** $V_1 = \dfrac{C_2V_2}{C_1}$

♦ Learning Exercise 1.4F

KEY MATH SKILL
Interpreting Graphs

Use the following graph to answer questions **a** and **b**:

Temperature Versus Time for Cooling of Coffee

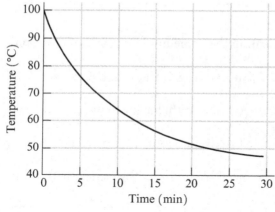

a. How many minutes elapse when coffee cools from 100 °C to 65 °C?

b. What is the temperature after the coffee has cooled for 15 min?

Answers

a. Temperature Versus Time for Cooling of Coffee

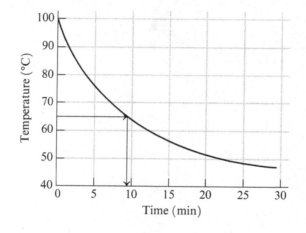

Draw a horizontal line from 65 °C on the temperature axis to intersect the curved line. Then drop a vertical line to the time axis and read the time where it crosses the x axis, which is estimated to be 9 min.

b.

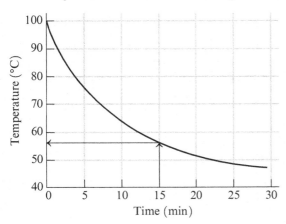

Draw a vertical line from 15 min on the time axis to intersect the curved line. Draw a horizontal line from the intersect to the temperature axis and read the temperature where it crosses the y axis, which is estimated to be 57 °C.

♦ Learning Exercise 1.4G

Use the following graph to answer questions **a** and **b**:

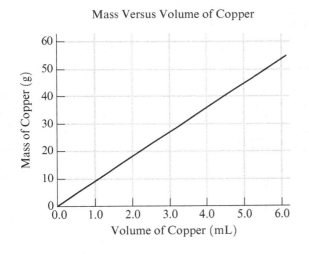

a. What is the volume, in milliliters, of 35 g of copper?

b. What is the mass, in grams, of 5 mL of copper?

Answers

a.

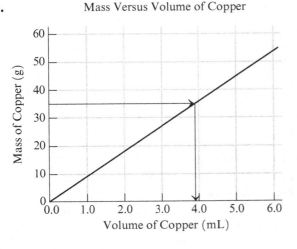

Draw a horizontal line from 35 g on the mass axis to intersect the line. Then drop a vertical line to the volume axis and read the volume where it crosses the x axis, which is estimated to be 3.9 mL.

Chapter 1

b.

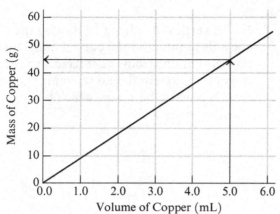
Mass Versus Volume of Copper

Draw a vertical line from 5.0 mL on the volume axis to intersect the line. Draw a horizontal line from the intersect to the mass axis and read the mass where it crosses the *y* axis, which is estimated to be 45 g.

1.5 Writing Numbers in Scientific Notation

Learning Goal: Write a number in scientific notation.

- Large and small numbers can be written using scientific notation in which the decimal point is moved to give a coefficient and the number of spaces moved is shown as a power of 10.
- A large number will have a positive power of 10, while a small number will have a negative power of 10.
- For numbers greater than 10, the decimal point is moved to the left to give a positive power of 10.
- For numbers less than 1, the decimal point is moved to the right to give a negative power of 10.

Key Terms for Sections 1.1 to 1.5

Match each of the following key terms with the correct description:

    a. scientific method   **b.** experiment   **c.** scientific notation
    d. conclusion/theory   **e.** chemistry   **f.** hypothesis

1. ____ an explanation of nature validated by many experiments
2. ____ the study of substances and how they interact
3. ____ a possible explanation of a natural phenomenon
4. ____ the process of making observations, writing a hypothesis, and testing with experiments
5. ____ a procedure used to test a hypothesis
6. ____ a form of writing large and small numbers using a coefficient that is at least 1 but less than 10, followed by a power of 10

Answers   **1.** d   **2.** e   **3.** f   **4.** a   **5.** b   **6.** c

Study Note

1. For a number greater than 10, the decimal point is moved to the left to give a number that is at least 1 but less than 10 and a positive power of 10. For a number less than 1, the decimal point is moved to the right to give a coefficient that is at least 1 but less than 10 and a negative power of 10.
2. To express 2500 in scientific notation, we can write it as 2.5×1000. Because 1000 is the same as 10^3, we can express 2500 in scientific notation as 2.5×10^3.
3. To express 0.082 in scientific notation, we move the decimal point two places to the right and write 8.2×0.01. Because 0.01 is the same as 10^{-2}, we can express 0.082 in scientific notation as 8.2×10^{-2}.

Chemistry in Our Lives

Solution Guide to Writing a Number in Scientific Notation	
STEP 1	Move the decimal point to obtain a coefficient that is at least 1 but less than 10.
STEP 2	Express the number of places moved as a power of 10.
STEP 3	Write the product of the coefficient multiplied by the power of 10.

♦ **Learning Exercise 1.5A**

KEY MATH SKILL
Writing Numbers in Scientific Notation

Write each of the following numbers in scientific notation:

a. 24 100 _____   b. 825 _____
c. 230 000 _____   d. 53 000 000 _____
e. 0.002 _____   f. 0.000 001 5 _____
g. 0.08 _____   h. 0.000 24 _____

Answers a. 2.41×10^4 b. 8.25×10^2 c. 2.3×10^5 d. 5.3×10^7
e. 2×10^{-3} f. 1.5×10^{-6} g. 8×10^{-2} h. 2.4×10^{-4}

♦ **Learning Exercise 1.5B**

Which number in each of the following pairs is larger?

a. 2500 or 2.5×10^2 _____   b. 0.04 or 4×10^{-3} _____
c. 65 000 or 6.5×10^5 _____   d. 0.000 35 or 3.5×10^{-3} _____
e. 300 000 or 3×10^6 _____   f. 0.002 or 2×10^{-4} _____

Answers a. 2500   b. 0.04   c. 6.5×10^5
d. 3.5×10^{-3} e. 3×10^6 f. 0.002

Checklist for Chapter 1

You are ready to take the Practice Test for Chapter 1. Be sure you have accomplished the following learning goals for this chapter. If not, review the Section listed at the end of the goal. Then apply your new skills and understanding to the Practice Test.

After studying Chapter 1, I can successfully:

_____ Identify a substance as a chemical. (1.1)

_____ Describe the steps of the scientific method. (1.2)

_____ Identify strategies that are effective for learning. (1.3)

_____ Develop a study plan for successfully learning chemistry. (1.3)

_____ Use math skills needed for chemistry: place values, positive and negative numbers, percentages, solving equations, and interpreting graphs. (1.4)

_____ Write numbers in scientific notation using a coefficient and a power of 10. (1.5)

Copyright © 2018 Pearson Education, Inc.

Chapter 1

Practice Test for Chapter 1

The chapter Sections to review are shown in parentheses at the end of each question.

1. Which of the following would be described as a chemical? (1.1)
   - A. sleeping
   - B. salt
   - C. singing
   - D. listening to a concert
   - E. energy

2. Which of the following is not a chemical? (1.1)
   - A. aspirin
   - B. sugar
   - C. feeling cold
   - D. glucose
   - E. vanilla

For questions 3 through 7, identify each statement as an observation (O), a hypothesis (H), an experiment (E), or a conclusion (C): (1.2)

3. ____ More sugar dissolves in 50 mL of hot water than in 50 mL of cold water.

4. ____ Samples containing 20 g of sugar each are placed separately in a glass of cold water and a glass of hot water.

5. ____ Sugar consists of white crystals.

6. ____ Water flows downhill due to gravity.

7. ____ Drinking 10 glasses of water a day will help me lose weight.

For questions 8 through 12, answer yes or no. (1.3)

To learn chemistry, I will:

8. ____ work Practice Problems as I read each Section of a chapter

9. ____ attend some classes but not all

10. ____ try to work each Sample Problem before I look at the Solution

11. ____ have a regular study time

12. ____ wait until the night before the exam to start studying

13. Evaluate $\frac{22 + 23}{-4 + 1}$. (1.4)
    - A. 5
    - B. −5
    - C. −7
    - D. −15
    - E. 15

14. In a big cats exhibit at the zoo, there are 8 lions, 6 tigers, and 2 leopards. What percentage of the animals in the exhibit are lions? (1.4)
    - A. 50%
    - B. 30%
    - C. 20%
    - D. 10%
    - E. 5%

15. The contents found in a container at a victim's home were identified as 65.0 g of ethylene glycol in 182 g of solution. What is the percentage of ethylene glycol in the solution? (1.4)
    - A. 15.3%
    - B. 26.3%
    - C. 35.7%
    - D. 48.0%
    - E. 65.0%

16. For breakfast, a person eats 55 g of eggs, 65 g of banana, and 85 g of milk. What percentage of the breakfast was milk? (1.4)
    - A. 85%
    - B. 65%
    - C. 41%
    - D. 32%
    - E. 27%

17. Solve for the value of a in the equation $5a + 10 = 30$. (1.4)
    - A. 2
    - B. 4
    - C. 8
    - D. 20
    - E. 40

18. Use the following graph to determine the solubility of carbon dioxide in water, in g CO_2/100 g water, at 40 °C: (1.4)

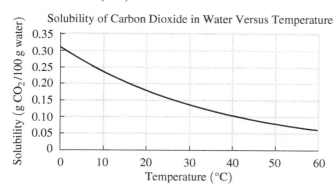

    A. 0.10   **B.** 0.15   **C.** 0.20   **D.** 0.25   **E.** 0.30

19. The number 42 000 written in scientific notation is (1.5)
    A. 42   **B.** 42×10^3 **C.** 4.2×10^3 **D.** 4.2×10^{-3} **E.** 4.2×10^4

20. The number 0.005 written in scientific notation is (1.5)
    A. 5   **B.** 5×10^{-3} **C.** 5×10^{-2} **D.** 0.5×10^{-4} **E.** 5×10^3

Answers to the Practice Test

1. B	**2.** C	**3.** O	**4.** E	**5.** O
6. C	**7.** H	**8.** yes	**9.** no	**10.** yes
11. yes	**12.** no	**13.** D	**14.** A	**15.** C
16. C	**17.** B	**18.** A	**19.** E	**20.** B

Chapter 1

Selected Answers and Solutions to Text Problems

1.1 **a.** Chemistry is the study of the composition, structure, properties, and reactions of matter.
   b. A chemical is a substance that has the same composition and properties wherever it is found.

1.3 Many chemicals are listed on a vitamin bottle such as vitamin A, vitamin B_3, vitamin B_{12}, vitamin C, and folic acid.

1.5 Typical items found in a medicine cabinet and some of the chemicals they contain are as follows:
   Antacid tablets: calcium carbonate, cellulose, starch, stearic acid, silicon dioxide
   Mouthwash: water, alcohol, thymol, glycerol, sodium benzoate, benzoic acid
   Cough suppressant: menthol, beta-carotene, sucrose, glucose

1.7 **a.** An observation is a description or measurement of a natural phenomenon.
   b. A hypothesis proposes a possible explanation for a natural phenomenon.
   c. An experiment is a procedure that tests the validity of a hypothesis.
   d. An observation is a description or measurement of a natural phenomenon.
   e. An observation is a description or measurement of a natural phenomenon.
   f. A conclusion is an explanation of an observation that has been validated by repeated experiments that support a hypothesis.

1.9 **a.** An observation is a description or measurement of a natural phenomenon.
   b. A hypothesis proposes a possible explanation for a natural phenomenon.
   c. An experiment is a procedure that tests the validity of a hypothesis.
   d. An experiment is a procedure that tests the validity of a hypothesis.

1.11 There are several things you can do that will help you successfully learn chemistry, including forming a study group, going to class, asking yourself questions as you read new material in the textbook, working *Sample Problems* and *Study Checks*, working *Practice Problems* and checking *Answers*, reading the assignment ahead of class, self-testing during and after reading each Section, retesting on new information a few days later, going to the instructor's office hours, and keeping a problem notebook.

1.13 Ways you can enhance your learning of chemistry include:
   a. forming a study group.
   c. asking yourself questions while reading the text.
   e. answering the Engage questions.

1.15 **a.** The bolded 8 is in the thousandths place.
   b. The bolded 6 is in the ones place.
   c. The bolded 6 is in the hundreds place.

1.17 **a.** $15 - (-8) = 15 + 8 = 23$
   b. $-8 + (-22) = -30$
   c. $4 \times (-2) + 6 = -8 + 6 = -2$

1.19 **a.** $\dfrac{21 \text{ flu shots}}{25 \text{ patients}} \times 100\% = 84\%$ received flu shots
   b. total grams of alloy = 56 g silver + 22 g copper = 78 g of alloy
   $\dfrac{56 \text{ g silver}}{78 \text{ g alloy}} \times 100\% = 72\%$ silver
   c. total number of coins = 11 nickels + 5 quarters + 7 dimes = 23 coins
   $\dfrac{7 \text{ dimes}}{23 \text{ coins}} \times 100\% = 30\%$ dimes

Chemistry in Our Lives

1.21 a.
$$4a + 4 = 40$$
$$4a + \cancel{4} - \cancel{4} = 40 - 4$$
$$4a = 36$$
$$\frac{\cancel{4}a}{\cancel{4}} = \frac{36}{4}$$
$$a = 9$$

b.
$$\frac{a}{6} = 7$$
$$\cancel{6}\left(\frac{a}{\cancel{6}}\right) = 6(7)$$
$$a = 42$$

1.23 a. The graph shows the relationship between the temperature of a cup of tea and time.
b. The vertical axis measures temperature, in °C.
c. The values on the vertical axis range from 20 °C to 80 °C.
d. As time increases, the temperature decreases.

1.25 a. Move the decimal point four places to the left to give 5.5×10^4.
b. Move the decimal point two places to the left to give 4.8×10^2.
c. Move the decimal point six places to the right to give 5×10^{-6}.
d. Move the decimal point four places to the right to give 1.4×10^{-4}.
e. Move the decimal point three places to the right to give 7.2×10^{-3}.
f. Move the decimal point five places to the left to give 6.7×10^5.

1.27 a. 7.2×10^3, which is also 72×10^2, is larger than 8.2×10^2.
b. 3.2×10^{-2}, which is also 320×10^{-4}, is larger than 4.5×10^{-4}.
c. 1×10^4 or 10 000 is larger than 1×10^{-4} or 0.0001.
d. 6.8×10^{-2} or 0.068 is larger than 0.000 52.

1.29 $\dfrac{120 \text{ g ethylene glycol}}{450 \text{ g liquid}} \times 100\% = 27\%$ ethylene glycol

1.31 No. All of these ingredients are chemicals.

1.33 Yes. Sherlock's investigation includes making observations (gathering data), formulating a hypothesis, testing the hypothesis, and modifying it until one of the hypotheses is validated.

1.35 a. When two negative numbers are added, the answer has a negative sign.
b. When a positive and negative number are multiplied, the answer has a negative sign.

1.37 a. Describing the appearance of a patient is an observation.
b. Formulating a reason for the extinction of dinosaurs is a hypothesis.
c. Measuring the completion time of a race is an observation.

1.39 If experimental results do not support your hypothesis, you should:
b. modify your hypothesis.
c. do more experiments.

1.41 A successful study plan would include:
b. working the *Sample Problems* as you go through a chapter.
c. self-testing.

1.43 a. $4 \times (-8) = -32$
b. $-12 - 48 = -12 + (-48) = -60$
c. $\dfrac{-168}{-4} = 42$

Chapter 1

1.45 total number of gumdrops = 16 orange + 8 yellow + 16 black = 40 gumdrops

a. $\dfrac{8 \text{ yellow gumdrops}}{40 \text{ total gumdrops}} \times 100\% = 20\%$ yellow gumdrops

b. $\dfrac{16 \text{ black gumdrops}}{40 \text{ total gumdrops}} \times 100\% = 40\%$ black gumdrops

1.47
a. Move the decimal point five places to the left to give 1.2×10^5.
b. Move the decimal point seven places to the right to give 3.4×10^{-7}.
c. Move the decimal point two places to the right to give 6.6×10^{-2}.
d. Move the decimal point three places to the left to give 2.7×10^3.

1.49
a. An observation is a description or measurement of a natural phenomenon.
b. A hypothesis proposes a possible explanation for a natural phenomenon.
c. A conclusion is an explanation of an observation that has been validated by repeated experiments that support a hypothesis.

1.51
a. An observation is a description or measurement of a natural phenomenon.
b. A hypothesis proposes a possible explanation for a natural phenomenon.
c. An experiment is a procedure that tests the validity of a hypothesis.
d. A conclusion is an explanation of an observation that has been validated by repeated experiments that support a hypothesis.

1.53
a.
$$2x + 5 = 41$$
$$2x + \cancel{5} - \cancel{5} = 41 - 5$$
$$2x = 36$$
$$\dfrac{2x}{\cancel{2}} = \dfrac{36}{2}$$
$$x = 18$$

b.
$$\dfrac{5x}{3} = 40$$
$$\cancel{3}\left(\dfrac{5x}{\cancel{3}}\right) = 3(40)$$
$$5x = 120$$
$$\dfrac{\cancel{5}x}{\cancel{5}} = \dfrac{120}{5}$$
$$x = 24$$

1.55
a. The graph shows the relationship between the solubility of carbon dioxide in water and temperature.
b. The vertical axis measures the solubility of carbon dioxide in water ($g\ CO_2/100\ g$ water).
c. The values on the vertical axis range from 0 to $0.35\ g\ CO_2/100\ g$ water.
d. As temperature increases, the solubility of carbon dioxide in water decreases.

2 Chemistry and Measurements

Greg has been taking Inderal to treat his high blood pressure. Every two months, he has his blood pressure taken by a registered nurse. During his last visit, his blood pressure was lower, 130/85. Thus, the dosage of Inderal was reduced to 45 mg to be taken twice daily. If one tablet contains 15 mg of Inderal, how many tablets does Greg need to take in one day?

Credit: AVAVA / Shutterstock

LOOKING AHEAD

2.1 Units of Measurement
2.2 Measured Numbers and Significant Figures
2.3 Significant Figures in Calculations
2.4 Prefixes and Equalities
2.5 Writing Conversion Factors
2.6 Problem Solving Using Unit Conversion
2.7 Density

 The Health icon indicates a question that is related to health and medicine.

2.1 Units of Measurement

Learning Goal: Write the names and abbreviations for the metric or SI units used in measurements of volume, length, mass, temperature, and time.

- In science, physical quantities are described in units of the metric or International System of Units (SI).
- Some important units of measurement are liter (L) for volume, meter (m) for length, gram (g) and kilogram (kg) for mass, degree Celsius (°C) and kelvin (K) for temperature, and second (s) for time.

Chapter 2

Key Terms for Section 2.1

Match each of the following key terms with the correct description:

a. mass
b. volume
c. gram
d. second
e. meter
f. liter
g. International System of Units

1. _____ the metric unit for volume that is slightly larger than a quart
2. _____ the metric unit used in measurements of mass
3. _____ the amount of space occupied by a substance
4. _____ the metric and SI unit for time
5. _____ a measure of the quantity of material in an object
6. _____ the official system of measurement used by scientists and used in most countries of the world
7. _____ the metric unit for length that is slightly longer than a yard

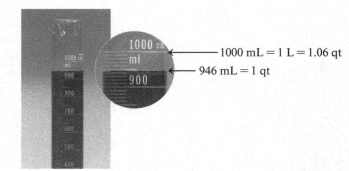

In the metric system, volume is based on the liter (L), which is equal to 1.06 quarts (qt).
Credit: Pearson Education

Answers   1. f   2. c   3. b   4. d   5. a   6. g   7. e

♦ **Learning Exercise 2.1**

Match the measurement with the quantity in each of the following:

a. length   b. volume   c. mass   d. temperature   e. time

1. _____ 45 g   2. _____ 8.2 m   3. _____ 215 °C   4. _____ 45 L
5. _____ 825 K   6. _____ 8.8 cm   7. _____ 140 kg   8. _____ 50 s

Answers   1. c   2. a   3. d   4. b
            5. d   6. a   7. c   8. e

2.2 Measured Numbers and Significant Figures

Learning Goal: Identify a number as measured or exact; determine the number of significant figures in a measured number.

On an electronic balance, the digital readout gives the mass of a nickel, which is 5.01 g.
Credit: Richard Megna/Fundamental Photographs, NYC

- A measured number is obtained when you use a measuring device.
- An exact number is obtained by counting items or from a definition within the same measuring system.
- There is uncertainty in every measured number but not in exact numbers.
- Significant figures in a measured number are all the digits, including the estimated digit.
- A number is a significant figure if it is not a zero.
- A zero is significant when it occurs between nonzero digits, at the end of a decimal number, or in the coefficient of a number written in scientific notation.

- A zero is not significant if it is at the beginning of a decimal number or used as a placeholder in a large number without a decimal point.

♦ Learning Exercise 2.2A

Identify the number(s) in each of the following statements as measured or exact and give a reason for your answer:

a. _____ There are 7 days in 1 week.

b. _____ A pulmonary treatment lasts for 25 min.

c. _____ There are 1000 g in 1 kg.

d. _____ The chef used 12 kg of potatoes.

e. _____ There are 4 books on the shelf.

f. _____ The height of a patient is 1.8 m.

Answers
a. exact (definition)
c. exact (metric definition)
e. exact (counted)
b. measured (use a watch)
d. measured (use a scale)
f. measured (use a metric ruler)

Study Note

Significant figures (abbreviated SFs) are all the numbers reported in a measurement, including the estimated digit. Zeros are significant unless they are placeholders appearing at the beginning of a decimal number or in a large number without a decimal point.

4.255 g (four SFs)   0.0040 m (two SFs)   46 500 L (three SFs)   150. mL (three SFs)

♦ Learning Exercise 2.2B

State the number of significant figures in each of the following measured numbers:

a. 35.24 g _____

b. 8.0×10^{-5} m _____

c. 55 000 m _____

d. 600. mL _____

e. 5.025 L _____

f. 0.006 kg _____

g. 2.680×10^5 mm _____

h. 25.0 °C _____

CORE CHEMISTRY SKILL
Counting Significant Figures

Answers
a. 4 SFs   b. 2 SFs   c. 2 SFs   d. 3 SFs
e. 4 SFs   f. 1 SF   g. 4 SFs   h. 3 SFs

2.3 Significant Figures in Calculations

Learning Goal: Adjust calculated answers to give the correct number of significant figures.

REVIEW
Identifying Place Values (1.4)
Using Positive and Negative Numbers in Calculations (1.4)

- In multiplication or division, the final answer is written so that it has the same number of significant figures (SFs) as the measurement with the fewest significant figures.

Chapter 2

- In addition or subtraction, the final answer is written so that it has the same number of decimal places as the measurement with the fewest decimal places.
- To evaluate a calculator answer, count the significant figures in the measurements and round off the calculator answer properly.
- Answers for calculations rarely use all the numbers that appear in the calculator display. Exact numbers are not included in the determination of the number of significant figures.

Study Note

1. To round off a number when the first digit to be dropped is *4 or less*, keep the digits you need and drop all the digits that follow.

    Round off 42.8254 to three SFs → 42.8 (drop 254)

2. To round off a number when the first digit to be dropped is *5 or greater*, keep the proper number of digits and increase the last retained digit by 1.

    Round off 8.4882 to two SFs → 8.5 (drop 882; increase the last retained digit by 1)

3. When rounding off large numbers without decimal points, maintain the value of the answer by adding nonsignificant zeros as placeholders.

    Round off 356 835 to three SFs → 357 000 (drop 835; increase the last retained digit by 1; add placeholder zeros)

♦ **Learning Exercise 2.3A**

KEY MATH SKILL
Rounding Off

Round off each of the following to two significant figures:

a. 88.75 m _____   b. 0.002 923 g _____   c. 50.525 s _____

d. 1.672 L _____   e. 0.001 055 8 kg _____   f. 82 080 mL _____

Answers   a. 89 m   b. 0.0029 g   c. 51 s   d. 1.7 L   e. 0.0011 kg   f. 82 000 mL

Study Note

1. An answer obtained from multiplying and dividing has the same number of significant figures as the measurement with the fewest significant figures.

    $\underset{\text{Two SFs}}{1.5} \times \underset{\text{Five SFs}}{32.546} = 48.819 \rightarrow 49$ *Answer rounded off to two SFs*

2. An answer obtained from adding or subtracting has the same number of decimal places as the measurement with the fewest decimal places.

    $\underset{\text{Thousandths place}}{82.223} + \underset{\text{Tenths place}}{4.1} = 86.323 \rightarrow 86.3$ *Answer rounded off to the tenths place*

♦ Learning Exercise 2.3B

Perform the following calculations with measured numbers. Write each answer with the correct number of significant figures or decimal places.

CORE CHEMISTRY SKILL
Using Significant Figures in Calculations

a. $1.3 \times 71.5 =$

b. $\dfrac{8.00}{4.00} =$

c. $\dfrac{(0.082)(25.4)}{(0.116)(3.4)} =$

d. $\dfrac{3.05 \times 1.86}{118.5} =$

e. $\dfrac{376}{0.0073} =$

f. $38.520 - 11.4 =$

g. $4.2 + 8.15 =$

h. $102.56 + 8.325 - 0.8825 =$

Answers
a. 93
b. 2.00
c. 5.3
d. 0.0479
e. 52 000 (5.2×10^4)
f. 27.1
g. 12.4
h. 110.00

2.4 Prefixes and Equalities

Learning Goal: Use the numerical values of prefixes to write a metric equality.

REVIEW
Writing Numbers in Scientific Notation (1.5)

- In the metric system, larger and smaller units use prefixes to change the size of the unit by factors of 10. For example, a prefix such as *centi* or *milli* preceding the unit *meter* gives a smaller length than a meter. A prefix such as *kilo* added to *gram* gives a unit that measures a mass that is 1000 times *greater* than a gram.
- An equality contains two units that measure the *same quantity* such as length, volume, mass, or time.
- Some common metric equalities are: 1 m = 100 cm; 1 L = 1000 mL; 1 kg = 1000 g.

Metric and SI Prefixes

Prefix	Symbol	Numerical Value	Scientific Notation	Equality
Prefixes That Increase the Size of the Unit				
tera	T	1 000 000 000 000	10^{12}	1 Ts = 1×10^{12} s 1 s = 1×10^{-12} Ts
giga	G	1 000 000 000	10^{9}	1 Gm = 1×10^{9} m 1 m = 1×10^{-9} Gm
mega	M	1 000 000	10^{6}	1 Mg = 1×10^{6} g 1 g = 1×10^{-6} Mg
kilo	k	1 000	10^{3}	1 km = 1×10^{3} m 1 m = 1×10^{-3} km

Chapter 2

Metric and SI Prefixes (continued)

Prefix	Symbol	Numerical Value	Scientific Notation	Equality
Prefixes That Decrease the Size of the Unit				
deci	d	0.1	10^{-1}	1 dL = 1×10^{-1} L 1 L = 10 dL
centi	c	0.01	10^{-2}	1 cm = 1×10^{-2} m 1 m = 100 cm
milli	m	0.001	10^{-3}	1 ms = 1×10^{-3} s 1 s = 1×10^{3} ms
micro	μ*	0.000 001	10^{-6}	1 μg = 1×10^{-6} g 1 g = 1×10^{6} μg
nano	n	0.000 000 001	10^{-9}	1 nm = 1×10^{-9} m 1 m = 1×10^{9} nm
pico	p	0.000 000 000 001	10^{-12}	1 ps = 1×10^{-12} s 1 s = 1×10^{12} ps

*In medicine, the abbreviation *mc* for the prefix *micro* is used because the symbol μ may be misread, which could result in a medication error. Thus, 1 μg would be written as 1 mcg.

♦ Learning Exercise 2.4A

CORE CHEMISTRY SKILL
Using Prefixes

Match the items in column A with those from column B.

A		B	
1. ___ megameter		**a.** nanometer	
2. ___ 0.1 m		**b.** decimeter	
3. ___ millimeter		**c.** 10^{-6} m	
4. ___ centimeter		**d.** 0.01 m	
5. ___ 10^{-9} m		**e.** 1000 m	
6. ___ micrometer		**f.** 10^{-3} m	
7. ___ kilometer		**g.** 10^{6} m	

Some Normal Laboratory Test Values

Substance in Blood	Normal Range
Albumin	3.5–5.4 g/dL
Ammonia	20–70 mcg/dL
Calcium	8.5–10.5 mg/dL
Cholesterol	105–250 mg/dL
Iron (male)	80–160 mcg/dL
Protein (total)	6.0–8.5 g/dL

Laboratory results for blood work are often reported in mass per deciliter (dL).

Answers   **1.** g   **2.** b   **3.** f   **4.** d
            5. a   **6.** c   **7.** e

♦ Learning Exercise 2.4B

Complete each of the following metric relationships:

a. 1 L = _____ mL   **b.** 1 L = _____ dL
c. 1 mm = _____ cm   **d.** 1 s = _____ ms
e. 1 kg = _____ g   **f.** 1 mm = _____ m
g. 1 g = _____ mcg   **h.** 1 dL = _____ L

Answers   **a.** 1000   **b.** 10   **c.** 0.1   **d.** 1000
            e. 1000   **f.** 0.001   **g.** 1 000 000 (10^6)   **h.** 0.1

During a physical examination, the metric length of an infant is measured.
Credit: Eric Schrader / Pearson Education

♦ **Learning Exercise 2.4C**

For each of the following pairs, which is the larger unit?

a. mL or mcL _____ b. g or kg _____

c. nm or m _____ d. centigram or decigram _____

e. kilosecond or second _____ f. Mm or mm _____

Answers   a. mL   b. kg   c. m   d. decigram   e. kilosecond   f. Mm

2.5 Writing Conversion Factors

Learning Goal: Write a conversion factor for two units that describe the same quantity.

> REVIEW
> Calculating Percentages (1.4)

- A conversion factor represents an equality expressed in the form of a fraction.
- Two conversion factors can be written for any relationship between equal quantities.
  For the metric–U.S. equality 2.54 cm = 1 in., the corresponding conversion factors are:

  $$\frac{2.54 \text{ cm}}{1 \text{ in.}} \quad \text{and} \quad \frac{1 \text{ in.}}{2.54 \text{ cm}}$$

- A percentage (%) is written as a conversion factor by expressing matching units in the relationship as the parts of a specific substance in 100 parts of the whole.
- An equality may be stated that applies only to a specific problem.

Some Common Equalities

Quantity	Metric (SI)	U.S.	Metric–U.S.
Length	1 km = 1000 m 1 m = 1000 mm 1 cm = 10 mm	1 ft = 12 in. 1 yd = 3 ft 1 mi = 5280 ft	2.54 cm = 1 in. (exact) 1 m = 39.4 in. 1 km = 0.621 mi
Volume	1 L = 1000 mL 1 dL = 100 mL 1 mL = 1 cm^3 1 mL = 1 cc*	1 qt = 4 cups 1 qt = 2 pt 1 gal = 4 qt	946 mL = 1 qt 1 L = 1.06 qt 473 mL = 1 pt 5 mL = 1 t (tsp)* 15 mL = 1 T (tbsp)*
Mass	1 kg = 1000 g 1 g = 1000 mg 1 mg = 1000 mcg*	1 lb = 16 oz	1 kg = 2.20 lb 454 g = 1 lb
Time	1 h = 60 min 1 min = 60 s	1 h = 60 min 1 min = 60 s	

*Used in medicine.

Study Note

Metric conversion factors are obtained from metric prefixes. For example, the metric equality 1 m = 100 cm is written as two conversion factors:

$$\frac{100 \text{ cm}}{1 \text{ m}} \quad \text{and} \quad \frac{1 \text{ m}}{100 \text{ cm}}$$

Chapter 2

♦ **Learning Exercise 2.5A**

> **CORE CHEMISTRY SKILL**
> Writing Conversion Factors from Equalities

Write the equality and two conversion factors for each of the following pairs of units:

a. millimeters and meters

b. kilograms and grams

c. kilograms and pounds

d. seconds and minutes

e. kilometers and meters

f. milliliters and quarts

g. deciliters and liters

Answers

a. $1 \text{ m} = 1000 \text{ mm}$

$$\frac{1000 \text{ mm}}{1 \text{ m}} \text{ and } \frac{1 \text{ m}}{1000 \text{ mm}}$$

b. $1 \text{ kg} = 1000 \text{ g}$

$$\frac{1000 \text{ g}}{1 \text{ kg}} \text{ and } \frac{1 \text{ kg}}{1000 \text{ g}}$$

c. $1 \text{ kg} = 2.20 \text{ lb}$

$$\frac{2.20 \text{ lb}}{1 \text{ kg}} \text{ and } \frac{1 \text{ kg}}{2.20 \text{ lb}}$$

d. $1 \text{ min} = 60 \text{ s}$

$$\frac{60 \text{ s}}{1 \text{ min}} \text{ and } \frac{1 \text{ min}}{60 \text{ s}}$$

e. $1 \text{ km} = 1000 \text{ m}$

$$\frac{1000 \text{ m}}{1 \text{ km}} \text{ and } \frac{1 \text{ km}}{1000 \text{ m}}$$

f. $1 \text{ qt} = 946 \text{ mL}$

$$\frac{946 \text{ mL}}{1 \text{ qt}} \text{ and } \frac{1 \text{ qt}}{946 \text{ mL}}$$

g. $1 \text{ L} = 10 \text{ dL}$

$$\frac{10 \text{ dL}}{1 \text{ L}} \text{ and } \frac{1 \text{ L}}{10 \text{ dL}}$$

🧍 Study Note

1. Sometimes, a statement within a problem gives an equality that is true only for that problem. Then conversion factors can be written that are true only for that problem. For example, a problem states that there are 50 mg of vitamin B in a tablet. The equality and its conversion factors are

   $$1 \text{ tablet} = 50 \text{ mg of vitamin B}$$

   $$\frac{50 \text{ mg vitamin B}}{1 \text{ tablet}} \text{ and } \frac{1 \text{ tablet}}{50 \text{ mg vitamin B}}$$

2. If a problem gives a percentage (%), it can be stated as parts per 100 parts. For example, a candy bar contains 45% by mass chocolate. This percentage (%) equality and its conversion factors are written using the same unit.

   $$45 \text{ g of chocolate} = 100 \text{ g of candy bar}$$

   $$\frac{45 \text{ g chocolate}}{100 \text{ g candy bar}} \text{ and } \frac{100 \text{ g candy bar}}{45 \text{ g chocolate}}$$

Chemistry and Measurements

♦ **Learning Exercise 2.5B**

Write the equality and two conversion factors for each of the following statements:

a. A cheese contains 55% fat by mass.

b. The Daily Value (DV) for zinc is 15 mg.

c. A 125-g steak contains 45 g of protein.

d. One iron tablet contains 65 mg of iron.

e. A car travels 14 miles on 1 gal of gasoline.

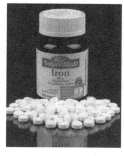

One tablet contains 65 mg of iron.
Credit: Editorial Image, LLC / Alamy

Answers

a.   55 g of fat = 100 g of cheese

$\dfrac{55 \text{ g fat}}{100 \text{ g cheese}}$ and $\dfrac{100 \text{ g cheese}}{55 \text{ g fat}}$

b. 15 mg of zinc = 1 day

$\dfrac{15 \text{ mg zinc}}{1 \text{ day}}$ and $\dfrac{1 \text{ day}}{15 \text{ mg zinc}}$

c. 125 g of steak = 45 g of protein

$\dfrac{45 \text{ g protein}}{125 \text{ g steak}}$ and $\dfrac{125 \text{ g steak}}{45 \text{ g protein}}$

d.   1 tablet = 65 mg of iron

$\dfrac{65 \text{ mg iron}}{1 \text{ tablet}}$ and $\dfrac{1 \text{ tablet}}{65 \text{ mg iron}}$

e.   1 gal = 14 mi

$\dfrac{14 \text{ mi}}{1 \text{ gal}}$ and $\dfrac{1 \text{ gal}}{14 \text{ mi}}$

♦ **Learning Exercise 2.5C**

Write the equality and two conversion factors, and identify the numbers as exact or give the number of significant figures for each of the following:

a. One tablet of calcium contains 315 mg of calcium.

b. The drug cured 37% of the patients.

c. One tablespoon of cough syrup contains 15 mL of cough syrup.

Answers

a.   1 tablet = 315 mg of calcium

$\dfrac{315 \text{ mg calcium}}{1 \text{ tablet}}$ and $\dfrac{1 \text{ tablet}}{315 \text{ mg calcium}}$

The 315 mg is measured: It has three SFs. The 1 tablet is exact.

Chapter 2

b.   37 cured = 100 patients

$$\frac{37 \text{ cured}}{100 \text{ patients}} \text{ and } \frac{100 \text{ patients}}{37 \text{ cured}}$$

The 37 patients cured and the 100 patients are both exact.

c. 1 tablespoon of syrup = 15 mL of syrup

$$\frac{15 \text{ mL syrup}}{1 \text{ tablespoon syrup}} \text{ and } \frac{1 \text{ tablespoon syrup}}{15 \text{ mL syrup}}$$

The 15 mL is measured: It has two SFs. The 1 tablespoon is exact.

2.6 Problem Solving Using Unit Conversion

Learning Goal: Use conversion factors to change from one unit to another.

- Conversion factors can be used to change a quantity expressed in one unit to a quantity expressed in another unit.
- In the process of problem solving, a given unit is multiplied by one or more conversion factors until the needed unit is obtained.

SAMPLE PROBLEM   Using Conversion Factors

A patient receives 2850 mL of saline solution. How many liters of the solution did the patient receive?

Solution Guide:

STEP 1 State the given and needed quantities.

Analyze the Problem	Given	Need	Connect
	2850 mL	liters	metric factor (L/mL)

STEP 2 Write a plan to convert the given unit to the needed unit.

milliliters $\xrightarrow{\text{Metric factor}}$ liters

STEP 3 State the equalities and conversion factors.

$$1 \text{ L} = 1000 \text{ mL}$$

$$\frac{1000 \text{ mL}}{1 \text{ L}} \text{ and } \frac{1 \text{ L}}{1000 \text{ mL}}$$

STEP 4 Set up the problem to cancel units and calculate the answer.

$$\underset{\text{Three SFs}}{2850 \text{ mL}} \times \underset{\text{Exact}}{\frac{1 \text{ L}}{1000 \text{ mL}}} = \underset{\text{Three SFs}}{2.85 \text{ L}}$$

♦ Learning Exercise 2.6A

Use metric conversion factors to solve each of the following problems. For **e** to **h**, complete boxes for Given, Need, Connect, and Answer.

CORE CHEMISTRY SKILL
Using Conversion Factors

    a. 189 mL = _____ L   b. 2.7 cm = _____ mm

Chemistry and Measurements

c. 0.274 m = _____ cm   d. 0.076 kg = _____ g

e. How many meters tall is a person whose height is 163 cm?

Analyze the Problem	Given	Need	Connect

f. There is 0.646 L of water in a teapot. How many milliliters is that?

Analyze the Problem	Given	Need	Connect

g. If a ring contains 13 500 mg of gold, how many grams of gold are in the ring?

Analyze the Problem	Given	Need	Connect

h. You walked a distance of 1.5 km on the treadmill at the gym. How many meters did you walk?

Analyze the Problem	Given	Need	Connect

Answers   a. 0.189 L   b. 27 mm   c. 27.4 cm   d. 76 g

e.
Analyze the Problem	Given	Need	Connect	Answer
	163 cm	meters	metric factor (m/cm)	1.63 m

f.
Analyze the Problem	Given	Need	Connect	Answer
	0.646 L	milliliters	metric factor (mL/L)	646 mL

g.
Analyze the Problem	Given	Need	Connect	Answer
	13 500 mg	grams	metric factor (g/mg)	13.5 g

h.
Analyze the Problem	Given	Need	Connect	Answer
	1.5 km	meters	metric factor (m/km)	1500 m

Chapter 2

♦ **Learning Exercise 2.6B**

Use metric–U.S. conversion factors to solve each of the following problems:

a. 880 g = _____ lb

b. 4.6 qt = _____ mL

c. 1.50 ft = _____ cm

d. 8.10 in. = _____ mm

Answers   **a.** 1.9 lb   **b.** 4400 mL
            c. 45.7 cm   **d.** 206 mm

♦ **Learning Exercise 2.6C**

Use conversion factors to solve each of the following problems.
For **c** to **f**, complete boxes for Given, Need, and Connect, and solve the problem.

a. 120 mm = _____ in.

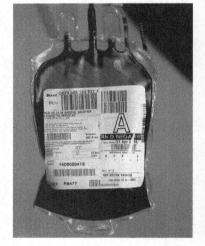

1 pt of blood contains 473 mL.
Credit: Syner-Comm/Alamy

b. 245 lb = _____ kg

c. In a triple-bypass surgery, a patient requires 3.00 pt of whole blood. How many milliliters of blood were given?

Analyze the Problem	Given	Need	Connect

d. Your friend has a height of 6 ft 3 in. What is your friend's height in meters?

Analyze the Problem	Given	Need	Connect

e. A doctor orders 0.450 g of a sulfa drug. On hand are 150-mg tablets. How many tablets of the sulfa drug are needed?

Analyze the Problem	Given	Need	Connect

f. A mouthwash contains 27% alcohol by volume. How many milliliters of alcohol are in a 1.18-L bottle of mouthwash?

Analyze the Problem	Given	Need	Connect

Answers a. 4.7 in. b. 111 kg

c.
Analyze the Problem	Given	Need	Connect	Answer
	3.00 pt	milliliters	factor (mL/pt)	1420 mL

d.
Analyze the Problem	Given	Need	Connect	Answer
	6 ft 3 in.	meters	factors (in./ft, m/in.)	1.9 m

e.
Analyze the Problem	Given	Need	Connect	Answer
	0.450 g, 150-mg tablet	tablets	factors (mg/g, tablet/mg)	3 tablets

f.
Analyze the Problem	Given	Need	Connect	Answer
	1.18 L, 27% (v/v)	milliliters of alcohol	factors (mL/L, % alcohol)	320 mL

2.7 Density

Learning Goal: Calculate the density of a substance; use the density to calculate the mass or volume of a substance.

- The density of a substance is the ratio of its mass to its volume, usually in units of g/cm^3 or g/mL.

  $$\text{Density} = \frac{\text{mass of substance}}{\text{volume of substance}}$$

- The volume of 1 mL is equal to the volume of 1 cm^3.
- Specific gravity is a relationship between the density of a substance and the density of water.

Chapter 2

Key Terms for Sections 2.2 to 2.7

Match each of the following key terms with the correct description:

    a. equality   **b.** conversion factor   **c.** measured number   **d.** exact number
    e. density   **f.** prefix   **g.** significant figures   **h.** specific gravity

1. _____ a number obtained by counting
2. _____ a part of the name of a metric unit that precedes the base unit and specifies the size of the measurement
3. _____ a ratio in which the numerator and denominator are quantities from an equality
4. _____ the relationship between two units that measure the same quantity
5. _____ the digits recorded in a measurement
6. _____ the number obtained when a quantity is determined by using a measuring device
7. _____ the relationship of the mass of an object to its volume expressed as grams per cubic centimeter (g/cm^3) or grams per milliliter (g/mL)
8. _____ the relationship between the density of a substance and the density of water

Answers   **1.** d **2.** f **3.** b **4.** a **5.** g **6.** c **7.** e **8.** h

♦ **Learning Exercise 2.7A**

a. Calculate the density, in grams per milliliter, of glycerol if a 200.0-mL sample has a mass of 252 g.

b. A person with diabetes may produce 5 to 12 L of urine per day. Calculate the density, in grams per milliliter, of a 100.0-mL urine sample that has a mass of 100.2 g.

c. A solid has a mass of 5.5 oz. When placed in a graduated cylinder with a water level of 55.2 mL, the object causes the water level to rise to 73.8 mL. What is the density of the object, in grams per milliliter?

Answers   **a.** 1.26 g/mL   **b.** 1.002 g/mL   **c.** 8.4 g/mL

Chemistry and Measurements

SAMPLE PROBLEM   Using Density

Density can be used as a factor to convert between the mass (g) and volume (mL) of a substance. The density of silver is 10.5 g/mL. What is the mass, in grams, of 6.0 mL of silver?

Solution:

STEP 1 State the given and needed quantities.

Analyze the Problem	Given	Need	Connect
	6.0 mL of silver, density 10.5 g/mL	grams of silver	density factor

STEP 2 Write a plan to calculate the needed quantity.

milliliters $\xrightarrow{\text{Density factor}}$ grams

STEP 3 Write the equalities and their conversion factors including density.

1 mL of silver = 10.5 g of silver

$$\frac{10.5 \text{ g}}{1 \text{ mL}} \quad \text{and} \quad \frac{1 \text{ mL}}{10.5 \text{ g}}$$

STEP 4 Set up the problem to calculate the needed quantity.

$$6.0 \text{ mL silver} \times \underbrace{\frac{10.5 \text{ g silver}}{1 \text{ mL silver}}}_{\text{Density factor}} = 63 \text{ g of silver}$$

♦ Learning Exercise 2.7B

Use density or specific gravity as a conversion factor to solve each of the following:

a. A sugar solution has a density of 1.20 g/mL. What is the mass, in grams, of 0.225 L of the solution?

> **CORE CHEMISTRY SKILL**
> Using Density as a Conversion Factor

b. A piece of pure gold weighs 0.26 lb. If gold has a density of 19.3 g/mL, what is the volume, in milliliters, of the piece of gold?

c. A urine sample has a specific gravity of 1.015 and a volume of 42.50 mL. What is the mass, in grams, of the solution?

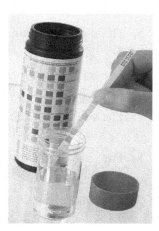

A dipstick is used to measure the specific gravity of a urine sample.
Credit: JPC-PROD/Fotolia

d. A 600.-g sample of an intravenous glucose solution has a density of 1.28 g/mL. What is the volume, in liters, of the sample?

Answers   **a.** 270. g   **b.** 6.1 mL   **c.** 43.14 g   **d.** 0.469 L

Chapter 2

Checklist for Chapter 2

You are ready to take the Practice Test for Chapter 2. Be sure you have accomplished the following learning goals for this chapter. If not, review the Section listed at the end of the goal. Then apply your new skills and understanding to the Practice Test.

After studying Chapter 2, I can successfully:

____ Write the names and abbreviations for the metric (SI) units of measurement. (2.1)

____ Identify a number as a measured number or an exact number. (2.2)

____ Determine the number of significant figures in measured numbers. (2.2)

____ Report a calculated answer with the correct number of significant figures. (2.3)

____ Write an equality from the relationship between two equal quantities. (2.4)

____ Write two conversion factors for an equality. (2.5)

____ Use conversion factors to change from one unit to another unit. (2.6)

____ Calculate the density or specific gravity of a substance or use density or specific gravity to calculate the mass or volume. (2.7)

Practice Test for Chapter 2

The chapter Sections to review are shown in parentheses at the end of each question.

1. Which of the following is a metric measurement of volume? (2.1)
   A. kilogram **B.** kilowatt **C.** kiloliter **D.** kilometer **E.** kiloquart

2. Which of the following is a metric measurement of length? (2.1)
   A. centiliter **B.** centigram **C.** centiyard **D.** centifoot **E.** centimeter

3. The exact number in the following is (2.2)
   A. 4.5 L **B.** 4 syringes **C.** 6 qt **D.** 15 g **E.** 2.0 m

4. The measured number in the following is (2.2)
   A. 1 book **B.** 2 cars **C.** 4 flowers **D.** 5 rings **E.** 4 g

5. The number of significant figures in 105.4 m is (2.2)
   A. 1 **B.** 2 **C.** 3 **D.** 4 **E.** 5

6. The number of significant figures in 0.000 82 g is (2.2)
   A. 1 **B.** 2 **C.** 3 **D.** 4 **E.** 5

7. The calculator answer 5.7805 rounded to two significant figures is (2.2)
   A. 5 **B.** 5.7 **C.** 5.8 **D.** 5.78 **E.** 6.0

8. The calculator answer 3486.512 rounded to three significant figures is (2.2)
   A. 4000 **B.** 3500 **C.** 349 **D.** 3487 **E.** 3490

9. The answer for $16.0 \div 8.0$ with the correct number of significant figures is (2.3)
   A. 2 **B.** 2.0 **C.** 2.00 **D.** 0.2 **E.** 5.0

10. The answer for $58.5 + 9.158$ with the correct number of decimal places is (2.3)
    A. 67 **B.** 67.6 **C.** 67.7 **D.** 67.66 **E.** 67.658

11. The answer for $\dfrac{2.5 \times 3.12}{4.6}$ with the correct number of significant figures is (2.3)
    A. 0.54 **B.** 7.8 **C.** 0.85 **D.** 1.7 **E.** 1.69

Chemistry and Measurements

12. Which of these prefixes has the largest value? (2.4)
    A. centi   B. deci   C. milli   D. kilo   E. micro
13. What is the decimal equivalent of the prefix *centi*? (2.4)
    A. 0.001   B. 0.01   C. 0.1   D. 10   E. 100
14. Which of the following is the smallest unit of mass measurement? (2.4)
    A. gram   B. milligram   C. kilogram   D. decigram   E. centigram
15. Which of the following is the largest unit of volume measurement? (2.4)
    A. mL   B. dL   C. cm^3 D. L   E. kL
16. Which of the following is a conversion factor? (2.5)
    A. 12 in.   B. 3 ft = 1 yd   C. 20 m   D. $\dfrac{1000 \text{ g}}{1 \text{ kg}}$ E. 2 cm^3
17. Which is the correct conversion factor for milliliters and liters? (2.5)
    A. $\dfrac{1000 \text{ mL}}{1 \text{ L}}$ B. $\dfrac{100 \text{ mL}}{1 \text{ L}}$ C. $\dfrac{10 \text{ mL}}{1 \text{ L}}$ D. $\dfrac{0.01 \text{ mL}}{1 \text{ L}}$ E. $\dfrac{0.001 \text{ mL}}{1 \text{ L}}$
18. Which is the correct conversion factor for millimeters and centimeters? (2.5)
    A. $\dfrac{1 \text{ mm}}{1 \text{ cm}}$ B. $\dfrac{10 \text{ mm}}{1 \text{ cm}}$ C. $\dfrac{100 \text{ cm}}{1 \text{ mm}}$ D. $\dfrac{100 \text{ mm}}{1 \text{ cm}}$ E. $\dfrac{10 \text{ cm}}{1 \text{ mm}}$
19. A doctor orders 20 mg of prednisone. If a tablet contains 5 mg of prednisone, how many tablets should be given? (2.6)
    A. 1   B. 2   C. 3   D. 4   E. 5
20. 294 mm is equal to (2.6)
    A. 2940 m   B. 29.4 m   C. 2.94 m   D. 0.294 m   E. 0.0294 m
21. A doctor orders 200 mg of penicillin to be injected. If it is available as 1000 mg of penicillin per 5 mL, how many milliliters should be given? (2.6)
    A. 1 mL   B. 2 mL   C. 3 mL   D. 4 mL   E. 5 mL
22. What is the volume, in liters, of 800. mL of blood plasma? (2.6)
    A. 0.8 L   B. 0.80 L   C. 0.800 L   D. 8.0 L   E. 80.0 L
23. What is the mass, in kilograms, of a 22-lb toddler? (2.6)
    A. 10. kg   B. 48 kg   C. 10 000 kg   D. 0.048 kg   E. 22 000 kg
24. The number of milliliters in 2 dL of an antiseptic liquid is (2.6)
    A. 20 mL   B. 200 mL   C. 2000 mL   D. 20 000 mL   E. 500 000 mL
25. What is the height, in meters, of a person who is 5 ft 4 in. tall? (2.6)
    A. 64 m   B. 25 m   C. 14 m   D. 1.6 m   E. 1.3 m
26. How many ounces are in 1500 grams of carrots? (2.6)
    A. 94 oz   B. 53 oz   C. 24 000 oz   D. 33 oz   E. 3.3 oz
27. How many quarts of orange juice are in 255 mL of juice? (2.6)
    A. 0.255 qt   B. 270 qt   C. 236 qt   D. 0.270 qt   E. 3.71 qt
28. Your doctor places you on a 2200-kcal diet, with 19% of the kilocalories from fat. How many kilocalories are you allowed from fat? (2.6)
    A. 19 kcal   B. 2200 kcal   C. 1900 kcal   D. 420 kcal   E. 4200 kcal

Chapter 2

29. 1.5 ft is the same length, in centimeters, as (2.6)
    A. 46 cm   B. 7.1 cm   C. 18 cm   D. 3.8 cm   E. 0.59 cm

30. How many milliliters of a salt solution with a density of 1.8 g/mL are needed to provide 450 g of salt solution? (2.7)
    A. 250 mL   B. 25 mL   C. 810 mL   D. 450 mL   E. 45 mL

31. Three liquids have densities of 1.15 g/mL, 0.79 g/mL, and 0.95 g/mL. When the liquids, which do not mix, are poured into a graduated cylinder, the liquid at the top has a density of (2.7)
    A. 1.15 g/mL   B. 1.00 g/mL   C. 0.95 g/mL   D. 0.79 g/mL   E. 0.16 g/mL

32. A sample of oil has a mass of 65 g and a volume of 80.0 mL. What is the specific gravity of the oil? (2.7)
    A. 1.5   B. 1.4   C. 1.2   D. 0.90   E. 0.81

33. What is the mass, in grams, of a 15.0-mL sample of an antihistamine solution with a density of 1.04 g/mL? (2.7)
    A. 104 g   B. 10.4 g   C. 15.6 g   D. 1.56 g   E. 14.4 g

34. Ethanol has a density of 0.785 g/mL. What is the mass, in grams, of 0.250 L of ethanol? (2.7)
    A. 196 g   B. 158 g   C. 318 g   D. 0.253 g   E. 0.160 g

35. A patient has a blood volume of 6.8 qt. If the density of blood is 1.06 g/mL, what is the mass, in grams, of the patient's blood? (2.7)
    A. 1060 g   B. 6400 g   C. 6800 g   D. 1500 g   E. 990 g

Answers to the Practice Test

1. C	2. E	3. B	4. E	5. D
6. B	7. C	8. E	9. B	10. C
11. D	12. D	13. B	14. B	15. E
16. D	17. A	18. B	19. D	20. D
21. A	22. C	23. A	24. B	25. D
26. B	27. D	28. D	29. A	30. A
31. D	32. E	33. C	34. A	35. C

Selected Answers and Solutions to Text Problems

2.1 a. The abbreviation for the unit gram is g.
b. The abbreviation for the unit liter is L.
c. The abbreviation for the unit degree Celsius is °C.
d. The abbreviation for the unit pound is lb.
e. The abbreviation for the unit second is s.

2.3 a. A liter is a unit of volume.
b. A centimeter is a unit of length.
c. A kilometer is a unit of length.
d. A second is a unit of time.

2.5 a. The unit is a meter, which is a unit of length.
b. The unit is a gram, which is a unit of mass.
c. The unit is a milliliter, which is a unit of volume.
d. The unit is a second, which is a unit of time.
e. The unit is a degree Celsius, which is a unit of temperature.

2.7 a. The unit is a second, which is a unit of time.
b. The unit is a kilogram, which is a unit of mass.
c. The unit is a gram, which is a unit of mass.
d. The unit is a degree Celsius, which is a unit of temperature.

2.9 a. All five numbers are significant figures (5 SFs).
b. Only the two nonzero numbers are significant (2 SFs); the preceding zeros are placeholders.
c. Only the two nonzero numbers are significant (2 SFs); the zeros that follow are placeholders.
d. All three numbers in the coefficient of a number written in scientific notation are significant (3 SFs).
e. All four numbers to the right of the decimal point, including the last zero in the decimal number, are significant (4 SFs).
f. All three numbers, including the zeros at the end of a decimal number, are significant (3 SFs).

2.11 Both measurements in part **b** have three significant figures, and both measurements in part **c** have two significant figures.

2.13 a. Zeros at the beginning of a decimal number are not significant.
b. Zeros between nonzero digits are significant.
c. Zeros at the end of a decimal number are significant.
d. Zeros in the coefficient of a number written in scientific notation are significant.
e. Zeros used as placeholders in a large number without a decimal point are not significant.

2.15 a. 5000 L is the same as 5×1000 L, which is written in scientific notation as 5.0×10^3 L with two significant figures.
b. 30 000 g is the same as $3 \times 10\,000$ g, which is written in scientific notation as 3.0×10^4 g with two significant figures.
c. 100 000 m is the same as $1 \times 100\,000$ m, which is written in scientific notation as 1.0×10^5 m with two significant figures.
d. 0.000 25 cm is the same as $2.5 \times \dfrac{1}{10\,000}$ cm, which is written in scientific notation as 2.5×10^{-4} cm.

2.17 Measured numbers are obtained using some type of measuring device. Exact numbers are numbers obtained by counting items or using a definition that compares two units in the same measuring system.
a. The value 67.5 kg is a measured number; measurement of mass requires a measuring device.
b. The value 2 tablets is obtained by counting, making it an exact number.

c. The values in the metric definition 1 L = 1000 mL are exact numbers.
d. The value 1720 km is a measured number; measurement of distance requires a measuring device.

2.19 Measured numbers are obtained using some type of measuring device. Exact numbers are numbers obtained by counting items or using a definition that compares two units in the same measuring system.
  a. 6 oz of hamburger meat is a measured number (3 hamburgers is a counted/exact number).
  b. Neither are measured numbers (both 1 table and 4 chairs are counted/exact numbers).
  c. Both 0.75 lb of grapes and 350 g of butter are measured numbers.
  d. Neither are measured numbers (the values in a definition are exact numbers).

2.21 a. 1.607 kg is a measured number and has 4 SFs.
  b. 130 mcg is a measured number and has 2 SFs.
  c. 4.02×10^6 red blood cells is a measured number and has 3 SFs.
  d. 23 babies is a counted/exact number.

2.23 a. 1.85 kg; the last digit is dropped since it is 4 or less.
  b. 88.2 L; since the fourth digit is 4 or less, the last three digits are dropped.
  c. 0.004 74 cm; since the fourth significant digit (the first digit to be dropped) is 5 or greater, the last retained digit is increased by 1 when the last four digits are dropped.
  d. 8810 m; since the fourth significant digit (the first digit to be dropped) is 5 or greater, the last retained digit is increased by 1 when the last digit is dropped (a nonsignificant zero is added at the end as a placeholder).
  e. 1.83×10^5 s; since the fourth digit is 4 or less, the last digit is dropped. The $\times 10^5$ is retained so that the magnitude of the answer is not changed.

2.25 a. To round off 56.855 m to three significant figures, drop the final digits 55 and increase the last retained digit by 1 to give 56.9 m.
  b. To round off 0.002 282 g to three significant figures, drop the final digit 2 to give 0.002 28 g.
  c. To round off 11 527 s to three significant figures, drop the final digits 27 and add two zeros as placeholders to give 11 500 s (1.15×10^4 s).
  d. To express 8.1 L to three significant figures, add a significant zero to give 8.10 L.

2.27 a. $45.7 \times 0.034 = 1.6$ Two significant figures are allowed since 0.034 has 2 SFs.
  b. $0.002\ 78 \times 5 = 0.01$ One significant figure is allowed since 5 has 1 SF.
  c. $\dfrac{34.56}{1.25} = 27.6$ Three significant figures are allowed since 1.25 has 3 SFs.
  d. $\dfrac{(0.2465)(25)}{1.78} = 3.5$ Two significant figures are allowed since 25 has 2 SFs.
  e. $(2.8 \times 10^4)(5.05 \times 10^{-6}) = 0.14$ or 1.4×10^{-1} Two significant figures are allowed since 2.8×10^4 has 2 SFs.
  f. $\dfrac{(3.45 \times 10^{-2})(1.8 \times 10^5)}{(8 \times 10^3)} = 0.8$ or 8×10^{-1} One significant figure is allowed since 8×10^3 has 1 SF.

2.29 a. 45.48 cm + 8.057 cm = 53.54 cm Two decimal places are allowed since 45.48 cm has two decimal places.
  b. 23.45 g + 104.1 g + 0.025 g = 127.6 g One decimal place is allowed since 104.1 g has one decimal place.
  c. 145.675 mL − 24.2 mL = 121.5 mL One decimal place is allowed since 24.2 mL has one decimal place.
  d. 1.08 L − 0.585 L = 0.50 L Two decimal places are allowed since 1.08 L has two decimal places.

2.31 a. mg **b.** dL **c.** km **d.** pg

2.33 a. centiliter **b.** kilogram **c.** millisecond **d.** gigameter

2.35 a. 0.01 **b.** 1 000 000 000 000 (or 1×10^{12})
c. 0.001 (or 1×10^{-3}) **d.** 0.1

2.37 a. decigram **b.** microgram
c. kilogram **d.** centigram

2.39 a. 1 m = 100 cm **b.** 1 m = 1×10^9 nm
c. 1 mm = 0.001 m **d.** 1 L = 1000 mL

2.41 a. kilogram, since 10^3 g is greater than 10^{-3} g
b. milliliter, since 10^{-3} L is greater than 10^{-6} L
c. km, since 10^3 m is greater than 10^0 m
d. kL, since 10^3 L is greater than 10^{-1} L
e. nanometer, since 10^{-9} m is greater than 10^{-12} m

2.43 A conversion factor can be inverted to give a second conversion factor: $\dfrac{1 \text{ m}}{100 \text{ cm}}$ and $\dfrac{100 \text{ cm}}{1 \text{ m}}$

2.45 a. 1 m = 100 cm; $\dfrac{100 \text{ cm}}{1 \text{ m}}$ and $\dfrac{1 \text{ m}}{100 \text{ cm}}$

b. 1 g = 1×10^9 ng; $\dfrac{1 \times 10^9 \text{ ng}}{1 \text{ g}}$ and $\dfrac{1 \text{ g}}{1 \times 10^9 \text{ ng}}$

c. 1 kL = 1000 L; $\dfrac{1000 \text{ L}}{1 \text{ kL}}$ and $\dfrac{1 \text{ kL}}{1000 \text{ L}}$

d. 1 s = 1000 ms; $\dfrac{1000 \text{ ms}}{1 \text{ s}}$ and $\dfrac{1 \text{ s}}{1000 \text{ ms}}$

2.47 a. 1 yd = 3 ft; $\dfrac{3 \text{ ft}}{1 \text{ yd}}$ and $\dfrac{1 \text{ yd}}{3 \text{ ft}}$; the 1 yd and 3 ft are both exact (U.S. definition).

b. 1 kg = 2.20 lb; $\dfrac{2.20 \text{ lb}}{1 \text{ kg}}$ and $\dfrac{1 \text{ kg}}{2.20 \text{ lb}}$; the 2.20 lb is measured: it has 3 SFs; the 1 kg is exact.

c. 1 gal of gasoline = 27 mi; $\dfrac{27 \text{ mi}}{1 \text{ gal gasoline}}$ and $\dfrac{1 \text{ gal gasoline}}{27 \text{ mi}}$; the 27 mi is measured: it has 2 SFs; the 1 gal is exact.

d. 100 g of sterling = 93 g of silver; $\dfrac{93 \text{ g silver}}{100 \text{ g sterling}}$ and $\dfrac{100 \text{ g sterling}}{93 \text{ g silver}}$; the 93 g is measured: it has 2 SFs; the 100 g is exact.

2.49 a. 1 s = 3.5 m; $\dfrac{3.5 \text{ m}}{1 \text{ s}}$ and $\dfrac{1 \text{ s}}{3.5 \text{ m}}$; the 3.5 m is measured: it has 2 SFs; the 1 s is exact.

b. 3.5 g of potassium = 1 day; $\dfrac{3.5 \text{ g potassium}}{1 \text{ day}}$ and $\dfrac{1 \text{ day}}{3.5 \text{ g potassium}}$; the 3.5 g is measured: it has 2 SFs; the 1 day is exact.

c. 1 L of gasoline = 26.0 km; $\dfrac{26.0 \text{ km}}{1 \text{ L gasoline}}$ and $\dfrac{1 \text{ L gasoline}}{26.0 \text{ km}}$; the 26.0 km is measured: it has 3 SFs; the 1 L is exact.

d. 100 g of crust = 28.2 g of silicon; $\dfrac{28.2 \text{ g silicon}}{100 \text{ g crust}}$ and $\dfrac{100 \text{ g crust}}{28.2 \text{ g silicon}}$; the 28.2 g is measured: it has 3 SFs; the 100 g is exact.

Chapter 2

2.51 a. 1 tablet = 630 mg of calcium; $\dfrac{630 \text{ mg calcium}}{1 \text{ tablet}}$ and $\dfrac{1 \text{ tablet}}{630 \text{ mg calcium}}$; the 630 mg is measured: it has 2 SFs; the 1 tablet is exact.

b. 60 mg of vitamin C = 1 day; $\dfrac{60 \text{ mg vitamin C}}{1 \text{ day}}$ and $\dfrac{1 \text{ day}}{60 \text{ mg vitamin C}}$; the 60 mg is measured: it has 1 SF; the 1 day is exact.

c. 1 tablet = 50 mg of atenolol; $\dfrac{50 \text{ mg atenolol}}{1 \text{ tablet}}$ and $\dfrac{1 \text{ tablet}}{50 \text{ mg atenolol}}$; the 50 mg is measured: it has 1 SF; the 1 tablet is exact.

d. 1 tablet = 81 mg of aspirin; $\dfrac{81 \text{ mg aspirin}}{1 \text{ tablet}}$ and $\dfrac{1 \text{ tablet}}{81 \text{ mg aspirin}}$; the 81 mg is measured: it has 2 SFs; the 1 tablet is exact.

2.53 a. 5 mL of syrup = 10 mg of Atarax; $\dfrac{10 \text{ mg Atarax}}{5 \text{ mL syrup}}$ and $\dfrac{5 \text{ mL syrup}}{10 \text{ mg Atarax}}$

b. 1 tablet = 0.25 g of Lanoxin; $\dfrac{0.25 \text{ g Lanoxin}}{1 \text{ tablet}}$ and $\dfrac{1 \text{ tablet}}{0.25 \text{ g Lanoxin}}$

c. 1 tablet = 300 mg of Motrin; $\dfrac{300 \text{ mg Motrin}}{1 \text{ tablet}}$ and $\dfrac{1 \text{ tablet}}{300 \text{ mg Motrin}}$

2.55 a. Given 44.2 mL   **Need** liters

    Plan mL → L $\dfrac{1 \text{ L}}{1000 \text{ mL}}$

    Set-up 44.2 m̶L̶ × $\dfrac{1 \text{ L}}{1000 \text{ m̶L̶}}$ = 0.0442 L (3 SFs)

b. Given 8.65 m   **Need** nanometers

    Plan m → nm $\dfrac{1 \times 10^9 \text{ nm}}{1 \text{ m}}$

    Set-up 8.65 m̶ × $\dfrac{1 \times 10^9 \text{ nm}}{1 \text{ m̶}}$ = 8.65 × 10^9 nm (3 SFs)

c. Given 5.2 × 10^8 g   **Need** megagrams

    Plan g → Mg $\dfrac{1 \text{ Mg}}{1 \times 10^6 \text{ g}}$

    Set-up 5.2 × 10^8 g̶ × $\dfrac{1 \text{ Mg}}{1 \times 10^6 \text{ g̶}}$ = 5.2 × 10^2 Mg (2 SFs)

d. Given 0.72 ks   **Need** milliseconds

    Plan ks → s → ms $\dfrac{1000 \text{ s}}{1 \text{ ks}}$ $\dfrac{1000 \text{ ms}}{1 \text{ s}}$

    Set-up 0.72 k̶s̶ × $\dfrac{1000 \text{ s̶}}{1 \text{ k̶s̶}}$ × $\dfrac{1000 \text{ ms}}{1 \text{ s̶}}$ = 7.2 × 10^5 ms (2 SFs)

2.57 a. Given 3.428 lb   **Need** kilograms

    Plan lb → kg $\dfrac{1 \text{ kg}}{2.20 \text{ lb}}$

    Set-up 3.428 l̶b̶ × $\dfrac{1 \text{ kg}}{2.20 \text{ l̶b̶}}$ = 1.56 kg (3 SFs)

b. Given 1.6 m   **Need** inches

   Plan m → in. $\dfrac{39.4 \text{ in.}}{1 \text{ m}}$

   Set-up 1.6 m × $\dfrac{39.4 \text{ in.}}{1 \text{ m}}$ = 63 in. (2 SFs)

c. Given 4.2 L   **Need** quarts

   Plan L → qt $\dfrac{1.06 \text{ qt}}{1 \text{ L}}$

   Set-up 4.2 L × $\dfrac{1.06 \text{ qt}}{1 \text{ L}}$ = 4.5 qt (2 SFs)

d. Given 0.672 ft   **Need** millimeters

   Plan ft → in. → cm → mm $\dfrac{12 \text{ in.}}{1 \text{ ft}}$ $\dfrac{2.54 \text{ cm}}{1 \text{ in.}}$ $\dfrac{10 \text{ mm}}{1 \text{ cm}}$

   Set-up 0.672 ft × $\dfrac{12 \text{ in.}}{1 \text{ ft}}$ × $\dfrac{2.54 \text{ cm}}{1 \text{ in.}}$ × $\dfrac{10 \text{ mm}}{1 \text{ cm}}$ = 205 mm (3 SFs)

2.59 a. Given 175 cm   **Need** meters

   Plan cm → m $\dfrac{1 \text{ m}}{100 \text{ cm}}$

   Set-up 175 cm × $\dfrac{1 \text{ m}}{100 \text{ cm}}$ = 1.75 m (3 SFs)

b. Given 5000 mL   **Need** liters

   Plan mL → L $\dfrac{1 \text{ L}}{1000 \text{ mL}}$

   Set-up 5000 mL × $\dfrac{1 \text{ L}}{1000 \text{ mL}}$ = 5 L (1 SF)

c. Given 0.0055 kg   **Need** grams

   Plan kg → g $\dfrac{1000 \text{ g}}{1 \text{ kg}}$

   Set-up 0.0055 kg × $\dfrac{1000 \text{ g}}{1 \text{ kg}}$ = 5.5 g (2 SFs)

d. Given 3500 cm³   **Need** liters

   Plan cm³ → mL → L $\dfrac{1 \text{ mL}}{1 \text{ cm}^3}$ $\dfrac{1 \text{ L}}{1000 \text{ mL}}$

   Set-up 3500 cm³ × $\dfrac{1 \text{ mL}}{1 \text{ cm}^3}$ × $\dfrac{1 \text{ L}}{1000 \text{ mL}}$ = 3.5 L (2 SFs)

2.61 a. Given 0.500 qt   **Need** milliliters

   Plan qt → mL $\dfrac{946 \text{ mL}}{1 \text{ qt}}$

   Set-up 0.500 qt × $\dfrac{946 \text{ mL}}{1 \text{ qt}}$ = 473 mL (3 SFs)

b. Given 175 lb   **Need** kilograms

   Plan lb → kg $\dfrac{1 \text{ kg}}{2.20 \text{ lb}}$

   Set-up 175 lb × $\dfrac{1 \text{ kg}}{2.20 \text{ lb}}$ = 79.5 kg (3 SFs)

c. Given 74 kg body mass, 15% body fat **Need** pounds of body fat
Plan kg of body mass → kg of body fat → lb of body fat
(percent equality: 100 kg of body mass = 15 g of body fat)

$$\frac{15 \text{ kg body fat}}{100 \text{ kg body mass}} \quad \frac{2.20 \text{ lb body fat}}{1 \text{ kg body fat}}$$

Set-up $74 \text{ kg body mass} \times \dfrac{15 \text{ kg body fat}}{100 \text{ kg body mass}} \times \dfrac{2.20 \text{ lb body fat}}{1 \text{ kg body fat}}$
= 24 lb of body fat (2 SFs)

d. Given 10.0 oz of fertilizer, 15% nitrogen **Need** grams of nitrogen
Plan oz of fertilizer → lb of fertilizer → g of fertilizer → g of nitrogen
(percent equality: 100 g of fertilizer = 15 g of nitrogen)

$$\frac{1 \text{ lb}}{16 \text{ oz}} \quad \frac{454 \text{ g}}{1 \text{ lb}} \quad \frac{15 \text{ g nitrogen}}{100 \text{ g fertilizer}}$$

Set-up $10.0 \text{ oz fertilizer} \times \dfrac{1 \text{ lb fertilizer}}{16 \text{ oz fertilizer}} \times \dfrac{454 \text{ g fertilizer}}{1 \text{ lb fertilizer}} \times \dfrac{15 \text{ g nitrogen}}{100 \text{ g fertilizer}}$
= 43 g of nitrogen (2 SFs)

2.63 a. Given 250 L of water **Need** gallons of water
Plan L → qt → gal $\dfrac{1.06 \text{ qt}}{1 \text{ L}} \quad \dfrac{1 \text{ gal}}{4 \text{ qt}}$

Set-up $250 \text{ L} \times \dfrac{1.06 \text{ qt}}{1 \text{ L}} \times \dfrac{1 \text{ gal}}{4 \text{ qt}} = 66 \text{ gal (2 SFs)}$

b. Given 0.024 g of sulfa drug, 8-mg tablets **Need** number of tablets
Plan g of sulfa drug → mg of sulfa drug → number of tablets

$$\frac{1000 \text{ mg}}{1 \text{ g}} \quad \frac{1 \text{ tablet}}{8 \text{ mg sulfa drug}}$$

Set-up $0.024 \text{ g sulfa drug} \times \dfrac{1000 \text{ mg}}{1 \text{ g}} \times \dfrac{1 \text{ tablet}}{8 \text{ mg sulfa drug}} = 3 \text{ tablets (1 SF)}$

c. Given 34-lb child, 115 mg of ampicillin/kg of body mass **Need** milligrams of ampicillin
Plan lb of body mass → kg of body mass → mg of ampicillin

$$\frac{1 \text{ kg}}{2.20 \text{ lb}} \quad \frac{115 \text{ mg ampicillin}}{1 \text{ kg body mass}}$$

Set-up $34 \text{ lb body mass} \times \dfrac{1 \text{ kg body mass}}{2.20 \text{ lb body mass}} \times \dfrac{115 \text{ mg ampicillin}}{1 \text{ kg body mass}}$
= 1800 mg of ampicillin (2 SFs)

d. Given 4.0 oz of ointment **Need** grams of ointment
Plan oz → lb → g $\dfrac{1 \text{ lb}}{16 \text{ oz}} \quad \dfrac{454 \text{ g}}{1 \text{ lb}}$

Set-up $4.0 \text{ oz} \times \dfrac{1 \text{ lb}}{16 \text{ oz}} \times \dfrac{454 \text{ g}}{1 \text{ lb}} = 110 \text{ g of ointment (2 SFs)}$

2.65 a. Given 500. mL of IV saline solution, 80. mL/h **Need** infusion time in hours
Plan mL of IV saline solution → hours $\dfrac{1 \text{ h}}{80. \text{ mL saline solution}}$

Set-up $500. \text{ mL saline solution} \times \dfrac{1 \text{ h}}{80. \text{ mL saline solution}} = 6.3 \text{ h (2 SFs)}$

b. Given 72.6-lb child, 1.5. mg of Medrol/kg of body mass, 20. mg of Medrol/mL of solution **Need** milliliters of Medrol solution

Plan lb of body mass → kg of body mass → mg of Medrol → mL of solution

$$\frac{1 \text{ kg}}{2.20 \text{ lb}} \quad \frac{1.5 \text{ mg Medrol}}{1 \text{ kg body mass}} \quad \frac{1 \text{ mL solution}}{20. \text{ mg Medrol}}$$

Set-up $72.6 \text{ lb body mass} \times \dfrac{1 \text{ kg body mass}}{2.20 \text{ lb body mass}} \times \dfrac{1.5 \text{ mg Medrol}}{1 \text{ kg body mass}} \times \dfrac{1 \text{ mL solution}}{20. \text{ mg Medrol}}$

$= 2.5 \text{ mL of Medrol solution (2 SFs)}$

2.67 Density is the mass of a substance divided by its volume. Density $= \dfrac{\text{mass (grams)}}{\text{volume (mL)}}$

The densities of solids and liquids are usually stated in g/mL or g/cm³, so in some problems the units will need to be converted.

a. Density $= \dfrac{\text{mass (grams)}}{\text{volume (mL)}} = \dfrac{24.0 \text{ g}}{20.0 \text{ mL}} = 1.20 \text{ g/mL (3 SFs)}$

b. Given 0.250 lb of butter, 130.3 mL   **Need** density (g/mL)

Plan lb → g, then calculate density $\dfrac{454 \text{ g}}{1 \text{ lb}}$

Set-up $0.250 \text{ lb} \times \dfrac{454 \text{ g}}{1 \text{ lb}} = 113.5 \text{ g (3 SFs allowed)}$

∴ Density $= \dfrac{\text{mass}}{\text{volume}} = \dfrac{113.5 \text{ g}}{130.3 \text{ mL}} = 0.871 \text{ g/mL (3 SFs)}$

c. Given 12.00 mL initial volume, 13.45 mL final volume, 4.50 g   **Need** density (g/mL)

Plan calculate volume by difference, then calculate density

Set-up volume of gem: 13.45 mL total − 12.00 mL water = 1.45 mL

∴ Density $= \dfrac{\text{mass}}{\text{volume}} = \dfrac{4.50 \text{ g}}{1.45 \text{ mL}} = 3.10 \text{ g/mL (3 SFs)}$

d. Density $= \dfrac{\text{mass}}{\text{volume}} = \dfrac{3.85 \text{ g}}{3.00 \text{ mL}} = 1.28 \text{ g/mL (3 SFs)}$

2.69 a. Given 514.1 g, 114 cm³   **Need** density (g/mL)

Plan convert volume cm³ → mL, then calculate density

Set-up $114 \text{ cm}^3 \times \dfrac{1 \text{ mL}}{1 \text{ cm}^3} = 114 \text{ mL}$

∴ Density $= \dfrac{\text{mass}}{\text{volume}} = \dfrac{514.1 \text{ g}}{114 \text{ mL}} = 4.51 \text{ g/mL (3 SFs)}$

b. Given 0.100 pt, 115.25 g initial, 182.48 g final   **Need** density (g/mL)

Plan pt → mL, then calculate mass by difference, then calculate density

$\dfrac{473 \text{ mL}}{1 \text{ pt}}$

Set-up $0.100 \text{ pt} \times \dfrac{473 \text{ mL}}{1 \text{ pt}} = 47.3 \text{ mL (3 SFs)}$

mass of syrup = 182.48 g − 115.25 g = 67.23 g

∴ Density $= \dfrac{\text{mass}}{\text{volume}} = \dfrac{67.23 \text{ g}}{47.3 \text{ mL}} = 1.42 \text{ g/mL (3 SFs)}$

Chapter 2

    c. Given 8.51 kg, 3.15 L   **Need** density (g/mL)

        Plan kg → g and L → mL, then calculate density $\dfrac{1000 \text{ g}}{1 \text{ kg}}$ $\dfrac{1000 \text{ mL}}{1 \text{ L}}$

        Set-up $8.51 \text{ kg} \times \dfrac{1000 \text{ g}}{1 \text{ kg}} = 8510 \text{ g}$ (3 SFs) and

                  $3.15 \text{ L} \times \dfrac{1000 \text{ mL}}{1 \text{ L}} = 3150 \text{ mL}$ (3 SFs)

                 ∴ Density = $\dfrac{\text{mass}}{\text{volume}} = \dfrac{8510 \text{ g}}{3150 \text{ mL}} = 2.70$ g/mL (3 SFs)

2.71 In these problems, the density is used as a conversion factor.

    a. Given 1.50 kg of ethanol   **Need** liters of ethanol

        Plan kg → g → mL → L $\dfrac{1000 \text{ g}}{1 \text{ kg}}$ $\dfrac{1 \text{ mL}}{0.79 \text{ g}}$ $\dfrac{1 \text{ L}}{1000 \text{ mL}}$

        Set-up $1.50 \text{ kg alcohol} \times \dfrac{1000 \text{ g}}{1 \text{ kg alcohol}} \times \dfrac{1 \text{ mL}}{0.79 \text{ g}} \times \dfrac{1 \text{ L}}{1000 \text{ mL}}$

                 = 1.9 L of ethanol (2 SFs)

    b. Given 6.5 mL of mercury   **Need** grams of mercury

        Plan mL → g $\dfrac{13.6 \text{ g}}{1 \text{ mL}}$

        Set-up $6.5 \text{ mL} \times \dfrac{13.6 \text{ g}}{1 \text{ mL}} = 88$ g of mercury (2 SFs)

    c. Given 225 cm³ of silver   **Need** ounces of silver

        Plan cm³ → mL → g → lb → oz $\dfrac{1 \text{ mL}}{1 \text{ cm}^3}$ $\dfrac{10.5 \text{ g}}{1 \text{ cm}^3}$ $\dfrac{1 \text{ lb}}{454 \text{ g}}$ $\dfrac{16 \text{ oz}}{1 \text{ lb}}$

        Set-up $225 \text{ cm}^3 \times \dfrac{1 \text{ mL}}{1 \text{ cm}^3} \times \dfrac{10.5 \text{ g}}{1 \text{ mL}} \times \dfrac{1 \text{ lb}}{454 \text{ g}} \times \dfrac{16 \text{ oz}}{1 \text{ lb}} = 83.3$ oz of silver (3 SFs)

2.73 a. Given 74.1 cm³ of copper   **Need** grams of copper

        Plan cm³ → mL → g $\dfrac{1 \text{ mL}}{1 \text{ cm}^3}$ $\dfrac{8.92 \text{ g}}{1 \text{ mL}}$

        Set-up $74.1 \text{ cm}^3 \times \dfrac{1 \text{ mL}}{1 \text{ cm}^3} \times \dfrac{8.92 \text{ g}}{1 \text{ mL}} = 661$ g of copper (3 SFs)

    b. Given 12.0 gal of gasoline   **Need** kilograms of gasoline

        Plan gal → qt → mL → g → kg $\dfrac{4 \text{ qt}}{1 \text{ gal}}$ $\dfrac{946 \text{ mL}}{1 \text{ qt}}$ $\dfrac{0.74 \text{ g}}{1 \text{ mL}}$ $\dfrac{1 \text{ kg}}{1000 \text{ g}}$

        Set-up $12.0 \text{ gal} \times \dfrac{4 \text{ qt}}{1 \text{ gal}} \times \dfrac{946 \text{ mL}}{1 \text{ qt}} \times \dfrac{0.74 \text{ g}}{1 \text{ mL}} \times \dfrac{1 \text{ kg}}{1000 \text{ g}} = 34$ kg of gasoline (2 SFs)

    c. Given 27 g of ice   **Need** cubic centimeters of ice

        Plan g → mL → cm³ $\dfrac{1 \text{ mL}}{0.92 \text{ g}}$ $\dfrac{1 \text{ cm}^3}{1 \text{ mL}}$

        Set-up $27 \text{ g} \times \dfrac{1 \text{ mL}}{0.92 \text{ g}} \times \dfrac{1 \text{ cm}^3}{1 \text{ mL}} = 29$ cm³ (2 SFs)

2.75 Because the density of aluminum is 2.70 g/cm³, silver is 10.5 g/cm³, and lead is 11.3 g/cm³, we can identify the unknown metal by calculating its density as follows:

    Density = $\dfrac{\text{mass of metal}}{\text{volume of metal}} = \dfrac{217 \text{ g}}{19.2 \text{ cm}^3} = 11.3$ g/cm³ (3 SFs)

    ∴ the metal is lead.

2.77 a. Specific gravity = $\dfrac{\text{density of substance}}{\text{density of water}} = \dfrac{1.030 \text{ g/mL}}{1.00 \text{ g/mL}} = 1.03$ (3 SFs)

b. Density = $\dfrac{\text{mass of glucose solution}}{\text{volume of glucose solution}} = \dfrac{20.6 \text{ g}}{20.0 \text{ mL}} = 1.03$ g/mL (3 SFs)

c. Specific gravity = $\dfrac{\text{density of substance}}{\text{density of water}}$

∴ Density of substance = specific gravity × density of water
  = 0.92 × 1.00 g/mL = 0.92 g/mL
Mass of substance = volume of substance × density of substance
  = 750 mL × 0.92 g/mL = 690 g (2 SFs)

d. Specific gravity = $\dfrac{\text{density of substance}}{\text{density of water}}$

∴ Density of substance = specific gravity × density of water
  = 0.850 × 1.00 g/mL = 0.850 g/mL

∴ Volume of solution = $\dfrac{\text{mass of solution}}{\text{density of solution}} = \dfrac{325 \text{ g}}{0.850 \text{ g/mL}} = 382$ mL (3 SFs)

2.79 a. 42 mcg of iron = 1 dL of blood; $\dfrac{42 \text{ mcg iron}}{1 \text{ dL blood}}$ and $\dfrac{1 \text{ dL blood}}{42 \text{ mcg iron}}$

b. Given 8.0-mL blood sample, 42 mcg of iron/dL   **Need** micrograms of iron
Plan mL of blood → dL of blood → mcg of iron $\dfrac{1 \text{ dL blood}}{100 \text{ mL blood}}$ $\dfrac{42 \text{ mcg iron}}{1 \text{ dL blood}}$
Set-up 8.0 mL blood × $\dfrac{1 \text{ dL blood}}{100 \text{ mL blood}}$ × $\dfrac{42 \text{ mcg iron}}{1 \text{ dL blood}}$ = 3.4 mcg of iron (2 SFs)

2.81 Both measurements in part **c** have two significant figures, and both measurements in part **d** have four significant figures.

2.83 a. The number of legs is a counted number; it is exact.
b. The height is measured with a ruler or tape measure; it is a measured number.
c. The number of chairs is a counted number; it is exact.
d. The area is measured with a ruler or tape measure; it is a measured number.

2.85 61.5 °C

2.87 a. length = 38.4 in. × $\dfrac{2.54 \text{ cm}}{1 \text{ in.}}$ = 97.5 cm (3 SFs)

b. length = 24.2 in. × $\dfrac{2.54 \text{ cm}}{1 \text{ in.}}$ = 61.5 cm (3 SFs)

c. There are three significant figures in the length measurement.
d. Area = length × width = 97.5 cm × 61.5 cm = 6.00×10^3 cm² (3 SFs)

2.89 a. Diagram 3; a cube that has a greater density than the water will sink to the bottom.
b. Diagram 4; a cube with a density of 0.80 g/mL will be about four-fifths submerged in the water.
c. Diagram 1; a cube with a density that is one-half the density of water will be one-half submerged in the water.
d. Diagram 2; a cube with the same density as water will float just at the surface of the water.

2.91 Since all three solids have a mass of 10.0 g, the one with the smallest volume must have the highest density; the one with the largest volume will have the lowest density.
A would be gold; it has the highest density (19.3 g/mL) and the smallest volume.
B would be silver; its density is intermediate (10.5 g/mL) and the volume is intermediate.
C would be aluminum; it has the lowest density (2.70 g/mL) and the largest volume.

Chapter 2

2.93 The green cube has the same volume as the gray cube. However, the green cube has a larger mass on the scale, which means that its mass/volume ratio is larger. Thus, the density of the green cube is higher than the density of the gray cube.

2.95 **a.** To round off 0.000 012 58 L to three significant figures, drop the final digit 8 and increase the last retained digit by 1 to give 0.000 012 6 L or 1.26×10^{-5} L.
  b. To round off 3.528×10^2 kg to three significant figures, drop the final digit 8 and increase the last retained digit by 1 to give 353 kg (3.53×10^2 kg).
  c. To express 125 111 m to three significant figures, drop the final digits 111 and add three zeros as placeholders to give 125 000 m (or 1.25×10^5 m).
  d. To express 34.9673 s to three significant figures, drop the final digits 673 and increase the last retained digit by 1 to give 35.0 s.

2.97 **a.** The total mass is the sum of the individual components of the dessert.
  137.25 g + 84 g + 43.7 g = 265 g. No places to the right of the decimal point are allowed since the mass of the fudge sauce (84 g) has no digits to the right of the decimal point.
  b. **Given** grams of dessert from part **a**   **Need** pounds of dessert
  Plan g → lb $\dfrac{1 \text{ lb}}{454 \text{ g}}$
  Set-up 265 g̶ dessert (total) $\times \dfrac{1 \text{ lb}}{454 \text{ g̶}}$ = 0.584 lb of dessert (3 SFs)

2.99 **Given** 1.95 euros/kg of grapes, \$1.14/euro   **Need** cost in dollars per pound
  Plan euros/kg → euros/lb → \$/lb $\dfrac{1.95 \text{ euros}}{1 \text{ kg grapes}}$ $\dfrac{1 \text{ kg}}{2.20 \text{ lb}}$ $\dfrac{\$1.14}{1 \text{ euro}}$
  Set-up $\dfrac{1.95 \text{ e̶u̶r̶o̶s̶}}{1 \text{ k̶g̶ g̶r̶a̶p̶e̶s̶}} \times \dfrac{1 \text{ k̶g̶}}{2.20 \text{ lb}} \times \dfrac{\$1.14}{1 \text{ e̶u̶r̶o̶}}$ = \$1.01/lb of grapes (3 SFs)

2.101 Given 4.0 lb of onions   **Need** number of onions
  Plan lb → g → number of onions $\dfrac{454 \text{ g}}{1 \text{ lb}}$ $\dfrac{1 \text{ onion}}{115 \text{ g}}$
  Set-up 4.0 l̶b̶ o̶n̶i̶o̶n̶s̶ $\times \dfrac{454 \text{ g̶}}{1 \text{ l̶b̶}} \times \dfrac{1 \text{ onion}}{115 \text{ g̶}}$ = 16 onions (2 SFs)

2.103 This problem requires several conversion factors. Let's take a look first at a possible unit plan. When you write out the unit plan, be sure you know a conversion factor you can use for each step.
  Given 7500 ft   **Need** minutes
  Plan ft → in. → cm → m → min $\dfrac{12 \text{ in.}}{1 \text{ ft}}$ $\dfrac{2.54 \text{ cm}}{1 \text{ in.}}$ $\dfrac{1 \text{ m}}{100 \text{ cm}}$ $\dfrac{1 \text{ min}}{55.0 \text{ m}}$
  Set-up 7500 f̶t̶ $\times \dfrac{12 \text{ i̶n̶.}}{1 \text{ f̶t̶}} \times \dfrac{2.54 \text{ c̶m̶}}{1 \text{ i̶n̶.}} \times \dfrac{1 \text{ m̶}}{100 \text{ c̶m̶}} \times \dfrac{1 \text{ min}}{55.0 \text{ m̶}}$ = 42 min (2 SFs)

2.105 Given 215 mL initial, 285 mL final volume, density of lead 11.3 g/mL   **Need** grams of lead
  Plan calculate the volume by difference and mL → g $\dfrac{11.3 \text{ g}}{1 \text{ mL}}$
  Set-up The difference between the initial volume of the water and its volume with the lead object will give us the volume of the lead object: 285 mL total − 215 mL water = 70. mL of lead, then
  70. m̶L̶ l̶e̶a̶d̶ $\times \dfrac{11.3 \text{ g lead}}{1 \text{ m̶L̶ l̶e̶a̶d̶}}$ = 790 g of lead (2 SFs)

2.107 Given 1.2 kg of gasoline **Need** milliliters of gasoline

Plan kg → g → mL $\dfrac{1000 \text{ g}}{1 \text{ mL}}$ $\dfrac{1 \text{ mL}}{0.74 \text{ g}}$

Set-up $1.2 \text{ kg} \times \dfrac{1000 \text{ g}}{1 \text{ kg}} \times \dfrac{1 \text{ mL}}{0.74 \text{ g}} = 1600 \text{ mL } (1.6 \times 10^3 \text{ mL})$ of gasoline (2 SFs)

2.109 a. Given 8.0 oz **Need** number of crackers

Plan oz → number of crackers $\dfrac{6 \text{ crackers}}{0.50 \text{ oz}}$

Set-up $8.0 \text{ oz} \times \dfrac{6 \text{ crackers}}{0.50 \text{ oz}} = 96$ crackers (2 SFs)

b. Given 10 crackers, 4 g of fat/serving **Need** ounces of fat

Plan number of crackers → servings → g of fat → lb of fat → oz of fat

$\dfrac{1 \text{ serving}}{6 \text{ crackers}}$ $\dfrac{4 \text{ g fat}}{1 \text{ serving}}$ $\dfrac{1 \text{ lb}}{454 \text{ g}}$ $\dfrac{16 \text{ oz}}{1 \text{ lb}}$

Set-up $10 \text{ crackers} \times \dfrac{1 \text{ serving}}{6 \text{ crackers}} \times \dfrac{4 \text{ g fat}}{1 \text{ serving}} \times \dfrac{1 \text{ lb}}{454 \text{ g}} \times \dfrac{16 \text{ oz}}{1 \text{ lb}} = 0.2$ oz of fat (1 SF)

c. Given 2.4 g of sodium, 140 mg of sodium/serving **Need** number of servings

Plan mg of sodium → servings $\dfrac{100 \text{ mg sodium}}{1 \text{ g sodium}}$ $\dfrac{1 \text{ serving}}{140 \text{ mg sodium}}$

Set-up $2.4 \text{ g sodium} \times \dfrac{1000 \text{ mg sodium}}{1 \text{ g sodium}} \times \dfrac{1 \text{ serving}}{140 \text{ mg sodium}} = 17$ servings (2 SFs)

2.111 Given 10 days, 4 tablets/day, 250-mg tablets **Need** ounces of amoxicillin

Plan days → tablets → mg of amoxicillin → g → lb → oz of amoxicillin

$\dfrac{4 \text{ tablets}}{1 \text{ day}}$ $\dfrac{250 \text{ mg amoxicillin}}{1 \text{ tablet}}$ $\dfrac{1 \text{ g}}{1000 \text{ mg}}$ $\dfrac{1 \text{ lb}}{454 \text{ g}}$ $\dfrac{16 \text{ oz}}{1 \text{ lb}}$

Set-up $10 \text{ days} \times \dfrac{4 \text{ tablets}}{1 \text{ day}} \times \dfrac{250 \text{ mg amoxicillin}}{1 \text{ tablet}} \times \dfrac{1 \text{ g}}{1000 \text{ mg}} \times \dfrac{1 \text{ lb}}{454 \text{ g}} \times \dfrac{16 \text{ oz}}{1 \text{ lb}}$
$= 0.35$ oz of amoxicillin (2 SFs)

2.113 Given 5.0 mL of elixir, 30. mg of phenobarbital/7.5 mL
Need milligrams of phenobarbital

Plan mL of elixir → mg of phenobarbital $\dfrac{30. \text{ mg phenobarbital}}{7.5 \text{ mL elixir}}$

Set-up $5.0 \text{ mL elixir} \times \dfrac{30. \text{ mg phenobarbital}}{7.5 \text{ mL elixir}} = 20.$ mg of phenobarbital (2 SFs)

2.115 Because the balance can measure mass to 0.001 g, the mass should be reported to 0.001 g. You should record the mass of the object as 31.075 g.

2.117 Given 3.0-h trip **Need** gallons of gasoline

Plan h → mi → km → L → qt → gal $\dfrac{55 \text{ mi}}{1 \text{ h}}$ $\dfrac{1 \text{ km}}{0.621 \text{ mi}}$ $\dfrac{1 \text{ L}}{11 \text{ km}}$ $\dfrac{1.06 \text{ qt}}{1 \text{ L}}$ $\dfrac{1 \text{ gal}}{4 \text{ qt}}$

Set-up $3.0 \text{ h} \times \dfrac{55 \text{ mi}}{1 \text{ h}} \times \dfrac{1 \text{ km}}{0.621 \text{ mi}} \times \dfrac{1 \text{ L}}{11 \text{ km}} \times \dfrac{1.06 \text{ qt}}{1 \text{ L}} \times \dfrac{1 \text{ gal}}{4 \text{ qt}}$
$= 6.4$ gal of gasoline (2 SFs)

2.119 Given 1.50 L of gasoline **Need** milliliters of olive oil
Plan L of gasoline → mL of gasoline → g of gasoline → g of olive oil → mL of olive oil
(equality from question: 1 g of olive oil = 1 g of gasoline)

$$\frac{1000 \text{ mL gasoline}}{1 \text{ L gasoline}} \quad \frac{0.74 \text{ g gasoline}}{1 \text{ mL gasoline}} \quad \frac{1 \text{ mL olive oil}}{0.92 \text{ g olive oil}}$$

Set-up $1.50 \text{ L gasoline} \times \dfrac{1000 \text{ mL}}{1 \text{ L}} \times \dfrac{0.74 \text{ g gasoline}}{1 \text{ mL gasoline}} = 1110 \text{ g of gasoline}$

$1110 \text{ g olive oil} \times \dfrac{1 \text{ mL olive oil}}{0.92 \text{ g olive oil}} = 1200 \text{ mL } (1.2 \times 10^3 \text{ mL}) \text{ of olive oil (2 SFs)}$

2.121 a. Given 65 kg of body mass, 3.0% fat **Need** pounds of fat
Plan kg of body mass → kg of fat → lb of fat
(percent equality: 100 kg of body mass = 3.0 kg of fat)

$$\frac{3.0 \text{ kg fat}}{100 \text{ kg body mass}} \quad \frac{2.20 \text{ lb fat}}{1 \text{ kg fat}}$$

Set-up $65 \text{ kg body mass} \times \dfrac{3.0 \text{ kg fat}}{100 \text{ kg body mass}} \times \dfrac{2.20 \text{ lb fat}}{1 \text{ kg fat}} = 4.3 \text{ lb of fat (2 SFs)}$

b. Given 3.0 L of fat **Need** pounds of fat
Plan L → mL → g → lb

$$\frac{1000 \text{ mL}}{1 \text{ L}} \quad \frac{0.909 \text{ g}}{1 \text{ mL}} \quad \frac{1 \text{ lb}}{454 \text{ g}}$$

Set-up $3.0 \text{ L} \times \dfrac{1000 \text{ mL}}{1 \text{ L}} \times \dfrac{0.909 \text{ g}}{1 \text{ mL}} \times \dfrac{1 \text{ lb}}{454 \text{ g}} = 6.0 \text{ lb of fat (2 SFs)}$

3
Matter and Energy

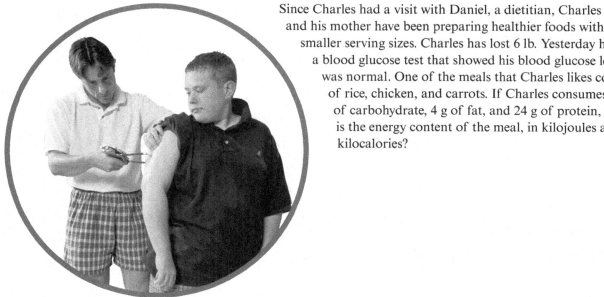

Since Charles had a visit with Daniel, a dietitian, Charles and his mother have been preparing healthier foods with smaller serving sizes. Charles has lost 6 lb. Yesterday he had a blood glucose test that showed his blood glucose level was normal. One of the meals that Charles likes consists of rice, chicken, and carrots. If Charles consumes 31 g of carbohydrate, 4 g of fat, and 24 g of protein, what is the energy content of the meal, in kilojoules and kilocalories?

Credit: Network Photographer/Alamy

LOOKING AHEAD

3.1 Classification of Matter
3.2 States and Properties of Matter
3.3 Temperature
3.4 Energy
3.5 Energy and Nutrition
3.6 Specific Heat
3.7 Changes of State

 The Health icon indicates a question that is related to health and medicine.

3.1 Classification of Matter

Learning Goal: Classify examples of matter as pure substances or mixtures.

- Matter is anything that has mass and occupies space.
- A pure substance, whether it is an element or compound, has a definite composition.
- Elements are the simplest type of matter.
- A compound consists of atoms of two or more elements always chemically combined in the same proportion.
- Mixtures contain two or more substances that are physically, not chemically, combined.
- Mixtures are classified as homogeneous or heterogeneous.

Chapter 3

Key Terms for Section 3.1

Match each of the following key terms with the correct description:

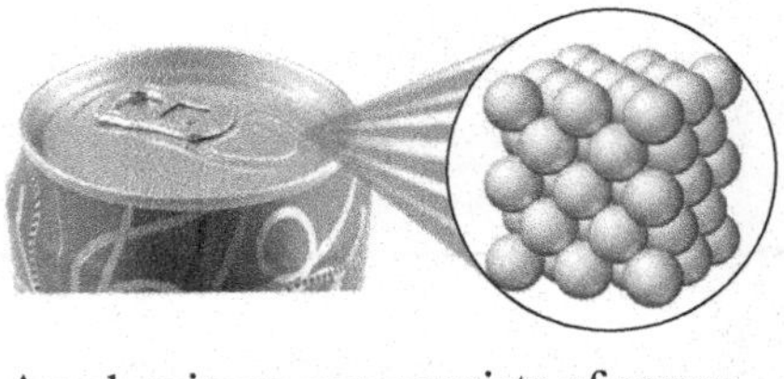

    a. element   **b.** mixture   **c.** pure substance
    d. matter   **e.** compound

1. ____ anything that has mass and occupies space
2. ____ an element or a compound that has a definite composition
3. ____ the physical combination of two or more substances that does not change the identities of the substances
4. ____ a pure substance consisting of two or more elements with a definite composition that can be broken down into simpler substances only by chemical methods
5. ____ a pure substance containing only one type of matter that cannot be broken down by chemical methods

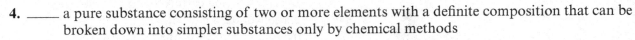

An aluminum can consists of many atoms of the element aluminum.
Credit: Norman Chan / Fotolia

Answers   **1.** d   **2.** c   **3.** b   **4.** e   **5.** a

♦ **Learning Exercise 3.1A**

Identify each of the following as an element or a compound:

a. _____ iron   **b.** _____ carbon dioxide
c. _____ potassium iodide   **d.** _____ gold
e. _____ aluminum   **f.** _____ table salt (sodium chloride)

Answers   **a.** element   **b.** compound   **c.** compound
    d. element   **e.** element   **f.** compound

♦ **Learning Exercise 3.1B**

Identify each of the following as a pure substance or a mixture:

a. _____ bananas and milk   **b.** _____ sulfur
c. _____ silver   **d.** _____ a bag of raisins and nuts
e. _____ pure water   **f.** _____ sand and water

Water, H_2O, consists of two atoms of H for one atom of O.
Credit: Pearson Education, Inc.

Answers   **a.** mixture   **b.** pure substance   **c.** pure substance
    d. mixture   **e.** pure substance   **f.** mixture

♦ **Learning Exercise 3.1C**

Identify each of the following mixtures as homogeneous or heterogeneous:

a. _____ chocolate milk   **b.** _____ sand and water
c. _____ orange soda   **d.** _____ a bag of raisins and nuts
e. _____ air   **f.** _____ vinegar

Answers   **a.** homogeneous   **b.** heterogeneous   **c.** homogeneous
    d. heterogeneous   **e.** homogeneous   **f.** homogeneous

Matter and Energy

3.2 States and Properties of Matter

Learning Goal: Identify the states and the physical and chemical properties of matter.

- The states of matter are solid, liquid, and gas.
- Physical properties are those characteristics of a substance that can change without affecting the identity of the substance.
- A substance undergoes a physical change when its shape, size, or state changes, but its composition does not change.
- Chemical properties are those characteristics of a substance that describe the ability of a substance to change into a new substance.
- A substance undergoes a chemical change when the original substance is converted into one or more new substances, which have different physical and chemical properties.
- Melting, freezing, boiling, and condensing are typical changes of state.

♦ **Learning Exercise 3.2A**

State whether each of the following statements describes a gas, a liquid, or a solid:

a. _____ There are no attractions among the particles.
b. _____ The particles are held close together in a definite pattern.
c. _____ This substance has a definite volume but no definite shape.
d. _____ The particles are moving extremely fast.
e. _____ This substance has no definite shape and no definite volume.
f. _____ The particles in this substance are vibrating slowly in fixed positions.
g. _____ This substance has a definite volume and a definite shape.

Answers   **a.** gas   **b.** solid   **c.** liquid   **d.** gas
           e. gas   **f.** solid   **g.** solid

♦ **Learning Exercise 3.2B**

Classify each of the following as a physical or chemical property:

a. _____ Silver is shiny.   b. _____ Water is a liquid at 25 °C.
c. _____ Wood burns.   d. _____ Mercury is a very dense liquid.
e. _____ Helium is not reactive. f. _____ Ice cubes float in water.

Answers   **a.** physical   **b.** physical   **c.** chemical
           d. physical   **e.** chemical   **f.** physical

CORE CHEMISTRY SKILL
Identifying Physical and Chemical Changes

♦ **Learning Exercise 3.2C**

Classify each of the following changes as physical or chemical:

a. _____ Sodium melts at 98 °C.   b. _____ Iron forms rust in air and water.
c. _____ Water condenses on a cold window. d. _____ Fireworks explode when ignited.
e. _____ Gasoline burns in a car engine.   f. _____ Paper is cut to make confetti.

Answers   **a.** physical   **b.** chemical   **c.** physical
           d. chemical   **e.** chemical   **f.** physical

Chapter 3

3.3 Temperature

Learning Goal: Given a temperature, calculate a corresponding temperature on another scale.

- Temperature is measured in degrees Celsius (°C) or kelvins (K). In the United States, the Fahrenheit scale (°F) is still in use.
- The equation $T_F = 1.8(T_C) + 32$ is used to convert a Celsius temperature to a Fahrenheit temperature.
- When rearranged for T_C, this equation is used to convert a Fahrenheit temperature to a Celsius temperature.

$$T_C = \frac{T_F - 32}{1.8}$$

- The temperature on the Celsius scale is related to the temperature on the Kelvin scale: $T_K = T_C + 273$.

	Solution Guide to Calculating Temperature
STEP 1	State the given and needed quantities.
STEP 2	Write a temperature equation.
STEP 3	Substitute in the known values and calculate the new temperature.

♦ **Learning Exercise 3.3**

Calculate the temperature in each of the following problems:

CORE CHEMISTRY SKILL
Converting between Temperature Scales

a. To prepare yogurt, milk is warmed to 185 °F. What Celsius temperature is needed to prepare the yogurt?

b. A heat-sensitive vaccine is stored at −12 °C. What is that temperature on a Fahrenheit thermometer?

c. A patient has a temperature of 39.5 °C. What is that temperature on the Fahrenheit scale?

d. A patient in the emergency room with hyperthermia has a temperature of 105 °F. What is the temperature on the Celsius scale?

e. The temperature in an autoclave used to sterilize surgical equipment is 253 °F. What is that temperature on the Celsius scale?

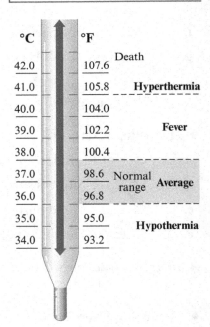

Very high temperatures (hyperthermia) can lead to convulsions and brain damage.
Credit: Digital Vision/Alamy

f. Liquid nitrogen at −196 °C is used to freeze and remove precancerous skin growths. What temperature will this be on the Kelvin scale?

Answers   **a.** 85.0 °C   **b.** 10 °F   **c.** 103.1 °F   **d.** 41 °C   **e.** 123 °C   **f.** 77 K

Matter and Energy

3.4 Energy

Learning Goal: Identify energy as potential or kinetic; convert between units of energy.
- Energy is the ability to do work.
- Potential energy is stored energy, which is determined by position or composition; kinetic energy is the energy of motion.
- The SI unit of energy is the joule (J), the metric unit is the calorie (cal).

$$1 \text{ cal} = 4.184 \text{ J}$$

$$\frac{4.184 \text{ J}}{1 \text{ cal}} \quad \text{and} \quad \frac{1 \text{ cal}}{4.184 \text{ J}}$$

Key Terms for Sections 3.2 to 3.4

Match each of the following key terms with the correct description:

a. potential energy   b. joule   c. kinetic energy
d. calorie   e. energy   f. physical change

1. _____ the SI unit of energy
2. _____ the energy of motion
3. _____ the amount of heat energy that raises the temperature of exactly 1 g of water exactly 1 °C
4. _____ a change in physical properties of a substance with no change in its identity
5. _____ the ability to do work
6. _____ a type of energy that is stored for future use

Answers   1. b   2. c   3. d   4. f   5. e   6. a

♦ Learning Exercise 3.4A

Indicate whether each of the following statements describes potential or kinetic energy:

a. _____ a potted plant sitting on a ledge   b. _____ water flowing down a stream
c. _____ logs sitting in a fireplace   d. _____ a piece of candy
e. _____ an arrow shot from a bow   f. _____ a ski jumper standing at the top of the ski jump
g. _____ a jogger running   h. _____ a skydiver waiting to jump
i. _____ your breakfast cereal   j. _____ a bowling ball striking the pins

Answers   a. potential   b. kinetic   c. potential   d. potential   e. kinetic
           f. potential   g. kinetic   h. potential   i. potential   j. kinetic

♦ Learning Exercise 3.4B

Match the words with the definitions below.

CORE CHEMISTRY SKILL
Using Energy Units

a. calorie   b. kilocalorie   c. joule

1. _____ the SI unit of work and energy
2. _____ the heat needed to raise 1 g of water by 1 °C
3. _____ 1000 cal

Answers   1. c   2. a   3. b

Chapter 3

♦ **Learning Exercise 3.4C**

Convert each of the following energy units:

a. 58 000 cal to kcal

b. 3450 J to cal

c. 2.8 kJ to cal

d. 15 200 cal to kJ

Answers   **a.** 58 kcal   **b.** 825 cal   **c.** 670 cal   **d.** 63.6 kJ

3.5 Energy and Nutrition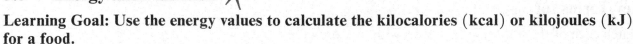

Learning Goal: Use the energy values to calculate the kilocalories (kcal) or kilojoules (kJ) for a food.

- The nutritional calorie (Cal) is the same amount of energy as 1 kcal, or 1000 calories.
- When a substance is burned in a calorimeter, the water that surrounds the reaction chamber absorbs the heat given off. The heat absorbed by the water is calculated, and the energy content for the substance (energy per gram) is determined.
- The energy values for three food types are: carbohydrate 4 kcal/g (17 kJ/g), fat 9 kcal/g (38 kJ/g), protein 4 kcal/g (17 kJ/g).
- The energy content of a food is the sum of kilocalories or kilojoules from carbohydrate, fat, and protein.

Solution Guide to Calculating the Energy from a Food	
STEP 1	State the given and needed quantities.
STEP 2	Use the energy value for each food type to calculate the kilocalories or kilojoules, rounded off to the tens place.
STEP 3	Add the energy for each food type to give the total energy from the food.

♦ **Learning Exercise 3.5A**

Calculate the kilojoules (kJ) for each food using the following data (round off the kilojoules for each food type to the tens place):

Food	Carbohydrate	Fat	Protein	kJ
a. green peas, cooked, 1 cup	19 g	1 g	9 g	_____
b. potato chips, 10 chips	10 g	8 g	1 g	_____
c. cream cheese, 8 oz	5 g	86 g	18 g	_____
d. lean hamburger, 3 oz	0 g	10 g	23 g	_____
e. banana, 1 medium	26 g	0 g	1 g	_____

Answers   **a.** 510 kJ   **b.** 490 kJ   **c.** 3670 kJ
          d. 770 kJ   **e.** 460 kJ

Matter and Energy

♦ **Learning Exercise 3.5B**

Use energy values to calculate each of the following:

a. Complete the following table listing ingredients for a peanut butter sandwich (round off the kilocalories for each food type to the tens place):

	Carbohydrate	Fat	Protein	kcal
Bread, 2 slices	30 g	0 g	5 g	_____
Peanut butter, 2 tbsp	10 g	13 g	8 g	_____
Jelly, 2 tsp	10 g	0 g	0 g	_____
Margarine, 1 tsp	0 g	7 g	0 g	_____

Total kcal in sandwich _____

b. How many kilocalories are in a single serving of pudding that contains 31 g of carbohydrate, 5 g of fat, and 5 g of protein (round off the kilocalories for each food type to the tens place)?

c. One bagel with light cream cheese has an energy content of 380 kcal. If there are 58 g of carbohydrate and 14 g of protein, how many grams of fat are in one bagel with light cream cheese (round off the kilocalories for each food type to the tens place)?

d. A serving of breakfast cereal provides 220 kcal. In this serving, there are 6 g of fat and 8 g of protein. How many grams of carbohydrate are in the cereal (round off the kilocalories for each food type to the tens place)?

Answers
a. bread = 140 kcal; peanut butter = 190 kcal; jelly = 40 kcal; margarine = 60 kcal
total kcal in sandwich = 430 kcal
b. carbohydrate = 120 kcal; fat = 50 kcal; protein = 20 kcal; total = 190 kcal
c. carbohydrate = 230 kcal; protein = 60 kcal
380 kcal − 230 kcal − 60 kcal = 90 kcal from fat

$$90 \text{ kcal} \times \frac{1 \text{ g fat}}{9 \text{ kcal}} = 10 \text{ g of fat}$$

d. fat = 50 kcal; protein = 30 kcal; 30 kcal + 50 kcal = 80 kcal from fat and protein
220 kcal − 80 kcal = 140 kcal from carbohydrate

$$140 \text{ kcal} \times \frac{1 \text{ g carbohydrate}}{4 \text{ kcal}} = 35 \text{ g of carbohydrate}$$

Chapter 3

3.6 Specific Heat

Learning Goal: Use specific heat to calculate heat loss or gain.

- Specific heat is the amount of energy required to raise the temperature of 1 g of a substance by 1 °C.
- The specific heat for liquid water is 1.00 cal/g °C or 4.184 J/g °C.
- The heat lost or gained can be calculated using the mass of the substance, temperature difference, and its specific heat (SH): Heat = mass $\times \Delta T \times SH$

	Solution Guide to Calculations Using Specific Heat
STEP 1	State the given and the needed quantities.
STEP 2	Calculate the temperature change (ΔT).
STEP 3	Write the heat equation and needed conversion factors.
STEP 4	Substitute in the given values and calculate the heat, making sure units cancel.

♦ **Learning Exercise 3.6**

CORE CHEMISTRY SKILL
Using the Heat Equation

The specific heat for water is 1.00 cal/g °C or 4.184 J/g °C. Calculate the kilocalories (kcal) and kilojoules (kJ) gained or released during the following:

a. heating 20.0 g of water from 22 °C to 77 °C

b. heating 10.0 g of water from 12.4 °C to 67.5 °C

c. cooling 0.450 kg of water from 80.0 °C to 35.0 °C

d. cooling 125 g of water from 72.0 °C to 45.0 °C

Answers
 a. 1.1 kcal, 4.6 kJ (gained) b. 0.551 kcal, 2.31 kJ (gained)
 c. 20.3 kcal, 84.7 kJ (released) d. 3.38 kcal, 14.1 kJ (released)

3.7 Changes of State

Learning Goal: Describe the changes of state between solids, liquids, and gases; calculate the energy released or absorbed.

CORE CHEMISTRY SKILL
Calculating Heat for Change of State

- A substance melts and freezes at its melting (freezing) point. During the process of melting or freezing, the temperature remains constant.
- The heat of fusion is the energy required to change 1 g of solid to liquid at the melting point. For ice to melt at 0 °C, 80. cal/g (or 334 J/g), is required. This is also the amount of heat lost when 1 g of water freezes at 0 °C.
- A substance boils and condenses at its boiling point. During the process of boiling or condensing, the temperature remains constant.
- The heat of vaporization is the energy required to change 1 g of liquid to 1 g of gas at the boiling point. For water to boil at 100 °C, 540 cal/g (or 2260 J/g), is required to change 1 g of liquid to 1 g of gas (steam); it is also the amount of heat released when 1 g of water vapor condenses at 100 °C.

- Evaporation is a surface phenomenon, while boiling occurs throughout the liquid.
- Sublimation is the change of state from a solid directly to a gas. Deposition is the opposite process.
- A heating or cooling curve illustrates the changes in temperature and state as heat is added to or removed from a substance.
- When a substance is heated or cooled, the energy gained or released is the total of the energy involved in temperature changes as well as the energy involved in changes of state.

Key Terms for Sections 3.5 to 3.7

Match each of the following key terms with the correct description:

    a. energy value   **b.** specific heat   **c.** Calorie (Cal)   **d.** sublimation   **e.** heat of vaporization

1. _____ a nutritional unit of energy equal to 1000 cal or 1 kcal
2. _____ a quantity of heat that changes the temperature of exactly 1 g of a substance by exactly 1 °C
3. _____ the number of kilocalories or kilojoules obtained per gram of carbohydrate, fat, or protein
4. _____ a substance changes directly from a solid to a gas
5. _____ the energy required to convert exactly 1 g of a liquid to vapor at its boiling point

Answers   **1.** c   **2.** b   **3.** a   **4.** d   **5.** e

♦ Learning Exercise 3.7A

Identify each of the following as

    a. melting   **b.** freezing   **c.** sublimation   **d.** condensation

1. _____ A liquid changes to a solid.
2. _____ Dry ice in an ice cream cart changes to a gas.
3. _____ Ice forms on the surface of a lake in winter.
4. _____ Butter in a hot pan turns to liquid.
5. _____ A gas changes to a liquid.

Answers   **1.** b   **2.** c   **3.** b   **4.** a   **5.** d

Melting and freezing are reversible processes.
Credit: John A. Rizzo / Photodisc / Getty Images

Solution Guide to Calculations Using a Heat Conversion Factor	
STEP 1	State the given and needed quantities.
STEP 2	Write a plan to convert the given quantity to the needed quantity.
STEP 3	Write the heat conversion factor and any metric factor.
STEP 4	Set up the problem and calculate the needed quantity.

Chapter 3

♦ Learning Exercise 3.7B

Calculate each of the following when a substance melts or freezes:

a. How many joules of heat are needed to melt 24.0 g of ice at 0 °C?

b. How much heat, in kilojoules, is released when 325 g of water freezes at 0 °C?

An ice bag is used to treat a sports injury.
Credit: wsphotos/Getty Images

c. How many kilocalories of heat are required to melt 120 g of ice in an ice bag at 0 °C?

d. How many grams of ice would melt when 1200 cal of heat is absorbed?

Answers   a. 8020 J   b. 109 kJ   c. 9.6 kcal   d. 15 g

♦ Learning Exercise 3.7C

Calculate each of the following when a substance boils or condenses:

a. How many joules are needed to completely change 7.20 g of water to vapor at 100 °C?

b. How many kilocalories are released when 42 g of steam at 100 °C condenses to form liquid water at 100 °C?

c. How many grams of water can be converted to steam at 100 °C when 155 kJ of energy is absorbed?

d. How many grams of steam condense if 24 kcal is released at 100 °C?

Answers   a. 16 300 J   b. 23 kcal   c. 68.6 g   d. 44 g

♦ Learning Exercise 3.7D

On each heating or cooling curve, indicate the portion that corresponds to a solid, liquid, or gas and the changes of state.

1. Draw a heating curve for water that begins at −20 °C and ends at 120 °C.

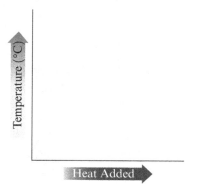

2. Draw a heating curve for bromine from −25 °C to 75 °C. Bromine has a melting point of −7 °C and a boiling point of 59 °C.

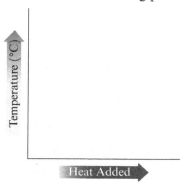

3. Draw a cooling curve for sodium from 1000 °C to 0 °C. Sodium has a freezing point of 98 °C and a boiling (condensation) point of 883 °C.

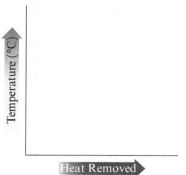

Answers

1.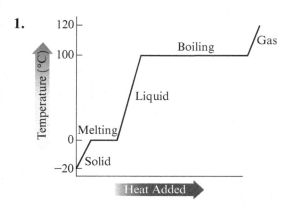

2.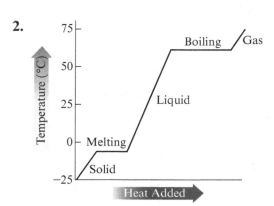

Chapter 3

3.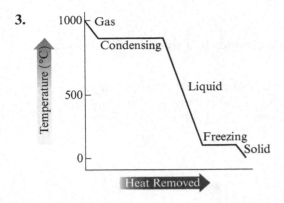

♦ **Learning Exercise 3.7E**

Calculate the energy change in each of the following:

a. How many kilojoules are released when 35 g of water at 65 °C cools to 0 °C and completely changes to solid? (*Hint*: Two steps are needed.)

b. How many kilocalories are needed to melt 15 g of ice at 0 °C, heat the water to 100 °C, and convert the water to gas at 100 °C? (*Hint*: Three steps are needed.)

Answers   a. 21 kJ   b. 11 kcal

Checklist for Chapter 3

You are ready to take the Practice Test for Chapter 3. Be sure you have accomplished the following learning goals for this chapter. If not, review the Section listed at the end of the goal. Then apply your new skills and understanding to the Practice Test.

After studying Chapter 3, I can successfully:

____ Classify matter as a pure substance or mixture. (3.1)

____ Classify a mixture as homogeneous or heterogeneous. (3.1)

____ Identify the physical state of a substance as a solid, liquid, or gas. (3.2)

____ Calculate a temperature in degrees Fahrenheit, degrees Celsius, or kelvins. (3.3)

____ Describe some forms of energy. (3.4)

____ Change a quantity in one energy unit to another energy unit. (3.4)

____ Using the energy values, calculate the energy, in kilocalories (kcal) or kilojoules (kJ), for a food sample. (3.5)

_____ Given the mass of a sample, specific heat, and the temperature change, calculate the heat lost or gained. (3.6)

_____ Calculate the heat change for a change of state for a specific amount of a substance. (3.7)

_____ Calculate the total heat for a combination of change of temperature and change of state for a specific amount of a substance. (3.7)

_____ Draw heating and cooling curves using the melting and boiling points of a substance. (3.7)

Practice Test for Chapter 3

The chapter Sections to review are shown in parentheses at the end of each question.

For questions 1 through 4, classify each of the following as a pure substance (P) *or a mixture* (M): (3.1)

1. _____ toothpaste
2. _____ platinum
3. _____ chromium
4. _____ mouthwash

For questions 5 through 8, classify each of the following mixtures as homogeneous (Ho) *or heterogeneous* (He): (3.1)

5. _____ noodle soup
6. _____ salt water
7. _____ chocolate chip cookie
8. _____ mouthwash

9. Which of the following is a chemical property? (3.2)
   A. dynamite explodes
   B. a shiny metal
   C. breaking up cement
   D. a melting point of 110 °C
   E. rain on a cool day

For questions 10 through 12, answer with solid (S), *liquid* (L), *or gas* (G): (3.2)

10. _____ has a definite volume but takes the shape of a container

11. _____ does not have a definite shape or definite volume

12. _____ has a definite shape and a definite volume

13. Which of the following is a chemical property of silver? (3.2)
    A. density of 10.5 g/mL
    B. shiny
    C. melts at 961 °C
    D. good conductor of heat
    E. reacts to form tarnish

14. Which of the following is a physical property of silicon? (3.2)
    A. burns in chlorine
    B. has a black to gray color
    C. reacts with nitric acid
    D. used to form silicone
    E. reacts with oxygen to form sand

For questions 15 through 19, answer as physical change (P) *or chemical change* (C): (3.2)

15. _____ Butter melts in a hot pan.
16. _____ Iron forms rust with oxygen.
17. _____ Baking powder forms bubbles (CO_2) as a cake is baking.
18. _____ Water boils.
19. _____ Propane burns in a camp stove.

20. 105 °F = _____ °C (3.3)
    A. 73 °C
    B. 41 °C
    C. 58 °C
    D. 90 °C
    E. 189 °C

Chapter 3

21. The melting point of gold is 1064 °C. The Fahrenheit temperature needed to melt gold would be (3.3)
    A. 129 °F
    B. 623 °F
    C. 1031 °F
    D. 1913 °F
    E. 1947 °F

22. The average daytime temperature on the planet Mercury is 683 K. What is this temperature on the Celsius scale? (3.3)
    A. 956 °C
    B. 715 °C
    C. 680 °C
    D. 410 °C
    E. 303 °C

23. Which of the following would be described as potential energy? (3.4)
    A. a car going around a racetrack
    B. a rabbit hopping
    C. oil in an oil well
    D. a moving merry-go-round
    E. a bouncing ball

24. Which of the following would be described as kinetic energy? (3.4)
    A. a car battery
    B. a can of tennis balls
    C. gasoline in a car fuel tank
    D. a box of matches
    E. a tennis ball crossing over the net

For questions 25 through 27, consider a glass of milk with an energy content of 170 kcal. In the milk, there are 12 g of carbohydrate, 9.0 g of fat, and protein: (round off the kilocalories for each food type to the tens place) (3.5)

25. The number of kilocalories provided by the carbohydrate is
    A. 4 kcal
    B. 9 kcal
    C. 40 kcal
    D. 50 kcal
    E. 80 kcal

26. The number of kilocalories provided by the fat is
    A. 9 kcal
    B. 40 kcal
    C. 60 kcal
    D. 70 kcal
    E. 80 kcal

27. Using your answers to questions 25 and 26, the number of kilocalories provided by the protein is
    A. 4 kcal
    B. 30 kcal
    C. 40 kcal
    D. 90 kcal
    E. 120 kcal

28. A patient on a diet has a lunch of 3 oz of tuna (8 g of fat, 16 g of protein), 1 cup of nonfat milk (12 g of carbohydrate, 9 g of protein), and 1 cup of raw carrots (11 g of carbohydrate, 2 g of protein). What is the total kilocalories for the lunch? (*Round off the kilocalories for each food type to the tens place.*) (3.5)

    A. 270 kcal
    B. 200 kcal
    C. 110 kcal
    D. 90 kcal
    E. 50 kcal

29. For the lunch in question 28, how many hours of swimming are needed to expend the same energy, in kilocalories, if swimming requires 2100 kJ/h? (3.5)
    A. 5.7 h
    B. 5.0 h
    C. 2.8 h
    D. 1.6 h
    E. 0.54 h

30. The number of joules needed to raise the temperature of 5.0 g of water from 25 °C to 55 °C is (3.6)
    A. 5.0 J
    B. 36 J
    C. 30.0 J
    D. 335 J
    E. 630 J

31. The number of kilojoules released when 15 g of water cools from 58 °C to 22 °C is (3.6)
    A. 0.13 kJ
    B. 0.54 kJ
    C. 2.3 kJ
    D. 63 kJ
    E. 150 kJ

For questions 32 through 36, match items A to E with one of the statements: (3.7)

   A. melting
   B. evaporation
   C. heat of fusion
   D. heat of vaporization
   E. boiling

32. _____ the energy required to convert a gram of solid to liquid

33. _____ the heat needed to convert a gram of liquid to gas

34. _____ the conversion of a liquid to gas at the surface of a liquid

35. _____ the conversion of solid to liquid

36. _____ the formation of a gas within the liquid as well as on the surface

37. The number of joules needed to convert 15.0 g of ice to liquid at 0 °C is (3.7)
   A. 73 J   B. 334 J   C. 2260 J   D. 4670 J   E. 5010 J

38. The number of calories released when 2.0 g of water at 50 °C is cooled and frozen at 0 °C is (3.7)
   A. 80. cal   B. 160 cal   C. 260 cal   D. 440 cal   E. 1100 cal

39. What is the total number of kilojoules required to convert 25 g of ice at 0 °C to gas at 100 °C? (3.7)
   A. 8.4 kJ   B. 19 kJ   C. 59 kJ   D. 67 kJ   E. 75 kJ

For questions 40 through 43, consider the heating curve for p-toluidine. Answer the following questions when heat is added to p-toluidine at −20 °C, where toluidine is below its melting point: (3.7)

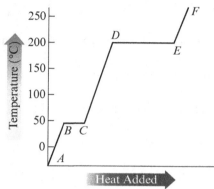

40. On the heating curve, segment **BC** indicates
   A. solid   B. melting   C. liquid   D. boiling   E. gas

41. On the heating curve, segment **CD** shows toluidine as
   A. solid   B. melting   C. liquid   D. boiling   E. gas

42. The boiling point of toluidine is
   A. 20 °C   B. 45 °C   C. 100 °C   D. 200 °C   E. 250 °C

43. On the heating curve, segment **EF** shows toluidine as
   A. solid   B. melting   C. liquid   D. boiling   E. gas

Answers to the Practice Test

1. M	2. P	3. P	4. M	5. He
6. Ho	7. He	8. Ho	9. A	10. L
11. G	12. S	13. E	14. B	15. P
16. C	17. C	18. P	19. C	20. B
21. E	22. D	23. C	24. E	25. D
26. E	27. C	28. A	29. E	30. E
31. C	32. C	33. D	34. B	35. A
36. E	37. E	38. C	39. E	40. B
41. C	42. D	43. E		

Chapter 3

Selected Answers and Solutions to Text Problems

3.1 *Elements* are the simplest type of pure substance, containing only one type of atom. *Compounds* contain two or more elements chemically combined in a specific proportion.
  a. A silicon chip is an element since it contains only one type of atom (Si).
  b. Hydrogen peroxide (H_2O_2) is a compound since it contains two elements (H, O) chemically combined.
  c. Oxygen (O_2) is an element since it contains only one type of atom (O).
  d. Rust (Fe_2O_3) is a compound since it contains two elements (Fe, O) chemically combined.
  e. Methane (CH_4) in natural gas is a compound since it contains two elements (C, H) chemically combined.

3.3 A *pure substance* is matter that has a fixed or definite composition: either elements or compounds. In a *mixture*, two or more substances are physically mixed but not chemically combined.
  a. Baking soda is composed of one type of matter ($NaHCO_3$), which makes it a pure substance.
  b. A blueberry muffin is composed of several substances mixed together, which makes it a mixture.
  c. Ice is composed of one type of matter (H_2O), which makes it a pure substance.
  d. Zinc is composed of one type of matter (Zn), which makes it a pure substance.
  e. Trimix is a physical mixture of oxygen, nitrogen, and helium gases, which makes it a mixture.

3.5 A *homogeneous mixture* has a uniform composition; a *heterogeneous mixture* does not have a uniform composition throughout the mixture.
  a. Vegetable soup is a heterogeneous mixture since it has chunks of vegetables.
  b. Tea is a homogeneous mixture since it has a uniform composition.
  c. Fruit salad is a heterogeneous mixture since it has chunks of fruit.
  d. Tea with ice and lemon slices is a heterogeneous mixture since it has chunks of ice and lemon.

3.7
  a. A gas has no definite volume or shape.
  b. In a gas, the particles do not interact with each other.
  c. In a solid, the particles are held in a rigid structure.

3.9 A *physical property* is a characteristic of the substance such as color, shape, odor, luster, size, melting point, and density. A *chemical property* is a characteristic that indicates the ability of a substance to change into a new substance.
  a. Color and physical state are physical properties.
  b. The ability to react with oxygen is a chemical property.
  c. The temperature of a substance is a physical property.
  d. Milk souring describes chemical reactions and is thus a chemical property.
  e. Burning butane gas in oxygen forms new substances, which makes it a chemical property.

3.11 When matter undergoes a *physical change*, its state or appearance changes, but its composition remains the same. When a *chemical change* occurs, the original substance is converted into a new substance, which has different physical and chemical properties.
  a. Water vapor condensing is a physical change since the physical form of the water changes, but the composition of the substance does not.
  b. Cesium metal reacting is a chemical change since new substances form.
  c. Gold melting is a physical change since the physical state changes, but not the composition of the substance.
  d. Cutting a puzzle results in a physical change since the size and shape change, but not the composition of the substance.
  e. Grating cheese results in a physical change since the size and shape change, but not the composition of the substance.

3.13
  a. The high reactivity of fluorine is a chemical property since it allows for the formation of new substances.
  b. The physical state of fluorine is a physical property.

Matter and Energy

c. The color of fluorine is a physical property.
d. The reactivity of fluorine with hydrogen is a chemical property since it allows for the formation of new substances.
e. The melting point of fluorine is a physical property.

3.15 The Fahrenheit temperature scale is still used in the United States. A normal body temperature is 98.6 °F on this scale. To convert her 99.8 °F temperature to the equivalent reading on the Celsius scale, the following calculation must be performed:

$$T_C = \frac{(99.8 - 32)}{1.8} = \frac{67.8}{1.8} = 37.7 \,°C \;(3 \text{ SFs}) \;(1.8 \text{ and } 32 \text{ are exact numbers})$$

Because a normal body temperature is 37.0 on the Celsius scale, her temperature of 37.7 °C would indicate a mild fever.

3.17 To convert Celsius to Fahrenheit: $T_F = 1.8(T_C) + 32$

To convert Fahrenheit to Celsius: $T_C = \dfrac{T_F - 32}{1.8}$ (1.8 and 32 are exact numbers)

To convert Celsius to Kelvin: $T_K = T_C + 273$
To convert Kelvin to Celsius: $T_C = T_K - 273$

a. $T_F = 1.8(T_C) + 32 = 1.8(37.0) + 32 = 66.6 + 32 = 98.6 \,°F$
b. $T_C = \dfrac{T_F - 32}{1.8} = \dfrac{(65.3 - 32)}{1.8} = \dfrac{33.3}{1.8} = 18.5 \,°C$
c. $T_K = T_C + 273 = -27 + 273 = 246 \text{ K}$
d. $T_K = T_C + 273 = 62 + 273 = 335 \text{ K}$
e. $T_C = \dfrac{T_F - 32}{1.8} = \dfrac{(114 - 32)}{1.8} = \dfrac{82}{1.8} = 46 \,°C$

3.19 a. $T_C = \dfrac{T_F - 32}{1.8} = \dfrac{(106 - 32)}{1.8} = \dfrac{74}{1.8} = 41 \,°C$

b. $T_C = \dfrac{T_F - 32}{1.8} = \dfrac{(103 - 32)}{1.8} = \dfrac{71}{1.8} = 39 \,°C$

No, there is no need to phone the doctor. The child's temperature is less than 40.0 °C.

3.21 When the roller-coaster car is at the top of the ramp, it has its maximum potential energy. As it descends, potential energy is converted into kinetic energy. At the bottom, all of its energy is kinetic.

3.23 a. Potential energy is stored in the water at the top of the waterfall.
b. Kinetic energy is displayed as the kicked ball moves.
c. Potential energy is stored in the chemical bonds in the coal.
d. Potential energy is stored when the skier is at the top of the hill.

3.25 a. **Given** 3500 cal   **Need** kilocalories
Plan cal → kcal $\dfrac{1 \text{ kcal}}{1000 \text{ cal}}$
Set-up $3500 \text{ cal} \times \dfrac{1 \text{ kcal}}{1000 \text{ cal}} = 3.5 \text{ kcal} \;(2 \text{ SFs})$

b. **Given** 415 J   **Need** calories
Plan J → cal $\dfrac{1 \text{ cal}}{4.184 \text{ J}}$
Set-up $415 \text{ J} \times \dfrac{1 \text{ cal}}{4.184 \text{ J}} = 99.2 \text{ cal} \;(3 \text{ SFs})$

Chapter 3

  c. **Given** 28 cal   **Need** joules
     Plan cal → J $\dfrac{4.184\ \text{J}}{1\ \text{cal}}$

     Set-up $28\ \cancel{\text{cal}} \times \dfrac{4.184\ \text{J}}{1\ \cancel{\text{cal}}} = 120\ \text{J}\ (2\ \text{SFs})$

  d. **Given** 4.5 kJ   **Need** calories
     Plan kJ → J → cal $\dfrac{1000\ \text{J}}{1\ \text{kJ}}\quad \dfrac{1\ \text{cal}}{4.184\ \text{J}}$

     Set-up $4.5\ \cancel{\text{kJ}} \times \dfrac{1000\ \cancel{\text{J}}}{1\ \cancel{\text{kJ}}} \times \dfrac{1\ \text{cal}}{4.184\ \cancel{\text{J}}} = 1100\ \text{cal}\ (2\ \text{SFs})$

3.27 a. **Given** 3.0 h, 270 kJ/h   **Need** joules
     Plan h → kJ → J $\dfrac{1000\ \text{J}}{1\ \text{kJ}}$

     Set-up $3.0\ \cancel{\text{h}} \times \dfrac{270\ \cancel{\text{kJ}}}{1.0\ \cancel{\text{h}}} \times \dfrac{1000\ \text{J}}{1\ \cancel{\text{kJ}}} = 8.1 \times 10^5\ \text{J}\ (2\ \text{SFs})$

  b. **Given** 3.0 h, 270 kJ/h   **Need** kilocalories
     Plan h → kJ → J → cal → kcal $\dfrac{1000\ \text{J}}{1\ \text{kJ}}\quad \dfrac{1\ \text{cal}}{4.184\ \text{J}}\quad \dfrac{1\ \text{kcal}}{1000\ \text{cal}}$

     Set-up $3.0\ \cancel{\text{h}} \times \dfrac{270\ \cancel{\text{kJ}}}{1.0\ \cancel{\text{h}}} \times \dfrac{1000\ \cancel{\text{J}}}{1\ \cancel{\text{kJ}}} \times \dfrac{1\ \cancel{\text{cal}}}{4.184\ \cancel{\text{J}}} \times \dfrac{1\ \text{kcal}}{1000\ \cancel{\text{cal}}} = 190\ \text{kcal}\ (2\ \text{SFs})$

3.29 a. **Given** 125 kJ   **Need** kilocalories
     Plan kJ → J → cal → kcal $\dfrac{1000\ \text{J}}{1\ \text{kJ}}\quad \dfrac{1\ \text{cal}}{4.184\ \text{J}}\quad \dfrac{1\ \text{kcal}}{1000\ \text{cal}}$

     Set-up $125\ \cancel{\text{kJ}} \times \dfrac{1000\ \cancel{\text{J}}}{1\ \cancel{\text{kJ}}} \times \dfrac{1\ \cancel{\text{cal}}}{4.184\ \cancel{\text{J}}} \times \dfrac{1\ \text{kcal}}{1000\ \cancel{\text{cal}}} = 29.9\ \text{kcal}\ (3\ \text{SFs})$

  b. **Given** 870. kJ   **Need** kilocalories
     Plan kJ → J → cal → kcal $\dfrac{1000\ \text{J}}{1\ \text{kJ}}\quad \dfrac{1\ \text{cal}}{4.184\ \text{J}}\quad \dfrac{1\ \text{kcal}}{1000\ \text{cal}}$

     Set-up $870.\ \cancel{\text{kJ}} \times \dfrac{1000\ \cancel{\text{J}}}{1\ \cancel{\text{kJ}}} \times \dfrac{1\ \cancel{\text{cal}}}{4.184\ \cancel{\text{J}}} \times \dfrac{1\ \text{kcal}}{1000\ \cancel{\text{cal}}} = 208\ \text{kcal}\ (3\ \text{SFs})$

3.31 a. **Given** one cup of orange juice that contains 26 g of carbohydrate, 2 g of protein, and no fat
     Need total energy in kilojoules

Food Type	Mass		Energy Value		Energy (rounded off to the tens place)
Carbohydrate	26 g	×	$\dfrac{17\ \text{kJ}}{1\ \text{g}}$	=	440 kJ
Protein	2 g	×	$\dfrac{17\ \text{kJ}}{1\ \text{g}}$	=	30 kJ
			Total energy content	=	470 kJ

  b. **Given** one apple that provides 72 kcal of energy and contains no fat or protein
     Need grams of carbohydrate

     Plan kcal → g of carbohydrate $\dfrac{1\ \text{g carbohydrate}}{4\ \text{kcal}}$

     Set-up $72\ \cancel{\text{kcal}} \times \dfrac{1\ \text{g carbohydrate}}{4\ \cancel{\text{kcal}}} = 18\ \text{g of carbohydrate}\ (2\ \text{SFs})$

c. Given one tablespoon of vegetable oil that contains 14 g of fat, and no carbohydrate or protein
Need total energy in kilocalories

Food Type	Mass		Energy Value		Energy (rounded off to the tens place)
Fat	14 g	×	$\dfrac{9 \text{ kcal}}{1 \text{ g}}$	=	130 kcal

d. Given one avocado that provides 405 kcal in total, and contains 13 g of carbohydrate and 5 g of protein
Need grams of fat

Food Type	Mass		Energy Value		Energy (rounded off to the tens place)
Carbohydrate	13 g	×	$\dfrac{4 \text{ kcal}}{1 \text{ g}}$	=	50 kcal
Fat	? g	×	$\dfrac{9 \text{ kcal}}{1 \text{ g}}$	=	? kcal
Protein	5 g	×	$\dfrac{4 \text{ kcal}}{1 \text{ g}}$	=	20 kcal
			Total energy content	=	405 kcal

∴ Energy from fat = 405 kcal − (50 kcal + 20 kcal) = 335 kcal

∴ 335 kcal × $\dfrac{1 \text{ g fat}}{9 \text{ kcal}}$ = 37 g of fat

3.33 Given one cup of clam chowder that contains 16 g of carbohydrate, 12 g of fat, and 9 g of protein
Need total energy in kilocalories and kilojoules

Food Type	Mass		Energy Value		Energy (rounded off to the tens place)
Carbohydrate	16 g	×	$\dfrac{4 \text{ kcal (or 17 kJ)}}{1 \text{ g}}$	=	60 kcal (or 270 kJ)
Fat	12 g	×	$\dfrac{9 \text{ kcal (or 38 kJ)}}{1 \text{ g}}$	=	110 kcal (or 460 kJ)
Protein	9 g	×	$\dfrac{4 \text{ kcal (or 17 kJ)}}{1 \text{ g}}$	=	40 kcal (or 150 kJ)
			Total energy content	=	210 kcal (or 880 kJ)

3.35 3.2 L glucose solution × $\dfrac{1000 \text{ mL solution}}{1 \text{ L solution}}$ × $\dfrac{5.0 \text{ g glucose}}{100. \text{ mL solution}}$ × $\dfrac{4 \text{ kcal}}{1 \text{ g glucose}}$
= 640 kcal (rounded off to the tens place)

3.37 Copper, which has the lowest specific heat of the samples, would reach the highest temperature.

3.39 a. Given $SH_{\text{water}} = 1.00$ cal/g °C; $m = 8.5$ g; $\Delta T = 36\,°C - 15\,°C = 21\,°C$
   Need heat in calories
   Plan Heat = $m \times \Delta T \times SH$

   Set-up Heat = 8.5 g × 21 °C × $\dfrac{1.00 \text{ cal}}{\text{g °C}}$ = 180 cal (2 SFs)

b. Given $SH_{water} = 4.184 \text{ J/g °C}; m = 25 \text{ g}; \Delta T = 86 \text{ °C} - 61 \text{ °C} = 25 \text{ °C}$
Need heat in joules
Plan Heat $= m \times \Delta T \times SH$

Set-up Heat $= 25 \text{ g} \times 25 \text{ °C} \times \dfrac{4.184 \text{ J}}{\text{g °C}} = 2600 \text{ J (2 SFs)}$

c. Given $SH_{water} = 1.00 \text{ cal/g °C}; m = 150 \text{ g}; \Delta T = 77 \text{ °C} - 15 \text{ °C} = 62 \text{ °C}$
Need heat in kilocalories

Plan Heat $= m \times \Delta T \times SH$ then cal $\to$ kcal $\dfrac{1 \text{ kcal}}{1000 \text{ cal}}$

Set-up Heat $= 150 \text{ g} \times 62 \text{ °C} \times \dfrac{1.00 \text{ cal}}{\text{g °C}} \times \dfrac{1 \text{ kcal}}{1000 \text{ cal}} = 9.3 \text{ kcal (2 SFs)}$

d. Given $SH_{copper} = 0.385 \text{ J/g °C}; m = 175 \text{ g}; \Delta T = 188 \text{ °C} - 28 \text{ °C} = 160. \text{ °C}$
Need heat in kilojoules

Plan Heat $= m \times \Delta T \times SH$ then J $\to$ kJ $\dfrac{1 \text{ kJ}}{1000 \text{ J}}$

Set-up Heat $= 175 \text{ g} \times 160. \text{ °C} \times \dfrac{0.385 \text{ J}}{\text{g °C}} \times \dfrac{1 \text{ kJ}}{1000 \text{ J}} = 10.8 \text{ kJ (3 SFs)}$

3.41 a. Given $SH_{water} = 4.184 \text{ J/g °C} = 1.00 \text{ cal/g °C}; m = 25.0 \text{ g};$
$\Delta T = 25.7 \text{ °C} - 12.5 \text{ °C} = 13.2 \text{ °C}$
Need heat in joules and calories
Plan Heat $= m \times \Delta T \times SH$

Set-up Heat $= 25.0 \text{ g} \times 13.2 \text{ °C} \times \dfrac{4.184 \text{ J}}{\text{g °C}} = 1380 \text{ J (3 SFs)}$

or Heat $= 25.0 \text{ g} \times 13.2 \text{ °C} \times \dfrac{1.00 \text{ cal}}{\text{g °C}} = 330. \text{ cal (3 SFs)}$

b. Given $SH_{copper} = 0.385 \text{ J/g °C} = 0.0920 \text{ cal/g °C}; m = 38.0 \text{ g};$
$\Delta T = 246 \text{ °C} - 122 \text{ °C} = 124 \text{ °C}$
Need heat in joules and calories
Plan Heat $= m \times \Delta T \times SH$

Set-up Heat $= 38.0 \text{ g} \times 124 \text{ °C} \times \dfrac{0.385 \text{ J}}{\text{g °C}} = 1810 \text{ J (3 SFs)}$

or Heat $= 38.0 \text{ g} \times 124 \text{ °C} \times \dfrac{0.0920 \text{ cal}}{\text{g °C}} = 434 \text{ cal (3 SFs)}$

c. Given $SH_{ethanol} = 2.46 \text{ J/g °C} = 0.588 \text{ cal/g °C}; m = 15.0 \text{ g};$
$\Delta T = 60.5 \text{ °C} - (-42.0 \text{ °C}) = 102.5 \text{ °C}$
Need heat in joules and calories
Plan Heat $= m \times \Delta T \times SH$

Set-up Heat $= 15.0 \text{ g} \times 102.5 \text{ °C} \times \dfrac{2.46 \text{ J}}{\text{g °C}} = 3780 \text{ J (3 SFs)}$

or Heat $= 15.0 \text{ g} \times 102.5 \text{ °C} \times \dfrac{0.588 \text{ cal}}{\text{g °C}} = 904 \text{ cal (3 SFs)}$

d. Given $SH_{iron} = 0.452 \text{ J/g °C} = 0.108 \text{ cal/g °C}; m = 125 \text{ g};$
$\Delta T = 118 \text{ °C} - 55 \text{ °C} = 63 \text{ °C}$
Need heat in joules and calories
Plan Heat $= m \times \Delta T \times SH$

Set-up Heat = 125 g × 63 °C × $\dfrac{0.452 \text{ J}}{\text{g °C}}$ = 3600 J (2 SFs)

or Heat = 125 g × 63 °C × $\dfrac{0.108 \text{ cal}}{\text{g °C}}$ = 850 cal (2 SFs)

3.43 a. The change from solid to liquid state is melting.
b. Coffee is freeze-dried using the process of sublimation.
c. Liquid water turning to ice is freezing.
d. Ice crystals form on a package of frozen corn due to deposition.

3.45 a. 65 g ice × $\dfrac{80. \text{ cal}}{1 \text{ g ice}}$ = 5200 cal (2 SFs); heat is absorbed

b. 17.0 g ice × $\dfrac{334 \text{ J}}{1 \text{ g ice}}$ = 5680 J (3 SFs); heat is absorbed

c. 225 g water × $\dfrac{80. \text{ cal}}{1 \text{ g water}}$ × $\dfrac{1 \text{ kcal}}{1000 \text{ cal}}$ = 18 kcal (2 SFs); heat is released

d. 50.0 g water × $\dfrac{334 \text{ J}}{1 \text{ g water}}$ × $\dfrac{1 \text{ kJ}}{1000 \text{ J}}$ = 16.7 kJ (3 SFs); heat is released

3.47 a. Water vapor in clouds changing to rain is an example of condensation.
b. Wet clothes drying on a clothesline involves evaporation.
c. Steam forming as lava flows into the ocean involves boiling.
d. Water droplets forming on a bathroom mirror after a hot shower involves condensation.

3.49 a. 10.0 g water × $\dfrac{540 \text{ cal}}{1 \text{ g water}}$ = 5400 cal (2 SFs); heat is absorbed

b. 5.00 g water × $\dfrac{2260 \text{ J}}{1 \text{ g water}}$ = 11 300 J (3 SFs); heat is absorbed

c. 8.0 kg steam × $\dfrac{1000 \text{ g}}{1 \text{ kg}}$ × $\dfrac{540 \text{ cal}}{1 \text{ g steam}}$ × $\dfrac{1 \text{ kcal}}{1000 \text{ cal}}$ = 4300 kcal (2 SFs); heat is released

d. 175 g steam × $\dfrac{2260 \text{ J}}{1 \text{ g steam}}$ × $\dfrac{1 \text{ kJ}}{1000 \text{ J}}$ = 396 kJ (3 SFs); heat is released

3.51

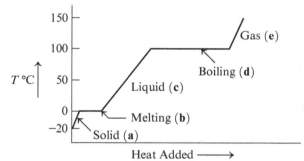

3.53 a. Two calculations are needed:

(1) ice 0 °C → water 0 °C: 50.0 g ice × $\dfrac{334 \text{ J}}{1 \text{ g ice}}$ = 16 700 J

(2) water 0 °C → 65.0 °C: ΔT = 65.0 °C − 0 °C = 65.0 °C

50.0 g water × 65.0 °C × $\dfrac{4.184 \text{ J}}{\text{g water °C}}$ = 13 600 J

∴ Total heat needed = 16 700 J + 13 600 J = 30 300 J (3 SFs)

Chapter 3

    b. Two calculations are needed:

        (1) steam 100 °C → water 100 °C: $15.0 \text{ g steam} \times \dfrac{540 \text{ cal}}{1 \text{ g steam}} \times \dfrac{1 \text{ kcal}}{1000 \text{ cal}} = 8.1 \text{ kcal}$

        (2) water 100 °C → 0 °C: $15.0 \text{ g water} \times 100. \text{ °C} \times \dfrac{1.00 \text{ cal}}{\text{g °C}} \times \dfrac{1 \text{ kcal}}{1000 \text{ cal}} = 1.50 \text{ kcal}$

        ∴ Total heat released = 8.1 kcal + 1.50 kcal = 9.6 kcal (2 SFs)

    c. Three calculations are needed:

        (1) ice 0 °C → water 0 °C: $24.0 \text{ g ice} \times \dfrac{334 \text{ J}}{1 \text{ g ice}} \times \dfrac{1 \text{ kJ}}{1000 \text{ J}} = 8.02 \text{ kJ}$

        (2) water 0 °C → 100 °C: $24.0 \text{ g water} \times 100. \text{ °C} \times \dfrac{4.184 \text{ J}}{\text{g °C}} \times \dfrac{1 \text{ kJ}}{1000 \text{ J}} = 10.0 \text{ kJ}$

        (3) water 100 °C → steam 100 °C: $24.0 \text{ g water} \times \dfrac{2260 \text{ J}}{1 \text{ g water}} \times \dfrac{1 \text{ kJ}}{1000 \text{ J}} = 54.2 \text{ kJ}$

        ∴ Total heat needed = 8.02 kJ + 10.0 kJ + 54.2 kJ = 72.2 kJ (3 SFs)

3.55 Two calculations are needed:

    (1) steam 100. °C → water 100. °C: $18.0 \text{ g steam} \times \dfrac{540 \text{ cal}}{1 \text{ g steam}} \times \dfrac{1 \text{ kcal}}{1000 \text{ cal}} = 9.7 \text{ kcal}$

    (2) water 100. °C → 37.0 °C: $\Delta T = 100 \text{ °C} - 37.0 \text{ °C} = 63 \text{ °C}$

    $18.0 \text{ g water} \times 63 \text{ °C} \times \dfrac{1.00 \text{ cal}}{\text{g °C}} \times \dfrac{1 \text{ kcal}}{1000 \text{ cal}} = 1.1 \text{ kcal}$

    ∴ Total heat released = 9.7 kcal + 1.1 kcal = 10.8 kcal (3 SFs)

3.57 **a.** Using the energy totals from Table 3.8:

Breakfast	Energy
Banana, 1 medium	110 kcal
Milk, nonfat, 1 cup	90 kcal
Egg, 1 large	70 kcal
	Total = 270 kcal

Lunch	Energy
Carrots, 1 cup	50 kcal
Beef, ground, 3 oz	220 kcal
Apple, 1 medium	60 kcal
Milk, nonfat, 1 cup	90 kcal
	Total = 420 kcal

Dinner	Energy
Chicken, no skin, 6 oz	2 × 110 kcal
Potato, baked	100 kcal
Broccoli, 3 oz	30 kcal
Milk, nonfat, 1 cup	90 kcal
	Total = 440 kcal

    b. total kilocalories for one day = 270 kcal + 420 kcal + 440 kcal = 1130 kcal
    c. Since Charles will maintain his weight if he consumes 1800 kcal per day, he will lose weight on this new diet (1130 kcal/day).
    d. For Charles, his energy balance is 1130 kcal/day − 1800 kcal/day = −670 kcal/day.

        $5.0 \text{ lb} \times \dfrac{3500 \text{ kcal}}{1.0 \text{ lb}} \times \dfrac{1 \text{ day}}{670 \text{ kcal}} = 26 \text{ days (2 SFs)}$

3.59 a. The diagram shows two different kinds of atoms chemically combined in a definite 2:1 ratio; it represents a compound.
  b. The diagram shows two different kinds of matter physically mixed, not chemically combined; it represents a mixture.
  c. The diagram contains only one kind of atom; it represents an element.

3.61 A *homogeneous mixture* has a uniform composition; a *heterogeneous mixture* does not have a uniform composition throughout the mixture.
  a. Lemon-flavored water is a homogeneous mixture since it has a uniform composition (as long as there are no lemon pieces).
  b. Stuffed mushrooms are a heterogeneous mixture since there are mushrooms and chunks of filling.
  c. Eye drops are a homogeneous mixture since they have a uniform composition.

3.63 41.5 °C
$$T_F = 1.8(T_C) + 32 = 1.8(41.5) + 32 = 74.7 + 32 = 107 °F \text{ (3 SFs)}$$

3.65 $T_C = \dfrac{T_F - 32}{1.8} = \dfrac{(155 - 32)}{1.8} = \dfrac{123}{1.8} = 68.3 °C$ (3 SFs)
(1.8 and 32 are exact numbers)
$T_K = T_C + 273 = 68.3 + 273 = 341 \text{ K (3 SFs)}$

3.67 Given 10.0 cm³ cubes of gold ($SH_{gold} = 0.129$ J/g °C), and aluminum ($SH_{aluminum} = 0.897$ J/g °C); $\Delta T = 25 °C - 15 °C = 10. °C$
  Need energy in joules and calories
  Plan cm³ → g   Heat = $m \times \Delta T \times SH$ $\dfrac{19.3 \text{ g gold}}{1 \text{ cm}^3}$ $\dfrac{2.70 \text{ g aluminum}}{1 \text{ cm}^3}$
  Set-up for gold: $10.0 \text{ cm}^3 \times \dfrac{19.3 \text{ g}}{1 \text{ cm}^3} \times 10. °C \times \dfrac{0.129 \text{ J}}{\text{g} °C} = 250 \text{ J (2 SFs)}$

  $10.0 \text{ cm}^3 \times \dfrac{19.3 \text{ g}}{1 \text{ cm}^3} \times 10. °C \times \dfrac{0.129 \text{ J}}{\text{g} °C} \times \dfrac{1 \text{ cal}}{4.184 \text{ J}} = 60. \text{ cal (2 SFs)}$

  for aluminum: $10.0 \text{ cm}^3 \times \dfrac{2.70 \text{ g}}{1 \text{ cm}^3} \times 10. °C \times \dfrac{0.897 \text{ J}}{\text{g} °C} = 240 \text{ J (2 SFs)}$

  $10.0 \text{ cm}^3 \times \dfrac{2.70 \text{ g}}{1 \text{ cm}^3} \times 10. °C \times \dfrac{0.897 \text{ J}}{\text{g} °C} \times \dfrac{1 \text{ cal}}{4.184 \text{ J}} = 58 \text{ cal (2 SFs)}$

  Thus, the energy needed to heat each of the metals is almost the same.

3.69 a. Given a meal consisting of: cheeseburger: 46 g of carbohydrate, 40. g of fat, 47 g of protein
  french fries: 47 g of carbohydrate, 16 g of fat, 4 g of protein
  chocolate shake: 76 g of carbohydrate, 10. g of fat, 10. g of protein
  Need energy from each food type in kilocalories
  Plan total grams of each food type, then g → kcal
  Set-up total carbohydrate = 46 g + 47 g + 76 g = 169 g
  total fat = 40. g + 16 g + 10. g = 66 g
  total protein = 47 g + 4 g + 10. g = 61 g

Food Type	Mass		Energy Value		Energy (rounded off to the tens place)
Carbohydrate	169 g	×	$\dfrac{4 \text{ kcal}}{1 \text{ g}}$	=	680 kcal
Fat	66 g	×	$\dfrac{9 \text{ kcal}}{1 \text{ g}}$	=	590 kcal
Protein	61 g	×	$\dfrac{4 \text{ kcal}}{1 \text{ g}}$	=	240 kcal

Chapter 3

b. total energy for the meal = 680 kcal + 590 kcal + 240 kcal = 1510 kcal

c. **Given** 1510 kcal from meal **Need** hours of sleeping to "burn off"

Plan kcal → h $\dfrac{1 \text{ h sleeping}}{60 \text{ kcal}}$

Set-up $1510 \text{ kcal} \times \dfrac{1 \text{ h sleeping}}{60 \text{ kcal}} = 25 \text{ h of sleeping (2 SFs)}$

d. **Given** 1510 kcal from meal **Need** hours of running to "burn off"

Plan kcal → h $\dfrac{1 \text{ h running}}{750 \text{ kcal}}$

Set-up $1510 \text{ kcal} \times \dfrac{1 \text{ h running}}{750 \text{ kcal}} = 2.0 \text{ h of running (2 SFs)}$

3.71 *Elements* are the simplest type of pure substance, containing only one type of atom. *Compounds* contain two or more elements chemically combined in a definite proportion. In a *mixture*, two or more substances are physically mixed but not chemically combined.
  a. Carbon in pencils is an element since it contains only one type of atom (C).
  b. Carbon monoxide is a compound since it contains two elements (C, O) chemically combined.
  c. Orange juice is composed of several substances physically mixed together (e.g. water, sugar, citric acid), which makes it a mixture.

3.73 A *homogeneous mixture* has a uniform composition; a *heterogeneous mixture* does not have a uniform composition throughout the mixture.
  a. A hot fudge sundae is a heterogeneous mixture since it has ice cream, fudge sauce, and perhaps a cherry.
  b. Herbal tea is a homogeneous mixture since it has a uniform composition.
  c. Vegetable oil is a homogeneous mixture since it has a uniform composition.

3.75 a. A vitamin tablet is a solid.
  b. Helium in a balloon is a gas.
  c. Milk is a liquid.
  d. Air is a mixture of gases.
  e. Charcoal is a solid.

3.77 A *physical property* is a characteristic of the substance such as color, shape, odor, luster, size, melting point, and density. A *chemical property* is a characteristic that indicates the ability of a substance to change into a new substance.
  a. The luster of gold is a physical property.
  b. The melting point of gold is a physical property.
  c. The ability of gold to conduct electricity is a physical property.
  d. The ability of gold to form a new substance with sulfur is a chemical property.

3.79 When matter undergoes a *physical change*, its state or appearance changes, but its composition remains the same. When a *chemical change* occurs, the original substance is converted into a new substance, which has different physical and chemical properties.
  a. Plant growth produces new substances, so it is a chemical change.
  b. A change of state from solid to liquid is a physical change.
  c. Chopping wood into smaller pieces is a physical change.
  d. Burning wood, which forms new substances, results in a chemical change.

3.81 a. $T_C = \dfrac{T_F - 32}{1.8} = \dfrac{(134 - 32)}{1.8} = \dfrac{102}{1.8} = 56.7 \,°C$

$T_K = T_C + 273 = 56.7 + 273 = 330. \text{ K}$

b. $T_C = \dfrac{T_F - 32}{1.8} = \dfrac{(-69.7 - 32)}{1.8} = \dfrac{-101.7}{1.8} = -56.50\ °C$

$T_K = T_C + 273 = -56.50 + 273 = 217\ K$

3.83 $T_C = \dfrac{T_F - 32}{1.8} = \dfrac{(-15 - 32)}{1.8} = \dfrac{-47}{1.8} = -26\ °C$

$T_K = T_C + 273 = -26 + 273 = 247\ K$

3.85 **Given** $m = 0.50$ g of oil; 18.9 kJ produced
Need energy value (kcal/g)
Plan kJ → kcal $\quad \dfrac{1000\ J}{1\ kJ} \quad \dfrac{1\ cal}{4.184\ J} \quad \dfrac{1\ kcal}{1000\ cal}$

then energy value $= \dfrac{\text{heat in kcal}}{\text{mass in g}}$

Set-up $18.9\ kJ \times \dfrac{1000\ J}{1\ kJ} \times \dfrac{1\ cal}{4.184\ J} \times \dfrac{1\ kcal}{1000\ cal} = 4.52$ kcal (3 SFs)

∴ energy value $= \dfrac{\text{heat produced}}{\text{mass of oil}} = \dfrac{4.52\ kcal}{0.50\ g\ oil} = 9.0$ kcal/g of vegetable oil (2 SFs)

3.87 Sand must have a lower specific heat than water since the same amount of heat causes a greater temperature change in the sand than in the water.

3.89 **a.** The melting point of chloroform is about $-60\ °C$.
b. The boiling point of chloroform is about $60\ °C$.
c. The diagonal line **A** represents the solid state as temperature increases. The horizontal line **B** represents the change from solid to liquid or melting of the substance. The diagonal line **C** represents the liquid state as temperature increases. The horizontal line **D** represents the change from liquid to gas or boiling of the liquid. The diagonal line **E** represents the gas state as temperature increases.
d. At $-80\ °C$, it is solid; at $-40\ °C$, it is liquid; at $25\ °C$, it is liquid; at $80\ °C$, it is gas.

3.91

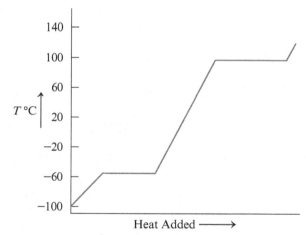

a. Dibromomethane is a solid at $-75\ °C$.
b. At $-53\ °C$, solid dibromomethane melts (solid → liquid).
c. Dibromomethane is a liquid at $-18\ °C$.
d. At $110\ °C$, dibromomethane is a gas.
e. Both solid and liquid dibromomethane will be present at the melting temperature of $-53\ °C$.

Chapter 3

3.93 Given 1 lb of body fat; 15% (m/m) water in body fat
Need kilocalories to "burn off"
Plan Because each gram of body fat contains 15% water, a person actually loses 85 grams of fat per hundred grams of body fat. (We considered 1 lb of fat as exactly 1 lb.)

lb of body fat → g of body fat → g of fat → kcal

$$\frac{454 \text{ g body fat}}{1 \text{ lb body fat}} \quad \frac{85 \text{ g fat}}{100 \text{ g body fat}} \quad \frac{9 \text{ kcal}}{1 \text{ g fat}}$$

Set-up $1 \text{ lb body fat} \times \dfrac{454 \text{ g body fat}}{1 \text{ lb body fat}} \times \dfrac{85 \text{ g fat}}{100 \text{ g body fat}} \times \dfrac{9 \text{ kcal}}{1 \text{ g fat}} = 3500 \text{ kcal (2 SFs)}$

3.95 Given $m = 725$ g of water; $SH_{water} = 4.184$ J/g °C; $\Delta T = 65\,°C - 37\,°C = 28\,°C$
Need heat in kilojoules

Plan Heat = $m \times \Delta T \times SH$, then J → kJ $\quad \dfrac{1 \text{ kJ}}{1000 \text{ J}}$

Set-up Heat = $725 \text{ g} \times 28\,°C \times \dfrac{4.184 \text{ J}}{\text{g °C}} \times \dfrac{1 \text{ kJ}}{1000 \text{ J}} = 85 \text{ kJ (2 SFs)}$

∴ 85 kJ are lost from the water bottle and are transferred to the muscles.

3.97 Given 0.66 g of olive oil; 370 g of water; $SH_{water} = 1.00$ cal/g °C;
$\Delta T = 38.8\,°C - 22.7\,°C = 16.1\,°C$
Need energy value (in kcal/g)

Plan Heat = $m \times \Delta T \times SH$ then cal → kcal $\quad \dfrac{1 \text{ kcal}}{1000 \text{ cal}}$

Set-up Heat = $370 \text{ g} \times 16.1\,°C \times \dfrac{1 \text{ cal}}{\text{g °C}} \times \dfrac{1 \text{ kcal}}{1000 \text{ cal}} = 6.0 \text{ kcal}$

∴ energy value = $\dfrac{\text{heat produced}}{\text{mass of oil}} = \dfrac{6.0 \text{ kcal}}{0.66 \text{ g oil}} = 9.1 \text{ kcal/g of olive oil (2 SFs)}$

3.99 a. Given 2.4×10^7 J/1.0 lb of oil; $m = 150$ kg of water; $SH_{water} = 4.184$ J/g °C;
$\Delta T = 100\,°C - 22\,°C = 78\,°C$
Need kilograms of oil needed

Plan Heat = $m \times \Delta T \times SH$ then J → lb of oil → kg of oil $\quad \dfrac{1 \text{ kg}}{2.20 \text{ lb}}$

Set-up Heat = $150 \text{ kg water} \times \dfrac{1000 \text{ g}}{1 \text{ kg}} \times 78\,°C \times \dfrac{4.184 \text{ J}}{\text{g °C}} = 4.9 \times 10^7 \text{ J (2 SFs)}$

∴ $m_{oil} = 4.9 \times 10^7 \text{ J} \times \dfrac{1.0 \text{ lb oil}}{2.4 \times 10^7 \text{ J}} \times \dfrac{1.0 \text{ kg oil}}{2.20 \text{ lb oil}} = 0.93 \text{ kg of oil (2 SFs)}$

b. Given 2.4×10^7 J/1.0 lb of oil; $m = 150$ kg of water; water 100 °C → steam 100 °C
Need kilograms of oil needed

Plan heat for water 100 °C → steam 100 °C; then J → lb of oil → kg of oil $\quad \dfrac{1 \text{ kg}}{2.20 \text{ lb}}$

Set-up Heat = $150 \text{ kg water} \times \dfrac{1000 \text{ g}}{1 \text{ kg}} \times \dfrac{2260 \text{ J}}{1 \text{ g water}} = 3.4 \times 10^8 \text{ J (2 SFs)}$

∴ $m_{oil} = 3.4 \times 10^8 \text{ J} \times \dfrac{1.0 \text{ lb oil}}{2.4 \times 10^7 \text{ J}} \times \dfrac{1.0 \text{ kg oil}}{2.20 \text{ lb oil}} = 6.4 \text{ kg of oil (2 SFs)}$

3.101 Two calculations are required:

(1) ice 0.0 °C → water 0.0 °C: $275 \text{ g ice} \times \dfrac{334 \text{ J}}{1 \text{ g ice}} \times \dfrac{1 \text{ kJ}}{1000 \text{ J}} = 91.9 \text{ kJ}$

(2) water 0.0 °C → 24.0 °C: $\Delta T = 24.0\text{ °C} - 0.0\text{ °C} = 24.0\text{ °C}$;

$275 \text{ g water} \times 24.0\text{ °C} \times \dfrac{4.184 \text{ J}}{\text{g °C}} \times \dfrac{1 \text{ kJ}}{1000 \text{ J}} = 27.6 \text{ kJ}$

∴ Total heat absorbed = 91.9 kJ + 27.6 kJ = 119.5 kJ

3.103 Given

Copper	Water
$m = 70.0$ g	$m = 50.0$ g
initial temperature = 54.0 °C	initial temperature = 26.0 °C
final temperature = 29.2 °C	final temperature = 29.2 °C
$SH = ?$ J/g °C	$SH = 4.184$ J/g °C

Need specific heat (SH) of copper (J/g °C)
Plan heat lost by copper = heat gained by water
For both, heat = $m \times \Delta T \times SH$
Set-up For water:
heat gained = $m \times \Delta T \times SH = 50.0 \text{ g} \times (29.2\text{ °C} - 26.0\text{ °C}) \times \dfrac{4.184 \text{ J}}{\text{g °C}} = 669 \text{ J}$

For copper:
∴ heat lost by copper = 669 J

$SH = \dfrac{\text{heat lost}}{m \times \Delta T} = \dfrac{669 \text{ J}}{70.0 \text{ g} \times (54.0\text{ °C} - 29.2\text{ °C})} = \dfrac{0.385 \text{ J}}{\text{g °C}}$ (3 SFs)

3.105 a. Given heat = 11 J; $m = 4.7$ g; $\Delta T = 4.5$ °C
Need specific heat (J/g °C)
Plan $SH = \dfrac{\text{heat}}{m \times \Delta T}$
Set-up $SH = \dfrac{11 \text{ J}}{4.7 \text{ g} \times 4.5\text{ °C}} = \dfrac{0.52 \text{ J}}{\text{g °C}}$ (2 SFs)

b. By comparing this calculated value to the specific heats given in Table 3.11 we would identify the unknown metal as titanium ($SH = 0.523$ J/g °C) rather than aluminum ($SH = 0.897$ J/g °C).

Chapter 3

Selected Answers to Combining Ideas from Chapters 1 to 3

CI.1 **a.** There are 4 significant figures in the measurement 20.17 lb.

b. $20.17 \text{ lb} \times \dfrac{1 \text{ kg}}{2.20 \text{ lb}} = 9.17 \text{ kg}$ (3 SFs)

c. $9.17 \text{ kg} \times \dfrac{1000 \text{ g}}{1 \text{ kg}} \times \dfrac{1 \text{ cm}^3}{19.3 \text{ g}} = 475 \text{ cm}^3$ (3 SFs)

d. $T_F = 1.8(T_C) + 32 = 1.8(1064) + 32 = 1947\,°F$ (4 SFs)
$T_K = T_C + 273 = 1064 + 273 = 1337 \text{ K}$ (4 SFs)

e. $\Delta T = 1064\,°C - 500.\,°C = 564\,°C$

$9.17 \text{ kg} \times \dfrac{1000 \text{ g}}{1 \text{ kg}} \times \dfrac{0.129 \text{ J}}{\text{g}\,°C} \times \dfrac{1 \text{ cal}}{4.184 \text{ J}} \times 564\,°C \times \dfrac{1 \text{ kcal}}{1000 \text{ cal}} = 159 \text{ kcal}$ (3 SFs)

f. $9.17 \text{ kg} \times \dfrac{1000 \text{ g}}{1 \text{ kg}} \times \dfrac{\$42.06}{1 \text{ g}} = \$386\,000$ (3 SFs)

CI.3 **a.** The water has its own shape in sample **B**.

b. Water sample **A** is represented by diagram **2**, which shows the particles in a random arrangement but close together.

c. Water sample **B** is represented by diagram **1**, which shows the particles fixed in a definite arrangement.

d. The state of matter indicated in diagram **1** is a <u>solid</u>; in diagram **2**, it is a <u>liquid</u>; and in diagram **3**, it is a <u>gas</u>.

e. The motion of the particles is slowest in diagram <u>1</u>.

f. The arrangement of particles is farthest apart in diagram <u>3</u>.

g. The particles fill the volume of the container in diagram <u>3</u>.

h. water $45\,°C \rightarrow 0\,°C$: $\Delta T = 45\,°C - 0\,°C = 45\,°C$
Heat $= m \times \Delta T \times SH$

$= 19 \text{ g} \times 45\,°C \times \dfrac{4.184 \text{ J}}{\text{g}\,°C} \times \dfrac{1 \text{ kJ}}{1000 \text{ J}} = 3.6 \text{ kJ}$ (2 SFs)

∴ 3.6 kJ of heat is removed.

CI.5 **a.** $0.250 \text{ lb} \times \dfrac{454 \text{ g}}{1 \text{ lb}} \times \dfrac{1 \text{ cm}^3}{7.86 \text{ g}} = 14.4 \text{ cm}^3$ (3 SFs)

b. $30 \text{ nails} \times \dfrac{14.4 \text{ cm}^3}{75 \text{ nails}} = 5.76 \text{ cm}^3 = 5.76 \text{ mL}$ (3 SFs)

New water level $= 17.6 \text{ mL water} + 5.76 \text{ mL} = 23.4 \text{ mL}$ (3 SFs)

c. $\Delta T = 125\,°C - 16\,°C = 109\,°C$
Heat $= m \times \Delta T \times SH$

$= 0.250 \text{ lb} \times \dfrac{454 \text{ g}}{1 \text{ lb}} \times 109\,°C \times \dfrac{0.452 \text{ J}}{\text{g}\,°C} = 5590 \text{ J}$ (3 SFs)

d. iron $25\,°C \rightarrow 1535\,°C$: $\Delta T = 1535\,°C - 25\,°C = 1510.\,°C$

$1 \text{ nail} \times \dfrac{0.250 \text{ lb Fe}}{75 \text{ nails}} \times \dfrac{454 \text{ g Fe}}{1 \text{ lb Fe}} \times \dfrac{0.452 \text{ J}}{\text{g}\,°C} \times 1510.\,°C = 1030 \text{ J}$ (3 SFs)

4
Atoms and Elements

John is preparing to plant a new crop in his fields. Although the element nitrogen in the soil is important, he will also check the soil levels of the elements phosphorus and potassium, which are also important for a good yield. Tests show that the level of phosphorus in the soil is 15 mg/kg, which is low. Tests also show that the potassium level in the soil is in the optimal range from 150 to 200 mg/kg. Thus, John will apply a fertilizer that contains phosphorus but no potassium. What are the symbols of the elements potassium, phosphorus, and nitrogen found in fertilizers?

Credit: Martin Harvey/Alamy

LOOKING AHEAD

4.1 Elements and Symbols
4.2 The Periodic Table
4.3 The Atom
4.4 Atomic Number and Mass Number
4.5 Isotopes and Atomic Mass
4.6 Electron Energy Levels
4.7 Trends in Periodic Properties

 The Health icon indicates a question that is related to health and medicine.

4.1 Elements and Symbols

Learning Goal: Given the name of an element, write its correct symbol; from the symbol, write the correct name.

- Chemical symbols are one- or two-letter abbreviations for the names of the elements; only the first letter of a chemical symbol is a capital letter.
- Element names are derived from planets, mythology, colors, minerals, and names of geographical locations and famous scientists.

Chapter 4

♦ **Learning Exercise 4.1A**

Write the symbol for each of the following elements found in the body:

a. carbon _____   b. iron _____   c. sodium _____
d. phosphorus _____ e. oxygen _____   f. nitrogen _____
g. iodine _____   h. sulfur _____   i. potassium _____
j. cobalt _____   k. calcium _____   l. selenium _____

Answers   a. C   b. Fe   c. Na   d. P   e. O   f. N
            g. I   h. S   i. K   j. Co   k. Ca   l. Se

♦ **Learning Exercise 4.1B**

Write the name of the element represented by each of the following symbols:

a. Mg _____   b. Be _____
c. H _____   d. Si _____
e. Ag _____   f. Br _____
g. F _____   h. Zn _____
i. Cr _____   j. Al _____

The element silver is used for dental fillings.
Credit: Pearson Science/Pearson Education

Answers   a. magnesium   b. beryllium   c. hydrogen   d. silicon   e. silver
            f. bromine   g. fluorine   h. zinc   i. chromium j. aluminum

4.2 The Periodic Table

Learning Goal: Use the periodic table to identify the group and the period of an element; identify the element as a metal, a nonmetal, or a metalloid.

- The periodic table is an arrangement of the elements into vertical columns and horizontal rows.
- Each vertical column is a *group* (or *family*) of elements that have similar properties.
- A horizontal row of elements is called a *period*.
- On the periodic table, the metals are located on the left of the heavy zigzag line, the nonmetals are to the right, and metalloids are found along the heavy zigzag line.
- The element hydrogen (H) located on the top left of the periodic table is a nonmetal.
- Main group, or representative, elements are found in Groups 1A (1), 2A (2), and 3A (13) to 8A (18). Transition elements are B-group elements (3 to 12).
- Of all the elements, only about 20 are essential for the well-being of the human body. Oxygen, carbon, hydrogen, and nitrogen make up 96% of human body mass.

Study Note

1. The periodic table consists of horizontal rows called *periods* and vertical columns called *groups*, which are also called *families*.
2. Group 1A (1) contains the *alkali metals*, Group 2A (2) contains the *alkaline earth metals*, Group 7A (17) contains the *halogens*, and Group 8A (18) contains the *noble gases*.

Atoms and Elements

Periodic Table of Elements

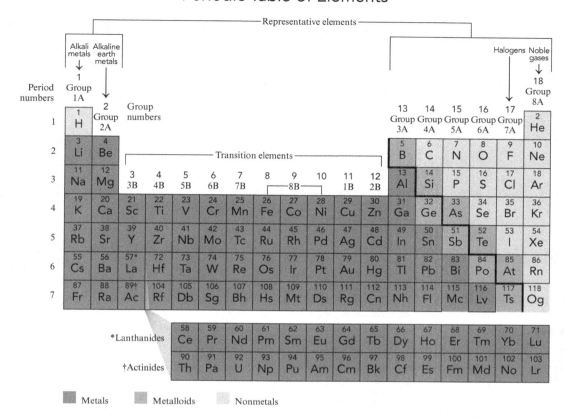

Key Terms for Sections 4.1 and 4.2

Match each of the following key terms with the correct description:

- **a.** nonmetal
- **b.** period
- **c.** chemical symbol
- **d.** metal
- **e.** group number
- **f.** halogen
- **g.** periodic table
- **h.** transition element

1. _____ a horizontal row of elements in the periodic table
2. _____ an element with little or no luster that is a poor conductor of heat and electricity
3. _____ a number that appears at the top of each vertical column (group) in the periodic table
4. _____ an element that is shiny, malleable, ductile, and a good conductor of heat and electricity
5. _____ an element in the center of the periodic table that has a group number of 3 through 12
6. _____ an arrangement of elements by increasing atomic number
7. _____ a one- or two-letter abbreviation that represents the name of an element
8. _____ an element in Group 7A (17)

Answers
1. b 2. a 3. e 4. d
5. h 6. g 7. c 8. f

Chapter 4

♦ Learning Exercise 4.2A

Indicate whether the following elements are part of the same group (G), the same period (P), or neither (N):

a. Li, O, and F _____
b. Br, Cl, and F _____
c. Al, Si, and Cl _____
d. Cl, Mg, and P _____
e. Mg, Ca, and Ba _____
f. C, S, and Br _____
g. Sb, P, and As _____
h. K, Ca, and Br _____

Answers   a. P   b. G   c. P   d. P
            e. G   f. N   g. G   h. P

♦ Learning Exercise 4.2B

Complete the list of some elements important in the body with the name, symbol, group number, period number, and whether it is a metal, nonmetal, or metalloid in the following table:

Name and Symbol	Group Number	Period Number	Type of Element
	2A (2)	3	
Carbon, C			
	5A (15)	2	
Sulfur, S			
	5A (15)	5	
	1A (1)	1	
		3	metalloid

Answers

Name and Symbol	Group Number	Period Number	Type of Element
Magnesium, Mg	2A (2)	3	metal
Carbon, C	4A (14)	2	nonmetal
Nitrogen, N	5A (15)	2	nonmetal
Sulfur, S	6A (16)	3	nonmetal
Antimony, Sb	5A (15)	5	metalloid
Hydrogen, H	1A (1)	1	nonmetal
Silicon, Si	4A (14)	3	metalloid

Atoms and Elements

♦ **Learning Exercise 4.2C**

Identify each of the following elements as a metal (M), a nonmetal (NM), or a metalloid (ML):

a. Cl ____   b. P ____   c. Fe ____   d. As ____   e. Al ____
f. O ____   g. Ca ____   h. Zn ____   i. Ge ____   j. Ag ____

Answers   a. NM   b. NM   c. M   d. ML   e. M
            f. NM   g. M   h. M   i. ML   j. M

♦ **Learning Exercise 4.2D**

Match the name of each of the following chemical groups with the elements Cs, F, He, Cr, Sr, Ne, Li, Cu, and Br:

a. halogen _____

b. noble gas _____

c. alkali metal _____

d. alkaline earth metal _____

e. transition element _____

Answers   a. F, Br   b. He, Ne   c. Cs, Li   d. Sr   e. Cr, Cu

4.3 The Atom

Learning Goal: Describe the electrical charge and location in an atom for a proton, a neutron, and an electron.

- An atom is the smallest particle that retains the characteristics of an element.
- Atoms are composed of three types of subatomic particles. Protons have a positive charge (+), electrons carry a negative charge (−), and neutrons are electrically neutral.
- The protons and neutrons, with masses of about 1 amu, are found in the tiny, dense nucleus.
- The electrons are located outside the nucleus and have masses much smaller than the mass of a neutron or proton.

Dalton's Atomic Theory

1. All matter is made up of tiny particles called atoms.
2. All atoms of a given element are the same and different from atoms of other elements.
3. Atoms of two or more different elements combine to form compounds. A particular compound is always made up of the same kinds of atoms and the same number of each kind of atom.
4. A chemical reaction involves the rearrangement, separation, or combination of atoms. Atoms are neither created nor destroyed during a chemical reaction.

Chapter 4

♦ **Learning Exercise 4.3A**

True (T) or False (F): Each of the following statements is consistent with current atomic theory:

a. ____ All matter is composed of atoms.

b. ____ All atoms of an element are similar.

c. ____ Atoms can combine to form compounds.

d. ____ Most of the mass of the atom is outside of the nucleus.

Answers   a. T   b. T   c. T   d. F

Aluminum foil consists of atoms of aluminum.
Credit: Russ Lappa/Pearson Education/Prentice Hall

♦ **Learning Exercise 4.3B**

Match the following terms with the correct descriptions:

   a. proton   b. neutron   c. electron   d. nucleus

1. ____ is found in the nucleus of an atom
2. ____ has a 1− charge
3. ____ is found outside the nucleus
4. ____ has a mass of about 1 amu
5. ____ is the small, dense center of the atom
6. ____ is neutral
7. ____ is attracted to a proton

Answers   1. a and b   2. c   3. c   4. a and b   5. d   6. b   7. c

4.4 Atomic Number and Mass Number

Learning Goal: Given the atomic number and the mass number of an atom, state the number of protons, neutrons, and electrons.

REVIEW
Using Positive and Negative Numbers in Calculations (1.4)

- The atomic number, which can be found on the periodic table, is the number of protons in a given atom of an element.
- In a neutral atom, the number of protons is equal to the number of electrons.
- The mass number is the total number of protons and neutrons in an atom.

Study Note

1. The *atomic number* is the number of protons in every atom of an element. In neutral atoms, the number of electrons equals the number of protons.
2. The *mass number* is the total number of neutrons and protons in the nucleus of an atom.
3. The number of neutrons is *mass number − atomic number*.

Example: How many protons and neutrons are in the nucleus of a strontium atom with a mass number of 89?

Solution: The atomic number of Sr is 38. Thus, an atom of Sr has 38 protons.
Mass number − atomic number = number of neutrons: 89 − 38 = 51 neutrons

---

78   Copyright © 2018 Pearson Education, Inc.

Atoms and Elements

♦ Learning Exercise 4.4A

CORE CHEMISTRY SKILL
Counting Protons and Neutrons

Give the number of protons in each of the following neutral atoms:

a. an atom of carbon _____

b. an atom of the element with atomic number 15 _____

c. an atom with a mass number of 40 and atomic number 19 _____

d. an atom with 9 neutrons and a mass number of 19 _____

e. a neutral atom that has 18 electrons _____

Answers a. 6   b. 15   c. 19   d. 10   e. 18

♦ Learning Exercise 4.4B

Determine the number of neutrons in each of the following atoms:

a. a mass number of 33 and atomic number 15 _____

b. a mass number of 10 and 5 protons _____

c. a selenium atom with a mass number of 80 _____

d. a mass number of 9 and atomic number 4 _____

e. a mass number of 22 and 10 protons _____

f. a titanium atom with a mass number of 46 _____

Answers a. 18   b. 5   c. 46   d. 5   e. 12   f. 24

4.5 Isotopes and Atomic Mass

Learning Goal: Determine the number of protons, neutrons, and electrons in one or more of the isotopes of an element; identify the most abundant isotope of an element.

- *Isotopes* are atoms of the same element that have the same number of protons but different numbers of neutrons.
- The atomic mass of an element is the average mass of all the isotopes in a naturally occurring sample of that element.

Key Terms for Sections 4.3 to 4.5

Match each of the following key terms with the correct description:

a. electron   b. atom   c. atomic mass   d. mass number   e. neutron
f. isotope   g. atomic number   h. nucleus   i. proton

1. _____ the total number of neutrons and protons in the nucleus of an atom

2. _____ the smallest particle of an element that retains the characteristics of the element

3. _____ a negatively charged subatomic particle having a minute mass

4. _____ a neutral subatomic particle found in the nucleus of an atom

5. _____ a number that is equal to the number of protons in an atom
6. _____ the compact, extremely dense center of an atom where the protons and neutrons are located
7. _____ a positively charged subatomic particle found in the nucleus of an atom
8. _____ the average mass of all the naturally occurring isotopes of an element
9. _____ an atom that differs in mass number from another atom of the same element

Answers   1. d   2. b   3. a   4. e   5. g
            6. h   7. i   8. c   9. f

Study Note

In the atomic symbol for a particular atom, the mass number appears in the upper left corner and the atomic number in the lower left corner of the atomic symbol.   Mass number → $^{32}_{16}$S   Atomic number →

This isotope is also written several other ways including sulfur-32, S-32, and ^{32}S.

♦ Learning Exercise 4.5A

CORE CHEMISTRY SKILL
Writing Atomic Symbols for Isotopes

Complete the following table for neutral atoms:

Atomic Symbol	Atomic Number	Mass Number	Number of Protons	Number of Neutrons	Number of Electrons
	15			16	
			47	60	
		55		30	
	35			45	
		35	17		
$^{120}_{50}$Sn					

Atoms of Mg

Answers

Atomic Symbol	Atomic Number	Mass Number	Number of Protons	Number of Neutrons	Number of Electrons
$^{31}_{15}$P	15	31	15	16	15
$^{107}_{47}$Ag	47	107	47	60	47
$^{55}_{25}$Mn	25	55	25	30	25
$^{80}_{35}$Br	35	80	35	45	35
$^{35}_{17}$Cl	17	35	17	18	17
$^{120}_{50}$Sn	50	120	50	70	50

Isotopes of Mg

$^{24}_{12}$Mg $^{25}_{12}$Mg $^{26}_{12}$Mg

The nuclei of three naturally occurring magnesium isotopes have the same number of protons but different numbers of neutrons.

Credit: Pearson Science/Pearson Education

♦ Learning Exercise 4.5B

Identify the sets of atoms that are isotopes:

a. $^{20}_{10}X$ b. $^{20}_{11}X$ c. $^{21}_{11}X$ d. $^{19}_{10}X$ e. $^{19}_{9}X$

Answer   Atoms **a** and **d** are isotopes (atomic number 10); atoms **b** and **c** are isotopes (atomic number 11).

♦ Learning Exercise 4.5C

Copper has two naturally occurring isotopes, $^{63}_{29}Cu$ and $^{65}_{29}Cu$. If that is the case, why is the atomic mass of copper listed as 63.55 on the periodic table?

Answer   Copper in nature consists of two isotopes that have different masses. The atomic mass is the average of the individual masses of the two isotopes, which takes into consideration their percent abundance in the sample. The atomic mass does not represent the mass of any individual atom.

♦ Learning Exercise 4.5D

Rubidium is the element with atomic number 37.

a. What is the symbol for rubidium? _____

b. What is the group number for rubidium? _____

c. What is the name of the family that includes rubidium? _____

d. Rubidium has two naturally occurring isotopes, $^{85}_{37}Rb$ and $^{87}_{37}Rb$. How many protons, neutrons, and electrons are in each isotope?

e. There are two naturally occurring isotopes of rubidium, $^{85}_{37}Rb$ and $^{87}_{37}Rb$. Which isotope of rubidium is more abundant?

Answers
a. Rb
b. Group 1A (1)
c. alkali metals
d. The isotope $^{85}_{37}Rb$ has 37 protons, 48 neutrons, and 37 electrons. The isotope $^{87}_{37}Rb$ has 37 protons, 50 neutrons, and 37 electrons.
e. On the periodic table, the atomic mass of rubidium is 85.47. The isotope Rb-85, which has a mass closer to the atomic mass of rubidium, is more abundant.

Chapter 4

4.6 Electron Energy Levels

Learning Goal: Given the name or symbol of one of the first 20 elements in the periodic table, write the electron arrangement.

- Electromagnetic radiation is energy that travels as waves at the speed of light.
- The electromagnetic spectrum is all forms of electromagnetic radiation, arranged in order of decreasing wavelength.
- An atomic spectrum is a series of colored lines that correspond to the specific energies emitted by a heated element.
- In an atom, the electrons of similar energy are grouped in specific energy levels. The first level can hold 2 electrons, the second level can hold 8 electrons, and the third level will take up to 18 electrons.
- The electron arrangement specifies the number of electrons in each energy level beginning with the lowest energy level.

♦ **Learning Exercise 4.6**

CORE CHEMISTRY SKILL
Writing Electron Arrangements

Write the electron arrangements for the following elements:

Element	Energy Level 1 2 3 4	Element	Energy Level 1 2 3 4
a. beryllium	_____	b. phosphorus	_____
c. carbon	_____	d. nitrogen	_____
e. potassium	_____	f. chlorine	_____
g. sodium	_____	h. silicon	_____

Answers a. 2,2 b. 2,8,5 c. 2,4 d. 2,5
 e. 2,8,8,1 f. 2,8,7 g. 2,8,1 h. 2,8,4

4.7 Trends in Periodic Properties

Learning Goal: Use the electron arrangement of elements to explain the trends in periodic properties.

CORE CHEMISTRY SKILL
Identifying Trends in Periodic Properties

- The physical and chemical properties of the elements change in a periodic manner going across each period and are repeated in each successive period.
- Representative elements in a group have similar behavior.
- The group number of a representative element gives the number of valence electrons in the atoms of that group.
- The Lewis symbol is a convenient way to represent the valence electrons, which are shown as dots on the sides, top, or bottom of the symbol for an element. For beryllium, the Lewis symbol is:

    Be·
- The atomic size of representative elements increases going down a group and decreases going from left to right across a period.
- The ionization energy of representative elements decreases going down a group and increases going from left to right across a period.
- The metallic character of representative elements increases going down a group and decreases going from left to right across a period.

Atoms and Elements

Key Terms for Sections 4.6 and 4.7

Match each of the following key terms with the correct description:

    a. energy level   **b.** electron arrangement   **c.** valence electrons
    d. ionization energy   **e.** group number   **f.** Lewis symbol

1. _____ the electrons within an atom arranged by increasing energy
2. _____ the representation of an atom that shows valence electrons as dots around the symbol of the element
3. _____ a measure of how easily an element loses a valence electron
4. _____ the electrons in the highest energy level of an atom
5. _____ the number at the top of each vertical column in the periodic table
6. _____ a group of electrons of similar energy

Answers   **1.** b   **2.** f   **3.** d   **4.** c   **5.** e   **6.** a

♦ **Learning Exercise 4.7A**

CORE CHEMISTRY SKILL
Drawing Lewis Symbols

State the number of valence electrons, the group number of each element, and draw the Lewis symbol for each of the following:

Element	Valence Electrons	Group Number	Lewis Symbol
Sulfur			
Silicon			
Magnesium			
Hydrogen			
Fluorine			
Aluminum			

Copyright © 2018 Pearson Education, Inc.

Chapter 4

Answers

Element	Valence Electrons	Group Number	Lewis Symbol
Sulfur	6 e^-	Group 6A (16)	:S̈:
Silicon	4 e^-	Group 4A (14)	·S̈i·
Magnesium	2 e^-	Group 2A (2)	Ṁg·
Hydrogen	1 e^-	Group 1A (1)	H·
Fluorine	7 e^-	Group 7A (17)	:F̈·
Aluminum	3 e^-	Group 3A (13)	·Ȧl·

♦ **Learning Exercise 4.7B**

Select the larger atom in each pair.

a. ____ Mg or Ca b. ____ Si or Cl
c. ____ Sr or Rb d. ____ Br or Cl
e. ____ Li or Cs f. ____ Li or N
g. ____ N or P   h. ____ As or Ca

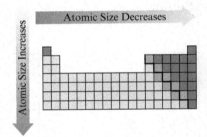

Atomic size increases going down a group and decreases going from left to right across a period.

Answers   a. Ca   b. Si   c. Rb   d. Br
             e. Cs   f. Li   g. P   h. Ca

♦ **Learning Exercise 4.7C**

Select the element in each pair with the lower ionization energy.

a. ____ Mg or Na b. ____ P or Cl
c. ____ K or Rb   d. ____ Br or F
e. ____ Li or O   f. ____ Sb or N
g. ____ K or Br   h. ____ S or Na

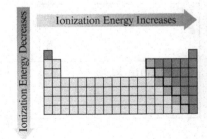

Ionization energy decreases going down a group and increases going from left to right across a period.

Answers   a. Na   b. P   c. Rb   d. Br
             e. Li   f. Sb   g. K   h. Na

Atoms and Elements

♦ Learning Exercise 4.7D

Select the element in each pair that has more metallic character.

a. _____ K or Na   b. _____ O or Se
c. _____ Ca or Br   d. _____ I or F
e. _____ Li or N   f. _____ Pb or Ba

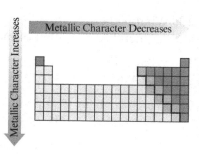

Answers   a. K   b. Se   c. Ca
            d. I   e. Li   f. Ba

Metallic character increases going down a group and decreases going from left to right across a period.

Checklist for Chapter 4

You are ready to take the Practice Test for Chapter 4. Be sure you have accomplished the following learning goals for this chapter. If not, review the Section listed at the end of the goal. Then apply your new skills and understanding to the Practice Test.

After studying Chapter 4, I can successfully:

_____ Write the correct symbol or name for an element. (4.1)

_____ Use the periodic table to identify the group and period of an element. (4.2)

_____ Identify an element as a metal, nonmetal, or metalloid. (4.2)

_____ State the electrical charge, mass, and location of the protons, neutrons, and electrons in an atom. (4.3)

_____ Given the atomic number and mass number of an atom, state the number of protons, neutrons, and electrons. (4.4)

_____ Identify an isotope and describe the atomic mass of an element. (4.5)

_____ Write the electron arrangement for the elements with atomic number 1 to 20. (4.6)

_____ Draw the Lewis symbol for a representative element. (4.7)

_____ Predict the periodic trends in valence electrons, group numbers, atomic size, ionization energy, and metallic character going down a group or across a period. (4.7)

Practice Test for Chapter 4

The chapter Sections to review are shown in parentheses at the end of each question.

For questions 1 through 5, write the correct symbol for each of the elements listed: (4.1)

1. lithium _____   2. hydrogen _____   3. barium _____
4. gold _____   5. magnesium _____

For questions 6 through 10, write the correct element name for each of the symbols listed: (4.1)

6. Fe _____   7. Cu _____   8. Cl _____
9. Pb _____   10. Ag _____

11. Some important elements in the body, C, N, and O, are part of a (4.2)
    A. period   B. group   C. neither

12. Some important elements in the body, N, P, and As are part of a (4.2)
    A. period   B. group   C. neither

Chapter 4

13. What is the classification of an atom with 15 protons and 17 neutrons? (4.2)
    A. metal   B. nonmetal   C. transition element   D. noble gas   E. halogen

14. What is the group number of the element with atomic number 3? (4.2)
    A. 1A (1)   B. 2A (2)   C. 3B (3)   D. 3A (13)   E. 5A (15)

15. Which element would you expect to have properties most like oxygen? (4.2)
    A. nitrogen   B. carbon   C. chlorine   D. argon   E. sulfur

For questions 16 through 18, indicate whether the statement is True or False: (4.3)

16. _____ The electrons of an atom are located in the nucleus.

17. _____ Most of the mass of an atom is due to the protons and neutrons.

18. _____ All matter is composed of atoms.

For questions 19 through 22, consider an atom with 12 protons and 13 neutrons: (4.4)

19. This atom has an atomic number of
    A. 12   B. 13   C. 23   D. 24   E. 25

20. This atom has a mass number of
    A. 12   B. 13   C. 23   D. 24   E. 25

21. This is an atom of
    A. carbon   B. sodium   C. magnesium   D. aluminum   E. manganese

22. The number of electrons in this atom is
    A. 12   B. 13   C. 23   D. 24   E. 25

For questions 23 through 26, consider an atom of calcium with a mass number of 44: (4.4)

23. This atom of calcium has an atomic number of
    A. 20   B. 24   C. 40   D. 41   E. 44

24. The number of protons in this atom of calcium is
    A. 20   B. 24   C. 40   D. 41   E. 44

25. The number of neutrons in this atom of calcium is
    A. 20   B. 24   C. 40   D. 41   E. 44

26. The number of electrons in this atom of calcium is
    A. 20   B. 24   C. 40   D. 41   E. 44

27. A mercury atom, $^{201}_{80}Hg$ has (4.4)
    A. 80 p^+, 80 n^0, 80 e^-
    B. 201 p^+, 201 n^0, 201 e^-
    C. 80 p^+, 201 n^0, 80 e^-
    D. 80 p^+, 121 n^0, 80 e^-
    E. 80 p^+, 121 n^0, 121 e^-

For questions 28 and 29, use the following list of atoms: (4.5)

$^{14}_{7}G$ $^{16}_{8}J$ $^{19}_{9}M$ $^{16}_{7}Q$ $^{18}_{8}R$

28. Which atom(s) is (are) isotopes of an atom with 8 protons and 9 neutrons?
    A. J   B. J, R   C. M, Q   D. M   E. Q

29. Which atom(s) is (are) isotopes of an atom with 7 protons and 8 neutrons?
    A. G   B. J   C. G, Q   D. J, R   E. none

30. Which of the following is an isotope of nitrogen? (4.5)
    A. $^{14}_{8}N$ B. $^{7}_{3}N$ C. $^{10}_{5}N$ D. $^{4}_{2}He$ E. $^{15}_{7}N$

31. The electron arrangement for an oxygen atom is (4.6)
    A. 2,4   B. 2,8   C. 2,6   D. 2,4,2   E. 2,6,2

32. The electron arrangement for an aluminum atom is (4.6)
   A. 2,11   B. 2,8,5   C. 2,8,3   D. 2,10,1   E. 2,2,6,3

33. The number of valence electrons in gallium is (4.7)
   A. 1   B. 2   C. 3   D. 13   E. 31

34. An atom of oxygen is larger than an atom of (4.7)
   A. lithium   B. sulfur   C. fluorine   D. boron   E. argon

35. Of Li, Na, K, Rb, and Cs, the element with the most metallic character is (4.7)
   A. Li   B. Na   C. K   D. Rb   E. Cs

36. Of C, Si, Ge, Sn, and Pb, the element with the highest ionization energy is (4.7)
   A. C   B. Si   C. Ge   D. Sn   E. Pb

37. The Lewis symbol •X• would be correct for an element in (4.7)
   A. Group 1A (1)   B. Group 2A (2)   C. Group 4A (14)
   D. Group 6A (16)   E. Group 7A (17)

38. The Lewis symbol X• would be correct for (4.7)
   A. magnesium   B. chlorine   C. sulfur
   D. cesium   E. nitrogen

Answers to the Practice Test

1. Li	**2.** H	**3.** Ba	**4.** Au	**5.** Mg
6. iron	**7.** copper	**8.** chlorine	**9.** lead	**10.** silver
11. A	**12.** B	**13.** B	**14.** A	**15.** E
16. False	**17.** True	**18.** True	**19.** A	**20.** E
21. C	**22.** A	**23.** A	**24.** A	**25.** B
26. A	**27.** D	**28.** B	**29.** C	**30.** E
31. C	**32.** C	**33.** C	**34.** C	**35.** E
36. A	**37.** B	**38.** D		

Chapter 4

Selected Answers and Solutions to Text Problems

4.1 **a.** Cu is the symbol for copper.
  b. Pt is the symbol for platinum.
  c. Ca is the symbol for calcium.
  d. Mn is the symbol for manganese.
  e. Fe is the symbol for iron.
  f. Ba is the symbol for barium.
  g. Pb is the symbol for lead.
  h. Sr is the symbol for strontium.

4.3 **a.** Carbon is the element with the symbol C.
  b. Chlorine is the element with the symbol Cl.
  c. Iodine is the element with the symbol I.
  d. Selenium is the element with the symbol Se.
  e. Nitrogen is the element with the symbol N.
  f. Sulfur is the element with the symbol S.
  g. Zinc is the element with the symbol Zn.
  h. Cobalt is the element with the symbol Co.

4.5 **a.** Sodium (Na) and chlorine (Cl) are in NaCl.
  b. Calcium (Ca), sulfur (S), and oxygen (O) are in $CaSO_4$.
  c. Carbon (C), hydrogen (H), chlorine (Cl), nitrogen (N), and oxygen (O) are in $C_{15}H_{22}ClNO_2$.
  d. Lithium (Li), carbon (C), and oxygen (O) are in Li_2CO_3.

4.7 **a.** C, N, and O are in Period 2.
  b. He is the element at the top of Group 8A (18).
  c. The alkali metals are the elements in Group 1A (1).
  d. Period 2 is the horizontal row of elements that ends with neon (Ne).

4.9 **a.** C is Group 4A (14), Period 2.
  b. He is the noble gas in Period 1.
  c. Na is the alkali metal in Period 3.
  d. Ca is Group 2A (2), Period 4.
  e. Al is Group 3A (13), Period 3.

4.11 On the periodic table, *metals* are located to the left of the heavy zigzag line, *nonmetals* are elements to the right, and *metalloids* (B, Si, Ge, As, Sb, Te, Po, and At) are located along the line.
  a. Ca is a metal.
  b. S is a nonmetal.
  c. Metals are shiny.
  d. An element that is a gas at room temperature is a nonmetal.
  e. Group 8A (18) elements are nonmetals.
  f. Br is a nonmetal.
  g. B is a metalloid.
  h. Ag is a metal.

4.13 **a.** Ca, an alkaline earth metal, is needed for bones and teeth, muscle contraction, and nerve impulses.
  b. Fe, a transition element, is a component of the oxygen carrier hemoglobin.
  c. K, an alkali metal, is the most common ion (K^+) in cells, and is needed for muscle contraction and nerve impulses.
  d. Cl, a halogen, is the most common negative ion (Cl^-) in fluids outside cells, and is a component of stomach acid (HCl).

Atoms and Elements

4.15 **a.** A macromineral is an element essential to health that is present in the human body in amounts from 5 to 1000 g.
b. Sulfur is a component of proteins, liver, vitamin B_1, and insulin.
c. 86 g is a typical amount of sulfur in a 60.-kg adult.

4.17 **a.** The electron has the smallest mass.
b. The proton has a 1+ charge.
c. The electron is found outside the nucleus.
d. The neutron is electrically neutral.

4.19 Rutherford determined that an atom contains a small, compact nucleus that is positively charged.

4.21 **a.** True
b. True
c. True
d. False; since a neutron has no charge, it is not attracted to a proton (a proton is attracted to an electron).

4.23 In the process of brushing your hair, strands of hair become charged with like charges that repel each other.

4.25 **a.** The atomic number is the same as the number of protons in an atom.
b. Both are needed since the number of neutrons is the (mass number) − (atomic number).
c. The mass number is the number of particles (protons + neutrons) in the nucleus.
d. The atomic number is the same as the number of electrons in a neutral atom.

4.27 The atomic number defines the element and is found above the symbol of the element in the periodic table.
a. Lithium, Li, has an atomic number of 3.
b. Fluorine, F, has an atomic number of 9.
c. Calcium, Ca, has an atomic number of 20.
d. Zinc, Zn, has an atomic number of 30.
e. Neon, Ne, has an atomic number of 10.
f. Silicon, Si, has an atomic number of 14.
g. Iodine, I, has an atomic number of 53.
h. Oxygen, O, has an atomic number of 8.

4.29 The atomic number gives the number of protons in the nucleus of an atom. Since atoms are neutral, the atomic number also gives the number of electrons in the neutral atom.
a. There are 18 protons and 18 electrons in a neutral argon atom.
b. There are 25 protons and 25 electrons in a neutral manganese atom.
c. There are 53 protons and 53 electrons in a neutral iodine atom.
d. There are 48 protons and 48 electrons in a neutral cadmium atom.

4.31 The atomic number is the same as the number of protons in an atom and the number of electrons in a neutral atom; the atomic number defines the element. The number of neutrons is the (mass number) − (atomic number).

Name of the Element	Symbol	Atomic Number	Mass Number	Number of Protons	Number of Neutrons	Number of Electrons
Zinc	Zn	30	66	30	66 − 30 = 36	30
Magnesium	Mg	12	12 + 12 = 24	12	12	12
Potassium	K	19	19 + 20 = 39	19	20	19
Sulfur	S	16	16 + 15 = 31	16	15	16
Iron	Fe	26	56	26	56 − 26 = 30	26

Chapter 4

4.33 **a.** Since the atomic number of strontium is 38, every Sr atom has 38 protons. An atom of strontium (mass number 89) has 51 neutrons ($89 - 38 = 51\,n$). Neutral atoms have the same number of protons and electrons. Therefore, 38 protons, 51 neutrons, 38 electrons.
b. Since the atomic number of chromium is 24, every Cr atom has 24 protons. An atom of chromium (mass number 52) has 28 neutrons ($52 - 24 = 28\,n$). Neutral atoms have the same number of protons and electrons. Therefore, 24 protons, 28 neutrons, 24 electrons.
c. Since the atomic number of sulfur is 16, every S atom has 16 protons. An atom of sulfur (mass number 34) has 18 neutrons ($34 - 16 = 18\,n$). Neutral atoms have the same number of protons and electrons. Therefore, 16 protons, 18 neutrons, 16 electrons.
d. Since the atomic number of bromine is 35, every Br atom has 35 protons. An atom of bromine (mass number 81) has 46 neutrons ($81 - 35 = 46\,n$). Neutral atoms have the same number of protons and electrons. Therefore, 35 protons, 46 neutrons, 35 electrons.

4.35 **a.** Since the number of protons is 15, the atomic number is 15 and the element symbol is P. The mass number is the sum of the number of protons and the number of neutrons, $15 + 16 = 31$. The atomic symbol for this isotope is $^{31}_{15}\text{P}$.
b. Since the number of protons is 35, the atomic number is 35 and the element symbol is Br. The mass number is the sum of the number of protons and the number of neutrons, $35 + 45 = 80$. The atomic symbol for this isotope is $^{80}_{35}\text{Br}$.
c. Since the number of electrons is 50, there must be 50 protons in a neutral atom. Since the number of protons is 50, the atomic number is 50, and the element symbol is Sn. The mass number is the sum of the number of protons and the number of neutrons, $50 + 72 = 122$. The atomic symbol for this isotope is $^{122}_{50}\text{Sn}$.
d. Since the element is chlorine, the element symbol is Cl, the atomic number is 17, and the number of protons is 17. The mass number is the sum of the number of protons and the number of neutrons, $17 + 18 = 35$. The atomic symbol for this isotope is $^{35}_{17}\text{Cl}$.
e. Since the element is mercury, the element symbol is Hg, the atomic number is 80, and the number of protons is 80. The mass number is the sum of the number of protons and the number of neutrons, $80 + 122 = 202$. The atomic symbol for this isotope is $^{202}_{80}\text{Hg}$.

4.37 **a.** Since the element is argon, the element symbol is Ar, the atomic number is 18, and the number of protons is 18. The atomic symbols for the isotopes with mass numbers of 36, 38, and 40 are $^{36}_{18}\text{Ar}$, $^{38}_{18}\text{Ar}$, and $^{40}_{18}\text{Ar}$, respectively.
b. They all have the same atomic number (the same number of protons and electrons).
c. They have different numbers of neutrons, which gives them different mass numbers.
d. The atomic mass of argon listed on the periodic table is the average atomic mass of all the naturally occurring isotopes of argon.
e. The isotope Ar-40 ($^{40}_{18}\text{Ar}$) is the most abundant in a sample of argon because its mass is closest to the average atomic mass of argon listed on the periodic table (39.95 amu).

4.39 Since the atomic mass of gallium is 69.72 amu, the more abundant isotope of gallium is $^{69}_{31}\text{Ga}$.

4.41 Since the atomic mass of copper (63.55 amu) is closer to 63 amu, there are more atoms of $^{63}_{29}\text{Cu}$ in a sample of copper.

4.43 Since the atomic mass of thallium (204.4 amu) is closer to 205 amu, the more abundant isotope of thallium is $^{205}_{81}\text{Tl}$.

4.45 Electrons can move to higher energy levels when they <u>absorb</u> energy.

4.47 **a.** Green light has greater energy than yellow light.
b. Blue light has greater energy than microwaves.

4.49 **a.** Carbon with atomic number 6 has 6 electrons, which are arranged with 2 electrons in energy level 1 and 4 electrons in energy level 2:   2,4
b. Argon with atomic number 18 has 18 electrons, which are arranged with 2 electrons in energy level 1, 8 electrons in energy level 2, and 8 electrons in energy level 3:   2,8,8

c. Potassium with atomic number 19 has 19 electrons, which are arranged with 2 electrons in energy level 1, 8 electrons in energy level 2, 8 electrons in energy level 3, and 1 electron in energy level 4: 2,8,8,1

d. Silicon with atomic number 14 has 14 electrons, which are arranged with 2 electrons in energy level 1, 8 electrons in energy level 2, and 4 electrons in energy level 3: 2,8,4

e. Helium with atomic number 2, has 2 electrons which are in energy level 1: 2

f. Nitrogen with atomic number 7 has 7 electrons, which are arranged with 2 electrons in energy level 1 and 5 electrons in energy level 2: 2,5

4.51
a. 2,1 is the electron arrangement of lithium (Li).
b. 2,8,2 is the electron arrangement of magnesium (Mg).
c. 1 is the electron arrangement of hydrogen (H).
d. 2,8,7 is the electron arrangement of chlorine (Cl).
e. 2,6 is the electron arrangement of oxygen (O).

4.53
a. Magnesium, in Group 2A (2), has 2 valence electrons.
b. Iodine, in Group 7A (17), has 7 valence electrons.
c. Oxygen, in Group 6A (16), has 6 valence electrons.
d. Phosphorus, in Group 5A (15), has 5 valence electrons.
e. Tin, in Group 4A (14), has 4 valence electrons.
f. Boron, in Group 3A (13), has 3 valence electrons.

4.55
a. Sulfur is in Group 6A (16);
$\cdot \ddot{\underset{..}{S}} :$

b. Nitrogen is in Group 5A (15);
$\cdot \underset{.}{\dot{N}} \cdot$

c. Calcium is in Group 2A (2);
$\dot{Ca} \cdot$

d. Sodium is in Group 1A (1);
$Na \cdot$

e. Gallium is in Group 3A (13);
$\cdot \dot{Ga} \cdot$

4.57 The atomic size of representative elements decreases going across a period from Group 1A to 8A and increases going down a group.
a. In Period 3, Na, which is on the left, is larger than Cl.
b. In Group 1A (1), Rb, which is farther down the group, is larger than Na.
c. In Period 3, Na, which is on the left, is larger than Mg.
d. In Period 5, Rb, which is on the left, is larger than I.

4.59
a. The atomic size of representative elements decreases from Group 1A to 8A: Mg, Al, Si.
b. The atomic size of representative elements increases going down a group: I, Br, Cl.
c. The atomic size of representative elements decreases from Group 1A to 8A: Sr, Sb, I.
d. The atomic size of representative elements decreases from Group 1A to 8A: Na, Si, P.

4.61
a. In Br, the valence electrons are closer to the nucleus, so Br has a higher ionization energy than I.
b. In Mg, the valence electrons are closer to the nucleus, so Mg has a higher ionization energy than Sr.
c. Attraction for the valence electrons increases going from left to right across a period, giving P a higher ionization energy than Si.
d. The noble gases have the highest ionization energy in each period, which gives Xe a higher ionization energy than I.

4.63 **a.** Br, Cl, F; ionization energy decreases going down a group.
  b. Na, Al, Cl; going across a period from left to right, ionization energy increases.
  c. Cs, K, Na; ionization energy decreases going down a group.
  d. Ca, As, Br; going across a period from left to right, ionization energy increases.

4.65 Na has a <u>larger</u> atomic size and is <u>more metallic</u> than P.

4.67 Ca, Ga, Ge, Br; since metallic character decreases from left to right across a period, Ca, in Group 2A (2), will be the most metallic, followed by Ga, then Ge, and finally Br, in Group 7A (17).

4.69 Sr has a <u>lower</u> ionization energy and <u>more</u> metallic character than Sb.

4.71 Going down Group 6A (16),
  a. the ionization energy <u>decreases</u>
  b. the atomic size <u>increases</u>
  c. the metallic character <u>increases</u>
  d. the number of valence electrons <u>remains the same</u>

4.73 **a.** False; an atom of N has a smaller atomic size than an atom of Li.
  b. True; an atom of N has a greater ionization energy than an atom of Li.
  c. True; an atom of N has a greater number of protons than an atom of Li.
  d. False; an atom of N has a less metallic character than an atom of Li.
  e. True; an atom of N has a greater number of valence electrons than an atom of Li.

4.75 **a.** Group 1A (1), alkali metals
  b. metal
  c. 19 protons
  d. Since the atomic mass of potassium is 39.10 amu, the most abundant isotope is K-39, $^{39}_{19}K$.

4.77 **a.** False; all atoms of a given element are identical to one another but different from atoms of other elements.
  b. True
  c. True
  d. False; atoms are never created or destroyed during a chemical reaction.

4.79 **a.** The atomic mass is the average of the masses of all of the naturally occurring isotopes of the element. Isotope masses are based on the masses of all subatomic particles in the atom: protons (1), neutrons (2), and electrons (3), although almost all of that mass comes from the protons (1) and neutrons (2).
  b. The number of protons (1) is the atomic number.
  c. The protons (1) are positively charged.
  d. The electrons (3) are negatively charged.
  e. The number of neutrons (2) is the (mass number) − (atomic number).

4.81 **a.** $^{16}_{8}X$, $^{17}_{8}X$, and $^{18}_{8}X$ all have an atomic number of 8, so all have 8 protons.
  b. $^{16}_{8}X$, $^{17}_{8}X$, and $^{18}_{8}X$ all have an atomic number of 8, so all are isotopes of oxygen.
  c. $^{16}_{8}X$ and $^{16}_{9}X$ have mass numbers of 16, whereas $^{18}_{8}X$ and $^{18}_{10}X$ have mass numbers of 18.
  d. $^{16}_{8}X$ (16 − 8 = 8 n) and $^{18}_{10}X$ (18 − 10 = 8 n) both have 8 neutrons.

4.83 **a.** $^{37}_{17}Cl$ and $^{38}_{18}Ar$ both have 20 neutrons (37 − 17 = 20 n; 38 − 18 = 20 n).
  b. Since both $^{36}_{16}S$ and $^{34}_{16}S$ have an atomic number of 16, they will both have 16 protons in the nucleus. Since the number of protons is 16, there must be 16 electrons in the neutral atom for both. They have different numbers of neutrons (36 − 16 = 20 n; 34 − 16 = 18 n).
  c. $^{40}_{18}Ar$ and $^{39}_{17}Cl$ both have 22 neutrons (40 − 18 = 22 n; 39 − 17 = 22 n).

4.85 a. The diagram shows 4 protons and 5 neutrons in the nucleus of the element. Since the number of protons is 4, the atomic number is 4, and the element symbol is Be. The mass number is the sum of the number of protons and the number of neutrons, $4 + 5 = 9$. The atomic symbol is $^{9}_{4}\text{Be}$.

  b. The diagram shows 5 protons and 6 neutrons in the nucleus of the element. Since the number of protons is 5, the atomic number is 5, and the element symbol is B. The mass number is the sum of the number of protons and the number of neutrons, $5 + 6 = 11$. The atomic symbol is $^{11}_{5}\text{B}$.

  c. The diagram shows 6 protons and 7 neutrons in the nucleus of the element. Since the number of protons is 6, the atomic number is 6, and the element symbol is C. The mass number is the sum of the number of protons and the number of neutrons, $6 + 7 = 13$. The atomic symbol is $^{13}_{6}\text{C}$.

  d. The diagram shows 5 protons and 5 neutrons in the nucleus of the element. Since the number of protons is 5, the atomic number is 5, and the element symbol is B. The mass number is the sum of the number of protons and the number of neutrons, $5 + 5 = 10$. The atomic symbol is $^{10}_{5}\text{B}$.

  e. The diagram shows 6 protons and 6 neutrons in the nucleus of the element. Since the number of protons is 6, the atomic number is 6, and the element symbol is C. The mass number is the sum of the number of protons and the number of neutrons, $6 + 6 = 12$. The atomic symbol is $^{12}_{6}\text{C}$.

  Representations **B** ($^{11}_{5}\text{B}$) and **D** ($^{10}_{5}\text{B}$) are isotopes of boron; **C** ($^{13}_{6}\text{C}$) and **E** ($^{12}_{6}\text{C}$) are isotopes of carbon.

4.87 Atomic size increases going down a group. Li is **D** because it would be smallest. Na is **A**, K is **C**, and Rb is **B**.

4.89 a. Na has the largest atomic size.
  b. Cl is a halogen.
  c. Si has the electron arrangement 2,8,4.
  d. Ar, a noble gas, has the highest ionization energy.
  e. S is in Group 6A (16).
  f. Na, in Group 1A (1), has the most metallic character.
  g. Mg, in Group 2A (2), has two valence electrons.

4.91 a. Br is in Group 7A (17), Period 4.
  b. Ar is in Group 8A (18), Period 3.
  c. K is in Group 1A (1), Period 4.
  d. Ra is in Group 2A (2), Period 7.

4.93 a. False; a proton is a positively charged particle.
  b. False; the neutron has about the same mass as a proton.
  c. True
  d. False; the nucleus is the tiny, dense central core of an atom.
  e. True

4.95 a. The atomic number gives the number of <u>protons</u> in the nucleus.
  b. In an atom, the number of electrons is equal to the number of <u>protons</u>.
  c. Sodium and potassium are examples of elements called <u>alkali metals</u>.

4.97 The atomic number defines the element and is found above the symbol of the element in the periodic table.
  a. Nickel, Ni, has an atomic number of 28.
  b. Barium, Ba, has an atomic number of 56.
  c. Radium, Ra, has an atomic number of 88.
  d. Arsenic, As, has an atomic number of 33.

Chapter 4

  e. Tin, Sn, has an atomic number of 50.
  f. Cesium, Cs, has an atomic number of 55.
  g. Gold, Au, has an atomic number of 79.
  h. Mercury, Hg, has an atomic number of 80.

4.99 a. Since the atomic number of cadmium is 48, every Cd atom has 48 protons. An atom of cadmium (mass number 114) has 66 neutrons ($114 - 48 = 66\,n$). Neutral atoms have the same number of protons and electrons. Therefore, 48 protons, 66 neutrons, 48 electrons.
  b. Since the atomic number of technetium is 43, every Tc atom has 43 protons. An atom of technetium (mass number 98) has 55 neutrons ($98 - 43 = 55\,n$). Neutral atoms have the same number of protons and electrons. Therefore, 43 protons, 55 neutrons, 43 electrons.
  c. Since the atomic number of gold is 79, every Au atom has 79 protons. An atom of gold (mass number 199) has 120 neutrons ($199 - 79 = 120\,n$). Neutral atoms have the same number of protons and electrons. Therefore, 79 protons, 120 neutrons, 79 electrons.
  d. Since the atomic number of radon is 86, every Rn atom has 86 protons. An atom of radon (mass number 222) has 136 neutrons ($222 - 86 = 136\,n$). Neutral atoms have the same number of protons and electrons. Therefore, 86 protons, 136 neutrons, 86 electrons.
  e. Since the atomic number of xenon is 54, every Xe atom has 54 protons. An atom of xenon (mass number 136) has 82 neutrons ($136 - 54 = 82\,n$). Neutral atoms have the same number of protons and electrons. Therefore, 54 protons, 82 neutrons, 54 electrons.

4.101

Name	Atomic Symbol	Number of Protons	Number of Neutrons	Number of Electrons
Sulfur	$^{34}_{16}\text{S}$	16	$34 - 16 = 18$	16
Nickel	$^{28+34}_{28}\text{Ni}$ or $^{62}_{28}\text{Ni}$	28	34	28
Magnesium	$^{12+14}_{12}\text{Mg}$ or $^{26}_{12}\text{Mg}$	12	14	12
Radon	$^{220}_{86}\text{Rn}$	86	$220 - 86 = 134$	86

4.103 a. Since the number of protons is 4, the atomic number is 4, and the element symbol is Be. The mass number is the sum of the number of protons and the number of neutrons, $4 + 5 = 9$. The atomic symbol for this isotope is $^{9}_{4}\text{Be}$.
  b. Since the number of protons is 12, the atomic number is 12, and the element symbol is Mg. The mass number is the sum of the number of protons and the number of neutrons, $12 + 14 = 26$. The atomic symbol for this isotope is $^{26}_{12}\text{Mg}$.
  c. Since the element is calcium, the element symbol is Ca, and the atomic number is 20. The mass number is given as 46. The atomic symbol for this isotope is $^{46}_{20}\text{Ca}$.
  d. Since the number of electrons is 30, there must be 30 protons in a neutral atom. Since the number of protons is 30, the atomic number is 30, and the element symbol is Zn. The mass number is the sum of the number of protons and the number of neutrons, $30 + 40 = 70$. The atomic symbol for this isotope is $^{70}_{30}\text{Zn}$.

4.105 a. Since the element is lead (Pb), the atomic number is 82, and the number of protons is 82. In a neutral atom, the number of electrons is equal to the number of protons, so there will be 82 electrons. The number of neutrons is the (mass number) − (atomic number) = $208 - 82 = 126\,n$. Therefore, 82 protons, 126 neutrons, 82 electrons.
  b. Since the element is lead (Pb), the atomic number is 82, and the number of protons is 82. The mass number is the sum of the number of protons and the number of neutrons, $82 + 132 = 214$. The atomic symbol for this isotope is $^{214}_{82}\text{Pb}$.
  c. Since the mass number is 214 (as in part **b**) and the number of neutrons is 131, the number of protons is the (mass number) − (number of neutrons) = $214 - 131 = 83\,p$. Since there are 83 protons, the atomic number is 83, and the element symbol is Bi (bismuth). The atomic symbol for this isotope is $^{214}_{83}\text{Bi}$.

4.107 a. Oxygen is in Group 6A (16) and has an electron arrangement of 2,6.
  b. Sodium is in Group 1A (1) and has an electron arrangement of 2,8,1.
  c. Neon is in Group 8A (18) and has an electron arrangement of 2,8.
  d. Boron is in Group 3A (13) and has an electron arrangement of 2,3.

4.109 Ca has a greater number of protons than K. This increase in positive charge increases the attraction for electrons which means that it is more difficult to remove the valence electrons. The valence electrons are farther from the nucleus in Ca than in Mg and less energy is needed to remove them.

4.111 a. Be is an alkaline earth metal.
  b. Li has the largest atomic size.
  c. F, in Group 7A (17), has the highest ionization energy.
  d. N is in Group 5A (15).
  e. Li has the most metallic character.

4.113 a. Since the element is strontium (Sr), which has atomic number 38, the number of protons is 38. In a neutral atom, the number of electrons is equal to the number of protons, so there will be 38 electrons. The number of neutrons is the (mass number) − (atomic number) = 87 − 38 = 49 n. In Sr-87, there are 38 protons, 49 neutrons, 38 electrons.
  b. The isotope Sr-88 ($^{88}_{38}$Sr) is the most abundant in a sample of strontium because its mass is closest to the average atomic mass of strontium listed on the periodic table (87.62 amu).
  c. The number of neutrons in Sr-84 is the (mass number) − (atomic number) = 84 − 38 = 46 n.
  d. The atomic mass of strontium (87.62 amu) listed on the periodic table is the average atomic mass of all the naturally occurring isotopes of strontium.

4.115 a. Atomic size increases going down a group, which gives O the smallest atomic size in Group 6A (16).
  b. Atomic size decreases going from left to right across a period, which gives Ar the smallest atomic size in Period 3.
  c. Ionization energy decreases going down a group, which gives N the highest ionization energy in Group 5A (15).
  d. Ionization energy increases going from left to right across a period, which gives Na the lowest ionization energy in Period 3.
  e. Metallic character increases going down a group, which gives Ra the most metallic character in Group 2A (2).

5
Nuclear Chemistry

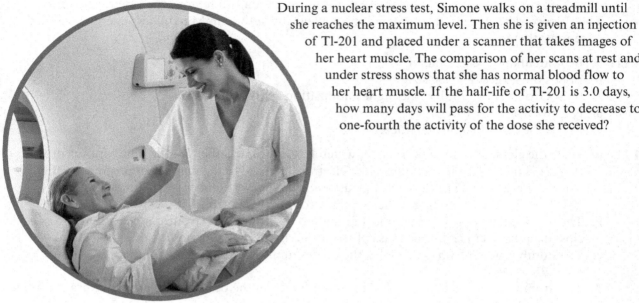

During a nuclear stress test, Simone walks on a treadmill until she reaches the maximum level. Then she is given an injection of Tl-201 and placed under a scanner that takes images of her heart muscle. The comparison of her scans at rest and under stress shows that she has normal blood flow to her heart muscle. If the half-life of Tl-201 is 3.0 days, how many days will pass for the activity to decrease to one-fourth the activity of the dose she received?

Credit: Tyler Olson/Shutterstock

LOOKING AHEAD

5.1 Natural Radioactivity
5.2 Nuclear Reactions
5.3 Radiation Measurement
5.4 Half-Life of a Radioisotope
5.5 Medical Applications Using Radioactivity
5.6 Nuclear Fission and Fusion

 The Health icon indicates a question that is related to health and medicine.

5.1 Natural Radioactivity

Learning Goal: Describe alpha, beta, positron, and gamma radiation.

> **REVIEW**
> Writing Atomic Symbols for Isotopes (4.5)

- Radioactive isotopes have unstable nuclei that break down (decay), spontaneously emitting alpha (α), beta (β), positron (β^+), or gamma (γ) radiation.
- An alpha particle is the same as a helium nucleus; it contains two protons and two neutrons.
- A beta particle is a high-energy electron, and a positron is a high-energy positive particle. A gamma ray is high-energy radiation.
- Because radiation can damage cells in the body, proper protection must be used: shielding, time limitation, and distance.

Nuclear Chemistry

> **Study Note**
>
> It is important to learn the symbols for the radiation particles in order to describe the different types of radiation:
>
> | $^{4}_{2}He$ or α | $^{0}_{-1}e$ or β | $^{0}_{+1}e$ or β^+ | $^{0}_{0}\gamma$ or γ | $^{1}_{1}H$ or p | $^{1}_{0}n$ or n |
> |---|---|---|---|---|---|
> | alpha particle | beta particle | positron | gamma ray | proton | neutron |

♦ **Learning Exercise 5.1A**

Match the terms in column A with the descriptions in column B.

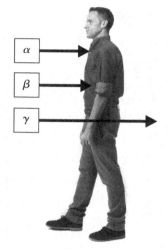

A		B
1. ____ $^{18}_{8}O$		a. symbol for a beta particle
2. ____ $^{0}_{0}\gamma$		b. symbol for an alpha particle
3. ____ $^{0}_{+1}e$		c. an atom that emits radiation
4. ____ radioisotope		d. symbol for a positron
5. ____ $^{4}_{2}He$		e. symbol for an isotope of oxygen
6. ____ $^{0}_{-1}e$		f. symbol for gamma radiation

Answers 1. e   2. f   3. d   4. c   5. b   6. a

Different types of radiation penetrate the body to different depths.
Credit: Celig/Shutterstock

♦ **Learning Exercise 5.1B**

Discuss some things you can do to minimize the amount of radiation received if you work with a radioactive substance. Describe how each method helps to limit the amount of radiation you would receive.

Answer

Three ways to minimize exposure to radiation are (1) use shielding, (2) keep time short in the radioactive area, and (3) keep as much distance as possible from the radioactive materials. Shielding such as clothing and gloves stop alpha and beta particles from reaching your skin, whereas lead or concrete will absorb gamma rays. Limiting the time spent near radioactive samples reduces exposure time. Increasing the distance from a radioactive source reduces the intensity of radiation. Wearing a film badge will monitor the amount of radiation you receive.

Chapter 5

♦ Learning Exercise 5.1C

What type(s) of radiation (alpha, beta, and/or gamma) would each of the following shielding materials protect you from?

a. heavy clothing _____   b. skin _____

c. paper _____   d. concrete _____

e. lead wall _____

Answers   a. alpha, beta   b. alpha   c. alpha
            d. alpha, beta, gamma e. alpha, beta, gamma

5.2 Nuclear Reactions

Learning Goal: Write a balanced nuclear equation for radioactive decay, showing mass numbers and atomic numbers.

> **REVIEW**
> Using Positive and Negative Numbers in Calculations (1.4)
> Solving Equations (1.4)
> Counting Protons and Neutrons (4.4)

- A balanced nuclear equation represents the changes in the nuclei of radioisotopes.

- In all nuclear equations, the sum of the mass numbers for the reactants and the products is equal and the sum of the atomic numbers for the reactants and the products is equal.

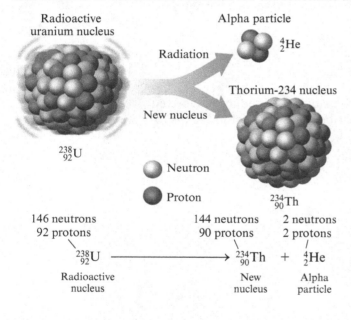

In the nuclear equation for alpha decay, the mass number of the new nucleus decreases by 4 and its atomic number decreases by 2.

- The new isotopes and the type of radiation emitted can be determined from the symbols that show the mass numbers and atomic numbers of the isotopes in the nuclear reaction.

    Radioisotope ⟶ new nucleus + radiation

    Total of the mass numbers is equal on both sides of the equation.

    $^{11}_{6}C \longrightarrow\ ^{7}_{4}Be + ^{4}_{2}He$

    Total of the atomic numbers is equal on both sides of the equation.

- A new radioactive isotope is produced when a nonradioactive isotope is bombarded by a small particle such as alpha particles, protons, neutrons, or small nuclei.

    $^{4}_{2}He$ + $^{10}_{5}B$ ⟶ $^{13}_{7}N$ + $^{1}_{0}n$
    Bombarding particle (α)   *Stable nucleus*   *New nucleus*   *Neutron emitted*

Nuclear Chemistry

Solution Guide to Writing a Nuclear Equation	
STEP 1	Write the incomplete nuclear equation.
STEP 2	Determine the missing mass number.
STEP 3	Determine the missing atomic number.
STEP 4	Determine the symbol of the new nucleus.
STEP 5	Complete the nuclear equation.

♦ **Learning Exercise 5.2A**

CORE CHEMISTRY SKILL
Writing Nuclear Equations

Complete each of the following nuclear equations:

a. $^{213}_{83}Bi \longrightarrow \,^{209}_{81}Tl + ?$ _____

b. $^{127}_{53}I \longrightarrow ? + \,^{1}_{0}n$ _____

c. $^{238}_{92}U \longrightarrow ? + \,^{4}_{2}He$ _____

d. $^{24}_{11}Na \longrightarrow ? + \,^{0}_{-1}e$ _____

e. $? \longrightarrow \,^{30}_{14}Si + \,^{0}_{-1}e$ _____

Answers a. $^{4}_{2}He$ b. $^{126}_{53}I$ c. $^{234}_{90}Th$ d. $^{24}_{12}Mg$ e. $^{30}_{13}Al$

♦ **Learning Exercise 5.2B**

Complete each of the following bombardment reactions:

a. $? + \,^{40}_{20}Ca \longrightarrow \,^{40}_{19}K + \,^{1}_{1}H$ _____

b. $^{1}_{0}n + \,^{27}_{13}Al \longrightarrow \,^{24}_{11}Na + ?$ _____

c. $^{1}_{0}n + \,^{10}_{5}B \longrightarrow ? + \,^{4}_{2}He$ _____

d. $? + \,^{23}_{11}Na \longrightarrow \,^{23}_{12}Mg + \,^{1}_{0}n$ _____

e. $^{1}_{1}H + \,^{197}_{79}Au \longrightarrow ? + \,^{1}_{0}n$ _____

Answers a. $^{1}_{0}n$ b. $^{4}_{2}He$ c. $^{7}_{3}Li$ d. $^{1}_{1}H$ e. $^{197}_{80}Hg$

5.3 Radiation Measurement

Learning Goal: Describe the detection and measurement of radiation.

- A Geiger counter may be used to detect radiation. When radiation passes through the gas in the counter tube, some atoms of gas are ionized, producing an electrical current.
- The activity of a radioactive sample measures the number of nuclear transformations per second. The curie (Ci) is equal to 3.7×10^{10} disintegrations/s. The SI unit is the becquerel (Bq), which is equal to 1 disintegration/s.

Chapter 5

- The radiation dose absorbed by a gram of body tissue is measured in the unit rad (radiation absorbed dose) or the SI unit gray (Gy), which is equal to 100 rad.
- The biological damage to the body caused by different types of radiation is measured in the unit rem (radiation equivalent) or the SI unit sievert (Sv), which is equal to 100 rem.

♦ **Learning Exercise 5.3A**

Match each type of measurement unit with the radiation process measured.

    a. curie   **b.** becquerel   **c.** rad   **d.** gray   **e.** rem

1. ____ an activity of 1 disintegration/s
2. ____ the amount of radiation absorbed by 1 g of material
3. ____ an activity of 3.7×10^{10} disintegrations/s
4. ____ a unit measuring the biological damage caused by different kinds of radiation
5. ____ a unit that measures the absorbed dose of radiation, which is equal to 100 rad

Answers   **1.** b   **2.** c   **3.** a   **4.** e   **5.** d

♦ **Learning Exercise 5.3B**

A sample of Ir-192 used in brachytherapy to treat breast and prostate cancers has an activity of 35 μCi.

a. What is its activity in becquerels?

b. If an absorbed dose is 15 mrem, what is the absorbed dose in sieverts?

Answers   **a.** 1.3×10^6 Bq   **b.** 1.5×10^{-4} Sv

5.4 Half-Life of a Radioisotope

Learning Goal: Given the half-life of a radioisotope, calculate the amount of radioisotope remaining after one or more half-lives.

REVIEW
Interpreting Graphs (1.4)
Using Conversion Factors (2.6)

- The half-life of a radioactive sample is the time required for one-half of the sample to decay (emit radiation).
- Most radioisotopes used in medicine, such as Tc-99m and I-131, have short half-lives. By comparison, many naturally occurring radioisotopes, such as C-14, Ra-226, and U-238, have long half-lives. For example, potassium-42 has a half-life of 12 h, whereas potassium-40 takes 1.3×10^9 yr for one-half of the radioactive sample to decay.

Nuclear Chemistry

SAMPLE PROBLEM   Using Half-Lives of a Radioisotope

Chromium-51, which is used to measure blood volume, has a half-life of 28 days. How many micrograms of a 16-μg sample of chromium-51 will remain active after 84 days?

Solution Guide:

STEP 1 State the given and needed quantities.

	Given	Need	Connect
Analyze the Problem	16 μg of Cr-51, 84 days elapsed, half-life = 28 days	micrograms of Cr-51 remaining	number of half-lives

STEP 2 Write a plan to calculate the unknown quantity.

84 days $\xrightarrow{\text{Half-life}}$ number of half-lives

16 micrograms of Cr-51 $\xrightarrow{\text{Number of half-lives}}$ micrograms of Cr-51 remaining

STEP 3 Write the half-life equality and conversion factors.

1 half-life of Cr-51 = 28 days

$$\frac{28 \text{ days}}{1 \text{ half-life}} \quad \text{and} \quad \frac{1 \text{ half-life}}{28 \text{ days}}$$

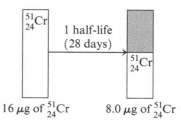

16 μg of $^{51}_{24}$Cr   8.0 μg of $^{51}_{24}$Cr

STEP 4 Set up the problem to calculate the needed quantity.

Number of half-lives = 84 days × $\frac{1 \text{ half-life}}{28 \text{ days}}$ = 3.0 half-lives

Now, we can calculate the quantity of Cr-51 that remains active.

16 μg $\xrightarrow{\text{1 half-life}}$ 8.0 μg $\xrightarrow{\text{2 half-lives}}$ 4.0 μg $\xrightarrow{\text{3 half-lives}}$ 2.0 μg

♦ **Learning Exercise 5.4A**

An 80.-mg sample of iodine-125 is used to treat prostate cancer. If iodine-125 has a half-life of 60. days, how many milligrams are radioactive

CORE CHEMISTRY SKILL
Using Half-Lives

a. after one half-life?

b. after two half-lives?

c. after 240 days?

Answers   a. 40. mg   b. 20. mg   c. 5.0 mg

Chapter 5

♦ **Learning Exercise 5.4B**

a. $^{99m}_{43}$Tc, which is used to image the heart, brain, and lungs, has a half-life of 6.0 h. If a technician picked up a 16-mg sample at 8 A.M., how many milligrams of the radioactive sample will remain active at 8 P.M. that same day?

b. Samarium-153, which is used to treat bone cancer, has a half-life of 46 h. How many micrograms of a 240-μg sample will be radioactive after 46 h?

c. Gallium-68, which is used for the detection of pancreatic cancer, has a half-life of 68 min. How many minutes will it take for 44 mg of Ga-68 to decay to 11 mg?

d. An archaeologist finds some pieces of a wooden boat at an ancient site. When a sample of the wood is analyzed for C-14, one-eighth of the original amount of C-14 remains. If the half-life of carbon-14 is 5730 yr, how many years ago was the boat made?

Answers   a. 4.0 mg   b. 120 μg   c. 136 min   d. 17 000 yr

5.5 Medical Applications Using Radioactivity

Learning Goal: Describe the use of radioisotopes in medicine.

- Nuclear medicine uses radioactive isotopes that go to specific sites in the body.
- For diagnostic work, radioisotopes are used that emit gamma rays and produce nonradioactive products.
- By detecting the radiation emitted by medical radioisotopes, evaluations can be made about the location and extent of an injury, disease, tumor, blood flow, or level of function of a particular organ.

♦ **Learning Exercise 5.5**

Write the atomic symbol for each of the following radioactive isotopes:

a. _____ bismuth-213, used to treat leukemia

b. _____ phosphorus-32, used to locate and treat brain tumors

c. _____ sodium-24, used to determine blood flow and to locate a blood clot or embolism

d. _____ nitrogen-13, used in positron emission tomography

Answers   a. $^{213}_{83}$Bi   b. $^{32}_{15}$P   c. $^{24}_{11}$Na   d. $^{13}_{7}$N

Nuclear Chemistry

5.6 Nuclear Fission and Fusion

Learning Goal: Describe the processes of nuclear fission and fusion.

- In fission, a large nucleus breaks apart into smaller nuclei, releasing one or more types of radiation and a great amount of energy.
- A chain reaction is a fission reaction that will continue once initiated.
- In fusion, small nuclei combine to form a larger nucleus, which releases great amounts of energy.
- Nuclear fission is currently used to produce electrical power while nuclear fusion applications are still in the experimental stage.

Key Terms for Sections 5.1 to 5.6

Match each of the following key terms with the correct description:

a. radiation	**b.** half-life	**c.** curie	**d.** fission
e. alpha particle	**f.** positron	**g.** rem	

1. _____ a particle identical to a helium nucleus produced in a radioactive nucleus
2. _____ the time required for one-half of a radioactive sample to undergo radioactive decay
3. _____ a unit of radiation measurement equal to 3.7×10^{10} disintegrations per second
4. _____ a process in which large nuclei split into smaller nuclei with the release of energy
5. _____ energy or particles released by radioactive atoms
6. _____ a measure of biological damage caused by radiation
7. _____ a particle produced when a proton is transformed into a neutron

Answers   **1.** e   **2.** b   **3.** c   **4.** d   **5.** a   **6.** g   **7.** f

♦ **Learning Exercise 5.6A**

Discuss the nuclear processes of fission and fusion for the production of energy.

Answer

Nuclear fission is a splitting of the atom into smaller nuclei accompanied by the release of large amounts of energy and radiation. In the process of *nuclear fusion*, two or more nuclei combine to form a heavier nucleus and release a large amount of energy. However, fusion requires a considerable amount of energy to initiate the process. Because the extremely high temperatures are difficult to maintain, fusion technology is still in the experimental stage.

♦ **Learning Exercise 5.6B**

Balance each of the following nuclear equations by writing the atomic symbol for the missing particle and state if each is a fission or fusion reaction:

a. $^{1}_{0}n + ^{235}_{92}U \longrightarrow ^{143}_{54}Xe + ? + 3^{1}_{0}n$ _____

b. $^{2}_{1}H + ^{2}_{1}H \longrightarrow ? + ^{1}_{0}n$ _____

c. $^{3}_{1}H + ? \longrightarrow ^{4}_{2}He + ^{1}_{0}n$ _____

d. $^{1}_{0}n + ^{235}_{92}U \longrightarrow ^{91}_{36}Kr + ? + 3^{1}_{0}n$ _____

Answers   **a.** $^{90}_{38}Sr$, fission   **b.** $^{3}_{2}He$, fusion   **c.** $^{2}_{1}H$, fusion   **d.** $^{142}_{56}Ba$, fission

Checklist for Chapter 5

You are ready to take the Practice Test for Chapter 5. Be sure you have accomplished the following learning goals for this chapter. If not, review the Section listed at the end of the goal. Then apply your new skills and understanding to the Practice Test.

After studying Chapter 5, I can successfully:

____ Describe alpha, beta, positron, and gamma radiation. (5.1)

____ Write a balanced nuclear equation showing mass numbers and atomic numbers for radioactive decay. (5.2)

____ Write a balanced nuclear equation for the formation of a radioactive isotope. (5.2)

____ Describe the detection and measurement of radiation. (5.3)

____ Calculate the amount of radioisotope remaining after one or more half-lives. (5.4)

____ Calculate the amount of time for a specific amount of a radioisotope to decay given its half-life. (5.4)

____ Describe the uses of radioisotopes in medicine. (5.5)

____ Describe the processes of nuclear fission and fusion. (5.6)

Practice Test for Chapter 5

The chapter Sections to review are shown in parentheses at the end of each question.

1. The correctly written symbol for an isotope of sulfur is (5.1)
   A. $^{30}_{16}Su$ B. $^{14}_{30}Si$ C. $^{30}_{16}S$ D. $^{30}_{16}Si$ E. $^{16}_{30}S$

2. Alpha particles are composed of (5.1)
   A. protons
   B. neutrons
   C. electrons
   D. protons and electrons
   E. protons and neutrons

3. Gamma radiation is a type of radiation that (5.1)
   A. originates in the electron shells
   B. is most dangerous
   C. is least dangerous
   D. is the heaviest
   E. travels the shortest distance

4. The charge on an alpha particle is (5.1)
   A. 1−   B. 1+   C. 2−   D. 2+   E. 4+

5. Beta particles are (5.1)
   A. protons
   B. neutrons
   C. electrons
   D. protons and electrons
   E. protons and neutrons

For questions 6 through 10, select from the following: (5.1)
   A. $^{0}_{-1}X$ B. $^{4}_{2}X$ C. $^{1}_{1}X$ D. $^{1}_{0}X$ E. $^{0}_{0}X$

6. an alpha particle

7. a beta particle

8. a gamma ray

9. a proton

10. a neutron

Nuclear Chemistry

11. Shielding from gamma rays is provided by (5.1)
    A. skin B. paper C. clothing D. lead E. air
12. The skin will provide shielding from (5.1)
    A. alpha particles B. beta particles C. gamma rays
    D. ultraviolet rays E. X-rays
13. When an atom emits an alpha particle, its mass number will (5.2)
    A. increase by 1 B. increase by 2 C. increase by 4
    D. decrease by 4 E. not change
14. When a nucleus emits a positron, the atomic number of the new nucleus (5.2)
    A. increases by 1 B. increases by 2 C. decreases by 1
    D. decreases by 2 E. does not change
15. When a nucleus emits a gamma ray, the atomic number of the new nucleus (5.2)
    A. increases by 1 B. increases by 2 C. decreases by 1
    D. decreases by 2 E. does not change

For questions 16 through 19, select the particle that completes each of the following equations: (5.2)

   A. neutron B. alpha particle C. beta particle D. gamma ray

16. $^{126}_{50}Sn \longrightarrow {}^{126}_{51}Sb + ?$
17. $^{99m}_{43}Tc \longrightarrow {}^{99}_{43}Tc + ?$
18. $^{69}_{30}Zn \longrightarrow {}^{69}_{31}Ga + ?$
19. $^{149}_{62}Sm \longrightarrow {}^{145}_{60}Nd + ?$

20. What symbol completes the following reaction? (5.2)

    $^{1}_{0}n + {}^{14}_{7}N \longrightarrow ? + {}^{1}_{1}H$

    A. $^{15}_{8}O$ B. $^{15}_{6}C$ C. $^{14}_{8}O$ D. $^{14}_{6}C$ E. $^{15}_{7}N$

21. To complete this nuclear equation, you need to write (5.2)

    $? + {}^{54}_{26}Fe \longrightarrow {}^{57}_{28}Ni + {}^{1}_{0}n$

    A. an alpha particle B. a beta particle C. gamma
    D. neutron E. proton

22. The name of the unit used to measure the number of disintegrations per second is (5.3)
    A. curie B. rad C. rem D. gray E. sievert
23. The rem and the sievert are units used to measure (5.3)
    A. activity of a radioactive sample
    B. biological damage of different types of radiation
    C. radiation absorbed
    D. background radiation
    E. half-life of a radioactive sample
24. Radiation can cause (5.3)
    A. nausea B. a lower white cell count C. fatigue
    D. hair loss E. all of these
25. The time required for a radioisotope to decay is measured by its (5.4)
    A. half-life B. protons C. activity D. fusion E. radioisotope
26. Oxygen-15, used in imaging brain function and blood flow, has a half-life of 2 min. How many half-lives have occurred in the 10 min it takes to prepare the sample? (5.4)
    A. 2 B. 3 C. 4 D. 5 E. 6

Chapter 5

27. Iodine-131, used to treat Graves' disease as well as thyroid and prostate cancers, has a half-life of 8.0 days. How many days will it take for a 80.-mg sample to decay to 10. mg? (5.4)

   A. 8.0 days   **B.** 16 days   **C.** 24 days   **D.** 32 days   **E.** 40. days

28. Phosphorus-32, used to treat leukemia and pancreatic cancer, has a half-life of 14.3 days. After 28.6 days, how many milligrams of a 120-mg sample will still be radioactive? (5.4)

   A. 240 mg   **B.** 120 mg   **C.** 60. mg   **D.** 30. mg   **E.** 15 mg

29. The radioisotope iodine-131 is used as a radioactive tracer for studying thyroid activity. The correctly written symbol for iodine-131 is (5.5)

   A. $_{53}I$ **B.** $_{131}I$ **C.** $_{53}^{131}I$ **D.** $_{131}^{53}I$ **E.** $_{53}^{78}I$

30. Radioisotopes used in medical diagnosis (5.5)
   A. have short half-lives   **B.** emit only gamma rays   **C.** locate in specific organs
   D. produce nonradioactive nuclei   **E.** all of these

31. The "splitting" of a large nucleus to form smaller particles accompanied by a release of energy is called (5.6)
   A. radioisotope   **B.** nuclear fission   **C.** nuclear fusion   **D.** bombardment   **E.** half-life

32. The process of combining small nuclei to form larger nuclei is (5.6)
   A. radioisotope   **B.** nuclear fission   **C.** nuclear fusion   **D.** bombardment   **E.** half-life

33. A fusion reaction (5.6)
   A. occurs in the Sun
   B. forms larger nuclei from smaller nuclei
   C. requires extremely high temperatures
   D. releases a large amount of energy
   E. all of these

Answers to the Practice Test

1. C	**2.** E	**3.** B	**4.** D	**5.** C
6. B	**7.** A	**8.** E	**9.** C	**10.** D
11. D	**12.** A	**13.** D	**14.** C	**15.** E
16. C	**17.** D	**18.** C	**19.** B	**20.** D
21. A	**22.** A	**23.** B	**24.** E	**25.** A
26. D	**27.** C	**28.** D	**29.** C	**30.** E
31. B	**32.** C	**33.** E		

Nuclear Chemistry

Selected Answers and Solutions to Text Problems

5.1 a. $^{4}_{2}He$ is the symbol for an alpha particle.
  b. $^{0}_{+1}e$ is the symbol for a positron.
  c. $^{0}_{0}\gamma$ is the symbol for gamma radiation.

5.3 a. Since the element is potassium, the element symbol is K and the atomic number is 19. The potassium isotopes with mass number 39, 40, and 41 will have the atomic symbol $^{39}_{19}K$, $^{40}_{19}K$, and $^{41}_{19}K$, respectively.
  b. Each isotope has 19 protons and 19 electrons, but they differ in the number of neutrons present. Potassium-39 has 39 − 19 = 20 neutrons, potassium-40 has 40 − 19 = 21 neutrons, and potassium-41 has 41 − 19 = 22 neutrons.

5.5 a. beta particle (β, $^{0}_{-1}e$)
  b. alpha particle (α, $^{4}_{2}He$)
  c. neutron (n, $^{1}_{0}n$)
  d. argon-38 ($^{38}_{18}Ar$)
  e. carbon-14 ($^{14}_{6}C$)

5.7 a. Since the element is copper, the element symbol is Cu and the atomic number is 29. The copper isotope with mass number 64 will have the atomic symbol $^{64}_{29}Cu$.
  b. Since the element is selenium, the element symbol is Se and the atomic number is 34. The selenium isotope with mass number 75 will have the atomic symbol $^{75}_{34}Se$.
  c. Since the element is sodium, the element symbol is Na and the atomic number is 11. The sodium isotope with mass number 24 will have the atomic symbol $^{24}_{11}Na$.
  d. Since the element is nitrogen, the element symbol is N and the atomic number is 7. The nitrogen isotope with mass number 15 will have the atomic symbol $^{15}_{7}N$.

5.9

Medical Use	Atomic Symbol	Mass Number	Number of Protons	Number of Neutrons
Heart imaging	$^{201}_{81}Tl$	201	81	120
Radiation therapy	$^{60}_{27}Co$	60	27	33
Abdominal scan	$^{67}_{31}Ga$	67	31	36
Hyperthyroidism	$^{131}_{53}I$	131	53	78
Leukemia treatment	$^{32}_{15}P$	32	15	17

5.11 a. 1. Alpha particles do not penetrate the skin.
  b. 3. Gamma radiation requires shielding protection that includes lead or thick concrete.
  c. 1. Alpha particles can be very harmful if ingested.

5.13 The mass number of the radioactive atom is reduced by four when an alpha particle ($^{4}_{2}He$) is emitted. The unknown product will have an atomic number that is two less than the atomic number of the radioactive atom.
  a. $^{208}_{84}Po \longrightarrow {}^{204}_{82}Pb + {}^{4}_{2}He$
  b. $^{232}_{90}Th \longrightarrow {}^{228}_{88}Ra + {}^{4}_{2}He$
  c. $^{251}_{102}No \longrightarrow {}^{247}_{100}Fm + {}^{4}_{2}He$
  d. $^{220}_{86}Rn \longrightarrow {}^{216}_{84}Po + {}^{4}_{2}He$

Chapter 5

5.15 The mass number of the radioactive atom is not changed when a beta particle ($_{-1}^{0}e$) is emitted. The unknown product will have an atomic number that is one greater than the atomic number of the radioactive atom.

    a. $_{11}^{25}\text{Na} \longrightarrow\ _{12}^{25}\text{Mg} +\ _{-1}^{0}e$
    b. $_{8}^{20}\text{O} \longrightarrow\ _{9}^{20}\text{F} +\ _{-1}^{0}e$
    c. $_{38}^{92}\text{Sr} \longrightarrow\ _{39}^{92}\text{Y} +\ _{-1}^{0}e$
    d. $_{26}^{60}\text{Fe} \longrightarrow\ _{27}^{60}\text{Co} +\ _{-1}^{0}e$

5.17 The mass number of the radioactive atom is not changed when a positron ($_{+1}^{0}e$) is emitted. The unknown product will have an atomic number that is one less than the atomic number of the radioactive atom.

    a. $_{14}^{26}\text{Si} \longrightarrow\ _{13}^{26}\text{Al} +\ _{+1}^{0}e$
    b. $_{27}^{54}\text{Co} \longrightarrow\ _{26}^{54}\text{Fe} +\ _{+1}^{0}e$
    c. $_{37}^{77}\text{Rb} \longrightarrow\ _{36}^{77}\text{Kr} +\ _{+1}^{0}e$
    d. $_{45}^{93}\text{Rh} \longrightarrow\ _{44}^{93}\text{Ru} +\ _{+1}^{0}e$

5.19 Balance the mass numbers and the atomic numbers in each nuclear equation.

    a. $_{13}^{28}\text{Al} \longrightarrow\ _{14}^{28}\text{Si} +\ _{-1}^{0}e$ $? =\ _{14}^{28}\text{Si}$ beta decay
    b. $_{73}^{180m}\text{Ta} \longrightarrow\ _{73}^{180}\text{Ta} +\ _{0}^{0}\gamma$ $? =\ _{0}^{0}\gamma$ gamma emission
    c. $_{29}^{66}\text{Cu} \longrightarrow\ _{30}^{66}\text{Zn} +\ _{-1}^{0}e$ $? =\ _{-1}^{0}e$ beta decay
    d. $_{92}^{238}\text{U} \longrightarrow\ _{90}^{234}\text{Th} +\ _{2}^{4}\text{He}$ $? =\ _{92}^{238}\text{U}$ alpha decay
    e. $_{80}^{188}\text{Hg} \longrightarrow\ _{79}^{188}\text{Au} +\ _{+1}^{0}e$ $? =\ _{79}^{188}\text{Au}$ positron emission

5.21 Balance the mass numbers and the atomic numbers in each nuclear equation.

    a. $_{0}^{1}n +\ _{4}^{9}\text{Be} \longrightarrow\ _{4}^{10}\text{Be}$ $? =\ _{4}^{10}\text{Be}$
    b. $_{0}^{1}n +\ _{52}^{131}\text{Te} \longrightarrow\ _{53}^{132}\text{I} +\ _{-1}^{0}e$ $? =\ _{53}^{132}\text{I}$
    c. $_{0}^{1}n +\ _{13}^{27}\text{Al} \longrightarrow\ _{11}^{24}\text{Na} +\ _{2}^{4}\text{He}$ $? =\ _{13}^{27}\text{Al}$
    d. $_{2}^{4}\text{He} +\ _{7}^{14}\text{N} \longrightarrow\ _{8}^{17}\text{O} +\ _{1}^{1}\text{H}$ $? =\ _{8}^{17}\text{O}$

5.23 **a.** 2. <u>Absorbed dose</u> can be measured in rad.
    b. 3. <u>Biological damage</u> can be measured in mrem.
    c. 1. <u>Activity</u> can be measured in mCi.
    d. 2. <u>Absorbed dose</u> can be measured in Gy.

5.25 $8 \text{ mGy} \times \dfrac{1 \text{ Gy}}{1000 \text{ mGy}} \times \dfrac{100 \text{ rad}}{1 \text{ Gy}} = 0.8 \text{ rad (1 SF)}$

Thus, a technician exposed to a 5-rad dose of radiation received more radiation than one exposed to 8 mGy (0.8 rad) of radiation.

5.27 **a.** $70.0 \text{ kg body mass} \times \dfrac{4.20 \text{ } \mu\text{Ci}}{1 \text{ kg body mass}} = 294 \text{ } \mu\text{Ci (3 SFs)}$

    b. From 50 rad of gamma radiation (we use a biological damage factor of 1, so 1 rad of gamma radiation = 1 rem):

    $50 \text{ rad} \times \dfrac{1 \text{ Gy}}{100 \text{ rad}} = 0.5 \text{ Gy (1 SF)}$

5.29 **a.** 2. two half-lives:

    $34 \text{ days} \times \dfrac{1 \text{ half-life}}{17 \text{ days}} = 2.0 \text{ half-lives}$

b. 1. one half-life:

$$20 \text{ min} \times \frac{1 \text{ half-life}}{20 \text{ min}} = 1 \text{ half-life}$$

c. 3. three half-lives:

$$21 \text{ h} \times \frac{1 \text{ half-life}}{7 \text{ h}} = 3 \text{ half-lives}$$

5.31 a. After one half-life, one-half of the sample would be radioactive:

80.0 mg of $^{99m}_{43}$Tc $\xrightarrow{1 \text{ half-life}}$ 40.0 mg of $^{99m}_{43}$Tc (3 SFs)

b. After two half-lives, one-fourth of the sample would still be radioactive:

80.0 mg of $^{99m}_{43}$Tc $\xrightarrow{1 \text{ half-life}}$ 40.0 mg of $^{99m}_{43}$Tc $\xrightarrow{2 \text{ half-lives}}$ 20.0 mg of $^{99m}_{43}$Tc (3 SFs)

c. $18 \text{ h} \times \dfrac{1 \text{ half-life}}{6.0 \text{ h}} = 3.0 \text{ half-lives}$

80.0 mg of $^{99m}_{43}$Tc $\xrightarrow{1 \text{ half-life}}$ 40.0 mg of $^{99m}_{43}$Tc $\xrightarrow{2 \text{ half-lives}}$ 20.0 mg of $^{99m}_{43}$Tc $\xrightarrow{3 \text{ half-lives}}$ 10.0 mg of $^{99m}_{43}$Tc (3 SFs)

d. $1.5 \text{ days} \times \dfrac{24 \text{ h}}{1 \text{ day}} \times \dfrac{1 \text{ half-life}}{6.0 \text{ h}} = 6.0 \text{ half-lives}$

80.0 mg of $^{99m}_{43}$Tc $\xrightarrow{1 \text{ half-life}}$ 40.0 mg of $^{99m}_{43}$Tc $\xrightarrow{2 \text{ half-lives}}$ 20.0 mg of $^{99m}_{43}$Tc $\xrightarrow{3 \text{ half-lives}}$ 10.0 mg of $^{99m}_{43}$Tc $\xrightarrow{4 \text{ half-lives}}$ 5.00 mg of $^{99m}_{43}$Tc $\xrightarrow{5 \text{ half-lives}}$ 2.50 mg of $^{99m}_{43}$Tc $\xrightarrow{6 \text{ half-lives}}$ 1.25 mg of $^{99m}_{43}$Tc (3 SFs)

5.33 The radiation level in a radioactive sample is cut in half with each half-life; the half-life of Sr-85 is 65 days.

a. For the radiation level to drop to one-fourth of its original level,

$\frac{1}{4} = \frac{1}{2} \times \frac{1}{2}$ or two half-lives

$$2 \text{ half-lives} \times \frac{65 \text{ days}}{1 \text{ half-life}} = 130 \text{ days} \text{ (2 SFs)}$$

b. For the radiation level to drop to one-eighth of its original level,

$\frac{1}{8} = \frac{1}{2} \times \frac{1}{2} \times \frac{1}{2}$ or three half-lives

$$3 \text{ half-lives} \times \frac{65 \text{ days}}{1 \text{ half-life}} = 195 \text{ days} \text{ (2 SFs)}$$

5.35 a. Since the elements calcium and phosphorus are part of bone, any calcium or phosphorus atom, regardless of isotope, will be carried to and become part of the bony structures of the body. Once there, the radiation emitted by any radioactive isotope can be used to diagnose or treat bone diseases.

b. Strontium (Sr) acts much like calcium (Ca) because both are Group 2A (2) elements. The body will accumulate radioactive strontium in bones in the same way that it incorporates calcium. Radioactive strontium is harmful to children because the radiation it produces causes more damage in cells that are dividing rapidly.

5.37 $4.0 \text{ mL solution} \times \dfrac{45 \ \mu\text{Ci}}{1 \text{ mL solution}} = 180 \ \mu\text{Ci}$ of selenium-75 (2 SFs)

5.39 a. $^{68}_{31}\text{Ga} \longrightarrow \ ^{68}_{30}\text{Mn} + \ ^{0}_{+1}e$

b. First, calculate the number of half-lives that have elapsed:

$$136 \text{ min} \times \frac{1 \text{ half-life}}{68 \text{ min}} = 2.0 \text{ half-lives}$$

Chapter 5

Now we can calculate the number of micrograms of gallium-68 that remain:

64 mcg of $^{68}_{31}$Ga $\xrightarrow{\text{1 half-life}}$ 32 mcg of $^{68}_{31}$Ga $\xrightarrow{\text{2 half-lives}}$ 16 mcg of $^{68}_{31}$Ga

5.41 Nuclear fission is the splitting of a large atom into smaller fragments with a simultaneous release of large amounts of energy.

5.43 $^{1}_{0}n + ^{235}_{92}U \longrightarrow ^{131}_{50}Sn + ^{103}_{42}Mo + 2^{1}_{0}n + \text{energy}$ $? = ^{103}_{42}Mo$

5.45 a. Neutrons bombard a nucleus in the <u>fission</u> process.
  b. The nuclear process that occurs in the Sun is <u>fusion</u>.
  c. <u>Fission</u> is the process in which a large nucleus splits into smaller nuclei.
  d. <u>Fusion</u> is the process in which small nuclei combine to form larger nuclei.

5.47 a. $74 \text{ MBq} \times \dfrac{1 \times 10^6 \text{ Bq}}{1 \text{ MBq}} \times \dfrac{1 \text{ Ci}}{3.7 \times 10^{10} \text{ Bq}} = 2.0 \times 10^{-3} \text{ Ci (2 SFs)}$

  b. $74 \text{ MBq} \times \dfrac{1 \times 10^6 \text{ Bq}}{1 \text{ MBq}} \times \dfrac{1 \text{ Ci}}{3.7 \times 10^{10} \text{ Bq}} \times \dfrac{1 \times 10^3 \text{ mCi}}{1 \text{ Ci}} = 2.0 \text{ mCi (2 SFs)}$

5.49 For the activity to drop to one-eighth of its original level, $\frac{1}{8} = \frac{1}{2} \times \frac{1}{2} \times \frac{1}{2}$ or three half-lives

$3 \text{ half-lives} \times \dfrac{3.0 \text{ days}}{1 \text{ half-life}} = 9.0 \text{ days (2 SFs)}$

5.51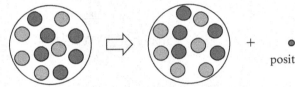

5.53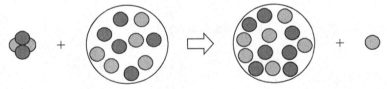

5.55 Half of a radioactive sample decays with each half-life:

$6.4 \, \mu\text{Ci of } ^{14}_{6}\text{C} \xrightarrow{\text{1 half-life}} 3.2 \, \mu\text{Ci of } ^{14}_{6}\text{C} \xrightarrow{\text{2 half-lives}} 1.6 \, \mu\text{Ci of } ^{14}_{6}\text{C} \xrightarrow{\text{3 half-lives}}$
$0.80 \, \mu\text{Ci of } ^{14}_{6}\text{C}$

∴ The activity of carbon-14 drops to 0.80 μCi in three half-lives or 3 × 5730 yr, which makes the age of the painting 17 000 yr (2 SFs).

5.57 a. $^{25}_{11}$Na has 11 protons and $25 - 11 = 14$ neutrons.
  b. $^{61}_{28}$Ni has 28 protons and $61 - 28 = 33$ neutrons.
  c. $^{84}_{37}$Rb has 37 protons and $84 - 37 = 47$ neutrons.
  d. $^{110}_{47}$Ag has 47 protons and $110 - 47 = 63$ neutrons.

5.59 a. gamma emission
  b. positron emission
  c. alpha decay

5.61 a. $^{225}_{90}\text{Th} \longrightarrow ^{221}_{88}\text{Ra} + ^{4}_{2}\text{He}$
  b. $^{210}_{83}\text{Bi} \longrightarrow ^{206}_{81}\text{Tl} + ^{4}_{2}\text{He}$
  c. $^{137}_{55}\text{Cs} \longrightarrow ^{137}_{56}\text{Ba} + ^{0}_{-1}e$
  d. $^{126}_{50}\text{Sn} \longrightarrow ^{126}_{51}\text{Sb} + ^{0}_{-1}e$
  e. $^{18}_{9}\text{F} \longrightarrow ^{18}_{8}\text{O} + ^{0}_{+1}e$

5.63 a. $^{4}_{2}He + ^{14}_{7}N \longrightarrow ^{17}_{8}O + ^{1}_{1}H$? = $^{17}_{8}O$
   b. $^{4}_{2}He + ^{27}_{13}Al \longrightarrow ^{30}_{14}Si + ^{1}_{1}H$? = $^{1}_{1}H$
   c. $^{1}_{0}n + ^{235}_{92}U \longrightarrow ^{90}_{38}Sr + 3^{1}_{0}n + ^{143}_{54}Xe$? = $^{143}_{54}Xe$
   d. $^{23m}_{12}Mg \longrightarrow ^{23}_{12}Mg + ^{0}_{0}\gamma$? = $^{23}_{12}Mg$

5.65 a. $^{16}_{8}O + ^{16}_{8}O \longrightarrow ^{28}_{14}Si + ^{4}_{2}He$
   b. $^{18}_{8}O + ^{249}_{98}Cf \longrightarrow ^{263}_{106}Sg + 4^{1}_{0}n$
   c. $^{222}_{86}Rn \longrightarrow ^{218}_{84}Po + ^{4}_{2}He$
   d. $^{80}_{38}Sr \longrightarrow ^{80}_{37}Rb + ^{0}_{+1}e$

5.67 First, calculate the number of half-lives that have elapsed:

$$24\ h \times \frac{1\ half\text{-}life}{6.0\ h} = 4.0\ half\text{-}lives$$

Now we can calculate the number of milligrams of technicium-99m that remain:

120 mg of $^{99m}_{43}Tc$ $\xrightarrow{1\ half\text{-}life}$ 60. mg of $^{99m}_{43}Tc$ $\xrightarrow{2\ half\text{-}lives}$ 30. mg of $^{99m}_{43}Tc$ $\xrightarrow{3\ half\text{-}lives}$
15 mg of $^{99m}_{43}Tc$ $\xrightarrow{4\ half\text{-}lives}$ 7.5 mg of $^{99m}_{43}Tc$ (2 SFs)

5.69 In the fission process, an atom splits into smaller nuclei with a simultaneous release of large amounts of energy. In fusion, two (or more) small nuclei combine (fuse) to form a larger nucleus, with a simultaneous release of large amounts of energy.

5.71 Fusion occurs naturally in the Sun and other stars.

5.73 $120\ nCi \times \dfrac{1 \times 10^{-9}\ Ci}{1\ nCi} \times \dfrac{3.7 \times 10^{10}\ Bq}{1\ Ci} = 4400\ Bq = 4.4 \times 10^{3}\ Bq$ (2 SFs)

5.75 Half of a radioactive sample decays with each half-life:

1.2 mg of $^{32}_{15}P$ $\xrightarrow{1\ half\text{-}life}$ 0.60 mg of $^{32}_{15}P$ $\xrightarrow{2\ half\text{-}lives}$ 0.30 mg of $^{32}_{15}P$

∴ Two half-lives must have elapsed during this time (28.6 days), yielding the half-life for phosphorus-32:

$$\frac{28.6\ days}{2\ half\text{-}lives} = 14.3\ days/half\text{-}life\ (3\ SFs)$$

5.77 a. $^{47}_{20}Ca \longrightarrow ^{47}_{21}Sc + ^{0}_{-1}e$
   b. First, calculate the number of half-lives that have elapsed:

$$18\ days \times \frac{1\ half\text{-}life}{4.5\ days} = 4.0\ half\text{-}lives$$

Now we can calculate the number of milligrams of calcium-47 that remain:

16 mg of $^{47}_{20}Ca$ $\xrightarrow{1\ half\text{-}life}$ 8.0 mg of $^{47}_{20}Ca$ $\xrightarrow{2\ half\text{-}lives}$ 4.0 mg of $^{47}_{20}Ca$ $\xrightarrow{3\ half\text{-}lives}$
2.0 mg of $^{47}_{20}Ca$ $\xrightarrow{4\ half\text{-}lives}$ 1.0 mg of $^{47}_{20}Ca$

   c. Half of a radioactive sample decays with each half-life:

4.8 mg of $^{47}_{20}Ca$ $\xrightarrow{1\ half\text{-}life}$ 2.4 mg of $^{47}_{20}Ca$ $\xrightarrow{2\ half\text{-}lives}$ 1.2 mg of $^{47}_{20}Ca$

∴ Two half-lives have elapsed.

$$2\ half\text{-}lives \times \frac{4.5\ days}{1\ half\text{-}life} = 9.0\ days\ (2\ SFs)$$

5.79 a. $^{180}_{80}Hg \longrightarrow ^{176}_{78}Pt + ^{4}_{2}He$
   b. $^{198}_{79}Au \longrightarrow ^{198}_{80}Hg + ^{0}_{-1}e$
   c. $^{82}_{37}Rb \longrightarrow ^{82}_{36}Kr + ^{0}_{+1}e$

5.81 $^{1}_{0}n + ^{238}_{92}U \longrightarrow ^{239}_{93}Np + ^{0}_{-1}e$

Chapter 5

5.83 Half of a radioactive sample decays with each half-life:

64 μCi of $^{201}_{81}$Tl $\xrightarrow{\text{1 half-life}}$ 32 μCi of $^{201}_{81}$Tl $\xrightarrow{\text{2 half-lives}}$ 16 μCi of $^{201}_{81}$Tl $\xrightarrow{\text{3 half-lives}}$

8.0 μCi of $^{201}_{81}$Tl $\xrightarrow{\text{4 half-lives}}$ 4.0 μCi of $^{201}_{81}$Tl

∴ Four half-lives must have elapsed during this time (12 days), yielding the half-life for thallium-201:

$$\frac{12 \text{ days}}{4 \text{ half-lives}} = 3.0 \text{ days/half-life (2 SFs)}$$

5.85 First, calculate the number of half-lives that have elapsed:

$$27 \text{ days} \times \frac{1 \text{ half-life}}{4.5 \text{ days}} = 6.0 \text{ half-lives}$$

Because the activity of a radioactive sample is cut in half with each half-life, the activity must have been double its present value before each half-life. For 6.0 half-lives, we need to double the value six times:

1.0 μCi of $^{47}_{20}$Ca $\xleftarrow{\text{1 half-life}}$ 2.0 μCi of $^{47}_{20}$Ca $\xleftarrow{\text{2 half-lives}}$ 4.0 μCi of $^{47}_{20}$Ca $\xleftarrow{\text{3 half-lives}}$ 8.0 μCi of $^{47}_{20}$Ca

$\xleftarrow{\text{4 half-lives}}$ 16 μCi of $^{47}_{20}$Ca $\xleftarrow{\text{5 half-lives}}$ 32 μCi of $^{47}_{20}$Ca $\xleftarrow{\text{6 half-lives}}$ 64 μCi of $^{47}_{20}$Ca

∴ The initial activity of the sample was 64 μCi (2 SFs).

5.87 $^{48}_{20}$Ca + $^{244}_{94}$Pu ⟶ $^{292}_{114}$Fl

5.89 First, calculate the number of half-lives that have elapsed since the technician was exposed:

$$36 \text{ h} \times \frac{1 \text{ half-life}}{12 \text{ h}} = 3.0 \text{ half-lives}$$

Because the activity of a radioactive sample is cut in half with each half-life, the activity must have been double its present value before each half-life. For 3.0 half-lives, we need to double the value three times:

2.0 μCi of $^{42}_{19}$K $\xleftarrow{\text{1 half-life}}$ 4.0 μCi of $^{42}_{19}$K $\xleftarrow{\text{2 half-lives}}$ 8.0 μCi of $^{42}_{19}$K $\xleftarrow{\text{3 half-lives}}$ 16 μCi of $^{42}_{19}$K

∴ The activity of the sample when the technician was exposed was 16 μCi (2 SFs).

6
Ionic and Molecular Compounds

Sarah, a pharmacist, talked with her customer regarding questions about stomach upset from the low-dose aspirin, $C_9H_8O_4$, which he is taking to prevent a heart attack or stroke. Sarah explained that when he takes a low-dose aspirin, he needs to drink a glass of water. If his stomach is still upset, he can take the aspirin with food or milk. A few days later, Sarah talked with her customer again. He told her he followed her suggestions and that he no longer had any upset stomach from the low-dose aspirin. Is aspirin an ionic or a molecular compound?

Credit: Design Pics Inc./Alamy

LOOKING AHEAD

6.1 Ions: Transfer of Electrons
6.2 Ionic Compounds
6.3 Naming and Writing Ionic Formulas
6.4 Polyatomic Ions

6.5 Molecular Compounds: Sharing Electrons
6.6 Lewis Structures for Molecules

6.7 Electronegativity and Bond Polarity
6.8 Shapes of Molecules
6.9 Polarity of Molecules and Intermolecular Forces

 The Health icon indicates a question that is related to health and medicine.

6.1 Ions: Transfer of Electrons

Learning Goal: Write the symbols for the simple ions of the representative elements.

> **REVIEW**
> Using Positive and Negative Numbers in Calculations (1.4)
> Writing Electron Arrangements (4.6)

- The stability of the noble gases is associated with a stable electron arrangement of eight electrons, an octet, in their outermost energy level. Helium is stable with two electrons in the $n = 1$ energy level.
- Atoms of elements other than the noble gases achieve stability by losing, gaining, or sharing valence electrons with other atoms in the formation of compounds.

Chapter 6

- Metals of the representative elements in Groups 1A (1), 2A (2), and 3A (13) achieve a noble gas electron arrangement by losing their valence electron(s) to form positively charged cations with a charge of 1+, 2+, or 3+, respectively.
- Nonmetals in Groups 5A (15), 6A (16), and 7A (17) gain valence electrons to achieve an octet forming negatively charged anions with a charge of 3−, 2−, or 1−, respectively.
- *Ionic bonds* occur when the valence electrons of atoms of a metal are transferred to atoms of nonmetals. For example, sodium atoms lose electrons and chlorine atoms gain electrons to form the ionic compound NaCl.
- *Covalent bonds* form when atoms of nonmetals share valence electrons. In the molecular compounds H_2O and C_3H_8, atoms share electrons.

Study Note

When an atom loses or gains electrons, it acquires the electron arrangement of the nearest noble gas. For example, sodium loses one electron, which gives the Na^+ ion the same electron arrangement as neon. Oxygen gains two electrons to give an oxide ion, O^{2-}, the same electron arrangement as neon.

♦ Learning Exercise 6.1A

The following metals, which are essential to health, lose electrons when they form ions. Indicate the group number, the number of electrons lost, and the ion (symbol and charge) for each of the following:

CORE CHEMISTRY SKILL
Writing Positive and Negative Ions

Metal	Group Number	Electrons Lost	Ion Formed
Magnesium			
Sodium			
Calcium			
Potassium			

Answers

Metal	Group Number	Electrons Lost	Ion Formed
Magnesium	2A (2)	2	Mg^{2+}
Sodium	1A (1)	1	Na^+
Calcium	2A (2)	2	Ca^{2+}
Potassium	1A (1)	1	K^+

Ionic and Molecular Compounds

♦ **Learning Exercise 6.1B**

The following nonmetals, which are essential to health, gain electrons when they form ions. Indicate the group number, the number of electrons gained, and the ion (symbol and charge) for each of the following:

Nonmetal	Group Number	Electrons Gained	Ion Formed
Chlorine			
Nitrogen			
Iodine			
Sulfur			

Answers

Nonmetal	Group Number	Electrons Gained	Ion Formed
Chlorine	7A (17)	1	Cl^-
Nitrogen	5A (15)	3	N^{3-}
Iodine	7A (17)	1	I^-
Sulfur	6A (16)	2	S^{2-}

6.2 Ionic Compounds

Learning Goal: Using charge balance, write the correct formula for an ionic compound.

- In the formulas of ionic compounds, the total positive charge is equal to the total negative charge. For example, the compound magnesium chloride, $MgCl_2$, contains Mg^{2+} and two Cl^-. The sum of the charges is zero: $1(2+) + 2(1-) = 0$.
- When two or more ions are needed for charge balance, that number is indicated by subscripts in the formula.
- A subscript of 1 is understood and not written.

For the following exercises, you may want to cut pieces of paper that represent typical positive and negative ions, as shown. To determine an ionic formula, place the smallest number of positive and negative pieces together to complete a geometric shape. Write the number of positive ions and negative ions as the subscripts for the formula.

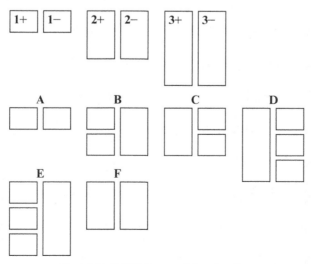

Chapter 6

♦ **Learning Exercise 6.2A**

CORE CHEMISTRY SKILL
Writing Ionic Formulas

Write the letter (A to F) that matches the arrangement of ions in the following compounds:

Compound	Combination	Compound	Combination
a. $MgCl_2$	_____	b. Na_2S	_____
c. $LiCl$	_____	d. CaO	_____
e. K_3N	_____	f. $AlBr_3$	_____
g. MgS	_____	h. $BaCl_2$	_____

Answers   a. C   b. B   c. A   d. F   e. E   f. D   g. F   h. C

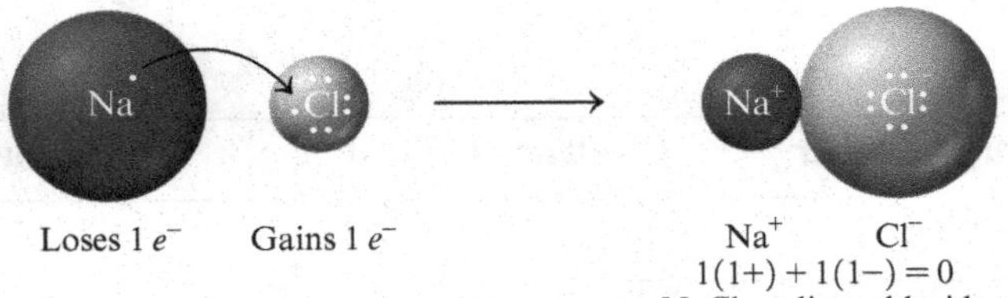

Na Loses 1 e^- Cl Gains 1 e^- → Na^+ Cl^-
Na^+ Cl^-
$1(1+) + 1(1-) = 0$
NaCl, sodium chloride

> **Study Note**
>
> You can check that the ionic formula you write is electrically neutral by multiplying each ionic charge by its subscript. When added together, their sum should equal zero. For example, the formula Na_2O gives $2(1+) + 1(2-) = (2+) + (2-) = 0$.

♦ **Learning Exercise 6.2B**

Write the correct formula for the ionic compound formed from each of the following pairs of ions:

a. Na^+ and Cl^- _____   b. K^+ and S^{2-} _____
c. Al^{3+} and O^{2-} _____   d. Sr^{2+} and Cl^- _____
e. Ca^{2+} and S^{2-} _____   f. Al^{3+} and Cl^- _____
g. Li^+ and N^{3-} _____   h. Ba^{2+} and P^{3-} _____

Answers   a. NaCl   b. K_2S c. Al_2O_3 d. $SrCl_2$
           e. CaS   f. $AlCl_3$ g. Li_3N h. Ba_3P_2

♦ **Learning Exercise 6.2C**

Write the symbols for the ions and the correct formula for the ionic compound formed by each of the following:

a. lithium and sulfur   b. potassium and phosphorus
c. barium and fluorine   d. gallium and sulfur

Answers   a. Li^+ and S^{2-}, Li_2S b. K^+ and P^{3-}, K_3P
           c. Ba^{2+} and F^-, BaF_2 d. Ga^{3+} and S^{2-}, Ga_2S_3

6.3 Naming and Writing Ionic Formulas

Learning Goal: Given the formula of an ionic compound, write the correct name; given the name of an ionic compound, write the correct formula.

REVIEW Solving Equations (1.4)

- In naming ionic compounds, the positive ion is named first, followed by the name of the negative ion. The name of a representative metal ion in Group 1A (1), 2A (2), or 3A (13) is the same as its element name. The name of a nonmetal ion is obtained by replacing the end of its element name with *ide*.
- Most transition elements form cations with two or more ionic charges. Then the ionic charge must be written as a Roman numeral in parentheses after the name of the metal. For example, the cations of iron, Fe^{2+} and Fe^{3+}, are named iron(II) and iron(III). The ions of copper are Cu^+, copper(I), and Cu^{2+}, copper(II).
- The only transition elements with fixed charges are zinc, Zn^{2+}; silver, Ag^+; and cadmium, Cd^{2+}; no Roman numerals are used when naming their cations in ionic compounds.

Solution Guide to Naming Ionic Compounds	
STEP 1	Identify the cation and anion.
STEP 2	Name the cation by its element name.
STEP 3	Name the anion by using the first syllable of its element name followed by *ide*.
STEP 4	Write the name for the cation first and the name for the anion second.

♦ **Learning Exercise 6.3A**

Write the symbols for the ions and the correct name for each of the following ionic compounds:

Formula	Ions		Name
a. Cs_2O	_____	_____	_____
b. $BaBr_2$	_____	_____	_____
c. Ca_3P_2	_____	_____	_____
d. Na_2S	_____	_____	_____

Answers
a. Cs^+, O^{2-}, cesium oxide
b. Ba^{2+}, Br^-, barium bromide
c. Ca^{2+}, P^{3-}, calcium phosphide
d. Na^+, S^{2-}, sodium sulfide

Solution Guide to Naming Ionic Compounds with Variable Charge Metal Ions	
STEP 1	Determine the charge of the cation from the anion.
STEP 2	Name the cation by its element name and use a Roman numeral in parentheses for the charge.
STEP 3	Name the anion by using the first syllable of its element name followed by *ide*.
STEP 4	Write the name for the cation first and the name for the anion second.

Chapter 6

♦ **Learning Exercise 6.3B**

Write the name for each of the following ions:

a. Cl^- _____   b. Fe^{2+} _____   c. Cu^+ _____

d. Ag^+ _____   e. O^{2-} _____   f. Ca^{2+} _____

g. S^{2-} _____   h. Al^{3+} _____   i. Fe^{3+} _____

j. Ba^{2+} _____   k. Cu^{2+} _____   l. N^{3-} _____

Answers
a. chloride   b. iron(II)   c. copper(I)
d. silver   e. oxide   f. calcium
g. sulfide   h. aluminum   i. iron(III)
j. barium   k. copper(II)   l. nitride

Some Metals That Form More Than One Positive Ion

Element	Possible Ions	Name of Ion
Bismuth	Bi^{3+}	Bismuth(III)
	Bi^{5+}	Bismuth(V)
Chromium	Cr^{2+}	Chromium(II)
	Cr^{3+}	Chromium(III)
Cobalt	Co^{2+}	Cobalt(II)
	Co^{3+}	Cobalt(III)
Copper	Cu^+	Copper(I)
	Cu^{2+}	Copper(II)
Gold	Au^+	Gold(I)
	Au^{3+}	Gold(III)
Iron	Fe^{2+}	Iron(II)
	Fe^{3+}	Iron(III)
Lead	Pb^{2+}	Lead(II)
	Pb^{4+}	Lead(IV)
Manganese	Mn^{2+}	Manganese(II)
	Mn^{3+}	Manganese(III)
Mercury	Hg_2^{2+}	Mercury(I)*
	Hg^{2+}	Mercury(II)
Nickel	Ni^{2+}	Nickel(II)
	Ni^{3+}	Nickel(III)
Tin	Sn^{2+}	Tin(II)
	Sn^{4+}	Tin(IV)

*Mercury(I) ions form an ion pair with a 2+ charge.

♦ **Learning Exercise 6.3C**

Write the symbol for each of the following ions:

Name	Symbol
a. chromium(III) ion	_____
b. cobalt(II) ion	_____
c. zinc ion	_____
d. lead(IV) ion	_____
e. gold(I) ion	_____
f. silver ion	_____
g. potassium ion	_____
h. nickel(III) ion	_____

Answers
a. Cr^{3+} b. Co^{2+} c. Zn^{2+}
d. Pb^{4+} e. Au^+ f. Ag^+
g. K^+ h. Ni^{3+}

♦ **Learning Exercise 6.3D**

Write the symbols for the ions and the correct name for each of the following ionic compounds:

CORE CHEMISTRY SKILL
Naming Ionic Compounds

Formula	Ions	Name
a. $CrCl_2$	_____ _____	_____
b. $SnBr_4$	_____ _____	_____
c. Na_3P	_____ _____	_____
d. Ni_2O_3	_____ _____	_____
e. CuO	_____ _____	_____
f. Mg_3N_2	_____ _____	_____

Ionic and Molecular Compounds

Answers
a. Cr^{2+}, Cl^-, chromium(II) chloride
b. Sn^{4+}, Br^-, tin(IV) bromide
c. Na^+, P^{3-}, sodium phosphide
d. Ni^{3+}, O^{2-}, nickel(III) oxide
e. Cu^{2+}, O^{2-}, copper(II) oxide
f. Mg^{2+}, N^{3-}, magnesium nitride

Solution Guide to Writing Formulas for Ionic Compounds	
STEP 1	Identify the cation and anion.
STEP 2	Balance the charges.
STEP 3	Write the formula, cation first, using subscripts from the charge balance.

♦ **Learning Exercise 6.3E**

Write the symbols for the ions and the correct formula for each of the following ionic compounds:

Compound	Cation	Anion	Formula of Compound
Aluminum sulfide			
Lead(II) chloride			
Barium oxide			
Gold(III) bromide			
Silver oxide			

Answers

Compound	Cation	Anion	Formula of Compound
Aluminum sulfide	Al^{3+}	S^{2-}	Al_2S_3
Lead(II) chloride	Pb^{2+}	Cl^-	$PbCl_2$
Barium oxide	Ba^{2+}	O^{2-}	BaO
Gold(III) bromide	Au^{3+}	Br^-	$AuBr_3$
Silver oxide	Ag^+	O^{2-}	Ag_2O

6.4 Polyatomic Ions

Learning Goal: Write the name and formula for an ionic compound containing a polyatomic ion.

- A polyatomic ion is a group of nonmetal atoms that carries an electrical charge, usually negative, 1−, 2−, or 3−. The polyatomic ion NH_4^+, ammonium ion, has a positive charge.
- Polyatomic ions cannot exist alone but are combined with an ion of the opposite charge.
- The names of ionic compounds containing polyatomic anions often end with *ate* or *ite*.
- When there is more than one polyatomic ion in the formula for a compound, the entire polyatomic ion formula is enclosed in parentheses and the subscript written outside the parentheses.

Fertilizer
NH_4NO_3

NH_4^+ NO_3^-
Ammonium ion   Nitrate ion

Credit: Pearson Education / Pearson Science

Chapter 6

> **Study Note**
>
> By learning the most common polyatomic ions such as nitrate NO_3^-, carbonate CO_3^{2-}, sulfate SO_4^{2-}, and phosphate PO_4^{3-}, you can derive their related polyatomic ions. For example, the nitrite ion, NO_2^-, has one oxygen atom less than the nitrate ion: NO_3^-, nitrate, and NO_2^-, nitrite.

Names and Formulas of Some Common Polyatomic Ions

Nonmetal	Formula of Ion*	Name of Ion
Hydrogen	**OH^-**	**Hydroxide**
Nitrogen	NH_4^+	Ammonium
	NO_3^-	**Nitrate**
	NO_2^-	Nitrite
Chlorine	ClO_4^-	Perchlorate
	ClO_3^-	**Chlorate**
	ClO_2^-	Chlorite
	ClO^-	Hypochlorite
Carbon	**CO_3^{2-}**	**Carbonate**
	HCO_3^-	Hydrogen carbonate (or bicarbonate)
	CN^-	Cyanide
	$C_2H_3O_2^-$	Acetate
Sulfur	**SO_4^{2-}**	**Sulfate**
	HSO_4^-	Hydrogen sulfate (or bisulfate)
	SO_3^{2-}	Sulfite
	HSO_3^-	Hydrogen sulfite (or bisulfite)
Phosphorus	**PO_4^{3-}**	**Phosphate**
	HPO_4^{2-}	Hydrogen phosphate
	$H_2PO_4^-$	Dihydrogen phosphate
	PO_3^{3-}	Phosphite

*Formulas and names in bold type indicate the most common polyatomic ion for that element.

♦ **Learning Exercise 6.4A**

Write the polyatomic ion (symbol and charge) for each of the following:

a. sulfate ion _____   b. hydroxide ion _____

c. carbonate ion _____   d. sulfite ion _____

e. ammonium ion _____   f. phosphate ion _____

g. nitrate ion _____   h. nitrite ion _____

Answers   a. SO_4^{2-} b. OH^- c. CO_3^{2-}
         d. SO_3^{2-} e. NH_4^+ f. PO_4^{3-}
         g. NO_3^- h. NO_2^-

Plaster cast
$CaSO_4$

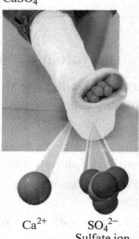

Ca^{2+} SO_4^{2-}
Sulfate ion

The sulfate ion in $CaSO_4$ is a polyatomic ion, which is a group of atoms that have an ionic charge.

Credit: Pearson Education/Pearson Science

Ionic and Molecular Compounds

	Solution Guide to Writing Formulas Containing Polyatomic Ions
STEP 1	Identify the cation and polyatomic ion (anion).
STEP 2	Balance the charges.
STEP 3	Write the formula, cation first, using the subscripts from charge balance.

◆ **Learning Exercise 6.4B**

Write the formulas for the ions and the correct formula for each of the following compounds:

Compound	Cation	Anion	Formula
Sodium phosphate			
Iron(II) hydroxide			
Ammonium carbonate			
Silver bicarbonate			
Chromium(III) sulfate			
Lead(II) nitrate			
Potassium sulfite			
Barium phosphate			

Answers

Compound	Cation	Anion	Formula
Sodium phosphate	Na^+	PO_4^{3-}	Na_3PO_4
Iron(II) hydroxide	Fe^{2+}	OH^-	$Fe(OH)_2$
Ammonium carbonate	NH_4^+	CO_3^{2-}	$(NH_4)_2CO_3$
Silver bicarbonate	Ag^+	HCO_3^-	$AgHCO_3$
Chromium(III) sulfate	Cr^{3+}	SO_4^{2-}	$Cr_2(SO_4)_3$
Lead(II) nitrate	Pb^{2+}	NO_3^-	$Pb(NO_3)_2$
Potassium sulfite	K^+	SO_3^{2-}	K_2SO_3
Barium phosphate	Ba^{2+}	PO_4^{3-}	$Ba_3(PO_4)_2$

◆ **Learning Exercise 6.4C**

Write the symbols for the ions and the correct name for each of the following ionic compounds:

Formula   **Ions**   **Name**

a. $Ba(NO_3)_2$ _____ _____   _____

b. $Fe_2(SO_4)_3$ _____ _____   _____

c. Na_3PO_3 _____ _____   _____

d. Al(ClO₃)₃ _____ _____ _____

e. NH₄NO₂ _____ _____ _____

f. Cr(OH)₂ _____ _____ _____

Answers   a. Ba^{2+}, NO_3^-, barium nitrate   b. Fe^{3+}, SO_4^{2-}, iron(III) sulfate
c. Na^+, PO_3^{3-}, sodium phosphite   d. Al^{3+}, ClO_3^-, aluminum chlorate
e. NH_4^+, NO_2^-, ammonium nitrite   f. Cr^{2+}, OH^-, chromium(II) hydroxide

6.5 Molecular Compounds: Sharing Electrons

Learning Goal: Given the formula of a molecular compound, write its correct name; given the name of a molecular compound, write its formula.

- Molecular compounds are composed of nonmetals bonded together to give discrete units called *molecules*.
- The name of a molecular compound is written using the element name of the first nonmetal and the first syllable of the second nonmetal name followed by the suffix *ide*.
- Prefixes are used to indicate the number of atoms of each nonmetal in the formula: mono (1), di (2), tri (3), tetra (4), penta (5), hexa (6), hepta (7), octa (8), nona (9), deca (10).
- In the name of a molecular compound, the prefix *mono* is usually omitted, as in NO, nitrogen oxide. Traditionally, however, CO is named carbon monoxide.
- If the vowels *o* and *o* or *a* and *o* appear together, the first vowel is omitted.
- The formula of a molecular compound is written using the symbols of the nonmetals in the name followed by subscripts determined from the prefixes.

Key Terms for Sections 6.1 to 6.5

Match each of the following key terms with the correct description:

a. ion   b. cation   c. anion
d. polyatomic ion   e. octet   f. covalent bond

1. _____ a positively charged ion
2. _____ a negatively charged ion
3. _____ the attraction between atoms formed by sharing electrons
4. _____ a group of covalently bonded nonmetal atoms that has an overall charge
5. _____ an atom or group of atoms having an electrical charge
6. _____ a set of eight valence electrons

Answers   1. b   2. c   3. f   4. d   5. a   6. e

Solution Guide to Naming Molecular Compounds	
STEP 1	Name the first nonmetal by its element name.
STEP 2	Name the second nonmetal by using the first syllable of its element name followed by *ide*.
STEP 3	Add prefixes to indicate the number of atoms (subscripts).

Ionic and Molecular Compounds

♦ Learning Exercise 6.5A

CORE CHEMISTRY SKILL
Writing the Names and Formulas for Molecular Compounds

Name the following molecular compounds:

a. CS_2 _____   b. PCl_3 _____

c. CO _____   d. SO_3 _____

e. N_2O_4 _____   f. CCl_4 _____

g. P_4S_6 _____   h. IF_7 _____

i. ClO_2 _____   j. S_2O _____

Answers
a. carbon disulfide
c. carbon monoxide
e. dinitrogen tetroxide
g. tetraphosphorus hexasulfide
i. chlorine dioxide
b. phosphorus trichloride
d. sulfur trioxide
f. carbon tetrachloride
h. iodine heptafluoride
j. disulfur oxide

Solution Guide to Writing Formulas for Molecular Compounds	
STEP 1	Write the symbols in the order of the elements in the name.
STEP 2	Write any prefixes as subscripts.

♦ Learning Exercise 6.5B

Write the formula for each of the following molecular compounds:

a. dinitrogen oxide _____   b. silicon tetrabromide _____

c. bromine pentafluoride _____   d. carbon dioxide _____

e. sulfur hexafluoride _____   f. oxygen difluoride _____

g. phosphorus octasulfide _____   h. selenium hexafluoride _____

i. iodine chloride _____   j. xenon trioxide _____

Answers
a. N_2O b. $SiBr_4$ c. BrF_5 d. CO_2 e. SF_6
f. OF_2 g. PS_8 h. SeF_6 i. ICl j. XeO_3

Summary of Writing Formulas and Names

- In ionic compounds containing two different elements, the first takes its element name. The ending of the name of the second element is replaced by *ide*. For example, $BaCl_2$ is named *barium chloride*. If the metal forms two or more positive ions, a Roman numeral is added to its name to indicate the ionic charge in the compound. For example, $FeCl_3$ is named *iron(III) chloride*.
- In ionic compounds with three or more elements, a group of atoms is named as a polyatomic ion. The names of negative polyatomic ions end in *ate* or *ite*, except for hydroxide and cyanide. For example, Na_2SO_4 is named *sodium sulfate*. No prefixes are used.
- When a polyatomic ion occurs two or more times in a formula, its formula is placed inside parentheses, and the number of ions is shown as a subscript after the parentheses as in $Ca(NO_3)_2$. If the subscript is 1, it is not written.
- In naming molecular compounds, a prefix before the name of one or both nonmetal names indicates the numerical value of a subscript. For example, N_2O_3 is named *dinitrogen trioxide*.

Chapter 6

♦ **Learning Exercise 6.5C**

Indicate the type of compound (ionic or molecular). Then give the formula and name for each.

Formula	Ionic or Molecular?	Name of Compound
$MgCl_2$		
		nitrogen trichloride
		potassium sulfate
Li_2O		
CBr_4		
Na_3PO_4		
		dihydrogen sulfide
		calcium hydrogen carbonate (bicarbonate)

Answers

Formula	Ionic or Molecular?	Name of Compound
$MgCl_2$	ionic	magnesium chloride
NCl_3	molecular	nitrogen trichloride
K_2SO_4	ionic	potassium sulfate
Li_2O	ionic	lithium oxide
CBr_4	molecular	carbon tetrabromide
Na_3PO_4	ionic	sodium phosphate
H_2S	molecular	dihydrogen sulfide
$Ca(HCO_3)_2$	ionic	calcium hydrogen carbonate (bicarbonate)

6.6 Lewis Structures for Molecules

REVIEW
Drawing Lewis Symbols (4.7)

Learning Goal: Draw the Lewis structures for molecular compounds with single and multiple bonds.

- In a covalent bond, atoms of nonmetals share electrons to achieve an octet (two for H).
- In a double bond, two pairs of electrons are shared between the same two atoms to complete octets.
- In a triple bond, three pairs of electrons are shared to complete octets.
- Bonding pairs of electrons are shared between atoms.
- Nonbonding electrons (lone pairs) are not shared between atoms.

♦ **Learning Exercise 6.6A**

List the number of bonds typically formed by the following atoms in molecular compounds:

a. N _____   b. S _____   c. P _____   d. C _____

e. Cl _____   f. O _____   g. H _____   h. F _____

Answers   a. 3   b. 2   c. 3   d. 4   e. 1   f. 2   g. 1   h. 1

SAMPLE PROBLEM Drawing Lewis Structures

Draw the Lewis structure for SCl_2.

Solution Guide:

Analyze the Problem	Given	Need	Connect
	SCl_2	Lewis structure	total valence electrons

STEP 1 Determine the arrangement of atoms.

Cl S Cl

STEP 2 Determine the total number of valence electrons.

$$\begin{aligned} &1 \text{ S atom } \times 6\,e^- &&= 6\,e^- \\ &2 \text{ Cl atoms} \times 7\,e^- &&= 14\,e^- \\ &\text{Total valence electrons for } SCl_2 &&= 20\,e^- \end{aligned}$$

STEP 3 Attach each bonded atom to the central atom with a pair of electrons. Four electrons are used to form two single bonds to connect the S atom and each Cl atom. Each bonding pair can be represented by a bond line.

Cl:S:Cl or Cl—S—Cl $\quad 20\,e^- - 4\,e^- = 16$ remaining valence electrons (8 pairs)

STEP 4 Use the remaining electrons to complete octets, using multiple bonds if needed. The remaining 16 electrons are placed as lone pairs around the S and Cl atoms to complete octets for all the atoms.

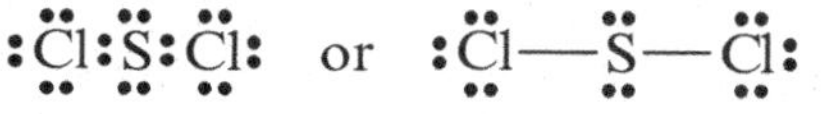

♦ Learning Exercise 6.6B

CORE CHEMISTRY SKILL
Drawing Lewis Structures

Draw the Lewis structure for each of the following:

a. H_2 **b.** NCl_3 **c.** HCl

d. Cl_2 **e.** H_2S **f.** CBr_4

Answers

a. H:H or H—H

b. 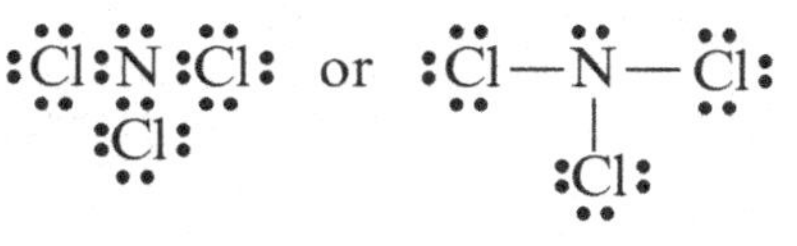

c. 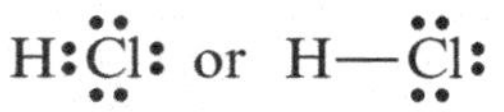

d. 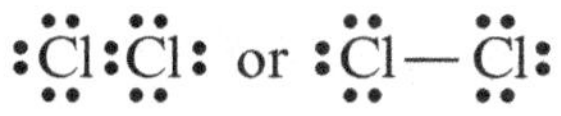

e. 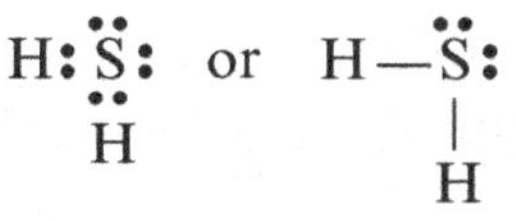

f. 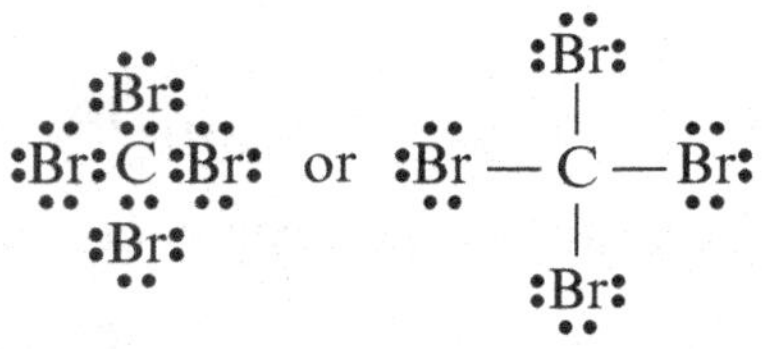

SAMPLE PROBLEM Drawing Lewis Structures with Multiple Bonds

Draw the Lewis structure for sufur dioxide, SO_2, in which the central atom is S.

Solution Guide:

Analyze the Problem	Given	Need	Connect
	SO_2	Lewis structure	total valence electrons

STEP 1 Determine the arrangement of atoms.

O   S   O

STEP 2 Determine the total number of valence electrons.

1 S atom × 6 e^- = 6 e^-
2 O atoms × 6 e^- = 12 e^-
Total valence electrons for SO_2 = 18 e^-

STEP 3 Attach each bonded atom to the central atom with a pair of electrons. Each bonding pair can be represented by a pair of electrons or a bond line.

O:S:O or O—S—O   18 e^- − 4 e^- = 14 remaining valence electrons (7 pairs)

STEP 4 Use the remaining electrons to complete octets, using multiple bonds if needed.

:Ö:S:Ö: or :Ö—S—Ö:

Because the octet for S is not complete, a lone pair from one O atom is shared with the central S atom to form a double bond.

:Ö:S::Ö: or :Ö—S=Ö:

♦ Learning Exercise 6.6C

Draw the Lewis structure for each of the following molecules:

a. CS_2

b. HCN

c. H_2CCH_2

d. HONO

Answers

a. :S̈::C::S̈: or :S̈=C=S̈:

b. H:C⋮⋮N: or H—C≡N:

c. H:C̈::C̈:H or H—C=C—H (with H H on top and H H on bottom)

d. H:Ö:N̈::Ö: or H—Ö—N̈=Ö:

6.7 Electronegativity and Bond Polarity

Learning Goal: Use electronegativity to determine the polarity of a bond.

- Electronegativity values indicate the ability of an atom to attract electrons in a chemical bond.
- Electronegativity values are low for metals and high for nonmetals.

Ionic and Molecular Compounds

[Electronegativity chart omitted in detail — values as shown: H 2.1; Li 1.0, Be 1.5; Na 0.9, Mg 1.2; K 0.8, Ca 1.0; Rb 0.8, Sr 1.0; Cs 0.7, Ba 0.9; B 2.0, C 2.5, N 3.0, O 3.5, F 4.0; Al 1.5, Si 1.8, P 2.1, S 2.5, Cl 3.0; Ga 1.6, Ge 1.8, As 2.0, Se 2.4, Br 2.8; In 1.7, Sn 1.8, Sb 1.9, Te 2.1, I 2.5; Tl 1.8, Pb 1.9, Bi 1.9, Po 2.0, At 2.1.]

- When atoms sharing electrons have the same or similar electronegativity values, electrons are shared equally and the bond is *nonpolar covalent*.
- Electrons are shared unequally in *polar covalent* bonds because they are attracted to the more electronegative atom.
- An electronegativity difference of 0.0 to 0.4 indicates a nonpolar covalent bond, whereas a difference of 0.5 to 1.8 indicates a polar covalent bond.
- An electronegativity difference of 1.9 to 3.3 indicates a bond that is ionic.
- A polar bond has a charge separation, which is called a *dipole*; the positive end of the dipole is marked as δ^+ and the negative end as δ^-.

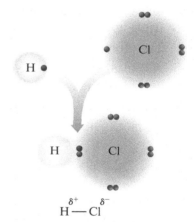

Electrons are shared unequally in the polar covalent bond of HCl.

Credit: Kameel4u/Shutterstock

♦ Learning Exercise 6.7A

Using the electronegativity values, determine the following:

1. the electronegativity difference for each pair
2. the type of bonding as ionic (I), polar covalent (PC), or nonpolar covalent (NC)

CORE CHEMISTRY SKILL
Using Electronegativity

Elements	Electronegativity Difference	Type of Bonding	Elements	Electronegativity Difference	Type of Bonding
a. H and O	_____	_____	b. N and S	_____	_____
c. Al and O	_____	_____	d. Cl and Cl	_____	_____
e. H and Cl	_____	_____	f. Li and F	_____	_____
g. S and F	_____	_____	h. H and C	_____	_____

Answers a. 1.4, PC   b. 0.5, PC   c. 2.0, I   d. 0.0, NC
e. 0.9, PC   f. 3.0, I   g. 1.5, PC   h. 0.4, NC

Chapter 6

♦ **Learning Exercise 6.7B**

For each of the following, write the symbols δ⁺ and δ⁻ over the appropriate atoms.

a. H—O   b. N—H   c. C—Cl
d. O—F   e. N—F   f. P—Cl

Answers

a. $\overset{\delta^+}{H}$—$\overset{\delta^-}{O}$ b. $\overset{\delta^-}{N}$—$\overset{\delta^+}{H}$ c. $\overset{\delta^+}{C}$—$\overset{\delta^-}{Cl}$

d. $\overset{\delta^+}{O}$—$\overset{\delta^-}{F}$ e. $\overset{\delta^+}{N}$—$\overset{\delta^-}{F}$ f. $\overset{\delta^+}{P}$—$\overset{\delta^-}{Cl}$

6.8 Shapes of Molecules

Learning Goal: Predict the three-dimensional structure of a molecule.

- The VSEPR theory predicts the geometry of a molecule by placing the electron groups around a central atom as far apart as possible.
- One electron group may consist of the following: a lone pair, a single bond, a double bond, or a triple bond.
- The atoms bonded to the central atom determine the three-dimensional geometry of a molecule.
- A central atom with two electron groups has a linear geometry (180°); three electron groups give a trigonal planar geometry (120°); and four electron pairs give a tetrahedral geometry (109°).
- A linear molecule has a central atom bonded to two atoms and no lone pairs.

Linear shape

- A trigonal planar molecule has a central atom bonded to three atoms and no lone pairs. A bent molecule with a bond angle of 120° has a central atom bonded to two atoms and one lone pair.

Trigonal planar shape   Bent shape (120°)

- A tetrahedral molecule has a central atom bonded to four atoms and no lone pairs. In a trigonal pyramidal molecule, a central atom is bonded to three atoms and one lone pair. In a bent molecule with a bond angle of 109°, a central atom is bonded to two atoms and two lone pairs.

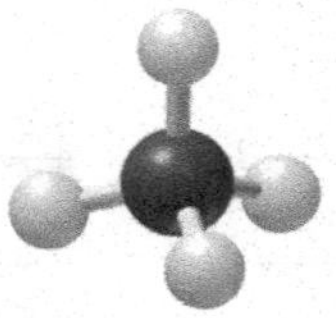

Tetrahedral shape 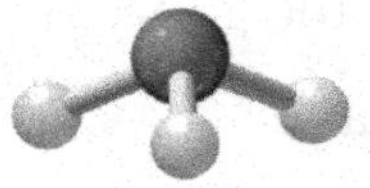   Trigonal pyramidal shape   Bent shape (109°)

♦ **Learning Exercise 6.8A**

Match the shape of a molecule with the following descriptions:

    a. linear   **b.** trigonal planar   **c.** tetrahedral
    d. trigonal pyramidal   **e.** bent (120°)   **f.** bent (109°)

1. has a central atom with three electron groups and three bonded atoms   _____
2. has a central atom with two electron groups and two bonded atoms   _____
3. has a central atom with four electron groups with three bonded atoms   _____
4. has a central atom with three electron groups and two bonded atoms   _____
5. has a central atom with four electron groups and four bonded atoms   _____
6. has a central atom with four electron groups and two bonded atoms   _____

Answers   **1.** b   **2.** a   **3.** d   **4.** e   **5.** c   **6.** f

Solution Guide to Predicting Shape of Molecules	
STEP 1	Draw the Lewis structure.
STEP 2	Arrange the electron groups around the central atom to minimize repulsion.
STEP 3	Use the atoms bonded to the central atom to determine the shape.

♦ **Learning Exercise 6.8B**

CORE CHEMISTRY SKILL
Predicting Shape

For each molecule, state the total number of valence electrons, draw the Lewis structure, state the number of electron groups and bonded atoms, and predict the shape.

Molecule or Ion	Valence Electrons	Lewis Structure	Electron Groups	Bonded Atoms	Shape
CH_4					
PCl_3					
SO_3					
H_2S					
SeO_2					

Chapter 6

Answers

Molecule or Ion	Valence Electrons	Lewis Structure	Electron Groups	Bonded Atoms	Shape
CH_4	8	H:C(H)(H):H	4	4	Tetrahedral
PCl_3	26	:Cl:P:Cl: :Cl:	4	3	Trigonal pyramidal
SO_3	24	:O:S::O: :O:	3	3	Trigonal planar
H_2S	8	H:S:H	4	2	Bent, 109°
SeO_2	18	:O::Se:O:	3	2	Bent, 120°

6.9 Polarity of Molecules and Intermolecular Forces

Learning Goal: Use the three-dimensional structure of a molecule to classify it as polar or nonpolar. Describe the intermolecular forces between ions, polar covalent molecules, and nonpolar covalent molecules.

- Nonpolar molecules can have polar bonds if the dipoles are in a symmetric arrangement and the bond dipoles cancel each other.
- In polar molecules, the bond dipoles do not cancel each other.
- The intermolecular forces between particles in solids and liquids determine their melting points.
- Ionic solids have high melting points due to strong ionic interactions between positive and negative ions.
- In polar substances, dipole–dipole attractions occur between the positive end of one molecule and the negative end of another.

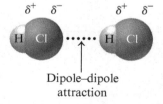

Dipole–dipole attraction

- Hydrogen bonding, a particularly strong type of dipole–dipole attraction, occurs between partially positive hydrogen atoms and the strongly electronegative atoms N, O, or F.

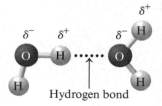

Hydrogen bond

Ionic and Molecular Compounds

- Dispersion forces occur when temporary dipoles form within the nonpolar molecules, causing weak attractions to other nonpolar molecules. Dispersion forces are the only intermolecular forces that occur between nonpolar molecules.

Key Terms for Sections 6.6 to 6.9

Match each of the following key terms with the correct description:

a. polar covalent bond
b. double bond
c. VSEPR theory
d. nonpolar covalent bond
e. dipole–dipole attraction
f. dispersion forces

1. _____ the equal sharing of valence electrons by two atoms
2. _____ the unequal attraction for shared electrons in a covalent bond
3. _____ the atoms in a molecule are arranged to minimize repulsion between electrons
4. _____ the sharing of two pairs of electrons by two atoms
5. _____ interaction between the positive end of a polar molecule and the negative end of another
6. _____ results from momentary polarization of nonpolar molecules

Answers 1. d   2. a   3. c   4. b   5. e   6. f

Solution Guide to Determining the Polarity of a Molecule	
STEP 1	Determine if the bonds are polar covalent or nonpolar covalent.
STEP 2	If the bonds are polar covalent, draw the Lewis structure and determine if the dipoles cancel.

♦ **Learning Exercise 6.9A**

Draw the bond dipoles in each of the following molecules and determine whether the molecule is polar or nonpolar:

CORE CHEMISTRY SKILL
Identifying Polarity of Molecules and Intermolecular Forces

a. CF_4

b. HBr

c. NH_3

d. SF_2

Answers

a.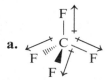

bond dipoles cancel, nonpolar

b. 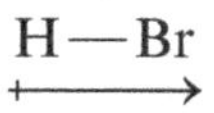 H—Br

bond dipole does not cancel, polar

c.
bond dipoles do not cancel, polar

d.
bond dipoles do not cancel, polar

♦ **Learning Exercise 6.9B**

Match the major type of intermolecular force with each of the following:
   a. ionic   **b.** dipole–dipole attractions   **c.** hydrogen bonding   **d.** dispersion forces

1. ____ KCl   2. ____ NF_3 3. ____ PF_3 4. ____ Cl_2
5. ____ HF   6. ____ H_2O 7. ____ C_4H_{10} 8. ____ Na_2O

Answers   **1.** a   **2.** b   **3.** b   **4.** d
            5. c   **6.** c   **7.** d   **8.** a

Checklist for Chapter 6

You are ready to take the Practice Test for Chapter 6. Be sure you have accomplished the following learning goals for this chapter. If not, review the Section listed at the end of the goal. Then apply your new skills and understanding to the Practice Test.

After studying Chapter 6, I can successfully:

_____ Illustrate the octet rule for the formation of ions. (6.1)

_____ Write the formulas for ionic compounds containing the ions of metals and nonmetals of representative elements. (6.2)

_____ Use charge balance to write an ionic formula. (6.2)

_____ Write the name for an ionic compound. (6.3)

_____ Write the formula for a compound containing a polyatomic ion. (6.4)

_____ Write the names and formulas for molecular compounds. (6.5)

_____ Draw the Lewis structures for molecular compounds. (6.6)

_____ Classify a bond as nonpolar covalent, polar covalent, or ionic. (6.7)

_____ Use the VSEPR theory to determine the shape of a molecule. (6.8)

_____ Classify a molecule as polar or nonpolar. (6.9)

_____ Identify the intermolecular forces between particles. (6.9)

Practice Test for Chapter 6

The chapter Sections to review are shown in parentheses at the end of each question.

For questions 1 through 4, consider an atom of phosphorus: (6.1)

1. Phosphorus is in Group
   A. 2A (2)   **B.** 3A (13)   **C.** 5A (15)   **D.** 7A (17)   **E.** 8A (18)

2. How many valence electrons does an atom of phosphorus have?
   A. 2   **B.** 3   **C.** 5   **D.** 8   **E.** 15

Ionic and Molecular Compounds

3. To achieve an octet, a phosphorus atom in an ionic compound will
   A. lose 1 electron   B. lose 2 electrons   C. lose 5 electrons
   D. gain 2 electrons   E. gain 3 electrons

4. A phosphide ion has a charge of
   A. 1+   B. 2+   C. 5+   D. 2−   E. 3−

5. To achieve an octet, a calcium atom (6.1)
   A. loses 1 electron   B. loses 2 electrons   C. loses 3 electrons
   D. gains 1 electron   E. gains 2 electrons

6. The correct ionic charge for a calcium ion is (6.1)
   A. 1+   B. 2+   C. 1−   D. 2−   E. 3−

7. Another name for a positive ion is (6.1)
   A. anion   B. cation   C. proton   D. positron   E. sodium

8. To achieve an octet, a chlorine atom (6.1)
   A. loses 1 electron   B. loses 2 electrons   C. loses 3 electrons
   D. gains 1 electron   E. gains 2 electrons

9. The correct ionic charge for a cesium ion is (6.1)
   A. 1+   B. 2+   C. 1−   D. 2−   E. 3−

10. The correct ionic charge for a fluoride ion is (6.1)
    A. 1+   B. 2+   C. 1−   D. 2−   E. 3−

11. When the elements magnesium and sulfur react, (6.2)
    A. an ionic compound forms
    B. a molecular compound forms
    C. no reaction occurs
    D. the two repel each other and will not combine
    E. none of the above

12. An ionic bond typically occurs between (6.2)
    A. two different nonmetals   B. two of the same type of nonmetals
    C. two noble gases   D. two different metals
    E. a metal and a nonmetal

13. The correct formula for a compound between sodium and sulfur is (6.2)
    A. SoS   B. NaS   C. Na_2S D. NaS_2 E. Na_2SO_4

14. The correct formula for a compound between aluminum and oxygen is (6.2)
    A. AlO   B. Al_2O C. AlO_3 D. Al_2O_3 E. Al_3O_2

15. The correct formula for a compound between barium and sulfur is (6.2)
    A. BaS   B. Ba_2S C. BaS_2 D. Ba_2S_2 E. $BaSO_4$

16. The correct formula for iron(III) fluoride is (6.2, 6.3)
    A. FeF   B. FeF_2 C. Fe_2F D. Fe_3F E. FeF_3

17. The correct formula for copper(II) chloride is (6.2, 6.3)
    A. CoCl   B. CuCl   C. $CoCl_2$ D. $CuCl_2$ E. Cu_2Cl

18. The correct formula for silver oxide is (6.2, 6.3)
    A. AgO   B. Ag_2O C. AgO_2 D. Ag_3O_2 E. Ag_3O

19. The correct ionic charge for a phosphate ion is (6.4)
    A. 1+   B. 2+   C. 1−   D. 2−   E. 3−

Chapter 6

20. The correct ionic charge for a sulfate ion is (6.4)
    A. 1+   B. 2+   C. 1−   D. 2−   E. 3−

21. The correct formula for ammonium sulfate is (6.2, 6.4)
    A. AmS   B. AmSO$_4$ C. (NH$_4$)$_2$S   D. NH$_4$SO$_4$ E. (NH$_4$)$_2$SO$_4$

22. The correct formula for lithium phosphate is (6.2, 6.4)
    A. LiPO$_4$ B. Li$_2$PO$_4$ C. Li$_3$PO$_4$ D. Li$_2$(PO$_4$)$_3$ E. Li$_3$(PO$_4$)$_2$

23. The correct formula for magnesium carbonate, used as a dietary magnesium supplement, is (6.2, 6.4)
    A. MgCO$_3$ B. Mg$_2$CO$_3$ C. Mg(CO$_3$)$_2$ D. MgCO   E. Mg$_2$(CO$_3$)$_3$

24. The correct formula for copper(I) sulfate, used as a fungicide, is (6.3, 6.4)
    A. CuSO$_3$ B. CuSO$_4$ C. Cu$_2$SO$_3$ D. Cu(SO$_4$)$_2$ E. Cu$_2$SO$_4$

25. The correct name of Al$_2$(HPO$_4$)$_3$ is (6.3, 6.4)
    A. aluminum hydrogen phosphite
    B. aluminum hydrogen phosphate
    C. aluminum hydrogen phosphorus
    D. aluminum hydrogen phosphorus oxide
    E. trialuminum dihydrogen phosphate

26. The correct name of CoS is (6.3, 6.4)
    A. copper sulfide   B. cobalt(II) sulfate   C. cobalt(I) sulfide
    D. cobalt sulfide   E. cobalt(II) sulfide

27. The correct name of MnCl$_2$, used as a supplement in intravenous solutions, is (6.3, 6.4)
    A. magnesium chloride   B. manganese(II) chlorine   C. manganese(II) chloride
    D. manganese chlorine   E. manganese(III) chloride

28. The correct name of ZnCO$_3$ is (6.3, 6.4)
    A. zinc(III) carbonate   B. zinc(II) carbonate   C. zinc bicarbonate
    D. zinc carbon trioxide   E. zinc carbonate

29. The correct name of Al$_2$O$_3$ is (6.3)
    A. aluminum oxide   B. aluminum(II) oxide   C. aluminum trioxide
    D. dialuminum trioxide   E. aluminum oxygenate

30. The correct name of Cr$_2$(SO$_3$)$_3$ is (6.3, 6.4)
    A. chromium sulfite   B. dichromium trisulfite   C. chromium(III) sulfite
    D. chromium(III) sulfate   E. chromium sulfate

31. The name of PF$_5$ is (6.5)
    A. potassium pentafluoride   B. phosphorus pentafluoride   C. phosphorus pentafluorine
    D. potassium pentafluorine   E. phosphorus heptafluoride

32. The correct name of NCl$_3$ is (6.5)
    A. nitrogen chloride   B. nitrogen trichloride   C. trinitrogen chloride
    D. nitrogen chlorine three   E. nitrogen trichlorine

33. The name of CO is (6.5)
    A. carbon monoxide   B. carbonic oxide   C. carbon oxide
    D. carbonious oxide   E. carboxide

34. The correct Lewis structure for CS$_2$ is: (6.6)
    A. $\ddot{S}=\ddot{C}=\ddot{S}$ B. $\ddot{S}=C=\ddot{S}$ C. $:\ddot{S}-\ddot{C}-\ddot{S}:$
    D. $:S-C-S:$ E. $:\ddot{S}-C\equiv\ddot{S}:$

Ionic and Molecular Compounds

35. A polar covalent bond occurs between (6.7)
   A. two different nonmetals
   B. two of the same type of nonmetals
   C. two noble gases
   D. two different metals
   E. a metal and a nonmetal

For questions 36 through 41, indicate the type of bonding expected to occur between the following elements: (6.7)

   ionic (I)   polar covalent (PC)   nonpolar covalent (NC)

36. _____ silicon and oxygen
37. _____ barium and chlorine
38. _____ aluminum and fluorine
39. _____ bromine and bromine
40. _____ sulfur and oxygen
41. _____ nitrogen and oxygen

For questions 42 through 47, match each of the following shapes with one of the descriptions: (6.8)

   A. linear   **B.** trigonal planar   **C.** tetrahedral
   D. trigonal pyramidal **E.** bent (120°)   **F.** bent (109°)

42. _____ has a central atom with three electron groups and three bonded atoms
43. _____ has a central atom with two electron groups and two bonded atoms
44. _____ has a central atom with four electron groups and three bonded atoms
45. _____ has a central atom with three electron groups and two bonded atoms
46. _____ has a central atom with four electron groups and four bonded atoms
47. _____ has a central atom with four electron groups and two bonded atoms

For questions 48 through 51, determine the shape for each of the following molecules: (6.8)

   A. linear   **B.** trigonal planar   **C.** bent, 120°
   D. tetrahedral   **E.** trigonal pyramidal   **F.** bent, 109°

48. PBr_3 **49.** CBr_4 **50.** H_2S **51.** CO

For questions 52 through 59, indicate the major type of intermolecular force (A to D) between particles of each of the following: (6.9)

   A. ionic   **B.** dipole–dipole attractions
   C. hydrogen bonds   **D.** dispersion forces

52. _____ CaO
53. _____ NCl_3
54. _____ SF_2
55. _____ Br_2
56. _____ H_2O_2
57. _____ NH_3
58. _____ C_3H_8
59. _____ BaF_2

For questions 60 through 64, draw the Lewis structure for CO_2: (6.6, 6.7, 6.8, 6.9)

60. The number of valence electrons needed in the Lewis structure of CO_2 is
   A. 10   **B.** 12   **C.** 14   **D.** 16   **E.** 18

61. The number of single bonds is
   A. 0   **B.** 1   **C.** 2   **D.** 3   **E.** 4

Chapter 6

62. The number of double bonds is
A. 0   B. 1   C. 2   D. 3   E. 4

63. The molecule is
A. nonpolar   B. polar   C. ionic

64. The shape of the molecule is
A. linear   B. bent (120°)   C. bent (109°)
D. trigonal planar   E. tetrahedral

Answers to the Practice Test

1. C	2. C	3. E	4. E	5. B
6. B	7. B	8. D	9. A	10. C
11. A	12. E	13. C	14. D	15. A
16. E	17. D	18. B	19. E	20. D
21. E	22. C	23. A	24. E	25. B
26. E	27. C	28. E	29. A	30. C
31. B	32. B	33. A	34. B	35. A
36. PC	37. I	38. I	39. NC	40. PC
41. PC	42. B	43. A	44. D	45. E
46. C	47. F	48. E	49. D	50. F
51. A	52. A	53. B	54. B	55. D
56. C	57. C	58. D	59. A	60. D
61. A	62. C	63. A	64. A	

Ionic and Molecular Compounds

Selected Answers and Solutions to Text Problems

6.1 Atoms with one, two, or three valence electrons will lose those electrons to acquire a noble gas electron configuration.
    a. Li loses 1 e^-.
    b. Ca loses 2 e^-.
    c. Ga loses 3 e^-.
    d. Cs loses 1 e^-.
    e. Ba loses 2 e^-.

6.3 Atoms form ions by losing or gaining valence electrons to achieve the same electron configuration as the nearest noble gas. Elements in Groups 1A (1), 2A (2), and 3A (13) lose valence electrons, whereas elements in Groups 5A (15), 6A (16), and 7A (17) gain valence electrons to complete octets.
    a. Sr loses 2 e^-.
    b. P gains 3 e^-.
    c. Elements in Group 7A (17) gain 1 e^-.
    d. Na loses 1 e^-.
    e. Br gains 1 e^-.

6.5
    a. The element with 3 protons is lithium. In a lithium ion with 2 electrons, the ionic charge would be 1+, (3+) + (2−) = 1+. The lithium ion is written as Li^+.
    b. The element with 9 protons is fluorine. In a fluorine ion with 10 electrons, the ionic charge would be 1−, (9+) + (10−) = 1−. The fluoride ion is written as F^-.
    c. The element with 12 protons is magnesium. In a magnesium ion with 10 electrons, the ionic charge would be 2+, (12+) + (10−) = 2+. The magnesium ion is written as Mg^{2+}.
    d. The element with 26 protons is iron. In an iron ion with 23 electrons, the ionic charge would be 3+, (26+) + (23−) = 3+. This iron ion is written as Fe^{3+}.

6.7
    a. Cu has an atomic number of 29, which means it has 29 protons. In a neutral atom, the number of electrons equals the number of protons. A copper ion with a charge of 2+ has lost 2 e^- to have 29 − 2 = 27 electrons. ∴ in a Cu^{2+} ion, there are 29 protons and 27 electrons.
    b. Se has an atomic number of 34, which means it has 34 protons. In a neutral atom, the number of electrons equals the number of protons. A selenium ion with a charge of 2− has gained 2 e^- to have 34 + 2 = 36 electrons. ∴ in a Se^{2-} ion, there are 34 protons and 36 electrons.
    c. Br has an atomic number of 35, which means it has 35 protons. In a neutral atom, the number of electrons equals the number of protons. A bromine ion with a charge of 1− has gained 1 e^- to have 35 + 1 = 36 electrons. ∴ in a Br^- ion, there are 35 protons and 36 electrons.
    d. Fe has an atomic number of 26, which means it has 26 protons. In a neutral atom, the number of electrons equals the number of protons. An iron ion with a charge of 3+ has lost 3 e^- to have 26 − 3 = 23 electrons. ∴ in a Fe^{3+} ion, there are 26 protons and 23 electrons.

6.9
    a. Chlorine in Group 7A (17) gains one electron to form chloride ion, Cl^-.
    b. Cesium in Group 1A (1) loses one electron to form cesium ion, Cs^+.
    c. Nitrogen in Group 5A (15) gains three electrons to form nitride ion, N^{3-}.
    d. Radium in Group 2A (2) loses two electrons to form radium ion, Ra^{2+}.

6.11
    a. Li^+ is called the lithium ion.
    b. Ca^{2+} is called the calcium ion.
    c. Ga^{3+} is called the gallium ion.
    d. P^{3-} is called the phosphide ion.

6.13
    a. There are 8 protons and 10 electrons in the O^{2-} ion; (8+) + (10−) = 2−.
    b. There are 19 protons and 18 electrons in the K^+ ion; (19+) + (18−) = 1+.
    c. There are 53 protons and 54 electrons in the I^- ion; (53+) + (54−) = 1−.
    d. There are 11 protons and 10 electrons in the Na^+ ion; (11+) + (10−) = 1+.

6.15 **a** (Li and Cl) and **c** (K and O) will form ionic compounds.

6.17
    a. Na^+ and O^{2-} → Na_2O
    b. Al^{3+} and Br^- → $AlBr_3$
    c. Ba^{2+} and N^{3-} → Ba_3N_2
    d. Mg^{2+} and F^- → MgF_2
    e. Al^{3+} and S^{2-} → Al_2S_3

Chapter 6

6.19 a. Ions: K^+ and $S^{2-} \to K_2S$ Check: $2(1+) + 1(2-) = 0$
      b. Ions: Na^+ and $N^{3-} \to Na_3N$ Check: $3(1+) + 1(3-) = 0$
      c. Ions: Al^{3+} and $I^- \to AlI_3$ Check: $1(3+) + 3(1-) = 0$
      d. Ions: Ga^{3+} and $O^{2-} \to Ga_2O_3$ Check: $2(3+) + 3(2-) = 0$

6.21 a. aluminum fluoride   b. calcium chloride
      c. sodium oxide   d. magnesium phosphide
      e. potassium iodide   f. barium fluoride

6.23 a. iron(II)   b. copper(II)
      c. zinc   d. lead(IV)
      e. chromium(III)   f. manganese(II)

6.25 a. Ions: Sn^{2+} and $Cl^- \to$ tin(II) chloride
      b. Ions: Fe^{2+} and $O^{2-} \to$ iron(II) oxide
      c. Ions: Cu^+ and $S^{2-} \to$ copper(I) sulfide
      d. Ions: Cu^{2+} and $S^{2-} \to$ copper(II) sulfide
      e. Ions: Cd^{2+} and $Br^- \to$ cadmium bromide
      f. Ions: Hg^{2+} and $Cl^- \to$ mercury(II) chloride

6.27 a. Au^{3+} b. Fe^{3+}
      c. Pb^{4+} d. Al^{3+}

6.29 a. Ions: Mg^{2+} and $Cl^- \to MgCl_2$ b. Ions: Na^+ and $S^{2-} \to Na_2S$
      c. Ions: Cu^+ and $O^{2-} \to Cu_2O$ d. Ions: Zn^{2+} and $P^{3-} \to Zn_3P_2$
      e. Ions: Au^{3+} and $N^{3-} \to AuN$ f. Ions: Co^{3+} and $F^- \to CoF_3$

6.31 a. Ions: Co^{3+} and $Cl^- \to CoCl_3$ b. Ions: Pb^{4+} and $O^{2-} \to PbO_2$
      c. Ions: Ag^+ and $I^- \to AgI$ d. Ions: Ca^{2+} and $N^{3-} \to Ca_3N_2$
      e. Ions: Cu^+ and $P^{3-} \to Cu_3P$ f. Ions: Cr^{2+} and $Cl^- \to CrCl_2$

6.33 a. Ions: K^+ and $P^{3-} \to K_3P$ b. Ions: Cu^{2+} and $Cl^- \to CuCl_2$
      c. Ions: Fe^{3+} and $Br^- \to FeBr_3$ d. Ions: Mg^{2+} and $O^{2-} \to MgO$

6.35 a. HCO_3^- b. NH_4^+ c. PO_3^{3-} d. ClO_3^-

6.37 a. sulfate   b. carbonate
      c. hydrogen sulfite (or bisulfite)   d. nitrate

6.39

	NO_2^-	CO_3^{2-}	HSO_4^-	PO_4^{3-}
Li^+	$LiNO_2$ Lithium nitrite	Li_2CO_3 Lithium carbonate	$LiHSO_4$ Lithium hydrogen sulfate	Li_3PO_4 Lithium phosphate
Cu^{2+}	$Cu(NO_2)_2$ Copper(II) nitrite	$CuCO_3$ Copper(II) carbonate	$Cu(HSO_4)_2$ Copper(II) hydrogen sulfate	$Cu_3(PO_4)_2$ Copper(II) phosphate
Ba^{2+}	$Ba(NO_2)_2$ Barium nitrite	$BaCO_3$ Barium carbonate	$Ba(HSO_4)_2$ Barium hydrogen sulfate	$Ba_3(PO_4)_2$ Barium phosphate

6.41 a. Ions: Ba^{2+} and $OH^- \to Ba(OH)_2$ b. Ions: Na^+ and $HSO_4^- \to NaHSO_4$
      c. Ions: Fe^{2+} and $NO_2^- \to Fe(NO_2)_2$ d. Ions: Zn^{2+} and $PO_4^{3-} \to Zn_3(PO_4)_2$
      e. Ions: Fe^{3+} and $CO_3^{2-} \to Fe_2(CO_3)_3$

6.43 a. CO_3^{2-}, sodium carbonate   b. NH_4^+, ammonium sulfide
      c. OH^-, calcium hydroxide   d. NO_2^-, tin(II) nitrite

Ionic and Molecular Compounds

6.45 a. Ions: $Zn^{2+} + C_2H_3O_2^- \rightarrow$ zinc acetate
   b. Ions: Mg^{2+} and $PO_4^{3-} \rightarrow$ magnesium phosphate
   c. Ions: $NH_4^+ + Cl^- \rightarrow$ ammonium chloride
   d. Ions: Na^+ and $HCO_3^- \rightarrow$ sodium bicarbonate or sodium hydrogen carbonate
   e. Ions: Na^+ and $NO_2^- \rightarrow$ sodium nitrite

6.47 When naming molecular compounds, prefixes are used to indicate the number of each atom as shown in the subscripts of the formula. The first nonmetal is named by its elemental name; the second nonmetal is named by using its elemental name with the ending changed to *ide*.
   a. 1 P and 3 Br $\rightarrow$ phosphorus tribromide   b. 2 Cl and 1 O $\rightarrow$ dichlorine oxide
   c. 1 C and 4 Br $\rightarrow$ carbon tetrabromide   d. 1 H and 1 F $\rightarrow$ hydrogen fluoride
   e. 1 N and 3 F $\rightarrow$ nitrogen trifluoride

6.49 When naming molecular compounds, prefixes are used to indicate the number of each atom as shown in the subscripts of the formula. The first nonmetal is named by its elemental name; the second nonmetal is named by using its elemental name with the ending changed to *ide*.
   a. 2 N and 3 O $\rightarrow$ dinitrogen trioxide
   b. 2 Si and 6 Br $\rightarrow$ disilicon hexabromide
   c. 4 P and 3 S $\rightarrow$ tetraphosphorus trisulfide
   d. 1 P and 5 Cl $\rightarrow$ phosphorus pentachloride
   e. 1 Se and 6 F $\rightarrow$ selenium hexafluoride

6.51 a. 1 C and 4 Cl $\rightarrow$ CCl_4 b. 1 C and 1 O $\rightarrow$ CO
   c. 1 P and 3 F $\rightarrow$ PF_3 d. 2 N and 4 O $\rightarrow$ N_2O_4

6.53 a. 1 O and 2 F $\rightarrow$ OF_2 b. 1 B and 3 Cl $\rightarrow$ BCl_3
   c. 2 N and 3 O $\rightarrow$ N_2O_3 d. 1 S and 6 F $\rightarrow$ SF_6

6.55 a. Ions: Al^{3+} and $SO_4^{2-} \rightarrow$ aluminum sulfate
   b. Ions: Ca^{2+} and $CO_3^{2-} \rightarrow$ calcium carbonate
   c. 2 N and 1 O $\rightarrow$ dinitrogen oxide
   d. Ions: $Mg^{2+} + OH^- \rightarrow$ magnesium hydroxide

6.57 a. $2\,H(1\,e^-) + 1\,S(6\,e^-) = 2 + 6 = 8$ valence electrons
   b. $2\,I(7\,e^-) = 14$ valence electrons
   c. $1\,C(4\,e^-) + 4\,Cl(7\,e^-) = 4 + 28 = 32$ valence electrons

6.59 a. $1\,H(1\,e^-) + 1\,F(7\,e^-) = 1 + 7 = 8$ valence electrons

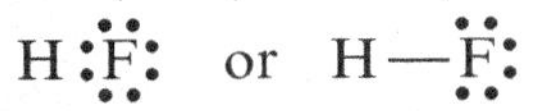

   H:F̈:   or   H—F̈:

   b. $1\,S(6\,e^-) + 2\,F(7\,e^-) = 6 + 14 = 20$ valence electrons

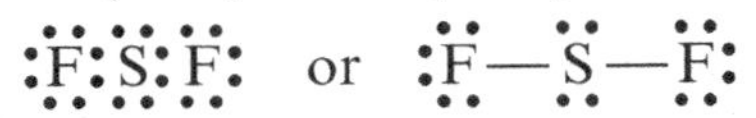

   :F̈:S̈:F̈:   or   :F̈—S̈—F̈:

   c. $1\,N(5\,e^-) + 3\,Br(7\,e^-) = 5 + 21 = 26$ valence electrons

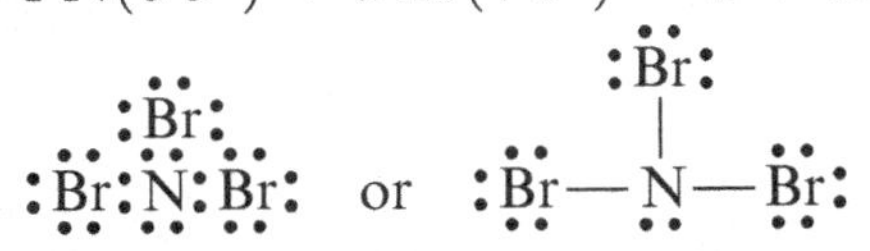

   :B̈r:
   :B̈r:N:B̈r: or   :B̈r—N—B̈r:
                              |
                             :B̈r:

   d. $1\,Cl(7\,e^-) + 1\,N(5\,e^-) + 2\,O(6\,e^-) = 7 + 5 + 12 = 24$ valence electrons

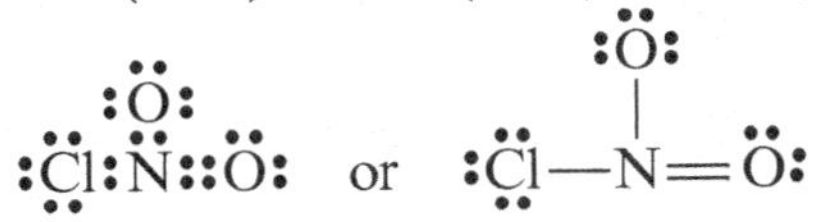

   :Ö:
   :C̈l:N::Ö: or   :C̈l—N=Ö:
                              |
                             :Ö:

6.61 a. The electronegativity values increase going from left to right across Period 2 from B to F.
   b. The electronegativity values decrease going down Group 2A (2) from Mg to Ba.
   c. The electronegativity values decrease going down Group 7A (17) from F to I.

Chapter 6

6.63 a. Electronegativity increases going up a group: K, Na, Li.
  b. Electronegativity increases going left to right across a period: Na, P, Cl.
  c. Electronegativity increases going across a period and at the top of a group: Ca, Se, O.

6.65 a; A nonpolar covalent bond would result from an electronegativity difference from 0.0 to 0.4.

6.67 a. electronegativity difference: 2.8 − 1.8 = 1.0, polar covalent
  b. electronegativity difference: 4.0 − 1.0 = 3.0, ionic
  c. electronegativity difference: 4.0 − 2.8 = 1.2, polar covalent
  d. electronegativity difference: 2.5 − 2.5 = 0.0, nonpolar covalent
  e. electronegativity difference: 3.0 − 2.1 = 0.9, polar covalent
  f. electronegativity difference: 2.5 − 2.1 = 0.4, nonpolar covalent

6.69 A dipole arrow points from the atom with the lower electronegativity value (more positive) to the atom in the bond that has the higher electronegativity value (more negative).

  a. $\overset{\delta^+}{N}$—$\overset{\delta^-}{F}$ →

  b. $\overset{\delta^+}{Si}$—$\overset{\delta^-}{Br}$ →

  c. $\overset{\delta^+}{C}$—$\overset{\delta^-}{O}$ →

  d. $\overset{\delta^+}{P}$—$\overset{\delta^-}{Br}$ →

  e. $\overset{\delta^-}{N}$—$\overset{\delta^+}{P}$ ←

6.71 a. 6, tetrahedral; Four electron groups around a central atom have a tetrahedral electron-group geometry. With four bonded atoms, the shape of the molecule is tetrahedral.
  b. 5, trigonal pyramidal; Four electron groups around a central atom have a tetrahedral electron-group geometry. With three bonded atoms (and one lone pair of electrons), the shape of the molecule is trigonal pyramidal.
  c. 3, trigonal planar; Three electron groups around a central atom have a trigonal planar electron-group geometry. With three bonded atoms, the shape of the molecule is trigonal planar.

6.73 SeO$_3$ 1 Se(6 e^-) + 3 O(6 e^-) = 6 + 18 = 24 valence electrons

$$\ddot{\text{:O:}}\ \|$$
:Ö—Se—Ö: (only 1 of the resonance structures is shown)

  a. There are <u>three</u> electron groups around the central Se atom.
  b. The electron-group geometry is <u>trigonal planar</u>.
  c. The number of atoms attached to the central Se atom is <u>three</u>.
  d. The shape of the molecule is <u>trigonal planar</u>.

6.75 In CF$_4$, the central atom C has four bonded atoms and no lone pairs of electrons, which gives CF$_4$ a tetrahedral shape. In NF$_3$, the central atom N has three bonded atoms and one lone pair of electrons, which gives NF$_3$ a trigonal pyramidal shape.

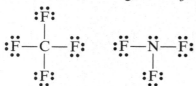

6.77 a. The central atom Ga has three electron groups bonded to three hydrogen atoms; GaH$_3$ has a trigonal planar shape.

140

b. The central O atom has four electron groups with two bonded atoms and two lone pairs of electrons, which gives OF_2 a bent shape (109°).

$$:\ddot{\underset{..}{F}}-\ddot{\underset{..}{O}}-\overset{\overset{..}{:\ddot{F}:}}{}$$

c. The central atom C has two electron groups bonded to two atoms; HCN is linear.

$$H-C\equiv N\!:$$

d. The central atom C has four electron groups bonded to four chlorine atoms; CCl_4 has a tetrahedral shape.

$$:\ddot{\underset{..}{Cl}}-\underset{\underset{:\ddot{\underset{..}{Cl}}:}{|}}{\overset{\overset{:\ddot{Cl}:}{|}}{C}}-\ddot{\underset{..}{Cl}}:$$

6.79 F_2 is a nonpolar molecule because there is a nonpolar covalent bond between F atoms, which have identical electronegativity values. In HF, the bond is a polar bond because there is a large electronegativity difference, which makes HF a polar molecule.

$:\ddot{\underset{..}{F}}-\ddot{\underset{..}{F}}:$ $H-\ddot{\underset{..}{F}}:$
nonpolar   polar

6.81 a. The molecule CS_2 contains two polar C—S bonds whose dipoles point in opposite directions. As a result of this linear geometry, the dipoles cancel, which makes CS_2 a nonpolar molecule.

$:\ddot{S}=C=\ddot{S}:$
nonpolar

b. The molecule NF_3 contains three polar covalent N—F bonds and a lone pair of electrons on the central N atom. This asymmetric trigonal pyramidal shape makes NF_3 a polar molecule.

polar

c. In the molecule CHF_3, there are three polar covalent C—F bonds and one nonpolar covalent C—H bond, which makes CHF_3 a polar molecule.

polar

d. In the molecule SO_3, there are three polar covalent S—O bonds arranged in a trigonal planar shape. As a result of this symmetric geometry, the dipoles cancel, which makes SO_3 a nonpolar molecule.

$$\underset{nonpolar}{:\ddot{\underset{..}{O}}=\underset{\underset{:\ddot{\underset{..}{O}}:}{\|}}{S}=\ddot{\underset{..}{O}}:}$$

Chapter 6

6.83 In the linear molecule CO_2, the two C—O dipoles cancel, resulting in a nonpolar molecule; in CO, there is only one dipole, making it a polar molecule.

$:\ddot{O}=C=\ddot{O}:$ $:C\equiv O:$

⟵⟶ ⟵⟶   ⟵⟶
   nonpolar   polar

6.85 **a.** BrF is a polar molecule. An attraction between the positive end of one polar molecule and the negative end of another polar molecule is called a dipole–dipole attraction.
b. An ionic bond is an attraction between a positive and negative ion, as in KCl.
c. NF_3 is a polar molecule. An attraction between the positive end of one polar molecule and the negative end of another polar molecule is called a dipole–dipole attraction.
d. Cl_2 is a nonpolar molecule. The weak attractions that occur between temporary dipoles in nonpolar molecules are called dispersion forces.

6.87 **a.** CH_3OH is a polar molecule. Hydrogen bonds are strong dipole–dipole attractions that occur between a partially positive hydrogen atom of one molecule and one of the strongly electronegative atoms N, O, or F in another, as is seen with CH_3OH molecules.
b. CO is a polar molecule. Dipole–dipole attractions occur between dipoles in polar molecules.
c. CF_4 is a nonpolar molecule. The weak attractions that occur between temporary dipoles in nonpolar molecules are called dispersion forces.
d. CH_3CH_3 is a nonpolar molecule. The weak attractions that occur between temporary dipoles in nonpolar molecules are called dispersion forces.

6.89 **a.** Ions: $Mg^{2+} + SO_4^{2-} \rightarrow MgSO_4$ **b.** Ions: $Sn^{2+} + F^- \rightarrow SnF_2$
c. Ions: $Al^{3+} + OH^- \rightarrow Al(OH)_3$

6.91 **a.** ionic   **b.** ionic   **c.** ionic

6.93 **a.** By losing two valence electrons from the third energy level, magnesium achieves an octet in the second energy level.
b. The magnesium ion Mg^{2+} has the same electron arrangement as the noble gas Ne (2,8).
c. Group 1A (1) and 2A (2) elements achieve octets by losing electrons when they form compounds. Group 8A (18) elements already have a stable octet of valence electrons (or two electrons for helium), so they are not normally found in compounds.

6.95 **a.** The element with 15 protons is phosphorus. In an ion of phosphorus with 18 electrons, the ionic charge would be 3−, (15+) + (18−) = 3−. The phosphide ion is written P^{3-}.
b. The element with 8 protons is oxygen. Since there are also 8 electrons, this is an oxygen (O) atom.
c. The element with 30 protons is zinc. In an ion of zinc with 28 electrons, the ionic charge would be 2+, (30+) + (28−) = 2+. The zinc ion is written Zn^{2+}.
d. The element with 26 protons is iron. In an ion of iron with 23 electrons, the ionic charge would be 3+, (26+) + (23−) = 3+. This iron ion is written Fe^{3+}.

6.97 **a.** X is in Group 1A (1); Y is in Group 6A (16).   **b.** ionic
c. X^+, Y^{2-} **d.** X_2Y
e. X^+ and $S^{2-} \rightarrow X_2S$ **f.** 1 Y and 2 Cl $\rightarrow YCl_2$
g. molecular

6.99

Electron Arrangements		Cation	Anion	Formula of Compound	Name of Compound
2,8,2	2,5	Mg^{2+}	N^{3-}	Mg_3N_2	Magnesium nitride
2,8,8,1	2,6	K^+	O^{2-}	K_2O	Potassium oxide
2,8,3	2,8,7	Al^{3+}	Cl^-	$AlCl_3$	Aluminum chloride

Ionic and Molecular Compounds

6.101 **a.** 2 valence electrons, 1 bonding pair, no lone pairs
b. 8 valence electrons, 1 bonding pair, 3 lone pairs
c. 14 valence electrons, 1 bonding pair, 6 lone pairs

6.103 **a.** 2; trigonal pyramidal shape, polar molecule
b. 1; bent shape (109°), polar molecule
c. 3; tetrahedral shape, nonpolar molecule

6.105 **a.** C—O (EN 3.5 − 2.5 = 1.0) and N—O (EN 3.5 − 3.0 = 0.5) are polar covalent bonds.
b. O—O (EN 3.5 − 3.5 = 0.0) is a nonpolar covalent bond.
c. Ca—O (EN 3.5 − 1.0 = 2.5) and K—O (EN 3.5 − 0.8 = 2.7) are ionic bonds.
d. C—O, N—O, O—O

6.107 **a.** PH_3 is a nonpolar molecule. Dispersion forces occur between temporary dipoles in nonpolar molecules.
b. NO_2 is a polar molecule. Dipole–dipole attractions occur between dipoles in polar molecules.
c. Hydrogen bonds are strong dipole–dipole attractions that occur between a partially positive hydrogen atom of one molecule and one of the strongly electronegative atoms N, O, or F in another.
d. Dispersion forces occur between temporary dipoles in Ar atoms.

6.109 **a.** N^{3-} is called the nitride ion. **b.** Mg^{2+} is called the magnesium ion.
c. O^{2-} is called the oxide ion. **d.** Al^{3+} is called the aluminum ion.

6.111 **a.** An element that forms an ion with a 2+ charge would be in Group 2A (2).
b. The Lewis symbol for an element in Group 2A (2) is $\dot{X}\cdot$
c. Mg is the Group 2A (2) element in Period 3.
d. Ions: X^{2+} and $N^{3-} \rightarrow X_3N_2$

6.113 **a.** Tin(IV) is Sn^{4+}.
b. The Sn^{4+} ion has 50 protons and 50 − 4 = 46 electrons.
c. Ions: Sn^{4+} and $O^{2-} \rightarrow SnO_2$
d. Ions: Sn^{4+} and $PO_4^{3-} \rightarrow Sn_3(PO_4)_4$

6.115 **a.** Ions: Sn^{2+} and $S^{2-} \rightarrow SnS$ **b.** Ions: Pb^{4+} and $O^{2-} \rightarrow PbO_2$
c. Ions: Ag^+ and $Cl^- \rightarrow AgCl$ **d.** Ions: Ca^{2+} and $N^{3-} \rightarrow Ca_3N_2$
e. Ions: Cu^+ and $P^{3-} \rightarrow Cu_3P$ **f.** Ions: Cr^{2+} and $Br^- \rightarrow CrBr_2$

6.117 **a.** 1 N and 3 Cl → nitrogen trichloride
b. 2 N and 3 S → dinitrogen trisulfide
c. 2 N and 1 O → dinitrogen oxide
d. 1 I and 1 F → iodine fluoride
e. 1 B and 3 F → boron trifluoride
f. 2 P and 5 O → diphosphorus pentoxide

6.119 **a.** 1 C and 1 S → CS **b.** 2 P and 5 O → P_2O_5
c. 2 H and 1 S → H_2S **d.** 1 S and 2 Cl → SCl_2

6.121 **a.** Ionic, ions are Fe^{3+} and $Cl^- \rightarrow$ iron(III) chloride
b. Ionic, ions are Na^+ and $SO_4^{2-} \rightarrow$ sodium sulfate
c. Molecular, 1 N and 2 O → nitrogen dioxide
d. Ionic, ions are Rb^+ and $S^{2-} \rightarrow$ rubidium sulfide
e. Molecular, 1 P and 5 F → phosphorus pentafluoride
f. Molecular, 1 C and 4 F → carbon tetrafluoride

6.123 **a.** Ions: Sn^{2+} and $CO_3^{2-} \rightarrow SnCO_3$ **b.** Ions: Li^+ and $P^{3-} \rightarrow Li_3P$
c. Molecular, 1 Si and 4 Cl → $SiCl_4$ **d.** Ions: Mn^{3+} and $O^{2-} \rightarrow Mn_2O_3$
e. Molecular, 4 P and 3 Se → P_4Se_3 **f.** Ions: Ca^{2+} and $Br^- \rightarrow CaBr_2$

6.125 a. $1\,H(1\,e^-) + 1\,N(5\,e^-) + 2\,O(6\,e^-) = 1 + 5 + 12 = 18$ valence electrons
b. $2\,C(4\,e^-) + 4\,H(1\,e^-) + 1\,O(6\,e^-) = 8 + 4 + 6 = 18$ valence electrons
c. $1\,C(4\,e^-) + 5\,H(1\,e^-) + 1\,N(5\,e^-) = 4 + 5 + 5 = 14$ valence electrons

6.127 a. $2\,Cl(7\,e^-) + 1\,O(6\,e^-) = 14 + 6 = 20$ valence electrons

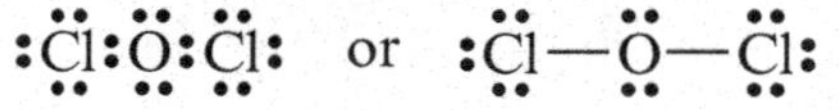

:C̈l:Ö:C̈l: or   :C̈l—Ö—C̈l:

b. $1\,N(5\,e^-) + 3\,H(1\,e^-) + 1\,O(6\,e^-) = 5 + 3 + 6 = 14$ valence electrons

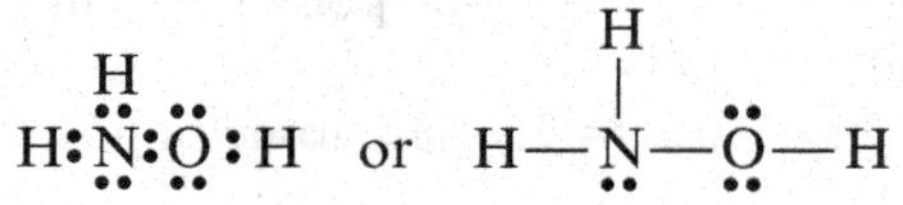

H:N̈:Ö:H   or   H—N—Ö—H
  H |
                     H

c. $2\,C(4\,e^-) + 2\,H(1\,e^-) + 2\,Cl(7\,e^-) = 8 + 2 + 14 = 24$ valence electrons

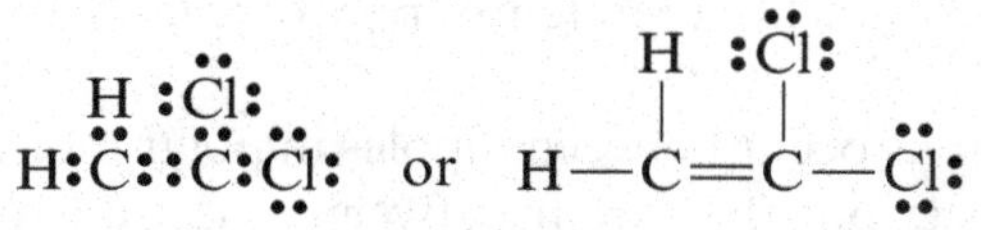

H:C::C:C̈l: or   H—C=C—C̈l:
  H :C̈l: | |
                    H :C̈l:

6.129 a. Electronegativity increases going up a group: I, Cl, F
b. Electronegativity increases going left to right across a period and at the top of a group: K, Li, S, Cl
c. Electronegativity increases going up a group: Ba, Sr, Mg, Be

6.131 a. C—O $(3.5 - 2.5 = 1.0)$ is more polar than C—N $(3.5 - 3.0 = 0.5)$.
b. N—F $(4.0 - 3.0 = 1.0)$ is more polar than N—Br $(3.0 - 2.8 = 0.2)$.
c. S—Cl $(3.0 - 2.5 = 0.5)$ is more polar than Br—Cl $(3.0 - 2.8 = 0.2)$.
d. Br—I $(2.8 - 2.5 = 0.3)$ is more polar than Br—Cl $(3.0 - 2.8 = 0.2)$.
e. N—F $(4.0 - 3.0 = 1.0)$ is more polar than N—O $(3.5 - 3.0 = 0.5)$.

6.133 A dipole arrow points from the atom with the lower electronegativity value (more positive) to the atom in the bond that has the higher electronegativity value (more negative).

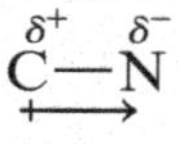

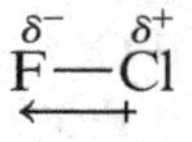

a. 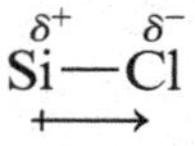 $\overset{\delta^+}{Si}—\overset{\delta^-}{Cl}$ →
b. $\overset{\delta^+}{C}—\overset{\delta^-}{N}$ →
c. $\overset{\delta^-}{F}—\overset{\delta^+}{Cl}$ ←
d. 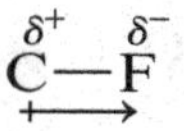 $\overset{\delta^+}{C}—\overset{\delta^-}{F}$ →
e. 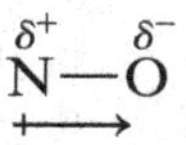 $\overset{\delta^+}{N}—\overset{\delta^-}{O}$ →

6.135 a. electronegativity difference: $3.0 - 1.8 = 1.2$, polar covalent
b. electronegativity difference: $2.5 - 2.5 = 0.0$, nonpolar covalent
c. electronegativity difference: $3.0 - 0.9 = 2.1$, ionic
d. electronegativity difference: $2.5 - 2.1 = 0.4$, nonpolar covalent
e. electronegativity difference: $4.0 - 4.0 = 0.0$, nonpolar covalent

6.137 a. $1\,N(5\,e^-) + 3\,F(7\,e^-) = 5 + 21 = 26$ valence electrons

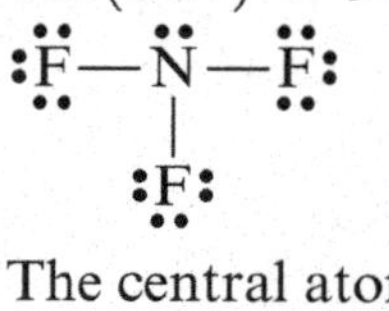

:F̈—N̈—F̈:
    |
   :F̈:

The central atom N has four electron groups with three bonded atoms and one lone pair of electrons, which gives NF₃ a trigonal pyramidal shape.

b. $1\,Si(4\,e^-) + 4\,Br(7\,e^-) = 4 + 28 = 32$ valence electrons

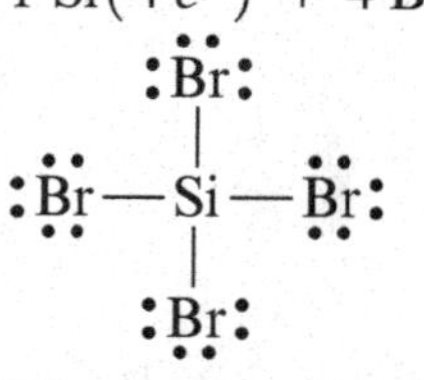

    :B̈r:
     |
:B̈r—Si—B̈r:
     |
    :B̈r:

The central atom Si has four electron groups bonded to four bromine atoms; SiBr₄ has a tetrahedral shape.

c. $1\,C(4\,e^-) + 2\,Se(6\,e^-) = 4 + 12 = 16$ valence electrons

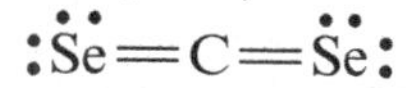

$$:\!\ddot{Se}\!=\!C\!=\!\ddot{Se}\!:$$

The central atom C has two electron groups bonded to two selenium atoms; CSe_2 has a linear shape.

6.139 a. $1\,C(4\,e^-) + 4\,Br(7\,e^-) = 4 + 28 = 32$ valence electrons

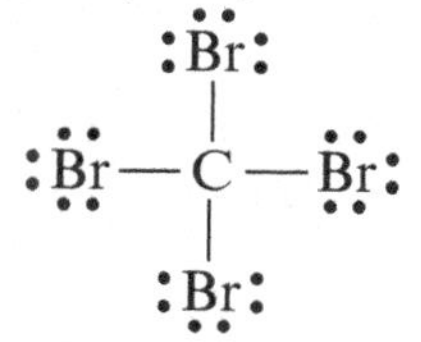

$$:\!\ddot{Br}\!-\!\underset{\underset{:\ddot{Br}:}{|}}{\overset{\overset{:\ddot{Br}:}{|}}{C}}\!-\!\ddot{Br}\!:$$

The central atom C has four electron groups bonded to four bromine atoms, which gives CBr_4 a tetrahedral shape.

b. $1\,O(6\,e^-) + 2\,H(1\,e^-) = 6 + 2 = 8$ valence electrons

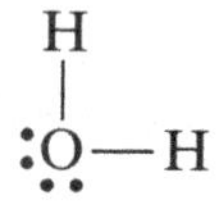

$$:\!\ddot{\underset{\underset{H}{|}}{O}}\!-\!H$$

The central atom O has four electron groups with two bonded atoms and two lone pairs of electrons, which gives H_2O a bent shape ($109°$).

6.141 a. A molecule that has a central atom with three bonded atoms and one lone pair will have a trigonal pyramidal shape. This asymmetric shape means the dipoles do not cancel, and the molecule will be polar.

b. A molecule that has a central atom with two bonded atoms and two lone pairs will have a bent shape ($109°$). This asymmetric shape means the dipoles do not cancel, and the molecule will be polar.

6.143 a. The molecule HBr contains only a single polar covalent H—Br bond (EN $2.8 - 2.1 = 0.7$); it is a polar molecule.

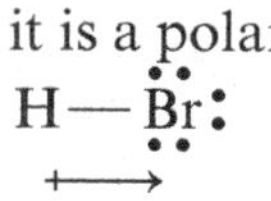

$$H\!-\!\ddot{Br}\!:$$
$\longleftrightarrow$

b. The molecule SiO_2 contains two polar covalent Si—O bonds (EN $3.5 - 1.8 = 1.7$) and has a linear shape. The two equal dipoles directed away from each other at $180°$ will cancel, resulting in a nonpolar molecule.

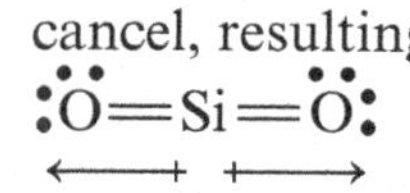

$$:\!\ddot{O}\!=\!Si\!=\!\ddot{O}\!:$$
$\longleftrightarrow \quad \longleftrightarrow$

c. The molecule NCl_3 contains three nonpolar covalent N—Cl bonds (EN $3.0 - 3.0 = 0.0$) and a lone pair of electrons on the central N atom, which gives the molecule a trigonal pyramidal shape. Since the molecule contains only nonpolar bonds, NCl_3 is a nonpolar molecule.

d. The molecule CH_3Cl consists of a central atom, C, with three nonpolar covalent C—H bonds (EN $2.5 - 2.1 = 0.4$) and one polar covalent C—Cl bond (EN $3.0 - 2.5 = 0.5$). The molecule has a tetrahedral shape, but the single dipole makes CH_3Cl a polar molecule.

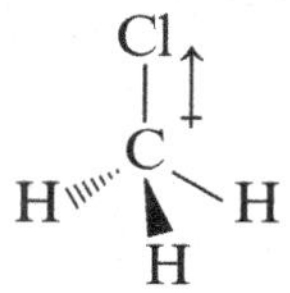

Chapter 6

e. The molecule NI₃ contains three polar covalent N—I bonds (EN 3.0 − 2.5 = 0.5) and a lone pair of electrons on the central N atom, which gives the molecule a trigonal pyramidal shape. This asymmetric shape does not allow the dipoles to cancel, which makes NI₃ a polar molecule.

f. The molecule H₂O contains two polar covalent O—H bonds (EN 3.5 − 2.1 = 1.4) and two lone pairs of electrons on the central O atom. The molecule has a bent shape (109°) and the asymmetric geometry of the O—H dipoles makes H₂O a polar molecule.

6.145 a. NF₃ is a polar molecule. Dipole–dipole attractions (2) occur between dipoles in polar molecules.
b. ClF is a polar molecule (EN 4.0 − 3.0 = 1.0). Dipole–dipole attractions (2) occur between dipoles in polar molecules.
c. Dispersion forces (4) occur between temporary dipoles in nonpolar Br₂ molecules.
d. Ionic bonds (1) are strong attractions between positive and negative ions, as in Cs₂O.
e. Dispersion forces (4) occur between temporary dipoles in nonpolar C₄H₁₀ molecules.
f. CH₃OH is a polar molecule and contains the polar O—H bond. Hydrogen bonding (3) involves strong dipole–dipole attractions that occur between a partially positive hydrogen atom of one polar molecule and one of the strongly electronegative atoms F, O, or N in another, as is seen with CH₃OH molecules.

6.147

Atom or Ion	Number of Protons	Number of Electrons	Electrons Lost/Gained
K⁺	19 p⁺	18 e⁻	1 e⁻ lost
Mg²⁺	12 p⁺	10 e⁻	2 e⁻ lost
O²⁻	8 p⁺	10 e⁻	2 e⁻ gained
Al³⁺	13 p⁺	10 e⁻	3 e⁻ lost

6.149 a. X as a X³⁺ ion would be in Group 3A (13).
b. X as a X²⁻ ion would be in Group 6A (16).
c. X as a X²⁺ ion would be in Group 2A (2).

6.151 Compounds with a metal and nonmetal are classified as ionic; compounds with two nonmetals are molecular.
a. Ionic, ions are Li⁺ and HPO₄²⁻ → lithium hydrogen phosphate
b. Molecular, 1 Cl and 3 F → chlorine trifluoride
c. Ionic, ions are Mg²⁺ and ClO₂⁻ → magnesium chlorite
d. Molecular, 1 N and 3 F → nitrogen trifluoride
e. Ionic, ions are Ca²⁺ and HSO₄⁻ → calcium bisulfate or calcium hydrogen sulfate
f. Ionic, ions are K⁺ and ClO₄⁻ → potassium perchlorate
g. Ionic, ions are Au³⁺ and SO₃²⁻ → gold(III) sulfite

6.153 a. 3 H(1 e⁻) + 1 N(5 e⁻) + 1 C(4 e⁻) + 1 O(6 e⁻) = 3 + 5 + 4 + 6 = 18 valence electrons

```
    H   :O:
    |   ‖
H—N—C—H
    ¨
```

b. $1\,\text{Cl}(7\,e^-) + 2\,\text{H}(1\,e^-) + 2\,\text{C}(4\,e^-) + 1\,\text{N}(5\,e^-) = 7 + 2 + 8 + 5 = 22$ valence electrons

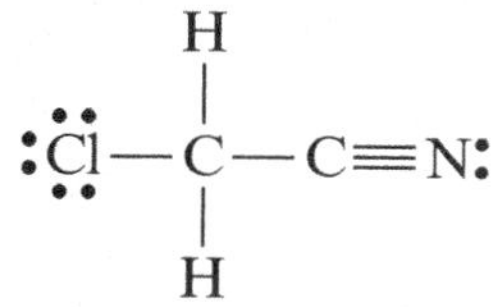

$$:\!\ddot{\underset{..}{\text{Cl}}}\!-\!\underset{\underset{\text{H}}{|}}{\overset{\overset{\text{H}}{|}}{\text{C}}}\!-\!\text{C}\!\equiv\!\text{N}\!:$$

c. $2\,\text{H}(1\,e^-) + 2\,\text{N}(5\,e^-) = 2 + 10 = 12$ valence electrons
$\text{H}-\ddot{\text{N}}=\ddot{\text{N}}-\text{H}$

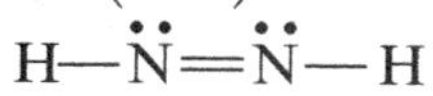

d. $1\,\text{Cl}(7\,e^-) + 2\,\text{C}(4\,e^-) + 2\,\text{O}(6\,e^-) + 3\,\text{H}(1\,e^-) = 7 + 8 + 12 + 3 = 30$ valence electrons

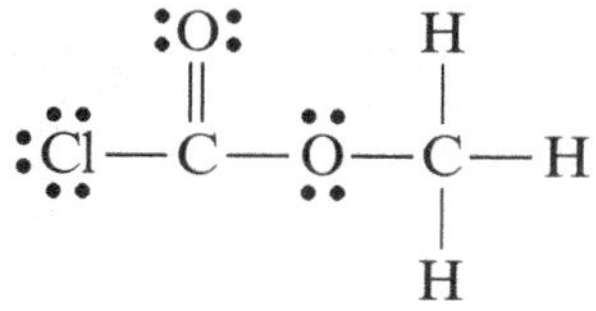

$$:\!\ddot{\underset{..}{\text{Cl}}}\!-\!\overset{\overset{:\text{O}:}{\|}}{\text{C}}\!-\!\ddot{\underset{..}{\text{O}}}\!-\!\underset{\underset{\text{H}}{|}}{\overset{\overset{\text{H}}{|}}{\text{C}}}\!-\!\text{H}$$

6.155 a. The central atom N has four electron groups with three bonded atoms and one lone pair of electrons, which gives NH_2Cl a trigonal pyramidal shape.

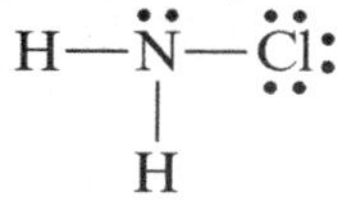

$\text{H}-\underset{\underset{\text{H}}{|}}{\ddot{\text{N}}}-\ddot{\underset{..}{\text{Cl}}}:$

b. The central atom Te has three electron groups with two bonded oxygen atoms and one lone pair of electrons, which gives the molecule TeO_2 a bent shape ($120°$).

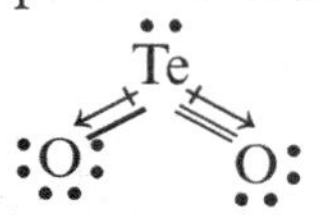

$:\!\ddot{\underset{..}{\text{O}}}\!\overset{\ddot{\text{Te}}}{\diagup\!\!\diagdown}\!\ddot{\underset{..}{\text{O}}}\!:$

Chapter 6

Selected Answers to Combining Ideas from Chapters 4 to 6

CI.7 **a.** Y has the higher electronegativity, since it is a nonmetal in Group 7A (17).
  b. X^{2+}, Y^-
  c. X has an electron arrangement of 2,8,2. Y has an electron arrangement of 2,8,7.
  d. X^{2+} has an electron arrangement of 2,8. Y^- has an electron arrangement of 2,8,8.
  e. X^{2+} has the same electron arrangement as the noble gas neon (Ne).
   Y^- has the same electron arrangement as the noble gas argon (Ar).
  f. $MgCl_2$, magnesium chloride

CI.9 **a.**

Isotope	Number of Protons	Number of Neutrons	Number of Electrons
$^{28}_{14}Si$	14	14	14
$^{29}_{14}Si$	14	15	14
$^{30}_{14}Si$	14	16	14

  b. Electron arrangement of Si: 2,8,4
  c. Since the atomic mass for Si is 28.02, the most abundant isotope of silicon is Si-28.
  d. $^{31}_{14}Si \longrightarrow {}^{31}_{15}P + {}^{0}_{-1}e$

  e. 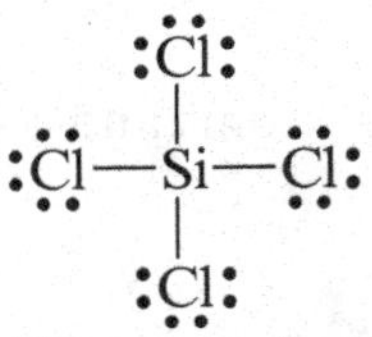

  $:\ddot{C}l-\underset{|}{\overset{|}{Si}}-\ddot{C}l:$ with $:\ddot{C}l:$ above and $:\ddot{C}l:$ below

  The central atom Si has four electron groups bonded to four chlorine atoms; $SiCl_4$ has a tetrahedral shape.

  f. Half of a radioactive sample decays with each half-life:

  16 μCi of $^{31}_{14}Si$ $\xrightarrow{\text{1 half-life}}$ 8.0 μCi of $^{31}_{14}Si$ $\xrightarrow{\text{2 half-lives}}$ 4.0 μCi of $^{31}_{14}Si$ $\xrightarrow{\text{3 half-lives}}$ 2.0 μCi of $^{31}_{14}Si$

  Therefore, 3 half-lives have passed.

  3 half-lives $\times \dfrac{2.6 \text{ h}}{1 \text{ half-life}} = 7.8$ h (2 SFs)

CI.11 **a.** $^{40}_{19}K \longrightarrow {}^{40}_{20}Ca + {}^{0}_{-1}e$, beta particle   and $^{40}_{19}K \longrightarrow {}^{40}_{18}Ar + {}^{0}_{+1}e$, positron

  b. 1.6 oz K $\times \dfrac{1 \text{ lb K}}{16 \text{ oz K}} \times \dfrac{454 \text{ g K}}{1 \text{ lb K}} \times \dfrac{0.012 \text{ g }^{40}_{19}K}{100 \text{ g K}} \times \dfrac{7.0 \text{ μCi}}{1 \text{ g }^{40}_{19}K} \times \dfrac{1 \text{ mCi}}{1000 \text{ μCi}}$

  $= 3.8 \times 10^{-5}$ mCi (2 SFs)

  1.6 oz K $\times \dfrac{1 \text{ lb K}}{16 \text{ oz K}} \times \dfrac{454 \text{ g K}}{1 \text{ lb K}} \times \dfrac{0.012 \text{ g }^{40}_{19}K}{100 \text{ g K}} \times \dfrac{7.0 \text{ μCi}}{1 \text{ g }^{40}_{19}K} \times \dfrac{1 \text{ Ci}}{10^6 \text{ μCi}} \times \dfrac{3.7 \times 10^{10} \text{ Bq}}{1 \text{ Ci}}$

  $= 1400$ Bq (2 SFs)

7 Chemical Quantities and Reactions

Since Natalie's diagnosis of mild emphysema due to secondhand cigarette smoke, she has been working with Angela, an exercise physiologist. Natalie's workout started with low-intensity exercises that use smaller muscles. Now that Natalie has increased the oxygen concentration in her blood, Angela has added some exercises that use larger muscles, which need more O_2. Today Natalie had a stress test and an EKG to assess and evaluate her progress. Her results show that her heart is functioning more efficiently and that she needs less oxygen. Her doctor told her that she has begun to reverse the progression of emphysema. How many moles of O_2 are in 48.0 g of O_2?

Credit: Javier Larrea/age fotostock

LOOKING AHEAD

7.1 The Mole
7.2 Molar Mass
7.3 Calculations Using Molar Mass
7.4 Equations for Chemical Reactions
7.5 Types of Chemical Reactions
7.6 Oxidation–Reduction Reactions
7.7 Mole Relationships in Chemical Equations
7.8 Mass Calculations for Chemical Reactions
7.9 Energy in Chemical Reactions

 The Health icon indicates a question that is related to health and medicine.

7.1 The Mole

Learning Goal: Use Avogadro's number to determine the number of particles in a given number of moles.

> **REVIEW**
> Writing Numbers in Scientific Notation (1.5)
> Counting Significant Figures (2.2)
> Writing Conversion Factors from Equalities (2.5)
> Using Conversion Factors (2.6)

- One mole of any compound contains Avogadro's number, 6.02×10^{23}, of particles (atoms, molecules, ions, formula units).
- Avogadro's number provides conversion factors between moles and the number of particles:

$$1 \text{ mole} = 6.02 \times 10^{23} \text{ particles}$$

$$\frac{6.02 \times 10^{23} \text{ particles}}{1 \text{ mole}} \quad \text{and} \quad \frac{1 \text{ mole}}{6.02 \times 10^{23} \text{ particles}}$$

Copyright © 2018 Pearson Education, Inc.

Chapter 7

	Solution Guide to Calculating the Number of Atoms or Molecules
STEP 1	State the given and needed quantities.
STEP 2	Write a plan to convert moles to atoms or molecules.
STEP 3	Use Avogadro's number to write conversion factors.
STEP 4	Set up the problem to calculate the number of particles.

♦ **Learning Exercise 7.1A**

CORE CHEMISTRY SKILL
Converting Particles to Moles

Use Avogadro's number to calculate each of the following:

a. number of Ca atoms in 3.00 moles of Ca

b. number of Zn atoms in 0.250 mole of Zn

c. number of SO_2 molecules in 0.118 mole of SO_2

d. number of moles of Ag in 4.88×10^{23} atoms of Ag

e. number of moles of NH_3 (ammonia) in 7.52×10^{23} molecules of NH_3

Answers
a. 1.81×10^{24} atoms of Ca b. 1.51×10^{23} atoms of Zn
c. 7.10×10^{22} molecules of SO_2 d. 0.811 mole of Ag
e. 1.25 moles of NH_3

Study Note

The subscripts in the formula of a compound indicate the number of moles of each element in 1 mole of that compound. For example, we can write the equalities and conversion factors for the moles of elements in the formula Mg_3N_2 as

1 mole of Mg_3N_2 = 3 moles of Mg atoms   1 mole of Mg_3N_2 = 2 moles of N atoms

$$\dfrac{3 \text{ moles Mg atoms}}{1 \text{ mole Mg}_3\text{N}_2} \text{ and } \dfrac{1 \text{ mole Mg}_3\text{N}_2}{3 \text{ moles Mg atoms}} \qquad \dfrac{2 \text{ moles N atoms}}{1 \text{ mole Mg}_3\text{N}_2} \text{ and } \dfrac{1 \text{ mole Mg}_3\text{N}_2}{2 \text{ moles N atoms}}$$

Chemical Quantities and Reactions

Solution Guide to Calculating the Moles of an Element in a Compound	
STEP 1	State the given and needed quantities.
STEP 2	Write a plan to convert moles of compound to moles of an element.
STEP 3	Write the equalities and conversion factors using subscripts.
STEP 4	Set up the problem to calculate the moles of an element.

♦ **Learning Exercise 7.1B**

Vitamin C (ascorbic acid) has the formula $C_6H_8O_6$.

a. How many moles of C are in 2.0 moles of vitamin C?

b. How many moles of H are in 5.0 moles of vitamin C?

c. How many moles of O are in 1.5 moles of vitamin C?

Answers   a. 12 moles of C   b. 40. moles of H   c. 9.0 moles of O

♦ **Learning Exercise 7.1C**

For the compound ibuprofen ($C_{13}H_{18}O_2$) used in Advil and Motrin, determine each of the following:

a. moles of C in 2.20 moles of ibuprofen

b. moles of O in 0.750 mole of ibuprofen

c. moles of ibuprofen that contain 15 moles of H

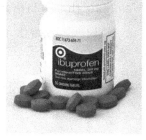

Ibuprofen is an anti-inflammatory.
Credit: Pearson Education / Pearson Science

Answers   a. 28.6 moles of C   b. 1.50 moles of O   c. 0.83 mole of ibuprofen

Chapter 7

7.2 Molar Mass

Learning Goal: Given the chemical formula of a substance, calculate its molar mass.

- The molar mass (g/mole) of an element is numerically equal to its atomic mass in grams.
- The molar mass (g/mole) of a compound is the sum of the mass (grams) for each element as indicated by the subscripts in the formula.

SAMPLE PROBLEM   Calculating Molar Mass

CORE CHEMISTRY SKILL
Calculating Molar Mass

What is the molar mass of silver nitrate, $AgNO_3$?

Solution Guide:

Analyze the Problem	Given	Need	Connect
	formula $AgNO_3$	molar mass of $AgNO_3$	periodic table

STEP 1 Obtain the molar mass of each element.

$$\frac{107.9 \text{ g Ag}}{1 \text{ mole Ag}} \quad \frac{14.01 \text{ g N}}{1 \text{ mole N}} \quad \frac{16.00 \text{ g O}}{1 \text{ mole O}}$$

STEP 2 Multiply each molar mass by the number of moles (subscript) in the formula.

$$1 \text{ mole Ag} \times \frac{107.9 \text{ g Ag}}{1 \text{ mole Ag}} = 107.9 \text{ g of Ag}$$

$$1 \text{ mole N} \times \frac{14.01 \text{ g N}}{1 \text{ mole N}} = 14.01 \text{ g of N}$$

$$3 \text{ moles O} \times \frac{16.00 \text{ g O}}{1 \text{ mole O}} = 48.00 \text{ g of O}$$

STEP 3 Calculate the molar mass by adding the masses of the elements.

Molar mass of $AgNO_3$ = 107.9 g + 14.01 g + 48.00 g = 169.9 g

♦ Learning Exercise 7.2

Calculate the molar mass for each of the following:

a. NaF, prevents dental caries

b. $FeC_4H_2O_4$, iron supplement

c. $C_{13}H_{18}O_2$, ibuprofen, anti-inflammatory

d. $KMnO_4$, used to treat fungus

e. CaF_2, source of fluoride in drinking water

f. $C_6H_8O_6$, vitamin C

$C_{12}H_{22}O_{11}$
1 mole of sucrose contains 342.3 g of sucrose (table sugar).
Credit: Pearson Education/Pearson Science

Answers   a. 41.99 g   b. 169.91 g   c. 206.27 g
             d. 158.04 g   e. 78.08 g   f. 176.12 g

7.3 Calculations Using Molar Mass

Learning Goal: Use molar mass to convert between grams and moles.

- The molar mass is used as a conversion factor to change a given quantity in moles to grams or grams to moles.
- The two conversion factors for the molar mass of NaOH (40.01 g/mole) are:

$$\frac{40.01 \text{ g NaOH}}{1 \text{ mole NaOH}} \quad \text{and} \quad \frac{1 \text{ mole NaOH}}{40.01 \text{ g NaOH}}$$

SAMPLE PROBLEM Converting Moles to Grams

What is the mass, in grams, of 0.254 mole of Na_2CO_3?

Solution Guide:

STEP 1 State the given and needed quantities.

Analyze the Problem	Given	Need	Connect
	0.254 mole of Na_2CO_3	grams of Na_2CO_3	molar mass

STEP 2 Write a plan to convert moles to grams.

moles of Na_2CO_3 $\xrightarrow{\text{Molar mass}}$ grams of Na_2CO_3

STEP 3 Determine the molar mass and write conversion factors.

1 mole of Na_2CO_3 = 105.99 g of Na_2CO_3

$$\frac{105.99 \text{ g } Na_2CO_3}{1 \text{ mole } Na_2CO_3} \quad \text{and} \quad \frac{1 \text{ mole } Na_2CO_3}{105.99 \text{ g } Na_2CO_3}$$

STEP 4 Set up the problem to convert moles to grams.

$$0.254 \text{ mole } Na_2CO_3 \times \frac{105.99 \text{ g } Na_2CO_3}{1 \text{ mole } Na_2CO_3} = 26.9 \text{ g of } Na_2CO_3$$

♦ **Learning Exercise 7.3A**

CORE CHEMISTRY SKILL
Using Molar Mass as a Conversion Factor

Calculate the mass, in grams, for each of the following:

a. 0.75 mole of S

b. 3.18 moles of K_2SO_4

c. 2.50 moles of NH_4Cl

d. 4.08 moles of $FeCl_3$

e. 2.28 moles of PCl_3

f. 0.815 mole of $Mg(NO_3)_2$

Answers   a. 24 g   b. 554 g   c. 134 g   d. 662 g   e. 313 g   f. 121 g

Chapter 7

♦ **Learning Exercise 7.3B**

Calculate the number of moles in each of the following quantities:

a. 108 g of C_5H_{12}

b. 6.12 g of CO_2

c. 38.7 g of $CaBr_2$

d. 236 g of Cl_2

e. 128 g of $Mg(OH)_2$

f. 172 g of Al_2O_3

g. The methane used to heat a home has a formula of CH_4. If 725 g of methane is used in one day, how many moles of methane are used during this time?

h. A vitamin tablet contains 18 mg of iron. If there are 100 tablets in a bottle, how many moles of iron are contained in the vitamins in the bottle?

Answers   a. 1.50 moles   b. 0.139 mole   c. 0.194 mole   d. 3.33 moles
            e. 2.19 moles   f. 1.69 moles   g. 45.2 moles   h. 0.032 mole

♦ **Learning Exercise 7.3C**

Octane, C_8H_{18}, is a component of gasoline.

a. How many moles of octane are in 34.7 g of octane?

b. How many grams of octane are in 0.737 mole of octane?

c. How many grams of C are in 135 g of octane?

d. How many grams of H are in 0.654 g of octane?

Answers   a. 0.304 mole   b. 84.2 g   c. 114 g   d. 0.104 g

7.4 Equations for Chemical Reactions

Learning Goal: Write a balanced chemical equation from the formulas of the reactants and products for a reaction; determine the number of atoms in the reactants and products.

REVIEW
Writing Ionic Formulas (6.2)
Naming Ionic Compounds (6.3)
Writing the Names and Formulas for Molecular Compounds (6.5)

- A chemical change occurs when a substance is converted into one or more new substances.
- When new substances form, a chemical reaction has taken place.

Chemical Quantities and Reactions

- Chemical change may be indicated by a formation of a gas (bubbles), change in color, formation of a solid, and/or heat produced or heat absorbed.
- A chemical equation shows the formulas of the reactants on the left side of the arrow and the formulas of the products on the right side.
- In a balanced equation, numbers called *coefficients*, which appear in front of the symbols or formulas, provide the same number of atoms for each kind of element on the reactant and product sides.
- Each formula in an equation is followed by an abbreviation, in parentheses, that gives the physical state of the substance: solid (s), liquid (l), or gas (g), and, if dissolved in water, an aqueous solution (aq).
- The Greek letter delta (Δ) over the arrow in an equation represents the application of heat to the reaction.

Bubbles (gas) form when $CaCO_3$ reacts with acid.
Credit: Sciencephotos/Alamy

◆ Learning Exercise 7.4A

Indicate the number of atoms of each element on the reactant side and on the product side for each of the following balanced equations:

a. $CaCO_3(s) \longrightarrow CaO(s) + CO_2(g)$

Element	Atoms on Reactant Side	Atoms on Product Side
Ca		
C		
O		

$2H_2(g) + O_2(g) \longrightarrow 2H_2O(g)$

Reactant atoms = Product atoms
When there are the same numbers of each type of atom in the reactants as in the products, the equation is balanced.

b. $2Na(s) + H_2O(l) \longrightarrow Na_2O(s) + H_2(g)$

Element	Atoms on Reactant Side	Atoms on Product Side
Na		
H		
O		

c. $C_5H_{12}(g) + 8O_2(g) \xrightarrow{\Delta} 5CO_2(g) + 6H_2O(g)$

Element	Atoms on Reactant Side	Atoms on Product Side
C		
H		
O		

d. $2AgNO_3(aq) + K_2S(aq) \longrightarrow Ag_2S(s) + 2KNO_3(aq)$

Element	Atoms on Reactant Side	Atoms on Product Side
Ag		
N		
O		
K		
S		

Chapter 7

e. $2Al(OH)_3(s) + 3H_2SO_4(aq) \longrightarrow 6H_2O(l) + Al_2(SO_4)_3(aq)$

Element	Atoms on Reactant Side	Atoms on Product Side
Al		
O		
H		
S		

Answers

a. $CaCO_3(s) \longrightarrow CaO(s) + CO_2(g)$

Element	Atoms on Reactant Side	Atoms on Product Side
Ca	1	1
C	1	1
O	3	3

b. $2Na(s) + H_2O(l) \longrightarrow Na_2O(s) + H_2(g)$

Element	Atoms on Reactant Side	Atoms on Product Side
Na	2	2
H	2	2
O	1	1

c. $C_5H_{12}(g) + 8O_2(g) \xrightarrow{\Delta} 5CO_2(g) + 6H_2O(g)$

Element	Atoms on Reactant Side	Atoms on Product Side
C	5	5
H	12	12
O	16	16

d. $2AgNO_3(aq) + K_2S(aq) \longrightarrow Ag_2S(s) + 2KNO_3(aq)$

Element	Atoms on Reactant Side	Atoms on Product Side
Ag	2	2
N	2	2
O	6	6
K	2	2
S	1	1

e. $2Al(OH)_3(s) + 3H_2SO_4(aq) \longrightarrow 6H_2O(l) + Al_2(SO_4)_3(aq)$

Element	Atoms on Reactant Side	Atoms on Product Side
Al	2	2
O	18	18
H	12	12
S	3	3

Chemical Quantities and Reactions

SAMPLE PROBLEM Balancing a Chemical Equation

CORE CHEMISTRY SKILL
Balancing a Chemical Equation

Balance the following equation:

$N_2(g) + H_2(g) \longrightarrow NH_3(g)$

Solution Guide:

Analyze the Problem	Given	Need	Connect
	reactants, products	balanced equation	equal numbers of atoms in reactants and products

STEP 1 Write an equation using the correct formulas for the reactants and products.

$N_2(g) + H_2(g) \longrightarrow NH_3(g)$

STEP 2 Count the atoms of each element in the reactants and products.

$N_2(g) + H_2(g) \longrightarrow NH_3(g)$
2N   2H   1N, 3H

STEP 3 Use coefficients to balance each element. Balance the N atoms by placing a coefficient of 2 in front of NH_3. (This increases the number of H atoms, too.) Recheck the number of N atoms and the number of H atoms.

$N_2(g) + H_2(g) \longrightarrow 2NH_3(g)$
2N   2H   2N, 6H

Balance the H atoms by placing a coefficient of 3 in front of H_2.

$N_2(g) + 3H_2(g) \longrightarrow 2NH_3(g)$

STEP 4 Check the final equation to confirm it is balanced.

$N_2(g) + 3H_2(g) \longrightarrow 2NH_3(g)$
2N   6H   2N, 6H   Balanced

♦ Learning Exercise 7.4B

Balance each of the following equations by placing coefficients in front of the formulas as needed:

a. ____ $MgO(s) \longrightarrow$ ____ $Mg(s) +$ ____ $O_2(g)$

b. ____ $Zn(s) +$ ____ $HCl(aq) \longrightarrow$ ____ $H_2(g) +$ ____ $ZnCl_2(aq)$

c. ____ $Al(s) +$ ____ $CuSO_4(aq) \longrightarrow$ ____ $Cu(s) +$ ____ $Al_2(SO_4)_3(aq)$

d. ____ $Al_2S_3(s) +$ ____ $H_2O(l) \longrightarrow$ ____ $Al(OH)_3(s) +$ ____ $H_2S(g)$

e. ____ $BaCl_2(aq) +$ ____ $Na_2SO_4(aq) \longrightarrow$ ____ $BaSO_4(s) +$ ____ $NaCl(aq)$

f. ____ $CO(g) +$ ____ $Fe_2O_3(s) \longrightarrow$ ____ $Fe(s) +$ ____ $CO_2(g)$

g. ____ $K(s) +$ ____ $H_2O(l) \longrightarrow$ ____ $H_2(g) +$ ____ $K_2O(aq)$

h. ____ $Fe(OH)_3(s) \longrightarrow$ ____ $H_2O(l) +$ ____ $Fe_2O_3(s)$

Chapter 7

    i. Galactose, $C_6H_{12}O_6(aq)$, is found in milk. In the body, galactose reacts with O_2 gas to form CO_2 gas and liquid water. Write a balanced chemical equation for the reaction.

    j. Maltose, $C_{12}H_{22}O_{11}(aq)$, is found in corn syrup. In the body, maltose reacts with O_2 gas to form CO_2 gas and liquid water. Write a balanced chemical equation for the reaction.

Answers
- a. $2MgO(s) \longrightarrow 2Mg(s) + O_2(g)$
- b. $Zn(s) + 2HCl(aq) \longrightarrow H_2(g) + ZnCl_2(aq)$
- c. $2Al(s) + 3CuSO_4(aq) \longrightarrow 3Cu(s) + Al_2(SO_4)_3(aq)$
- d. $Al_2S_3(s) + 6H_2O(l) \longrightarrow 2Al(OH)_3(s) + 3H_2S(g)$
- e. $BaCl_2(aq) + Na_2SO_4(aq) \longrightarrow BaSO_4(s) + 2NaCl(aq)$
- f. $3CO(g) + Fe_2O_3(s) \longrightarrow 2Fe(s) + 3CO_2(g)$
- g. $2K(s) + H_2O(l) \longrightarrow H_2(g) + K_2O(aq)$
- h. $2Fe(OH)_3(s) \longrightarrow 3H_2O(l) + Fe_2O_3(s)$
- i. $C_6H_{12}O_6(aq) + 6O_2(g) \longrightarrow 6CO_2(g) + 6H_2O(l)$
- j. $C_{12}H_{22}O_{11}(aq) + 12O_2(g) \longrightarrow 12CO_2(g) + 11H_2O(l)$

7.5 Types of Chemical Reactions

Learning Goal: Identify a chemical reaction as a combination, decomposition, single replacement, double replacement, or combustion.

- Reactions are classified as combination, decomposition, single replacement, double replacement, or combustion.
- In a *combination* reaction, two or more reactants form one product:

    $A + B \longrightarrow AB$

- In a *decomposition* reaction, a reactant splits into two or more simpler products:

    $AB \longrightarrow A + B$

- In a *single replacement* reaction, an uncombined element takes the place of an element in a compound:

    $A + BC \longrightarrow AC + B$

- In a *double replacement* reaction, the positive ions in the reactants switch places:

    $AB + CD \longrightarrow AD + CB$

- In a *combustion* reaction, a compound of carbon and hydrogen reacts with oxygen to form CO_2, H_2O, and energy.

    $C_XH_Y(g) + ZO_2(g) \xrightarrow{\Delta} XCO_2(g) + \frac{Y}{2}H_2O(g) + \text{energy}$

♦ Learning Exercise 7.5A

Match each of the following reactions with the type of reaction:

    a. combination   b. decomposition
    c. single replacement   d. double replacement   e. combustion

CORE CHEMISTRY SKILL
Classifying Types of Chemical Reactions

1. ____ $N_2(g) + 3H_2(g) \longrightarrow 2NH_3(g)$

2. ____ $CaCl_2(aq) + K_2CO_3(aq) \longrightarrow CaCO_3(s) + 2KCl(aq)$

3. ____ $2K_2O_2(aq) \longrightarrow 2K_2O(l) + O_2(g)$

4. ____ $CuO(s) + H_2(g) \longrightarrow Cu(s) + H_2O(l)$

5. ____ $N_2(g) + O_2(g) \longrightarrow 2NO(g)$

6. ____ $C_2H_4(g) + 3O_2(g) \xrightarrow{\Delta} 2CO_2(g) + 2H_2O(g) + \text{energy}$

7. ____ $PbCO_3(s) \longrightarrow PbO(s) + CO_2(g)$

8. ____ $2Al(s) + Fe_2O_3(s) \longrightarrow 2Fe(s) + Al_2O_3(s)$

In a combustion reaction, a candle burns using the oxygen in the air.
Credit: Sergiy Zavgorodny / Shutterstock

Answers   1. a   2. d   3. b   4. c
            5. a   6. e   7. b   8. c

♦ Learning Exercise 7.5B

One way to remove tarnish from silver is to place the silver object on a piece of aluminum foil and add boiling water and some baking soda. The unbalanced chemical equation is the following:

$$Al(s) + Ag_2S(s) \longrightarrow Ag(s) + Al_2S_3(s)$$

a. Write the balanced chemical equation for the reaction.

b. What type of reaction takes place?

Answers   **a.** $2Al(s) + 3Ag_2S(s) \longrightarrow 6Ag(s) + Al_2S_3(s)$
             b. single replacement

♦ Learning Exercise 7.5C

Octane, C_8H_{18}, a liquid compound in gasoline, burns in oxygen to produce the gases carbon dioxide and water, and energy.

a. Write the balanced chemical equation for the reaction.

b. What type of reaction takes place?

Answers   **a.** $2C_8H_{18}(l) + 25O_2(g) \xrightarrow{\Delta} 16CO_2(g) + 18H_2O(g) + \text{energy}$
             b. combustion

Chapter 7

7.6 Oxidation–Reduction Reactions

Learning Goal: Define the terms oxidation and reduction; identify the reactants oxidized and reduced.

- In an oxidation–reduction reaction, there is a loss and gain of electrons. In an oxidation, electrons are lost. In a reduction, electrons are gained.
- An oxidation must always be accompanied by a reduction. The number of electrons lost in the oxidation reaction is equal to the number of electrons gained in the reduction reaction.
- In biological systems, the term *oxidation* describes the gain of oxygen or the loss of hydrogen. The term *reduction* is used to describe a loss of oxygen or a gain of hydrogen.

Key Terms for Sections 7.1 to 7.6

Match each of the following key terms with the correct description:

    a. combustion reaction   **b.** molar mass   **c.** combination reaction
    d. decomposition reaction   **e.** single replacement reaction   **f.** oxidation–reduction reaction
    g. Avogadro's number

1. ____ a reaction in which a single reactant splits into two or more simpler substances
2. ____ a reaction in which one element replaces a different element in a compound
3. ____ a reaction in which a carbon-containing compound burns in oxygen from the air to produce carbon dioxide, water, and energy
4. ____ a reaction in which reactants combine to form a single product
5. ____ the number of items in a mole, equal to 6.02×10^{23}
6. ____ a reaction in which an equal number of electrons is lost from one reactant and gained by another reactant
7. ____ the mass, in grams, of an element that is numerically equal to atomic masses of the elements in a compound

Answers   **1.** d   **2.** e   **3.** a   **4.** c   **5.** g   **6.** f   **7.** b

♦ Learning Exercise 7.6

For each of the following reactions, indicate whether the underlined element is *oxidized* or *reduced*:

CORE CHEMISTRY SKILL
Identifying Oxidized and Reduced Substances

a. $\underline{Fe}^{3+}(aq) + e^- \longrightarrow Fe^{2+}(aq)$ Fe^{3+} is _____

b. $2\underline{Cl}^-(aq) \longrightarrow Cl_2(g) + 2\ e^-$ Cl^- is _____

c. $\underline{Ca}O(s) + H_2(g) \longrightarrow Ca(s) + H_2O(l)$ Ca^{2+} is _____

d. $4\underline{Al}(s) + 3O_2(g) \longrightarrow 2Al_2O_3(s)$ Al is _____

e. $\underline{Cu}Cl_2(aq) + Zn(s) \longrightarrow ZnCl_2(aq) + Cu(s)$ Cu^{2+} is _____

f. $2H\underline{Br}(aq) + Cl_2(g) \longrightarrow 2HCl(aq) + Br_2(g)$ Br⁻ is _____

g. $2\underline{Na}(s) + Cl_2(g) \longrightarrow 2NaCl(s)$ Na is _____

Answers   **a.** reduced   **b.** oxidized   **c.** reduced   **d.** oxidized
             e. reduced   **f.** oxidized   **g.** oxidized

7.7 Mole Relationships in Chemical Equations

Learning Goal: Use a mole–mole factor from a balanced chemical equation to calculate the number of moles of another substance in the reaction.

- The Law of Conservation of Mass states that there is no change in the total mass of the substances reacting in a chemical reaction.
- The coefficients in a balanced chemical equation indicate the moles of reactants and products in a reaction.
- Using the coefficients, mole–mole conversion factors can be written to relate any two substances in an equation.
- For the reaction of oxygen forming ozone, $3O_2(g) \longrightarrow 2O_3(g)$, the mole–mole factors are the following:

$$\frac{3 \text{ moles } O_2}{2 \text{ moles } O_3} \quad \text{and} \quad \frac{2 \text{ moles } O_3}{3 \text{ moles } O_2}$$

♦ **Learning Exercise 7.7A**

Write all of the possible mole–mole factors for the following equation:

$$N_2(g) + O_2(g) \longrightarrow 2NO(g)$$

Answers

For N_2 and O_2:

$$\frac{1 \text{ mole } N_2}{1 \text{ mole } O_2} \quad \text{and} \quad \frac{1 \text{ mole } O_2}{1 \text{ mole } N_2}$$

For N_2 and NO:

$$\frac{2 \text{ moles NO}}{1 \text{ mole } N_2} \quad \text{and} \quad \frac{1 \text{ mole } N_2}{2 \text{ moles NO}}$$

For O_2 and NO:

$$\frac{2 \text{ moles NO}}{1 \text{ mole } O_2} \quad \text{and} \quad \frac{1 \text{ mole } O_2}{2 \text{ moles NO}}$$

SAMPLE PROBLEM Calculating the Moles of Reactants and Products

For the equation $N_2(g) + O_2(g) \longrightarrow 2NO(g)$, calculate the number of moles of NO produced from 3.0 moles of N_2.

Solution Guide:

STEP 1 State the given and needed quantities (moles).

Analyze the Problem	Given	Need	Connect
	3.0 moles of N_2	moles of NO	mole–mole factor

Chapter 7

STEP 2 Write a plan to convert the given to the needed quantity (moles).

moles of N_2 $\xrightarrow{\text{Mole–mole factor}}$ moles of NO

STEP 3 Use coefficients to write mole–mole factors.

1 mole of N_2 = 2 moles of NO

$$\frac{2 \text{ moles NO}}{1 \text{ mole } N_2} \text{ and } \frac{1 \text{ mole } N_2}{2 \text{ moles NO}}$$

STEP 4 Set up the problem to give the needed quantity (moles).

$$3.0 \text{ moles } N_2 \times \frac{2 \text{ moles NO}}{1 \text{ mole } N_2} = 6.0 \text{ moles of NO}$$

♦ **Learning Exercise 7.7B**

Use the balanced chemical equation to answer the following questions:

CORE CHEMISTRY SKILL
Using Mole–Mole Factors

$$C_3H_8(g) + 5O_2(g) \xrightarrow{\Delta} 3CO_2(g) + 4H_2O(g)$$

a. How many moles of O_2 are needed to react with 2.00 moles of C_3H_8?

b. How many moles of CO_2 are produced when 4.00 moles of O_2 reacts?

c. How many moles of C_3H_8 react with 3.00 moles of O_2?

d. How many moles of H_2O are produced from 0.500 mole of C_3H_8?

Answers   a. 10.0 moles of O_2 b. 2.40 moles of CO_2 c. 0.600 mole of C_3H_8
d. 2.00 moles of H_2O

7.8 Mass Calculations for Chemical Reactions

REVIEW
Using Significant Figures in Calculations (2.3)

Learning Goal: Given the mass in grams of a substance in a reaction, calculate the mass in grams of another substance in the reaction.

- The grams of a substance in an equation are converted to grams of another substance using their molar masses and mole–mole factors.

Substance A   Substance B

grams of A $\xrightarrow{\text{Molar mass A}}$ moles of A $\xrightarrow{\text{Mole–mole factor B/A}}$ moles of B $\xrightarrow{\text{Molar mass B}}$ grams of B

Chemical Quantities and Reactions

SAMPLE PROBLEM   Mass of Product

How many grams of water are produced when 45.0 g of O_2 reacts with NH_3?

Solution Guide:

STEP 1 State the given and needed quantities (grams).

Analyze the Problem	Given	Need	Connect
	45.0 g of O_2	grams of H_2O	molar masses, mole–mole factor
Equation	$4NH_3(g) + 3O_2(g) \longrightarrow 2N_2(g) + 6H_2O(g)$		

STEP 2 Write a plan to convert the given to the needed quantity (grams).

grams of O_2 $\xrightarrow{\text{Molar mass } O_2}$ moles of O_2 $\xrightarrow{\text{Mole–mole factor}}$ moles of H_2O $\xrightarrow{\text{Molar mass } H_2O}$ grams of H_2O

STEP 3 Use coefficients to write mole–mole factors; write molar masses.

1 mole of O_2 = 32.00 g of O_2 1 mole of H_2O = 18.02 g of H_2O

$$\frac{32.00 \text{ g } O_2}{1 \text{ mole } O_2} \text{ and } \frac{1 \text{ mole } O_2}{32.00 \text{ g } O_2} \qquad \frac{18.02 \text{ g } H_2O}{1 \text{ mole } H_2O} \text{ and } \frac{1 \text{ mole } H_2O}{18.02 \text{ g } H_2O}$$

3 moles of O_2 = 6 moles of H_2O

$$\frac{3 \text{ moles } O_2}{6 \text{ moles } H_2O} \text{ and } \frac{6 \text{ moles } H_2O}{3 \text{ moles } O_2}$$

STEP 4 Set up the problem to give the needed quantity (grams).

$$45.0 \text{ g } O_2 \times \frac{1 \text{ mole } O_2}{32.00 \text{ g } O_2} \times \frac{6 \text{ moles } H_2O}{3 \text{ moles } O_2} \times \frac{18.02 \text{ g } H_2O}{1 \text{ mole } H_2O} = 50.7 \text{ g of } H_2O$$

♦ Learning Exercise 7.8

CORE CHEMISTRY SKILL
Converting Grams to Grams

Use the balanced chemical equation to answer the questions below:

$$2C_2H_6(g) + 7O_2(g) \xrightarrow{\Delta} 4CO_2(g) + 6H_2O(g)$$

a. How many grams of O_2 are needed to react with 120. g of C_2H_6?

b. How many grams of C_2H_6 are needed to react with 115 g of O_2?

c. How many grams of C_2H_6 react if 2.00 g of CO_2 is produced?

Chapter 7

d. How many grams of CO_2 are produced when 60.0 g of C_2H_6 reacts with sufficient oxygen?

e. How many grams of H_2O are produced when 82.5 g of O_2 reacts with sufficient C_2H_6?

Answers **a.** 447 g of O_2 **b.** 30.9 g of C_2H_6 **c.** 0.683 g of C_2H_6
d. 176 g of CO_2 **e.** 39.8 g of H_2O

7.9 Energy in Chemical Reactions

Learning Goal: Describe exothermic and endothermic reactions and factors that affect the rate of a reaction.

REVIEW
Using Energy Units (3.4)

- The heat of reaction is the energy released or absorbed when the reaction occurs.
- The heat of reaction is the energy difference between the reactants and the products.
- In exothermic reactions, the energy of the products is lower than that of the reactants, which means that energy is released. The heat of reaction for an exothermic reaction is written as a product.

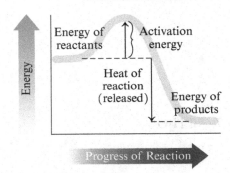

- In endothermic reactions, the energy of the products is higher than that of the reactants, which means that energy is absorbed. The heat of reaction for an endothermic reaction is written as a reactant.

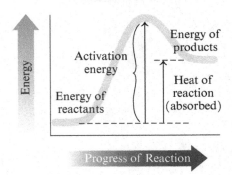

- The rate of a reaction is the speed at which the reactants are converted to products.
- Increasing the concentrations of reactants, raising the temperature, or adding a catalyst can increase the rate of a reaction.

Chemical Quantities and Reactions

Key Terms for Sections 7.7 to 7.9

Match each of the following key terms with the correct description:

　　a. catalyst　　　　　　　　b. exothermic reaction　　　　c. endothermic reaction
　　d. activation energy　　　e. mole–mole factor　　　　　f. conservation of mass

1. _____ a reaction in which the energy of the reactants is less than that of the products
2. _____ the amount of energy required to break the bonds between atoms of the reactants
3. _____ a reaction in which the energy of the reactants is greater than that of the products
4. _____ the law that states that there is no change in the mass of substances reacting
5. _____ a factor that relates the number of moles of two compounds derived from the coefficients in a balanced equation
6. _____ a substance that increases the rate of a reaction by decreasing the activation energy

Answers　　1. c　　2. d　　3. b　　4. f　　5. e　　6. a

♦ Learning Exercise 7.9A

Indicate whether each of the following is an endothermic or exothermic reaction:

a. $2H_2(g) + O_2(g) \longrightarrow 2H_2O(g) + 582 \text{ kJ}$　　　　_____

b. $C_2H_4(g) + 176 \text{ kJ} \longrightarrow H_2(g) + C_2H_2(g)$　　　_____

c. $2C(s) + O_2(g) \longrightarrow 2CO(g) + 220 \text{ kJ}$　　　　_____

d. $C_6H_{12}O_6(s) + 6O_2(g) \longrightarrow 6CO_2(g) + 6H_2O(l) + 2800 \text{ kJ}$　_____

e. $C_2H_6O(l) + 44 \text{ kJ} \longrightarrow C_2H_4(g) + H_2O(g)$　　　_____

Answers　　a. exothermic　　b. endothermic　　c. exothermic
　　　　　　　d. exothermic　　e. endothermic

♦ Learning Exercise 7.9B

Indicate how each of the following will affect (increase or decrease) the rate of the reaction for the combustion of C_3H_8:

$$C_3H_8(g) + 5O_2(g) \xrightarrow{\Delta} 3CO_2(g) + 4H_2O(g)$$

a. adding some C_3H_8　　　_____
b. adding a catalyst　　　　_____
c. removing some O_2　　　_____
d. decreasing the temperature　_____

Answers　　a. increase　　b. increase　　c. decrease　　d. decrease

Chapter 7

Checklist for Chapter 7

You are ready to take the Practice Test for Chapter 7. Be sure you have accomplished the following learning goals for this chapter. If not, review the Section listed at the end of the goal. Then apply your new skills and understanding to the Practice Test.

After studying Chapter 7, I can successfully:

____ Calculate the number of particles in a given number of moles of a substance. (7.1)

____ Calculate the molar mass given the chemical formula of a substance. (7.2)

____ Convert the grams of a substance to moles, and moles to grams. (7.3)

____ Write a balanced equation for a chemical reaction from the formulas of the reactants and products. (7.4)

____ Identify a reaction as a combination, decomposition, single replacement, double replacement, or combustion. (7.5)

____ Identify the substance oxidized and the substance reduced in an oxidation–reduction reaction. (7.6)

____ Use mole–mole factors from the mole relationships in an equation to calculate the moles of another substance in an equation for a chemical reaction. (7.7)

____ Calculate the mass of a substance in an equation using mole factors and molar masses. (7.8)

____ Identify exothermic and endothermic reactions from the heat in a chemical reaction. (7.9)

____ Determine how a change in conditions may affect the rate of a reaction. (7.9)

Practice Test for Chapter 7

The chapter Sections to review are shown in parentheses at the end of each question.

For questions 1 through 4, calculate the moles of each of the elements in the formula: (7.1)

Tagamet has the formula $C_{10}H_{16}N_6S$. Tagamet is used to inhibit the production of acid in the stomach. Select the correct answer for the number of moles of each of the following in 0.25 mole of Tagamet:

　　A. 0.25 mole　　B. 0.50 mole　　C. 1.5 moles　　D. 2.5 moles　　E. 4.0 moles

1. moles of C　　2. moles of H　　3. moles of N　　4. moles of S

5. The number of Na atoms in 0.0500 mole of Na is (7.1)

　　A. 0.500 atom　　B. 6.02×10^{23} atoms　　C. 3.01×10^{23} atoms
　　D. 3.01×10^{22} atoms　　E. 3.01×10^{25} atoms

6. What is the molar mass of K_2SO_4? (7.2)
　　A. 71.08 g　　B. 87.17 g　　C. 126.27 g　　D. 135.27 g　　E. 174.27 g

7. What is the molar mass of $NaNO_3$? (7.2)
　　A. 34.0 g　　B. 37.0 g　　C. 53.0 g　　D. 75.0 g　　E. 85.00 g

8. The number of grams in 0.600 mole of Cl_2 is (7.3)
　　A. 71.0 g　　B. 21.3 g　　C. 42.5 g　　D. 84.5 g　　E. 4.30 g

9. How many grams are in 4.00 moles of NH_3? (7.3)
　　A. 4.00 g　　B. 17.0 g　　C. 34.0 g　　D. 68.1 g　　E. 0.240 g

10. The number of moles of sodium hydroxide in 8.0 g of NaOH is (7.3)
　　A. 0.10 mole　　B. 0.20 mole　　C. 0.40 mole　　D. 2.0 moles　　E. 4.0 moles

Chemical Quantities and Reactions

11. The number of moles of water in 3.6 g of H_2O is (7.3)
    A. 0.20 mole   B. 1.0 mole   C. 2.0 moles   D. 3.0 moles   E. 4.0 moles

12. The number of moles of aluminum in 54 g of Al is (7.3)
    A. 0.20 mole   B. 1.0 mole   C. 2.0 moles   D. 3.0 moles   E. 4.0 moles

13. The number of moles of carbon dioxide in 2.2 g of CO_2 is (7.3)
    A. 2.0 moles   B. 1.0 mole   C. 0.20 mole   D. 0.050 mole   E. 0.010 mole

14. The number of moles of hydrogen in 0.20 g of H_2 is (7.3)
    A. 0.10 mole   B. 0.20 mole   C. 0.40 mole   D. 0.040 mole   E. 0.010 mole

For each of the unbalanced chemical equations in questions 15 through 19, balance the equation and indicate the correct coefficient for the component in the equation written in boldface type. (7.4)

    A. 1   B. 2   C. 3   D. 4   E. 5

15. ____ $Sn(s) + \mathbf{Cl_2}(g) \longrightarrow SnCl_4(s)$

16. ____ $Al(s) + H_2O(l) \longrightarrow \mathbf{H_2}(g) + \mathbf{Al_2O_3}(s)$

17. ____ $C_3H_4(g) + \mathbf{O_2}(g) \xrightarrow{\Delta} \mathbf{CO_2}(g) + \mathbf{H_2O}(g)$

18. ____ $\mathbf{NH_3}(g) + O_2(g) \longrightarrow N_2(g) + H_2O(g)$

19. ____ $N_2O(g) \longrightarrow N_2(g) + \mathbf{O_2}(g)$

For questions 20 through 25, classify each reaction as one of the following: (7.5)

    A. combination   B. decomposition   C. single replacement
    D. double replacement   E. combustion

20. ____ $S(s) + O_2(g) \longrightarrow SO_2(g)$

21. ____ $Fe_2O_3(s) + 3C(s) \longrightarrow 2Fe(s) + 3CO(g)$

22. ____ $CaCO_3(s) \longrightarrow CaO(s) + CO_2(g)$

23. ____ $Mg(s) + 2AgNO_3(aq) \longrightarrow Mg(NO_3)_2(aq) + 2Ag(s)$

24. ____ $C_6H_{12}(g) + 9O_2(g) \xrightarrow{\Delta} 6CO_2(g) + 6H_2O(g)$

25. ____ $Na_2S(aq) + Pb(NO_3)_2(aq) \longrightarrow PbS(s) + 2NaNO_3(aq)$

For questions 26 through 30, identify each reaction as an oxidation (O) *or a reduction* (R): (7.6)

26. $Ca(s) \longrightarrow Ca^{2+}(aq) + 2\,e^-$ _____   27. $Fe^{3+}(aq) + 3\,e^- \longrightarrow Fe(s)$ _____

28. $Al^{3+}(aq) + 3\,e^- \longrightarrow Al(s)$ _____   29. $Br_2(l) + 2\,e^- \longrightarrow 2Br^-(aq)$ _____

30. $Sn^{2+}(aq) \longrightarrow Sn^{4+}(aq) + 2\,e^-$ _____

For questions 31 through 35, use the reaction: (7.7, 7.8)

$C_2H_6O(g) + 3O_2(g) \xrightarrow{\Delta} 2CO_2(g) + 3H_2O(g)$
Ethanol

31. How many moles of oxygen are needed to react with 0.360 mole of ethanol?
    A. 0.120 mole   B. 1.08 moles   C. 1.20 moles   D. 3.00 moles   E. 6.00 moles

32. How many moles of water are produced when 12 moles of oxygen reacts?
    A. 3.0 moles   B. 6.0 moles   C. 8.0 moles   D. 12 moles   E. 36 moles

33. How many grams of carbon dioxide are produced when 92.0 g of ethanol reacts?
    A. 22.0 g   B. 44.0 g   C. 88.0 g   D. 92.0 g   E. 176 g

Chapter 7

34. How many grams of oxygen would be needed to produce 100. g of CO_2?
   A. 3.41 g   **B.** 48.5 g   **C.** 72.7 g   **D.** 100. g   **E.** 109 g

35. How many grams of water will be produced if 23 g of ethanol reacts?
   A. 54 g   **B.** 27 g   **C.** 18 g   **D.** 9.0 g   **E.** 6.0 g

For questions 36 through 38, indicate if the reactions are exothermic (**EX**) *or endothermic* (**EN**): (7.9)

36. $N_2(g) + 3H_2(g) \longrightarrow 2NH_3(g) + 84 \text{ kJ}$ _____

37. $2HCl(g) + 184 \text{ kJ} \longrightarrow H_2(g) + Cl_2(g)$ _____

38. $C_3H_8(g) + 5O_2(g) \xrightarrow{\Delta} 3CO_2(g) + 4H_2O(g) + 531 \text{ kcal}$ _____

For questions 39 and 40, indicate the effect of each of the following changes on the reaction: (7.9)

$C_2H_6O(g) + 3O_2(g) \xrightarrow{\Delta} 2CO_2(g) + 3H_2O(g)$

increase (**I**) *the rate of the reaction*   *decrease* (**D**) *the rate of the reaction*

39. adding O_2 ____

40. removing some $C_2H_6O(g)$ ____

Answers to the Practice Test

1. D	2. E	3. C	4. A	5. D
6. E	7. E	8. C	9. D	10. B
11. A	12. C	13. D	14. A	15. B
16. C	17. D	18. D	19. A	20. A
21. C	22. B	23. C	24. E	25. D
26. O	27. R	28. R	29. R	30. O
31. B	32. D	33. E	34. E	35. B
36. EX	37. EN	38. EX	39. I	40. D

Selected Answers and Solutions to Text Problems

7.1 One mole contains 6.02×10^{23} atoms of an element, molecules of a molecular substance, or formula units of an ionic substance.

7.3
a. $0.500 \text{ mole C} \times \dfrac{6.02 \times 10^{23} \text{ atoms C}}{1 \text{ mole C}} = 3.01 \times 10^{23}$ atoms of C (3 SFs)

b. $1.28 \text{ moles SO}_2 \times \dfrac{6.02 \times 10^{23} \text{ molecules SO}_2}{1 \text{ mole SO}_2} = 7.71 \times 10^{23}$ molecules of SO_2 (3 SFs)

c. $5.22 \times 10^{22} \text{ atoms Fe} \times \dfrac{1 \text{ mole Fe}}{6.02 \times 10^{23} \text{ atoms Fe}} = 0.0867$ mole of Fe (3 SFs)

d. $8.50 \times 10^{24} \text{ atoms C}_2\text{H}_6\text{O} \times \dfrac{1 \text{ mole C}_2\text{H}_6\text{O}}{6.02 \times 10^{23} \text{ atoms C}_2\text{H}_6\text{O}} = 14.1$ moles of C_2H_6O (3 SFs)

7.5 1 mole of H_3PO_4 contains 3 moles of H atoms, 1 mole of P atoms, and 4 moles of O atoms.

a. $2.00 \text{ moles H}_3\text{PO}_4 \times \dfrac{3 \text{ moles H}}{1 \text{ mole H}_3\text{PO}_4} = 6.00$ moles of H (3 SFs)

b. $2.00 \text{ moles H}_3\text{PO}_4 \times \dfrac{4 \text{ moles O}}{1 \text{ mole H}_3\text{PO}_4} = 8.00$ moles of O (3 SFs)

c. $2.00 \text{ moles H}_3\text{PO}_4 \times \dfrac{1 \text{ mole P}}{1 \text{ mole H}_3\text{PO}_4} \times \dfrac{6.02 \times 10^{23} \text{ atoms P}}{1 \text{ mole P}} = 1.20 \times 10^{24}$ atoms of P (3 SFs)

d. $2.00 \text{ moles H}_3\text{PO}_4 \times \dfrac{4 \text{ moles O}}{1 \text{ mole H}_3\text{PO}_4} \times \dfrac{6.02 \times 10^{23} \text{ atoms O}}{1 \text{ mole O}} = 4.82 \times 10^{24}$ atoms of O (3 SFs)

7.7 The subscripts indicate the moles of each element in one mole of that compound.

a. $1.5 \text{ moles quinine} \times \dfrac{24 \text{ moles H}}{1 \text{ mole quinine}} = 36$ moles of H (2 SFs)

b. $5.0 \text{ moles quinine} \times \dfrac{20 \text{ moles C}}{1 \text{ mole quinine}} = 1.0 \times 10^2$ moles of C (2 SFs)

c. $0.020 \text{ mole quinine} \times \dfrac{2 \text{ moles N}}{1 \text{ mole quinine}} = 0.040$ mole of N (2 SFs)

7.9
a. $2.30 \text{ moles naproxen} \times \dfrac{14 \text{ moles C}}{1 \text{ mole naproxen}} = 32.2$ moles of C (3 SFs)

b. $0.444 \text{ mole naproxen} \times \dfrac{14 \text{ moles H}}{1 \text{ mole naproxen}} = 6.22$ moles of H (3 SFs)

c. $0.0765 \text{ mole naproxen} \times \dfrac{3 \text{ moles O}}{1 \text{ mole naproxen}} = 0.230$ mole of O (3 SFs)

7.11
a. $2 \text{ moles Cl} \times \dfrac{35.45 \text{ g Cl}}{1 \text{ mole Cl}} = 70.90$ g of Cl

$\therefore$ Molar mass of Cl_2 = 70.90 g

b. $3 \text{ moles C} \times \dfrac{12.01 \text{ g C}}{1 \text{ mole C}} = 36.03$ g of C

$6 \text{ moles H} \times \dfrac{1.008 \text{ g H}}{1 \text{ mole H}} = 6.048$ g of H

$3 \text{ moles O} \times \dfrac{16.00 \text{ g O}}{1 \text{ mole O}} = 48.00$ g of O

Chapter 7

    3 moles of C   = 36.03 g of C
    6 moles of H   = 6.048 g of H
    3 moles of O   = 48.00 g of O
    ∴ Molar mass of $C_3H_6O_3$ = 90.08 g

c. $3 \text{ moles Mg} \times \dfrac{24.31 \text{ g Mg}}{1 \text{ mole Mg}} = 72.93$ g of Mg

    $2 \text{ moles P} \times \dfrac{30.97 \text{ g P}}{1 \text{ mole P}} = 61.94$ g of P

    $8 \text{ moles O} \times \dfrac{16.00 \text{ g O}}{1 \text{ mole O}} = 128.0$ g of O

    3 moles of Mg   = 72.93 g of Mg
    2 moles of P   = 61.94 g of P
    8 moles of O   = 128.0 g of O
    ∴ Molar mass of $Mg_3(PO_4)_2$ = 262.9 g

7.13 a. $1 \text{ mole Al} \times \dfrac{26.98 \text{ g Al}}{1 \text{ mole Al}} = 26.98$ g of Al

    $3 \text{ moles F} \times \dfrac{19.00 \text{ g F}}{1 \text{ mole F}} = 57.00$ g of F

    1 mole of Al   = 26.98 g of Al
    3 moles of F   = 57.00 g of F
    ∴ Molar mass of AlF_3 = 83.98 g

b. $2 \text{ moles C} \times \dfrac{12.01 \text{ g C}}{1 \text{ mole C}} = 24.02$ g of C

    $4 \text{ moles H} \times \dfrac{1.008 \text{ g H}}{1 \text{ mole H}} = 4.032$ g of H

    $2 \text{ moles Cl} \times \dfrac{35.45 \text{ g Cl}}{1 \text{ mole Cl}} = 70.90$ g of Cl

    2 moles of C   = 24.02 g of C
    4 moles of H   = 4.032 g of H
    2 moles of Cl   = 70.90 g of Cl
    ∴ Molar mass of $C_2H_4Cl_2$ = 98.95 g

c. $1 \text{ mole Sn} \times \dfrac{118.7 \text{ g Sn}}{1 \text{ mole Sn}} = 118.7$ g of Sn

    $2 \text{ moles F} \times \dfrac{19.00 \text{ g F}}{1 \text{ mole F}} = 38.00$ g of F

    1 mole of Sn   = 118.7 g of Sn
    2 moles of F   = 38.00 g of F
    ∴ Molar mass of SnF_2 = 156.7 g

7.15 a. $1 \text{ mole K} \times \dfrac{39.10 \text{ g K}}{1 \text{ mole K}} = 39.10$ g of K

    $1 \text{ mole Cl} \times \dfrac{35.45 \text{ g Cl}}{1 \text{ mole Cl}} = 35.45$ g of Cl

    1 mole of K   = 39.10 g of K
    1 mole of Cl   = 35.45 g of Cl
    ∴ Molar mass of KCl = 74.55 g

b. $6 \text{ moles C} \times \dfrac{12.01 \text{ g C}}{1 \text{ mole C}} = 72.06 \text{ g of C}$

$5 \text{ moles H} \times \dfrac{1.008 \text{ g H}}{1 \text{ mole H}} = 5.04 \text{ g of H}$

$1 \text{ mole N} \times \dfrac{14.01 \text{ g N}}{1 \text{ mole N}} = 14.01 \text{ g of N}$

$2 \text{ moles O} \times \dfrac{16.00 \text{ g O}}{1 \text{ mole O}} = 32.00 \text{ g of O}$

6 moles of C = 72.06 g of C
5 moles of H = 5.04 g of H
1 mole of N = 14.01 g of N
2 moles of O = 32.00 g of O
∴ Molar mass of $C_6H_5NO_2$ = 123.11 g

c. $19 \text{ moles C} \times \dfrac{12.01 \text{ g C}}{1 \text{ mole C}} = 228.2 \text{ g of C}$

$20 \text{ moles H} \times \dfrac{1.008 \text{ g H}}{1 \text{ mole H}} = 20.16 \text{ g of H}$

$1 \text{ mole F} \times \dfrac{19.00 \text{ g F}}{1 \text{ mole F}} = 19.00 \text{ g of F}$

$1 \text{ mole N} \times \dfrac{14.01 \text{ g N}}{1 \text{ mole N}} = 14.01 \text{ g of N}$

$3 \text{ moles O} \times \dfrac{16.00 \text{ g O}}{1 \text{ mole O}} = 48.00 \text{ g of O}$

19 moles of C = 228.2 g of C
20 moles of H = 20.16 g of H
1 mole of F = 19.00 g of F
1 mole of N = 14.01 g of N
3 moles of O = 48.00 g of O
∴ Molar mass of $C_{19}H_{20}FNO_3$ = 329.4 g

7.17 a. $2 \text{ moles Al} \times \dfrac{26.98 \text{ g Al}}{1 \text{ mole Al}} = 53.96 \text{ g of Al}$

$3 \text{ moles S} \times \dfrac{32.07 \text{ g S}}{1 \text{ mole S}} = 96.21 \text{ g of S}$

$12 \text{ moles O} \times \dfrac{16.00 \text{ g O}}{1 \text{ mole O}} = 192.0 \text{ g of O}$

2 moles of Al = 53.96 g of Al
3 moles of S = 96.21 g of S
12 moles of O = 192.0 g of O
∴ Molar mass of $Al_2(SO_4)_3$ = 342.2 g

b. $1 \text{ mole K} \times \dfrac{39.10 \text{ g K}}{1 \text{ mole K}} = 39.10 \text{ g of K}$

$4 \text{ moles C} \times \dfrac{12.01 \text{ g C}}{1 \text{ mole C}} = 48.04 \text{ g of C}$

$5 \text{ moles H} \times \dfrac{1.008 \text{ g H}}{1 \text{ mole H}} = 5.040 \text{ g of H}$

$6 \text{ moles O} \times \dfrac{16.00 \text{ g O}}{1 \text{ mole O}} = 96.00 \text{ g of O}$

Chapter 7

$$\begin{aligned}
1 \text{ mole of K} &= 39.10 \text{ g of K} \\
4 \text{ moles of C} &= 48.04 \text{ g of C} \\
5 \text{ moles of H} &= 5.040 \text{ g of H} \\
6 \text{ moles of O} &= \underline{96.00 \text{ g of O}} \\
\therefore \text{ Molar mass of } KC_4H_5O_6 &= 188.18 \text{ g}
\end{aligned}$$

c. $16 \text{ moles C} \times \dfrac{12.01 \text{ g C}}{1 \text{ mole C}} = 192.2 \text{ g of C}$

$19 \text{ moles H} \times \dfrac{1.008 \text{ g H}}{1 \text{ mole H}} = 19.15 \text{ g of H}$

$3 \text{ moles N} \times \dfrac{14.01 \text{ g N}}{1 \text{ mole N}} = 42.03 \text{ g of N}$

$5 \text{ moles O} \times \dfrac{16.00 \text{ g O}}{1 \text{ mole O}} = 80.00 \text{ g of O}$

$1 \text{ mole S} \times \dfrac{32.07 \text{ g S}}{1 \text{ mole S}} = 32.07 \text{ g of S}$

$$\begin{aligned}
16 \text{ moles of C} &= 192.2 \text{ g of C} \\
19 \text{ moles of H} &= 19.15 \text{ g of H} \\
3 \text{ moles of N} &= 42.03 \text{ g of N} \\
5 \text{ moles of O} &= 80.00 \text{ g of O} \\
1 \text{ mole of S} &= \underline{32.07 \text{ g of S}} \\
\therefore \text{ Molar mass of } C_{16}H_{19}N_3O_5S &= 365.5 \text{ g}
\end{aligned}$$

7.19 a. $8 \text{ moles C} \times \dfrac{12.01 \text{ g C}}{1 \text{ mole C}} = 96.08 \text{ g of C}$

$9 \text{ moles H} \times \dfrac{1.008 \text{ g H}}{1 \text{ mole H}} = 9.072 \text{ g of H}$

$1 \text{ mole N} \times \dfrac{14.01 \text{ g N}}{1 \text{ mole N}} = 14.01 \text{ g of N}$

$2 \text{ moles O} \times \dfrac{16.00 \text{ g O}}{1 \text{ mole O}} = 32.00 \text{ g of O}$

$$\begin{aligned}
8 \text{ moles of C} &= 96.08 \text{ g of C} \\
9 \text{ moles of H} &= 9.072 \text{ g of H} \\
1 \text{ mole of N} &= 14.01 \text{ g of N} \\
2 \text{ moles of O} &= \underline{32.00 \text{ g of O}} \\
\therefore \text{ Molar mass of } C_8H_9NO_2 &= 151.16 \text{ g}
\end{aligned}$$

b. $3 \text{ moles Ca} \times \dfrac{40.08 \text{ g Ca}}{1 \text{ mole Ca}} = 120.2 \text{ g of Ca}$

$12 \text{ moles C} \times \dfrac{12.01 \text{ g C}}{1 \text{ mole C}} = 144.1 \text{ g of C}$

$10 \text{ moles H} \times \dfrac{1.008 \text{ g H}}{1 \text{ mole H}} = 10.08 \text{ g of H}$

$14 \text{ moles O} \times \dfrac{16.00 \text{ g O}}{1 \text{ mole O}} = 224.0 \text{ g of O}$

$$\begin{aligned}
3 \text{ moles of Ca} &= 120.2 \text{ g of Ca} \\
12 \text{ moles of C} &= 144.1 \text{ g of C} \\
10 \text{ moles of H} &= 10.08 \text{ g of H} \\
14 \text{ moles of O} &= \underline{224.0 \text{ g of O}} \\
\therefore \text{ Molar mass of } Ca_3(C_6H_5O_7)_2 &= 498.4 \text{ g}
\end{aligned}$$

c. $17 \text{ moles C} \times \dfrac{12.01 \text{ g C}}{1 \text{ mole C}} = 204.2 \text{ g of C}$

$18 \text{ moles H} \times \dfrac{1.008 \text{ g H}}{1 \text{ mole H}} = 18.14 \text{ g of H}$

$1 \text{ mole F} \times \dfrac{19.00 \text{ g F}}{1 \text{ mole F}} = 19.00 \text{ g of F}$

$3 \text{ moles N} \times \dfrac{14.01 \text{ g N}}{1 \text{ mole N}} = 42.03 \text{ g of N}$

$3 \text{ moles O} \times \dfrac{16.00 \text{ g O}}{1 \text{ mole O}} = 48.00 \text{ g of O}$

17 moles of C = 204.2 g of C
18 moles of H = 18.14 g of H
1 mole of F = 19.00 g of F
3 moles of N = 42.03 g of N
3 moles of O = 48.00 g of O
∴ Molar mass of $C_{17}H_{18}FN_3O_3$ = 331.4 g

7.21 a. $1.50 \text{ moles Na} \times \dfrac{22.99 \text{ g Na}}{1 \text{ mole Na}} = 34.5 \text{ g of Na (3 SFs)}$

b. $2.80 \text{ moles Ca} \times \dfrac{40.08 \text{ g Ca}}{1 \text{ mole Ca}} = 112 \text{ g of Ca (3 SFs)}$

c. Molar mass of CO_2
= 1 mole of C + 2 moles of O = 12.01 g + 2(16.00 g) = 44.01 g

$0.125 \text{ mole } CO_2 \times \dfrac{44.01 \text{ g } CO_2}{1 \text{ mole } CO_2} = 5.50 \text{ g of } CO_2 \text{ (3 SFs)}$

d. Molar mass of Na_2CO_3
= 2 moles of Na + 1 mole of C + 3 moles of O = 2(22.99 g) + 12.01 g + 3(16.00 g)
= 105.99 g

$0.0485 \text{ mole } Na_2CO_3 \times \dfrac{105.99 \text{ g } Na_2CO_3}{1 \text{ mole } Na_2CO_3} = 5.14 \text{ g of } Na_2CO_3 \text{ (3 SFs)}$

e. Molar mass of PCl_3
= 1 mole of P + 3 moles of Cl = 30.97 g + 3(35.45 g) = 137.32 g

$7.14 \times 10^2 \text{ moles } PCl_3 \times \dfrac{137.32 \text{ g } PCl_3}{1 \text{ mole } PCl_3} = 9.80 \times 10^4 \text{ g of } PCl_3 \text{ (3 SFs)}$

7.23 a. $0.150 \text{ mole Ne} \times \dfrac{20.18 \text{ g Ne}}{1 \text{ mole Ne}} = 3.03 \text{ g of Ne}$

b. Molar mass of I_2
= 2 moles of I = 2(126.9 g) = 253.8 g

$0.150 \text{ mole } I_2 \times \dfrac{253.8 \text{ g } I_2}{1 \text{ mole } I_2} = 38.1 \text{ g of } I_2 \text{ (3 SFs)}$

c. Molar mass of Na_2O
= 2 moles of Na + 1 mole of O = 2(22.99 g) + 16.00 g = 61.98 g

$0.150 \text{ mole } Na_2O \times \dfrac{61.98 \text{ g } Na_2O}{1 \text{ mole } Na_2O} = 9.30 \text{ g of } Na_2O \text{ (3 SFs)}$

Chapter 7

 d. Molar mass of $Ca(NO_3)_2$
  = 1 mole of Ca + 2 moles of N + 6 moles of O = 40.08 g + 2(14.01 g) + 6(16.00 g) = 164.10 g

$$0.150 \text{ mole } Ca(NO_3)_2 \times \frac{164.10 \text{ g } Ca(NO_3)_2}{1 \text{ mole } Ca(NO_3)_2} = 24.6 \text{ g of } Ca(NO_3)_2 \text{ (3 SFs)}$$

 e. Molar mass of C_6H_{14}
  = 6 moles of C + 14 moles of H = 6(12.01 g) + 14(1.008 g) = 86.17 g

$$0.150 \text{ mole } C_6H_{14} \times \frac{86.17 \text{ g } C_6H_{14}}{1 \text{ mole } C_6H_{14}} = 12.9 \text{ g of } C_6H_{14} \text{ (3 SFs)}$$

7.25 **a.** $82.0 \text{ g Ag} \times \dfrac{1 \text{ mole Ag}}{107.9 \text{ g Ag}} = 0.760 \text{ mole of Ag (3 SFs)}$

 b. $0.288 \text{ g C} \times \dfrac{1 \text{ mole C}}{12.01 \text{ g C}} = 0.0240 \text{ mole of C (3 SFs)}$

 c. Molar mass of NH_3
  = 1 mole of N + 3 moles of H = 14.01 g + 3(1.008 g) = 17.03 g

$$15.0 \text{ g } NH_3 \times \frac{1 \text{ mole } NH_3}{17.03 \text{ g } NH_3} = 0.881 \text{ mole of } NH_3 \text{ (3 SFs)}$$

 d. Molar mass of CH_4
  = 1 mole of C + 4 moles of H = 12.01 g + 4(1.008 g) = 16.04 g

$$7.25 \text{ g } CH_4 \times \frac{1 \text{ mole } CH_4}{16.04 \text{ g } CH_4} = 0.452 \text{ mole of } CH_4 \text{ (3 SFs)}$$

 e. Molar mass of Fe_2O_3
  = 2 moles of Fe + 3 moles of O = 2(55.85 g) + 3(16.00 g) = 159.7 g

$$245 \text{ g } Fe_2O_3 \times \frac{1 \text{ mole } Fe_2O_3}{159.7 \text{ g } Fe_2O_3} = 1.53 \text{ moles of } Fe_2O_3 \text{ (3 SFs)}$$

7.27 **a.** $25.0 \text{ g He} \times \dfrac{1 \text{ mole He}}{4.003 \text{ g He}} = 6.25 \text{ moles of He (3 SFs)}$

 b. Molar mass of O_2
  = 2 moles of O = 2(16.00 g) = 32.00 g

$$25.0 \text{ g } O_2 \times \frac{1 \text{ mole } O_2}{32.00 \text{ g } O_2} = 0.781 \text{ mole of } O_2 \text{ (3 SFs)}$$

 c. Molar mass of $Al(OH)_3$
  = 1 mole of Al + 3 moles of O + 3 moles of H = 26.98 g + 3(16.00 g) + 3(1.008 g)
  = 78.00 g

$$25.0 \text{ g } Al(OH)_3 \times \frac{1 \text{ mole } Al(OH)_3}{78.00 \text{ g } Al(OH)_3} = 0.321 \text{ mole of } Al(OH)_3 \text{ (3 SFs)}$$

 d. Molar mass of Ga_2S_3
  = 2 moles of Ga + 3 moles of S = 2(69.72 g) + 3(32.07 g) = 235.7 g

$$25.0 \text{ g } Ga_2S_3 \times \frac{1 \text{ mole } Ga_2S_3}{235.7 \text{ g } Ga_2S_3} = 0.106 \text{ mole of } Ga_2S_3 \text{ (3 SFs)}$$

 e. Molar mass of C_4H_{10}
  = 4 moles of C + 10 moles of H = 4(12.01 g) + 10(1.008 g) = 58.12 g

$$25.0 \text{ g } C_4H_{10} \times \frac{1 \text{ mole } C_4H_{10}}{58.12 \text{ g } C_4H_{10}} = 0.430 \text{ mole of } C_4H_{10} \text{ (3 SFs)}$$

7.29 Molar mass of C_2H_5Cl
= 2 moles of C + 5 moles of H + 1 mole of Cl = 2(12.01 g) + 5(1.008 g) + 35.45 g = 64.51 g

a. $34.0 \text{ g } C_2H_5Cl \times \dfrac{1 \text{ mole } C_2H_5Cl}{64.51 \text{ g } C_2H_5Cl} = 0.527$ mole of C_2H_5Cl (3 SFs)

b. $1.50 \text{ moles } C_2H_5Cl \times \dfrac{64.51 \text{ g } C_2H_5Cl}{1 \text{ mole } C_2H_5Cl} = 96.8$ g of C_2H_5Cl (3 SFs)

7.31 a. Molar mass of $MgSO_4$
= 1 mole of Mg + 1 mole of S + 4 moles of O = 24.31 g + 32.07 g + 4(16.00 g)
= 120.38 g

$5.00 \text{ moles } MgSO_4 \times \dfrac{120.38 \text{ g } MgSO_4}{1 \text{ mole } MgSO_4} = 602$ g of $MgSO_4$ (3 SFs)

b. Molar mass of KI
= 1 mole of K + 1 mole of I = 39.10 g + 126.9 g = 166.0 g

$0.450 \text{ mole KI} \times \dfrac{166.0 \text{ g KI}}{1 \text{ mole KI}} = 74.7$ g of KI (3 SFs)

7.33 Molar mass of N_2O
= 2 moles of N + 1 mole of O = 2(14.01 g) + 16.00 g = 44.02 g

a. $1.50 \text{ moles } N_2O \times \dfrac{44.02 \text{ g } N_2O}{1 \text{ mole } N_2O} = 66.0$ g of N_2O (3 SFs)

b. $34.0 \text{ g } N_2O \times \dfrac{1 \text{ mole } N_2O}{44.02 \text{ g } N_2O} = 0.772$ mole of N_2O (3 SFs)

7.35 An equation is balanced when there are equal numbers of atoms of each element on the reactant side and on the product side.
a. not balanced (2 O ≠ 3 O)   **b.** balanced
c. not balanced (2 O ≠ 1 O)   **d.** balanced

7.37 Place coefficients in front of formulas until you make the atoms of each element equal on each side of the equation.
a. $N_2(g) + O_2(g) \longrightarrow 2NO(g)$
b. $2HgO(s) \longrightarrow 2Hg(l) + O_2(g)$
c. $4Fe(s) + 3O_2(g) \longrightarrow 2Fe_2O_3(s)$
d. $2Na(s) + Cl_2(g) \longrightarrow 2NaCl(s)$

7.39 a. $Mg(s) + 2AgNO_3(aq) \longrightarrow Mg(NO_3)_2(aq) + 2Ag(s)$
b. $2Al(s) + 3CuSO_4(aq) \longrightarrow Al_2(SO_4)_3(aq) + 3Cu(s)$
c. $Pb(NO_3)_2(aq) + 2NaCl(aq) \longrightarrow PbCl_2(s) + 2NaNO_3(aq)$
d. $2Al(s) + 6HCl(aq) \longrightarrow 3H_2(g) + 2AlCl_3(aq)$

7.41 a. This is a decomposition reaction because a single reactant splits into two simpler substances.
b. This is a single replacement reaction because one element in the reacting compound (I in BaI_2) is replaced by the other reactant (Br in Br_2).
c. This is a combustion reaction because a carbon-containing fuel burns in oxygen to produce carbon dioxide and water.
d. This is a double replacement reaction because the positive ions in the two reactants exchange places to form two products.
e. This is a combination reaction because atoms of two different elements bond to form one product.

7.43 a. combination   **b.** single replacement   **c.** decomposition
d. double replacement   **e.** combustion

Chapter 7

7.45 Oxidation is the loss of electrons; reduction is the gain of electrons.
  a. Na^+ gains an electron to form Na; this is a reduction.
  b. Ni loses electrons to form Ni^{2+}; this is an oxidation.
  c. Cr^{3+} gains electrons to form Cr; this is a reduction.
  d. $2H^+$ gain electrons to form H_2; this is a reduction.

7.47 An oxidized substance has lost electrons; a reduced substance has gained electrons.
  a. Zn loses electrons and is oxidized. Cl_2 gains electrons and is reduced.
  b. Br^- (in NaBr) loses electrons and is oxidized. Cl_2 gains electrons and is reduced.
  c. O^{2-} (in PbO) loses electrons and is oxidized. Pb^{2+} (in PbO) gains electrons and is reduced.
  d. Sn^{2+} loses electrons and oxidized. Fe^{3+} gains electrons and is reduced.

7.49 a. Fe^{3+} gains an electron to form Fe^{2+}; this is a reduction.
  b. Fe^{2+} loses an electron to form Fe^{3+}; this is an oxidation.

7.51 Linoleic acid gains hydrogen atoms and is reduced.

7.53 a. $\dfrac{2 \text{ moles } SO_2}{1 \text{ mole } O_2}$ and $\dfrac{1 \text{ mole } O_2}{2 \text{ moles } SO_2}$; $\dfrac{2 \text{ moles } SO_2}{2 \text{ moles } SO_3}$ and $\dfrac{2 \text{ moles } SO_3}{2 \text{ moles } SO_2}$; $\dfrac{1 \text{ mole } O_2}{2 \text{ moles } SO_3}$ and $\dfrac{2 \text{ moles } SO_3}{1 \text{ mole } O_2}$

  b. $\dfrac{4 \text{ moles } P}{5 \text{ moles } O_2}$ and $\dfrac{5 \text{ moles } O_2}{4 \text{ moles } P}$; $\dfrac{4 \text{ moles } P}{2 \text{ moles } P_2O_5}$ and $\dfrac{2 \text{ moles } P_2O_5}{4 \text{ moles } P}$; $\dfrac{5 \text{ moles } O_2}{2 \text{ moles } P_2O_5}$ and $\dfrac{2 \text{ moles } P_2O_5}{5 \text{ moles } O_2}$

7.55 a. $2.6 \text{ moles } H_2 \times \dfrac{1 \text{ mole } O_2}{2 \text{ moles } H_2} = 1.3 \text{ moles of } O_2$ (2 SFs)

  b. $5.0 \text{ moles } O_2 \times \dfrac{2 \text{ moles } H_2}{1 \text{ mole } O_2} = 10. \text{ moles of } H_2$ (2 SFs)

  c. $2.5 \text{ moles } O_2 \times \dfrac{2 \text{ moles } H_2O}{1 \text{ mole } O_2} = 5.0 \text{ moles of } H_2O$ (2 SFs)

7.57 a. $0.500 \text{ mole } SO_2 \times \dfrac{5 \text{ moles } C}{2 \text{ moles } SO_2} = 1.25 \text{ moles of } C$ (3 SFs)

  b. $1.2 \text{ moles } C \times \dfrac{4 \text{ moles } CO}{5 \text{ moles } C} = 0.96 \text{ mole of } CO$ (2 SFs)

  c. $0.50 \text{ mole } CS_2 \times \dfrac{2 \text{ moles } SO_2}{1 \text{ mole } CS_2} = 1.0 \text{ mole of } SO_2$ (2 SFs)

  d. $2.5 \text{ moles } C \times \dfrac{1 \text{ mole } CS_2}{5 \text{ moles } C} = 0.50 \text{ mole of } CS_2$ (2 SFs)

7.59 a. $57.5 \text{ g Na} \times \dfrac{1 \text{ mole Na}}{22.99 \text{ g Na}} \times \dfrac{2 \text{ moles } Na_2O}{4 \text{ moles Na}} \times \dfrac{61.98 \text{ g } Na_2O}{1 \text{ mole } Na_2O} = 77.5 \text{ g of } Na_2O$ (3 SFs)

  b. $18.0 \text{ g Na} \times \dfrac{1 \text{ mole Na}}{22.99 \text{ g Na}} \times \dfrac{1 \text{ mole } O_2}{4 \text{ moles Na}} \times \dfrac{32.00 \text{ g } O_2}{1 \text{ mole } O_2} = 6.26 \text{ g of } O_2$ (3 SFs)

  c. $75.0 \text{ g } Na_2O \times \dfrac{1 \text{ mole } Na_2O}{61.98 \text{ g } Na_2O} \times \dfrac{1 \text{ mole } O_2}{2 \text{ moles } Na_2O} \times \dfrac{32.00 \text{ g } O_2}{1 \text{ mole } O_2} = 19.4 \text{ g of } O_2$ (3 SFs)

7.61 a. $13.6 \text{ g } NH_3 \times \dfrac{1 \text{ mole } NH_3}{17.03 \text{ g } NH_3} \times \dfrac{3 \text{ moles } O_2}{4 \text{ moles } NH_3} \times \dfrac{32.00 \text{ g } O_2}{1 \text{ mole } O_2} = 19.2 \text{ g of } O_2$ (3 SFs)

  b. $6.50 \text{ g } O_2 \times \dfrac{1 \text{ mole } O_2}{32.00 \text{ g } O_2} \times \dfrac{2 \text{ moles } N_2}{3 \text{ moles } O_2} \times \dfrac{28.02 \text{ g } N_2}{1 \text{ mole } N_2} = 3.79 \text{ g of } N_2$ (3 SFs)

  c. $34.0 \text{ g } NH_3 \times \dfrac{1 \text{ mole } NH_3}{17.03 \text{ g } NH_3} \times \dfrac{6 \text{ moles } H_2O}{4 \text{ moles } NH_3} \times \dfrac{18.02 \text{ g } H_2O}{1 \text{ mole } H_2O} = 54.0 \text{ g of } H_2O$ (3 SFs)

7.63 a. $28.0 \text{ g NO}_2 \times \dfrac{1 \text{ mole NO}_2}{46.01 \text{ g NO}_2} \times \dfrac{1 \text{ mole H}_2\text{O}}{3 \text{ moles NO}_2} \times \dfrac{18.02 \text{ g H}_2\text{O}}{1 \text{ mole H}_2\text{O}} = 3.66 \text{ g of H}_2\text{O} \text{ (3 SFs)}$

b. $15.8 \text{ g H}_2\text{O} \times \dfrac{1 \text{ mole H}_2\text{O}}{18.02 \text{ g H}_2\text{O}} \times \dfrac{1 \text{ mole NO}}{1 \text{ mole H}_2\text{O}} \times \dfrac{30.01 \text{ g NO}}{1 \text{ mole NO}} = 26.3 \text{ g of NO (3 SFs)}$

c. $8.25 \text{ g NO}_2 \times \dfrac{1 \text{ mole NO}_2}{46.01 \text{ g NO}_2} \times \dfrac{2 \text{ moles HNO}_3}{3 \text{ moles NO}_2} \times \dfrac{63.02 \text{ g HNO}_3}{1 \text{ mole HNO}_3} = 7.53 \text{ g of HNO}_3 \text{ (3 SFs)}$

7.65 a. $2\text{PbS}(s) + 3\text{O}_2(g) \longrightarrow 2\text{PbO}(s) + 2\text{SO}_2(g)$

b. $29.9 \text{ g PbS} \times \dfrac{1 \text{ mole PbS}}{239.3 \text{ g PbS}} \times \dfrac{3 \text{ moles O}_2}{2 \text{ moles PbS}} \times \dfrac{32.00 \text{ g O}_2}{1 \text{ mole O}_2} = 6.00 \text{ g of O}_2 \text{ (3 SFs)}$

c. $65.0 \text{ g PbS} \times \dfrac{1 \text{ mole PbS}}{239.3 \text{ g PbS}} \times \dfrac{2 \text{ moles SO}_2}{2 \text{ moles PbS}} \times \dfrac{64.07 \text{ g SO}_2}{1 \text{ mole SO}_2} = 17.4 \text{ g of SO}_2 \text{ (3 SFs)}$

d. $128 \text{ g PbO} \times \dfrac{1 \text{ mole PbO}}{223.2 \text{ g PbO}} \times \dfrac{2 \text{ moles PbS}}{2 \text{ moles PbO}} \times \dfrac{239.3 \text{ g PbS}}{1 \text{ mole PbS}} = 137 \text{ g of PbS (3 SFs)}$

7.67 a. The energy of activation is the energy required to break the bonds of the reacting molecules.
b. In exothermic reactions, the energy of the products is lower than the energy of the reactants.
c.

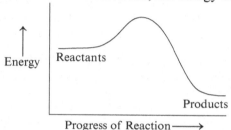

7.69 a. An exothermic reaction releases heat.
b. An endothermic reaction has a higher energy level for the products than the reactants.
c. The metabolism of glucose is an exothermic reaction, providing energy for the body.

7.71 a. Heat is a product; the reaction is exothermic.
b. Heat is a reactant; the reaction is endothermic.
c. Heat is a product; the reaction is exothermic.

7.73 a. The rate of a reaction tells how fast the products are formed or how fast the reactants are consumed.
b. Because more reactants will have the energy necessary to proceed to products (the activation energy) at room temperature than at refrigerator temperatures, the rate of formation of mold will be higher at room temperature.

7.75 a. Addition of a reactant increases the reaction rate.
b. Increasing the temperature increases the number of collisions with the energy of activation. The rate of reaction will be increased.
c. Addition of a catalyst increases the reaction rate.
d. Removal of reactant decreases the reaction rate.

7.77 a. $C_6H_{12}O_6(aq) + 6O_2(g) \longrightarrow 6CO_2(g) + 6H_2O(l)$
b. $6CO_2(g) + 6H_2O(l) \longrightarrow C_6H_{12}O_6(aq) + 6O_2(g)$

7.79 1. a. S_2Cl_2
  b. Molar mass of $S_2Cl_2 = 2(32.07 \text{ g}) + 2(35.45 \text{ g}) = 135.04 \text{ g (5 SFs)}$
  c. $10.0 \text{ g S}_2\text{Cl}_2 \times \dfrac{1 \text{ mole S}_2\text{Cl}_2}{135.04 \text{ g S}_2\text{Cl}_2} = 0.0741 \text{ mole of S}_2\text{Cl}_2 \text{ (3 SFs)}$

2. a. C_6H_6
b. Molar mass of C_6H_6 = 6(12.01 g) + 6(1.008 g) = 78.11 g (4 SFs)
c. 10.0 g C_6H_6 × $\dfrac{1 \text{ mole } C_6H_6}{78.11 \text{ g } C_6H_6}$ = 0.128 mole of C_6H_6 (3 SFs)

7.81 a. Molar mass of dipyrithione ($C_{10}H_8N_2O_2S_2$)
= 10(12.01 g) + 8(1.008 g) + 2(14.01 g) + 2(16.00 g) + 2(32.07 g) = 252.3 g (4 SFs)

b. 25.0 g $C_{10}H_8N_2O_2S_2$ × $\dfrac{1 \text{ mole } C_{10}H_8N_2O_2S_2}{252.3 \text{ g } C_{10}H_8N_2O_2S_2}$ = 0.0991 mole of $C_{10}H_8N_2O_2S_2$ (3 SFs)

c. 25.0 g $C_{10}H_8N_2O_2S_2$ × $\dfrac{1 \text{ mole } C_{10}H_8N_2O_2S_2}{252.3 \text{ g } C_{10}H_8N_2O_2S_2}$ × $\dfrac{10 \text{ moles C}}{1 \text{ mole } C_{10}H_8N_2O_2S_2}$
= 0.991 mole of C (3 SFs)

d. 8.2×10^{24} N atoms × $\dfrac{1 \text{ mole N}}{6.02 \times 10^{23} \text{ N atoms}}$ × $\dfrac{1 \text{ mole } C_{10}H_8O_2N_2S_2}{2 \text{ moles N}}$
= 6.8 moles of $C_{10}H_8O_2N_2S_2$ (2 SFs)

7.83 a. 1, 1, 2 combination **b.** 2, 2, 1 decomposition

7.85 a. reactants NO and O_2; product NO_2 **b.** $2NO(g) + O_2(g) \longrightarrow 2NO_2(g)$
c. combination

7.87 a. reactant NI_3; products N_2 and I_2 **b.** $2NI_3(s) \longrightarrow N_2(g) + 3I_2(g)$
c. decomposition

7.89 a. reactants Cl_2 and O_2; product OCl_2
b. $2Cl_2(g) + O_2(g) \longrightarrow 2OCl_2(g)$
c. combination

7.91 a. 1 mole Zn × $\dfrac{65.41 \text{ g Zn}}{1 \text{ mole Zn}}$ = 65.41 g of Zn

1 mole S × $\dfrac{32.07 \text{ g S}}{1 \text{ mole S}}$ = 32.07 g of S

4 moles O × $\dfrac{16.00 \text{ g O}}{1 \text{ mole O}}$ = 64.00 g of O

1 mole of Zn   = 65.41 g of Zn
1 mole of S   = 32.07 g of S
4 moles of O   = 64.00 g of O
∴ Molar mass of $ZnSO_4$ = 161.48 g

b. 1 mole Ca × $\dfrac{40.08 \text{ g Ca}}{1 \text{ mole Ca}}$ = 40.08 g of Ca

2 moles I × $\dfrac{126.9 \text{ g I}}{1 \text{ mole I}}$ = 253.8 g of I

6 moles O × $\dfrac{16.00 \text{ g O}}{1 \text{ mole O}}$ = 96.00 g of O

1 mole of Ca   = 40.08 g of Ca
2 moles of I   = 253.8 g of I
6 moles of O   = 96.00 g of O
∴ Molar mass of $Ca(IO_3)_2$ = 389.9 g

c. $5 \text{ moles C} \times \dfrac{12.01 \text{ g C}}{1 \text{ mole C}} = 60.05 \text{ g of C}$

$8 \text{ moles H} \times \dfrac{1.008 \text{ g H}}{1 \text{ mole H}} = 8.064 \text{ g of H}$

$1 \text{ mole N} \times \dfrac{14.01 \text{ g N}}{1 \text{ mole N}} = 14.01 \text{ g of N}$

$1 \text{ mole Na} \times \dfrac{22.99 \text{ g Na}}{1 \text{ mole Na}} = 22.99 \text{ g of Na}$

$4 \text{ moles O} \times \dfrac{16.00 \text{ g O}}{1 \text{ mole O}} = 64.00 \text{ g of O}$

5 moles of C   = 60.05 g of C
8 moles of H   = 8.064 g of H
1 mole of N   = 14.01 g of N
1 mole of Na   = 22.99 g of Na
4 moles of O   = 64.00 g of O
∴ Molar mass of $C_5H_8NNaO_4$ = 169.11 g

d. $6 \text{ moles C} \times \dfrac{12.01 \text{ g C}}{1 \text{ mole C}} = 72.06 \text{ g of C}$

$12 \text{ moles H} \times \dfrac{1.008 \text{ g H}}{1 \text{ mole H}} = 12.10 \text{ g of H}$

$2 \text{ moles O} \times \dfrac{16.00 \text{ g O}}{1 \text{ mole O}} = 32.00 \text{ g of O}$

6 moles of C   = 72.06 g of C
12 moles of H   = 12.10 g of H
2 moles of O   = 32.00 g of O
∴ Molar mass of $C_6H_{12}O_2$ = 116.16 g

7.93 a. $0.150 \text{ mole K} \times \dfrac{39.10 \text{ g K}}{1 \text{ mole K}} = 5.87 \text{ g of K (3 SFs)}$

b. $0.150 \text{ mole Cl}_2 \times \dfrac{70.90 \text{ g Cl}_2}{1 \text{ mole Cl}_2} = 10.6 \text{ g of Cl}_2 \text{ (3 SFs)}$

c. $0.150 \text{ mole Na}_2\text{CO}_3 \times \dfrac{105.99 \text{ g Na}_2\text{CO}_3}{1 \text{ mole Na}_2\text{CO}_3} = 15.9 \text{ g of Na}_2\text{CO}_3 \text{ (3 SFs)}$

7.95 a. $25.0 \text{ g CO}_2 \times \dfrac{1 \text{ mole CO}_2}{44.01 \text{ g CO}_2} = 0.568 \text{ mole of CO}_2 \text{ (3 SFs)}$

b. $25.0 \text{ g Al}_2\text{O}_3 \times \dfrac{1 \text{ mole Al}_2\text{O}_3}{101.96 \text{ g Al}_2\text{O}_3} = 0.245 \text{ mole of Al}_2\text{O}_3 \text{ (3 SFs)}$

c. $25.0 \text{ g MgCl}_2 \times \dfrac{1 \text{ mole MgCl}_2}{95.21 \text{ g MgCl}_2} = 0.263 \text{ mole of MgCl}_2 \text{ (3 SFs)}$

7.97 a. Atoms of a metal and a nonmetal forming an ionic compound is a combination reaction.
b. When a compound of hydrogen and carbon reacts with oxygen, it is a combustion reaction.
c. When calcium carbonate is heated to produce calcium oxide and carbon dioxide, it is a decomposition reaction.
d. Zinc replacing copper in $Cu(NO_3)_2$ is a single replacement reaction.

7.99
a. $NH_3(g) + HCl(g) \longrightarrow NH_4Cl(s)$ — combination
b. $C_4H_8(g) + 6O_2(g) \xrightarrow{\Delta} 4CO_2(g) + 4H_2O(g)$ — combustion
c. $2Sb(s) + 3Cl_2(g) \longrightarrow 2SbCl_3(s)$ — combination
d. $2NI_3(s) \longrightarrow N_2(g) + 3I_2(g)$ — decomposition
e. $2KBr(aq) + Cl_2(aq) \longrightarrow 2KCl(aq) + Br_2(l)$ — single replacement
f. $2Fe(s) + 3H_2SO_4(aq) \longrightarrow 3H_2(g) + Fe_2(SO_4)_3(aq)$ — single replacement
g. $Al_2(SO_4)_3(aq) + 6NaOH(aq) \longrightarrow 2Al(OH)_3(s) + 3Na_2SO_4(aq)$ — double replacement

7.101
a. Zn^{2+} gains electrons to form Zn; this is a reduction.
b. Al loses electrons to form Al^{3+}; this is an oxidation.
c. Pb loses electrons to form Pb^{2+}; this is an oxidation.
d. Cl_2 gains electrons to form $2Cl^-$; this is a reduction.

7.103
a. $2NH_3(g) + 5F_2(g) \longrightarrow N_2F_4(g) + 6HF(g)$

b. $4.00 \text{ moles HF} \times \dfrac{2 \text{ moles } NH_3}{6 \text{ moles HF}} = 1.33 \text{ moles of } NH_3$ (3 SFs)

$4.00 \text{ moles HF} \times \dfrac{5 \text{ moles } F_2}{6 \text{ moles HF}} = 3.33 \text{ moles of } F_2$ (3 SFs)

c. $25.5 \text{ g } NH_3 \times \dfrac{1 \text{ mole } NH_3}{17.03 \text{ g } NH_3} \times \dfrac{5 \text{ moles } F_2}{2 \text{ moles } NH_3} \times \dfrac{38.00 \text{ g } F_2}{1 \text{ mole } F_2} = 142 \text{ g of } F_2$ (3 SFs)

d. $3.40 \text{ g } NH_3 \times \dfrac{1 \text{ mole } NH_3}{17.03 \text{ g } NH_3} \times \dfrac{1 \text{ mole } N_2F_4}{2 \text{ moles } NH_3} \times \dfrac{104.02 \text{ g } N_2F_4}{1 \text{ mole } N_2F_4} = 10.4 \text{ g of } N_2F_4$ (3 SFs)

7.105
a. $C_5H_{12}(g) + 8O_2(g) \xrightarrow{\Delta} 5CO_2(g) + 6H_2O(g) + \text{energy}$

b. $72 \text{ g } H_2O \times \dfrac{1 \text{ mole } H_2O}{18.02 \text{ g } H_2O} \times \dfrac{1 \text{ mole } C_5H_{12}}{6 \text{ moles } H_2O} \times \dfrac{72.15 \text{ g } C_5H_{12}}{1 \text{ mole } C_5H_{12}} = 48 \text{ g of } C_5H_{12}$ (2 SFs)

c. $32.0 \text{ g } O_2 \times \dfrac{1 \text{ mole } O_2}{32.00 \text{ g } O_2} \times \dfrac{5 \text{ moles } CO_2}{8 \text{ moles } O_2} \times \dfrac{44.01 \text{ g } CO_2}{1 \text{ mole } CO_2} = 27.5 \text{ g of } CO_2$ (3 SFs)

7.107
a. Since heat is produced during the formation of $SiCl_4$, this is an exothermic reaction.
b. Heat is given off; the energy of the product is lower than the energy of the reactants.

7.109
a. $124 \text{ g } C_2H_6O \times \dfrac{1 \text{ mole } C_2H_6O}{46.07 \text{ g } C_2H_6O} \times \dfrac{1 \text{ mole } C_6H_{12}O_6}{2 \text{ moles } C_2H_6O} \times \dfrac{180.16 \text{ g } C_6H_{12}O_6}{1 \text{ mole } C_6H_{12}O_6}$
$= 242 \text{ g of } C_6H_{12}O_6$ (3 SFs)

b. $0.240 \text{ kg } C_6H_{12}O_6 \times \dfrac{1000 \text{ g}}{1 \text{ kg}} \times \dfrac{1 \text{ mole } C_6H_{12}O_6}{180.16 \text{ g } C_6H_{12}O_6} \times \dfrac{2 \text{ moles } C_2H_6O}{1 \text{ mole } C_6H_{12}O_6} \times \dfrac{46.07 \text{ g } C_2H_6O}{1 \text{ mole } C_2H_6O}$
$= 123 \text{ g of } C_2H_6O$ (3 SFs)

7.111
a. $4Al(s) + 3O_2(g) \longrightarrow 2Al_2O_3(s)$
b. This is a combination reaction.

c. $4.50 \text{ moles Al} \times \dfrac{3 \text{ moles } O_2}{4 \text{ moles Al}} = 3.38 \text{ moles of } O_2$ (3 SFs)

d. $50.2 \text{ g Al} \times \dfrac{1 \text{ mole Al}}{26.98 \text{ g Al}} \times \dfrac{2 \text{ moles } Al_2O_3}{4 \text{ moles Al}} \times \dfrac{101.96 \text{ g } Al_2O_3}{1 \text{ mole } Al_2O_3} = 94.9 \text{ g of } Al_2O_3$ (3 SFs)

e. $8.00 \text{ g } O_2 \times \dfrac{1 \text{ mole } O_2}{32.00 \text{ g } O_2} \times \dfrac{2 \text{ moles } Al_2O_3}{3 \text{ moles } O_2} \times \dfrac{101.96 \text{ g } Al_2O_3}{1 \text{ mole } Al_2O_3} = 17.0 \text{ g of } Al_2O_3$ (3 SFs)

7.113
a. $0.500 \text{ mole } C_3H_6O_3 \times \dfrac{6.02 \times 10^{23} \text{ molecules}}{1 \text{ mole } C_3H_6O_3} = 3.01 \times 10^{23} \text{ molecules of } C_3H_6O_3$ (3 SFs)

b. $1.50 \text{ moles } C_3H_6O_3 \times \dfrac{3 \text{ moles C}}{1 \text{ mole } C_3H_6O_3} \times \dfrac{6.02 \times 10^{23} \text{ atoms}}{1 \text{ mole C}} = 2.71 \times 10^{24} \text{ atoms of C}$ (3 SFs)

c. $4.5 \times 10^{24} \text{ O atoms} \times \dfrac{1 \text{ mole O}}{6.02 \times 10^{23} \text{ O atoms}} \times \dfrac{1 \text{ mole } C_3H_6O_3}{3 \text{ moles O}} = 2.5$ moles of $C_3H_6O_3$ (2 SFs)

d. Molar mass of lactic acid $(C_3H_6O_3)$
= $3(12.01 \text{ g}) + 6(1.008 \text{ g}) + 3(16.00 \text{ g}) = 90.08$ g (4 SFs)

7.115 a. $2C_2H_2(g) + 5O_2(g) \xrightarrow{\Delta} 4CO_2(g) + 2H_2O(g) + 2600$ kJ

b. Since heat is produced during the combustion of acetylene, the reaction is exothermic.

c. $2.00 \text{ moles } O_2 \times \dfrac{2 \text{ moles } H_2O}{5 \text{ moles } O_2} = 0.800$ mole of H_2O (3 SFs)

d. $9.80 \text{ g } C_2H_2 \times \dfrac{1 \text{ mole } C_2H_2}{26.04 \text{ g } C_2H_2} \times \dfrac{5 \text{ moles } O_2}{2 \text{ moles } C_2H_2} \times \dfrac{32.00 \text{ g } O_2}{1 \text{ mole } O_2} = 30.1$ g of O_2 (3 SFs)

8
Gases

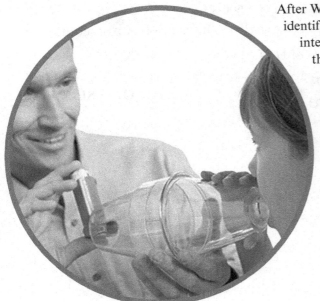

After Whitney's asthma attack, she and her parents learned to identify and treat her asthma. Because her asthma is mild and intermittent, they follow a management plan to control the symptoms. Whitney participates in physical activity and exercise to maintain lung function and to prevent recurrent attacks. If her symptoms worsen, she uses a bronchodilator to help her breathing. If Whitney has a lung capacity of 3.2 L at 37 °C and 756 mmHg, what is her lung capacity, in liters, at STP?

Credit: Adam Gault/Alamy

LOOKING AHEAD

8.1 Properties of Gases
8.2 Pressure and Volume (Boyle's Law)
8.3 Temperature and Volume (Charles's Law)
8.4 Temperature and Pressure (Gay-Lussac's Law)
8.5 The Combined Gas Law
8.6 Volume and Moles (Avogadro's Law)
8.7 Partial Pressures (Dalton's Law)

 The Health icon indicates a question that is related to health and medicine.

8.1 Properties of Gases

Learning Goal: Describe the kinetic molecular theory of gases and the units of measurement used for gases.

> **REVIEW**
> Using Significant Figures in Calculations (2.3)
> Writing Conversion Factors from Equalities (2.5)
> Using Conversion Factors (2.6)

- In a gas, small particles (atoms or molecules) are far apart and moving randomly with high velocities.
- The attractive forces between the particles of a gas are usually very small.
- The actual volume occupied by gas molecules is extremely small compared with the volume that the gas occupies.
- Gas particles are in constant motion, moving rapidly in straight paths.
- The average kinetic energy of gas molecules is proportional to the Kelvin temperature.
- A gas is described by the physical properties of pressure (P), volume (V), temperature (T), and amount (n).
- A gas exerts pressure, which is the force that gas particles exert on the walls of a container.

Gases

- Units of gas pressure include torr (Torr), millimeters of mercury (mmHg), pascals (Pa), kilopascals (kPa), lb/in.² (psi), and atmospheres (atm).
- Common units of volume are liters (L) and milliliters (mL).
- Temperature is usually measured in °C. In calculations involving gases, temperatures must use the Kelvin scale.
- Amount of gas is usually measured in grams but in calculations involving gases, moles must be used.

♦ **Learning Exercise 8.1A**

Answer *true* (T) or *false* (F) for each of the following:

a. _____ Gases are composed of small particles.
b. _____ Gas molecules are usually close together.
c. _____ Gas molecules move rapidly because they are strongly attracted.
d. _____ Distances between gas molecules are large.
e. _____ Gas molecules travel in straight lines until they collide.
f. _____ Pressure is the force of gas particles striking the walls of a container.
g. _____ Kinetic energy decreases with increasing temperature.
h. _____ 450 K is a measure of temperature.
i. _____ 3.56 g is a measure of pressure.

Answers   a. T   b. F   c. F   d. T   e. T
            f. T   g. F   h. T   i. F

Units for Measuring Pressure

Unit	Abbreviation	Unit Equivalent to 1 atm
atmosphere	atm	1 atm (exact)
millimeters of Hg	mmHg	760 mmHg (exact)
torr	Torr	760 Torr (exact)
inches of Hg	inHg	29.9 inHg
pounds per square inch	lb/in.² (psi)	14.7 lb/in.²
pascal	Pa	101 325 Pa
kilopascal	kPa	101.325 kPa

♦ **Learning Exercise 8.1B**

Complete the following:

a. 1.50 atm = _____ Torr

b. 550 mmHg = _____ atm

c. 0.725 atm = _____ kPa

d. 1520 Torr = _____ atm

e. 30.5 psi = _____ mmHg

f. 87.6 kPa = _____ Torr

g. During the weather report on TV, the pressure is given as 29.4 inHg. What is this pressure in millimeters of mercury and in atmospheres?

Chapter 8

Answers   a. 1140 Torr   b. 0.72 atm   c. 73.5 kPa   d. 2.00 atm
          e. 1580 mmHg   f. 657 Torr   g. 747 mmHg, 0.983 atm

8.2 Pressure and Volume (Boyle's Law)

Learning Goal: Use the pressure–volume relationship (Boyle's law) to calculate the unknown pressure or volume when the temperature and amount of gas do not change.

REVIEW
Solving Equations (1.4)

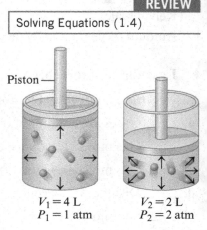

- According to Boyle's law, pressure increases if volume decreases and pressure decreases if volume increases, when the temperature and amount of gas do not change.
- If two properties change in opposite directions, the properties have an inverse relationship.
- The volume (V) of a gas changes inversely with the pressure (P) of the gas when T and n do not change:

$$P_1 V_1 = P_2 V_2$$ The subscripts 1 and 2 represent initial and final conditions.

Key Terms for Sections 8.1 and 8.2

Match each of the following key terms with the correct description:

a. kinetic molecular theory   b. inverse relationship   c. Boyle's law
d. pressure   e. atmospheric pressure

1. _____ the pressure exerted by the atmosphere
2. _____ a force exerted by gas particles when they collide with the sides of a container
3. _____ a relationship in which two properties change in opposite directions
4. _____ the volume of a gas varies inversely with the pressure of a gas, if there is no change in temperature and amount of gas
5. _____ the model that explains the behavior of gas particles

Answers   1. e   2. d   3. b   4. c   5. a

SAMPLE PROBLEM   Using Boyle's Law

A balloon has a volume of 3.00 L and a pressure of 755 mmHg. What is the volume, in liters, of the balloon at a final pressure of 638 mmHg, if the temperature and amount of gas do not change?

Solution Guide:

STEP 1 State the given and needed quantities.

Analyze the Problem	Given	Need	Connect
	$P_1 = 755$ mmHg $P_2 = 638$ mmHg $V_1 = 3.00$ L	V_2	Boyle's law $P_1 V_1 = P_2 V_2$
	Factors that do not change: T and n		**Predict:** P decreases, V increases

STEP 2 Rearrange the gas law equation to solve for the unknown quantity.

$$V_2 = V_1 \times \frac{P_1}{P_2}$$

STEP 3 Substitute values into the gas law equation and calculate.

$$V_2 = 3.00 \text{ L} \times \frac{755 \text{ mmHg}}{638 \text{ mmHg}} = 3.55 \text{ L}$$

♦ **Learning Exercise 8.2**

CORE CHEMISTRY SKILL
Using the Gas Laws

Solve each of the following problems using Boyle's law:

a. A sample of 4.0 L of He has a pressure of 800. mmHg. What is the pressure, in millimeters of mercury, when the volume is reduced to 1.0 L, if there is no change in temperature and amount of gas?

b. The maximum volume of air that can fill a person's lungs is 6250 mL at a pressure of 728 mmHg. Inhalation occurs as the pressure in the lungs drops to 721 mmHg with no change in temperature and amount of gas. To what volume, in milliliters, do the lungs expand, if there is no change in temperature and amount of gas?

c. A sample of O_2 at a pressure of 5.00 atm has a volume of 3.00 L. If the gas pressure is changed to 760. mmHg, what volume, in liters, will the O_2 occupy, if there is no change in temperature and amount of gas?

d. A sample of 250. mL of N_2 is initially at a pressure of 2.50 atm. If the pressure changes to 825 mmHg, what is the volume, in milliliters, if there is no change in temperature and amount of gas?

Answers   a. 3200 mmHg   b. 6310 mL   c. 15.0 L   d. 576 mL

8.3 Temperature and Volume (Charles's Law)

Learning Goal: Use the temperature–volume relationship (Charles's law) to calculate the unknown temperature or volume when the pressure and amount of gas do not change.

- According to Charles's law, when the pressure and quantity do not change, the volume of a gas increases when the temperature of the gas increases, and the volume decreases when the temperature decreases.
- In a direct relationship, two properties increase or decrease together.
- The volume (V) of a gas is directly related to its Kelvin temperature (T) when there is no change in the pressure or moles of the gas:

$$\frac{V_1}{T_1} = \frac{V_2}{T_2}$$ The subscripts 1 and 2 represent initial and final conditions.

$T_1 = 200 \text{ K}$
$V_1 = 1 \text{ L}$

$T_2 = 400 \text{ K}$
$V_2 = 2 \text{ L}$

Chapter 8

♦ **Learning Exercise 8.3**

Solve each of the following problems using Charles's law:

a. A balloon has a volume of 2.5 L at a temperature of 0 °C. What is the volume, in liters, of the balloon when the temperature rises to 120 °C, if there is no change in pressure and amount of gas?

b. A balloon is filled with He to a volume of 6600 L at a temperature of 223 °C. To what temperature, in degrees Celsius, must the gas be cooled to decrease the volume to 4800 L, if there is no change in pressure and amount of gas?

c. A sample of 750. mL of Ne is heated from 120. °C to 350. °C. What is the volume, in milliliters, if there is no change in pressure and amount of gas?

d. What is the temperature, in degrees Celsius, of 350. mL of O_2 at 22 °C when its volume increases to 0.800 L, if there is no change in pressure and amount of gas?

Answers a. 3.6 L b. 88 °C c. 1190 mL d. 401 °C

8.4 Temperature and Pressure (Gay-Lussac's Law)

Learning Goal: Use the temperature–pressure relationship (Gay-Lussac's law) to calculate the unknown temperature or pressure when the volume and amount of gas do not change.

- According to Gay-Lussac's law, when volume and amount do not change, the gas pressure increases when the temperature of the gas increases, and the pressure decreases when the temperature decreases.

- The pressure (P) of a gas is directly related to its Kelvin temperature (T), when there is no change in the volume or moles of the gas:

$$\frac{P_1}{T_1} = \frac{P_2}{T_2}$$ The subscripts 1 and 2 represent initial and final conditions.

$T_1 = 200$ K $T_2 = 400$ K
$P_1 = 1$ atm $P_2 = 2$ atm

♦ **Learning Exercise 8.4**

Solve each of the following problems using Gay-Lussac's law:

a. A sample of He has a pressure of 860. mmHg at a temperature of 225 K. What is the pressure, in millimeters of mercury, when the He reaches a temperature of 675 K, if there is no change in volume or amount of gas?

186

b. A balloon contains a gas with a pressure of 580. mmHg and a temperature of 227 °C. What is the pressure, in millimeters of mercury, of the gas when the temperature drops to 27 °C, if there is no change in volume or amount of gas?

c. A spray can contains a gas with a pressure of 3.0 atm at a temperature of 17 °C. What is the pressure, in atmospheres, inside the container if the temperature rises to 110 °C, if there is no change in volume or amount of gas?

d. A gas has a pressure of 1200. mmHg at 300. °C. What will the temperature, in degrees Celsius, be when the pressure falls to 1.10 atm, if there is no change in volume or amount of gas?

Answers **a.** 2580 mmHg   **b.** 348 mmHg   **c.** 4.0 atm   **d.** 126 °C

8.5 The Combined Gas Law

Learning Goal: Use the combined gas law to calculate the unknown pressure, volume, or temperature of a gas when changes in two of these properties are given and the amount of gas does not change.

- When the amount of a gas does not change, the gas laws can be combined into a relationship of pressure (P), volume (V), and temperature (T):

$$\frac{P_1 V_1}{T_1} = \frac{P_2 V_2}{T_2}$$ The subscripts 1 and 2 represent initial and final conditions.

Key Terms for Sections 8.3 to 8.5

Match each of the following key terms with the correct description:

   a. Gay-Lussac's law   **b.** direct relationship   **c.** Charles's law
   d. combined gas law

1. ____ the gas law stating that the volume of a gas changes directly with a change in Kelvin temperature when pressure and amount (moles) of the gas do not change
2. ____ the gas law stating that the pressure of a gas changes directly with a change in Kelvin temperature when the number of moles of a gas and its volume do not change
3. ____ a relationship in which two properties change in the same direction
4. ____ a relationship that combines several gas laws relating pressure, volume, and temperature when the amount of gas does not change

Answers   **1.** c   **2.** a   **3.** b   **4.** d

Chapter 8

♦ Learning Exercise 8.5

Solve each of the following problems using the combined gas law:

a. A 4.0-L sample of N_2 has a pressure of 1200 mmHg at 220 K. What is the pressure, in millimeters of mercury, of the sample when the volume increases to 20. L at 440 K, if there is no change in the amount of gas?

b. A 25.0-mL bubble forms at the ocean depths when the pressure is 10.0 atm and the temperature is 5.0 °C. What is the volume, in milliliters, of that bubble at the ocean surface when the pressure is 760. mmHg and the temperature is 25 °C, if there is no change in the amount of gas?

c. A 35.0-mL sample of Ar has a pressure of 1.0 atm and a temperature of 15 °C. What is the volume, in milliliters, if the pressure goes to 2.0 atm and the temperature to 45 °C, if there is no change in the amount of gas?

d. A weather balloon is launched from the Earth's surface with a volume of 315 L, a temperature of 12 °C, and pressure of 0.930 atm. What is the volume, in liters, of the balloon in the upper atmosphere when the pressure is 116 mmHg and the temperature is −35 °C, if there is no change in the amount of gas?

e. A 10.0-L sample of gas is emitted from a volcano with a pressure of 1.20 atm and a temperature of 150. °C. What is the volume, in liters, of the gas when its pressure is 0.900 atm and the temperature is −40. °C, if there is no change in the amount of gas?

Answers **a.** 480 mmHg **b.** 268 mL **c.** 19 mL **d.** 1.60×10^3 L **e.** 7.34 L

8.6 Volume and Moles (Avogadro's Law)

Learning Goal: Use Avogadro's law to calculate the unknown amount or volume of a gas when the pressure and temperature do not change.

> REVIEW
> Using Molar Mass as a Conversion Factor (7.3)

- If the number of moles of gas increases, the volume increases; if the number of moles of gas decreases, the volume decreases.
- Avogadro's law states that equal volumes of gases at the same temperature and pressure contain the same number of moles. The volume (V) of a gas is directly related to the number of moles of the gas when the pressure and temperature of the gas do not change:

$$\frac{V_1}{n_1} = \frac{V_2}{n_2}$$ The subscripts 1 and 2 represent initial and final conditions.

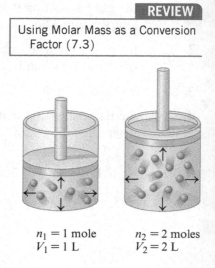

$n_1 = 1$ mole
$V_1 = 1$ L

$n_2 = 2$ moles
$V_2 = 2$ L

- At STP conditions, standard temperature (0 °C) and pressure (1 atm), 1.00 mole of a gas occupies a volume of 22.4 L, the *molar volume*.

♦ **Learning Exercise 8.6A**

Solve each of the following problems using Avogadro's law:

a. A balloon containing 0.50 mole of He has a volume of 4.0 L. What is the volume, in liters, when 1.0 mole of N_2 is added to the balloon, if there is no change in pressure and temperature?

b. A balloon containing 1.0 mole of O_2 has a volume of 15 L. What is the volume, in liters, of the balloon when 3.0 moles of He is added, if there is no change in pressure and temperature?

Answers   **a.** 12 L   **b.** 60. L

SAMPLE PROBLEM   Using Molar Volume

How many liters would 2.00 moles of N_2 occupy at STP?

Solution Guide:

STEP 1 State the given and needed quantities.

Analyze the Problem	Given	Need	Connect
	2.00 moles of N_2 at STP	liters of N_2 at STP	molar volume (STP)

STEP 2 Write a plan to calculate the needed quantity.

moles of N_2 $\xrightarrow{\text{Molar volume}}$ liters of N_2 (STP)

STEP 3 Write the equalities and conversion factors including 22.4 L/mole at STP.

$$1 \text{ mole of } N_2 = 22.4 \text{ L of } N_2 \text{ (STP)}$$

$$\frac{22.4 \text{ L } N_2 \text{ (STP)}}{1 \text{ mole } N_2} \quad \text{and} \quad \frac{1 \text{ mole } N_2}{22.4 \text{ L } N_2 \text{ (STP)}}$$

STEP 4 Set up the problem with factors to cancel units.

$$2.00 \text{ moles } N_2 \times \frac{22.4 \text{ L } N_2 \text{ (STP)}}{1 \text{ mole } N_2} = 44.8 \text{ L of } N_2 \text{ (STP)}$$

♦ **Learning Exercise 8.6B**

Solve each of the following problems using molar volume:

a. What is the volume, in liters, occupied by 18.5 g of N_2 at STP?

b. What is the mass, in grams, of 4.48 L of O_2 at STP?

Answers   **a.** 14.8 L   **b.** 6.40 g

8.7 Partial Pressures (Dalton's Law)

Learning Goal: Use Dalton's law of partial pressures to calculate the total pressure of a mixture of gases.

- In a mixture of two or more gases, the total pressure is the sum of the partial pressures (subscripts 1, 2, 3 . . . represent the partial pressures of the individual gases).

    $$P_{total} = P_1 + P_2 + P_3 + \cdots$$

- The partial pressure of a gas in a mixture is the pressure it would exert if it were the only gas in the container.

Key Terms for Sections 8.6 and 8.7

Match each of the following key terms with the correct description:

    a. molar volume   **b.** STP   **c.** Dalton's law
    d. partial pressure   **e.** Avogadro's law

1. ____ the gas law stating that the total pressure exerted by a mixture of gases in a container is the sum of the pressures that each gas would exert alone
2. ____ the pressure exerted by a single gas in a gas mixture
3. ____ the volume occupied by 1 mole of a gas at STP
4. ____ the gas law stating that the volume of a gas is directly related to the number of moles of gas when pressure and temperature do not change
5. ____ the standard conditions of 0 °C (273 K) and 1 atm, used for the comparison of gases

Oxygen therapy increases the amount of oxygen available to the tissues of the body.
Credit: Levent Konuk/Shutterstock

Answers   **1.** c   **2.** d   **3.** a   **4.** e   **5.** b

Solution Guide to Calculating Partial Pressure of a Gas in a Mixture	
STEP 1	Write the equation for the sum of the partial pressures.
STEP 2	Rearrange the equation to solve for the unknown pressure.
STEP 3	Substitute known pressures into the equation and calculate the unknown pressure.

Gases

♦ **Learning Exercise 8.7A**

> **CORE CHEMISTRY SKILL**
> Calculating Partial Pressure

Solve each of the following problems using Dalton's law:

a. What is the total pressure, in millimeters of mercury, of a gas sample that contains O_2 at 0.500 atm, N_2 at 132 Torr, and He at 224 mmHg?

b. What is the total pressure, in atmospheres, of a gas sample that contains He at 285 mmHg and O_2 at 1.20 atm?

c. A gas sample containing N_2 and O_2 has a pressure of 1500. mmHg. If the partial pressure of the N_2 is 0.900 atm, what is the partial pressure, in millimeters of mercury, of the O_2 in the mixture?

Answers   a. 736 mmHg   b. 1.58 atm   c. 816 mmHg

♦ **Learning Exercise 8.7B**

Fill in the blanks by writing Increases or Decreases for a gas in a closed container:

	Pressure	Volume	Moles	Temperature
a.	_____	Increases	Constant	Constant
b.	Increases	Constant	_____	Constant
c.	Constant	Decreases	_____	Constant
d.	Increases	Constant	Constant	_____
e.	Constant	_____	Constant	Decreases
f.	_____	Constant	Increases	Constant

Answers   a. Decreases   b. Increases   c. Decreases   d. Increases   e. Decreases   f. Increases

Checklist for Chapter 8

You are ready to take the Practice Test for Chapter 8. Be sure you have accomplished the following learning goals for this chapter. If not, review the Section listed at the end of the goal. Then apply your new skills and understanding to the Practice Test.

After studying Chapter 8, I can successfully:

_____ Describe the kinetic molecular theory of gases. (8.1)

_____ Change the units of pressure from one to another. (8.1)

_____ Use the pressure–volume relationship (Boyle's law) to calculate the unknown pressure or volume when the temperature and amount of gas do not change. (8.2)

_____ Use the temperature–volume relationship (Charles's law) to calculate the unknown temperature or volume when the pressure and amount of gas do not change. (8.3)

_____ Use the temperature–pressure relationship (Gay-Lussac's law) to calculate the unknown temperature or pressure when the volume and amount of gas do not change. (8.4)

Chapter 8

_____ Use the combined gas law to calculate the unknown pressure, volume, or temperature of a fixed amount of gas when changes in two of these properties are given when the amount of gas does not change. (8.5)

_____ Describe the relationship between the amount of a gas and its volume (Avogadro's law) when the pressure and temperature do not change and use this relationship in calculations. (8.6)

_____ Calculate the total pressure of a gas mixture from the partial pressures. (8.7)

Practice Test for Chapter 8

The chapter Sections to review are shown in parentheses at the end of each question.

For questions 1 through 5, use the kinetic molecular theory of gases to answer true (T) or false (F): (8.1)

1. _____ A gas does not have its own volume or shape.
2. _____ The molecules of a gas are moving extremely fast.
3. _____ The collisions of gas molecules with the walls of their container create pressure.
4. _____ Gas molecules are close together and move in straight lines.
5. _____ The attractive forces between molecules are very small.

6. What is the pressure, in atmospheres, of a gas with a pressure of 1200 mmHg? (8.1)
   A. 0.63 atm
   B. 0.79 atm
   C. 1.2 atm
   D. 1.6 atm
   E. 2.0 atm

7. The relationship that the volume of a gas is inversely related to its pressure (when temperature and moles of the gas do not change) is known as (8.2)
   A. Boyle's law
   B. Charles's law
   C. Gay-Lussac's law
   D. Dalton's law
   E. Avogadro's law

8. A 6.00-L sample of O_2 has a pressure of 660. mmHg. When the volume is reduced to 2.00 L with no change in temperature and amount of gas, it will have a pressure of (8.2)
   A. 1980 mmHg
   B. 1320 mmHg
   C. 330. mmHg
   D. 220. mmHg
   E. 110. mmHg

9. If the temperature and amount of a gas do not change, but its volume doubles, its pressure will (8.2)
   A. double
   B. triple
   C. decrease to one-half the original pressure
   D. decrease to one-fourth the original pressure
   E. not change

10. If the temperature of a gas is increased, (8.3, 8.4)
    A. the pressure will decrease
    B. the volume will increase
    C. the volume will decrease
    D. the number of molecules will increase
    E. none of these

11. When a gas is heated in a closed metal container, the (8.4)
    A. pressure increases
    B. pressure decreases
    C. volume increases
    D. volume decreases
    E. number of molecules increases

Gases

12. A sample of N_2 at 110 K has a pressure of 1.0 atm. When the temperature is increased to 360 K with no change in volume and amount of gas, what is the pressure? (8.4)
  A. 0.50 atm   B. 1.0 atm   C. 1.5 atm   D. 3.3 atm   E. 4.0 atm

13. A gas sample with a volume of 4.0 L has a pressure of 750 mmHg and a temperature of 77 °C. What is its volume at 277 °C and 250 mmHg, if there is no change in the amount of gas? (8.5)
  A. 7.6 L   B. 19 L   C. 2.1 L   D. 0.00056 L   E. 3.3 L

14. A sample of 2.00 moles of gas initially at STP is converted to a volume of 5.0 L and a temperature of 27 °C. What is the pressure, in atmospheres, if there is no change in the amount of gas? (8.5, 8.6)
  A. 0.12 atm   B. 5.5 atm   C. 7.5 atm   D. 9.8 atm   E. 10. atm

15. The conditions for standard temperature and pressure (STP) are (8.6)
  A. 0 K, 1 atm   B. 0 °C, 10 atm   C. 25 °C, 1 atm
  D. 273 K, 1 atm   E. 273 K, 0.5 atm

16. The volume occupied by 1.50 moles of CH_4 at STP is (8.6)
  A. 44.10 L   B. 33.6 L   C. 22.4 L   D. 11.2 L   E. 5.60 L

17. How many grams of O_2 are present in 44.1 L of O_2 at STP? (8.6)
  A. 10.0 g   B. 16.0 g   C. 32.0 g   D. 410.0 g   E. 63.0 g

18. The pressure of a gas will increase when (8.6)
  A. the volume increases
  B. the temperature decreases
  C. more molecules of gas are added
  D. molecules of gas are removed
  E. none of these

19. If two gases have the same volume, temperature, and pressure, they also have the same (8.6)
  A. density   B. number of molecules   C. molar mass
  D. speed   E. size molecules

20. A gas mixture contains He with a partial pressure of 0.100 atm, O_2 with a partial pressure of 445 mmHg, and N_2 with a partial pressure of 235 Torr. What is the total pressure, in atmospheres, for the gas mixture? (8.7)
  A. 0.995 atm   B. 1.39 atm   C. 1.69 atm   D. 2.00 atm   E. 10.0 atm

21. A mixture of O_2 and N_2 has a total pressure of 1040 mmHg. If the O_2 has a partial pressure of 510 mmHg, what is the partial pressure of the N_2? (8.7)
  A. 240 mmHg   B. 530 mmHg   C. 770 mmHg
  D. 1040 mmHg   E. 1350 mmHg

22. 3.00 moles of He in a steel container has a pressure of 12.0 atm. What is the pressure after 4.00 moles of Ne is added, if there is no change in temperature? (8.7)
  A. 5.14 atm   B. 16.0 atm   C. 28.0 atm   D. 32.0 atm   E. 45.0 atm

Answers to the Practice Test

1. T	**2.** T	**3.** T	**4.** F	**5.** T
6. D	**7.** A	**8.** A	**9.** C	**10.** B
11. A	**12.** D	**13.** B	**14.** D	**15.** D
16. B	**17.** E	**18.** C	**19.** B	**20.** A
21. B	**22.** C			

Chapter 8

Selected Answers and Solutions to Text Problems

8.1 **a.** At a higher temperature, gas particles have greater kinetic energy, which makes them move faster.
b. Because there are great distances between the particles of a gas, they can be pushed closer together and still remain a gas.
c. Gas particles are very far apart, meaning that the mass of a gas in a certain volume is very small, resulting in a low density.

8.3 **a.** The temperature of a gas can be expressed in kelvins.
b. The volume of a gas can be expressed in milliliters.
c. The amount of a gas can be expressed in grams.
d. Pressure can be expressed in millimeters of mercury (mmHg).

8.5 Statements **a**, **d**, and **e** describe the pressure of a gas.

8.7 **a.** $2.00 \text{ atm} \times \dfrac{760 \text{ Torr}}{1 \text{ atm}} = 1520 \text{ Torr (3 SFs)}$

b. $2.00 \text{ atm} \times \dfrac{14.7 \text{ lb/in.}^2}{1 \text{ atm}} = 29.4 \text{ lb/in.}^2 \text{ (3 SFs)}$

c. $2.00 \text{ atm} \times \dfrac{760 \text{ mmHg}}{1 \text{ atm}} = 1520 \text{ mmHg (3 SFs)}$

d. $2.00 \text{ atm} \times \dfrac{101.325 \text{ kPa}}{1 \text{ atm}} = 203 \text{ kPa (3 SFs)}$

8.9 As the scuba diver ascends to the surface, external pressure decreases. If the air in the lungs, which is at a higher pressure, were not exhaled, its volume would expand and severely damage the lungs. The pressure of the gas in the lungs must adjust to changes in the external pressure.

8.11 **a.** The pressure is greater in cylinder A. According to Boyle's law, a decrease in volume pushes the gas particles closer together, which will cause an increase in the pressure.
b. From Boyle's law, we know that pressure is inversely related to volume. The mmHg units must be converted to atm for unit cancellation in the calculation, and because the pressure increases, volume must decrease.
According to Boyle's law, $P_1 V_1 = P_2 V_2$, then

$V_2 = V_1 \times \dfrac{P_1}{P_2} = 220 \text{ mL} \times \dfrac{650 \text{ mmHg}}{1.2 \text{ atm}} \times \dfrac{1 \text{ atm}}{760 \text{ mmHg}} = 160 \text{ mL (2 SFs)}$

8.13 **a.** The pressure of the gas *increases* to two times the original pressure when the volume is halved.
b. The pressure *decreases* to one-third the original pressure when the volume expands to three times its initial value.
c. The pressure *increases* to 10 times the original pressure when the volume decreases to one-tenth of the initial volume.

8.15 From Boyle's law, we know that pressure is inversely related to volume (for example, pressure increases when volume decreases).
a. Volume increases; pressure must decrease.

$P_2 = P_1 \times \dfrac{V_1}{V_2} = 655 \text{ mmHg} \times \dfrac{10.0 \text{ L}}{20.0 \text{ L}} = 328 \text{ mmHg (3 SFs)}$

b. Volume decreases; pressure must increase.

$P_2 = P_1 \times \dfrac{V_1}{V_2} = 655 \text{ mmHg} \times \dfrac{10.0 \text{ L}}{2.50 \text{ L}} = 2620 \text{ mmHg (3 SFs)}$

c. The mL units must be converted to L for unit cancellation in the calculation, and because the volume increases, pressure must decrease.

$$P_2 = P_1 \times \frac{V_1}{V_2} = 655 \text{ mmHg} \times \frac{10.0 \text{ L}}{13\,800 \text{ mL}} \times \frac{1000 \text{ mL}}{1 \text{ L}} = 475 \text{ mmHg (3 SFs)}$$

d. The mL units must be converted to L for unit cancellation in the calculation, and because the volume decreases, pressure must increase.

$$P_2 = P_1 \times \frac{V_1}{V_2} = 655 \text{ mmHg} \times \frac{10.0 \text{ L}}{1250 \text{ mL}} \times \frac{1000 \text{ mL}}{1 \text{ L}} = 5240 \text{ mmHg (3 SFs)}$$

8.17 From Boyle's law, we know that pressure is inversely related to volume.
  a. Pressure decreases; volume must increase.

$$V_2 = V_1 \times \frac{P_1}{P_2} = 50.0 \text{ L} \times \frac{760. \text{ mmHg}}{725 \text{ mmHg}} = 52.4 \text{ L (3 SFs)}$$

  b. The mmHg units must be converted to atm for unit cancellation in the calculation, and because the pressure increases, volume must decrease.

$$P_1 = 760. \text{ mmHg} \times \frac{1 \text{ atm}}{760 \text{ mmHg}} = 1.00 \text{ atm}$$

$$V_2 = V_1 \times \frac{P_1}{P_2} = 50.0 \text{ L} \times \frac{1.00 \text{ atm}}{2.0 \text{ atm}} = 25 \text{ L (2 SFs)}$$

  c. The mmHg units must be converted to atm for unit cancellation in the calculation, and because the pressure decreases, volume must increase.

$$P_1 = 760. \text{ mmHg} \times \frac{1 \text{ atm}}{760 \text{ mmHg}} = 1.00 \text{ atm}$$

$$V_2 = V_1 \times \frac{P_1}{P_2} = 50.0 \text{ L} \times \frac{1.00 \text{ atm}}{0.500 \text{ atm}} = 100. \text{ L (3 SFs)}$$

  d. The mmHg units must be converted to torr for unit cancellation in the calculation, and because the pressure increases, volume must decrease.

$$P_1 = 760. \text{ mmHg} \times \frac{760 \text{ Torr}}{760 \text{ mmHg}} = 760. \text{ Torr}$$

$$V_2 = V_1 \times \frac{P_1}{P_2} = 50.0 \text{ L} \times \frac{760. \text{ Torr}}{850 \text{ Torr}} = 45 \text{ L (2 SFs)}$$

8.19 Volume increases; initial pressure must have been higher.

$$P_1 = P_2 \times \frac{V_2}{V_1} = 3.62 \text{ atm} \times \frac{9.73 \text{ L}}{5.40 \text{ L}} = 6.52 \text{ atm (3 SFs)}$$

8.21 Pressure decreases; volume must increase.

$$V_2 = V_1 \times \frac{P_1}{P_2} = 5.0 \text{ L} \times \frac{5.0 \text{ atm}}{1.0 \text{ atm}} = 25 \text{ L (2 SFs)}$$

8.23
  a. *Inspiration* begins when the diaphragm contracts, causing the lungs to expand. The increased volume of the thoracic cavity reduces the pressure in the lungs such that air flows into the lungs.
  b. *Expiration* occurs as the diaphragm relaxes, causing a decrease in the volume of the lungs. The pressure of the air in the lungs increases, and air flows out of the lungs.
  c. *Inspiration* occurs when the pressure within the lungs is less than that of the atmosphere.

8.25 According to Charles's law, there is a direct relationship between Kelvin temperature and volume (for example, volume increases when temperature increases, if the pressure and amount of gas do not change).
  a. Diagram C shows an increased volume corresponding to an increase in temperature.
  b. Diagram A shows a decreased volume corresponding to a decrease in temperature.
  c. Diagram B shows no change in volume, which corresponds to no net change in temperature.

8.27 According to Charles's law, the volume of a gas is directly related to the Kelvin temperature. In all gas law computations, temperatures must be in kelvins. (Temperatures in °C are converted to K by the addition of 273.) The initial temperature for all cases here is $T_1 = 15\,°C + 273 = 288\,K$.
  a. Volume increases; temperature must have increased.
  $$T_2 = T_1 \times \frac{V_2}{V_1} = 288\,K \times \frac{5.00\,L}{2.50\,L} = 576\,K \quad 576\,K - 273 = 303\,°C\;(3\;SFs)$$
  b. Volume decreases; temperature must have decreased.
  $$T_2 = T_1 \times \frac{V_2}{V_1} = 288\,K \times \frac{1250\,mL}{2.50\,L} \times \frac{1\,L}{1000\,mL} = 144\,K \quad 144\,K - 273 = -129\,°C\;(3\;SFs)$$
  c. Volume increases; temperature must have increased.
  $$T_2 = T_1 \times \frac{V_2}{V_1} = 288\,K \times \frac{7.50\,L}{2.50\,L} = 864\,K \quad 864\,K - 273 = 591\,°C\;(3\;SFs)$$
  d. Volume increases; temperature must have increased.
  $$T_2 = T_1 \times \frac{V_2}{V_1} = 288\,K \times \frac{3550\,mL}{2.50\,L} \times \frac{1\,L}{1000\,mL} = 409\,K \quad 409\,K - 273 = 136\,°C\;(3\;SFs)$$

8.29 According to Charles's law, the volume of a gas is directly related to the Kelvin temperature. In all gas law computations, temperatures must be in kelvins. (Temperatures in °C are converted to K by the addition of 273.) The initial temperature for all cases here is $T_1 = 75\,°C + 273 = 348\,K$.
  a. When temperature decreases, volume must also decrease.
  $T_2 = 55\,°C + 273 = 328\,K$
  $$V_2 = V_1 \times \frac{T_2}{T_1} = 2500\,mL \times \frac{328\,K}{348\,K} = 2400\,mL\;(2\;SFs)$$
  b. When temperature increases, volume must also increase.
  $$V_2 = V_1 \times \frac{T_2}{T_1} = 2500\,mL \times \frac{680.\,K}{348\,K} = 4900\,mL\;(2\;SFs)$$
  c. When temperature decreases, volume must also decrease.
  $T_2 = -25\,°C + 273 = 248\,K$
  $$V_2 = V_1 \times \frac{T_2}{T_1} = 2500\,mL \times \frac{248\,K}{348\,K} = 1800\,mL\;(2\;SFs)$$
  d. When temperature decreases, volume must also decrease.
  $$V_2 = V_1 \times \frac{T_2}{T_1} = 2500\,mL \times \frac{240.\,K}{348\,K} = 1700\,mL\;(2\;SFs)$$

8.31 Volume decreases; initial temperature must have been higher.
  $T_2 = 32\,°C + 273 = 305\,K$
  $$T_1 = T_2 \times \frac{V_1}{V_2} = 305\,K \times \frac{0.256\,L}{0.198\,L} = 394\,K \quad 394\,K - 273 = 121\,°C\;(3\;SFs)$$

8.33 According to Gay-Lussac's law, temperature is directly related to pressure. In all gas law computations, temperatures must be in kelvins. (Temperatures in °C are converted to K by the addition of 273.)

a. When temperature decreases, pressure must also decrease.

$T_1 = 155\,°C + 273 = 428\,K \qquad T_2 = 0\,°C + 273 = 273\,K$

$P_2 = P_1 \times \dfrac{T_2}{T_1} = 1200\,\text{Torr} \times \dfrac{1\,\text{mmHg}}{1\,\text{Torr}} \times \dfrac{273\,K}{428\,K} = 770\,\text{mmHg (2 SFs)}$

b. When temperature increases, pressure must also increase.

$T_1 = 12\,°C + 273 = 285\,K \qquad T_2 = 35\,°C + 273 = 308\,K$

$P_2 = P_1 \times \dfrac{T_2}{T_1} = 1.40\,\text{atm} \times \dfrac{760\,\text{mmHg}}{1\,\text{atm}} \times \dfrac{308\,K}{285\,K} = 1150\,\text{mmHg (3 SFs)}$

8.35 According to Gay-Lussac's law, pressure is directly related to temperature. In all gas law computations, temperatures must be in kelvins. (Temperatures in °C are converted to K by the addition of 273.)

a. Pressure decreases; temperature must have decreased.

$T_1 = 25\,°C + 273 = 298\,K$

$T_2 = T_1 \times \dfrac{P_2}{P_1} = 298\,K \times \dfrac{620.\,\text{mmHg}}{740.\,\text{mmHg}} = 250.\,K \quad 250.\,K - 273 = -23\,°C\ (2\,\text{SFs})$

b. Pressure increases; temperature must have increased.

$T_1 = -18\,°C + 273 = 255\,K$

$T_2 = T_1 \times \dfrac{P_2}{P_1} = 255\,K \times \dfrac{1250\,\text{Torr}}{0.950\,\text{atm}} \times \dfrac{1\,\text{atm}}{760\,\text{Torr}} = 441\,K \quad 441\,K - 273 = 168\,°C\ (3\,\text{SFs})$

8.37 Pressure increases; temperature must have increased.

$T_1 = 22\,°C + 273 = 295\,K$

$T_2 = \dfrac{T_1}{P_1} \times P_2 = 295\,K \times \dfrac{766\,\text{mmHg}}{744\,\text{mmHg}} = 304\,K \quad 304\,K - 273 = 31\,°C\ (2\,\text{SFs})$

8.39 According to Gay-Lussac's law, pressure is directly related to temperature. In all gas law computations, temperatures must be in kelvins. (Temperatures in °C are converted to K by the addition of 273.)

When temperature increases, pressure must also increase.

$T_1 = 5\,°C + 273 = 278\,K \quad T_2 = 22\,°C + 273 = 295\,K$

$P_2 = P_1 \times \dfrac{T_2}{T_1} = 1.8\,\text{atm} \times \dfrac{295\,K}{278\,K} = 1.9\,\text{atm (2 SFs)}$

8.41 $\dfrac{P_1 V_1}{T_1} = \dfrac{P_2 V_2}{T_2} \qquad \dfrac{P_1 V_1}{T_1} \times \dfrac{T_2 T_1}{P_1 V_1} = \dfrac{P_2 V_2}{T_2} \times \dfrac{T_2 T_1}{P_1 V_1} \qquad \therefore\ T_2 = T_1 \times \dfrac{P_2}{P_1} \times \dfrac{V_2}{V_1}$

8.43 $T_1 = 25\,°C + 273 = 298\,K;\qquad V_1 = 6.50\,L;\qquad P_1 = 845\,\text{mmHg}$

a. $T_2 = 325\,K;\qquad V_2 = 1850\,\text{mL} = 1.85\,L;\qquad P_2 =\ ?\ \text{atm}$

$P_2 = P_1 \times \dfrac{V_1}{V_2} \times \dfrac{T_2}{T_1} = 845\,\text{mmHg} \times \dfrac{6.50\,L}{1.85\,L} \times \dfrac{325\,K}{298\,K} \times \dfrac{1\,\text{atm}}{760\,\text{mmHg}} = 4.26\,\text{atm (3 SFs)}$

b. $T_2 = 12\,°C + 273 = 285\,K;\qquad V_2 = 2.25\,L;\qquad P_2 =\ ?\ \text{atm}$

$P_2 = P_1 \times \dfrac{V_1}{V_2} \times \dfrac{T_2}{T_1} = 845\,\text{mmHg} \times \dfrac{6.50\,L}{2.25\,L} \times \dfrac{285\,K}{298\,K} \times \dfrac{1\,\text{atm}}{760\,\text{mmHg}} = 3.07\,\text{atm (3 SFs)}$

c. $T_2 = 47\,°C + 273 = 320.\,K;\qquad V_2 = 12.8\,L;\qquad P_2 =\ ?\ \text{atm}$

$P_2 = P_1 \times \dfrac{V_1}{V_2} \times \dfrac{T_2}{T_1} = 845\,\text{mmHg} \times \dfrac{6.50\,L}{12.8\,L} \times \dfrac{320.\,K}{298\,K} \times \dfrac{1\,\text{atm}}{760\,\text{mmHg}} = 0.606\,\text{atm (3 SFs)}$

Chapter 8

8.45 $T_1 = 212\,°C + 273 = 485\,K;$ $\quad V_1 = 124\,mL;\quad P_1 = 1.80\,atm$
$T_2 = ?\,°C;\quad V_2 = 138\,mL;\quad P_2 = 0.800\,atm$
$T_2 = T_1 \times \dfrac{P_2}{P_1} \times \dfrac{V_2}{V_1} = 485\,K \times \dfrac{0.800\,\text{atm}}{1.80\,\text{atm}} \times \dfrac{138\,\text{mL}}{124\,\text{mL}} = 240.\,K$
$240.\,K - 273 = -33\,°C\ (2\,SFs)$

8.47 The volume increases because the number of gas particles in the tire or basketball is increased.

8.49 According to Avogadro's law, the volume of a gas is directly related to the number of moles of gas.
$n_1 = 1.50$ moles Ne;$\quad V_1 = 8.00\,L$

a. $V_2 = V_1 \times \dfrac{n_2}{n_1} = 8.00\,L \times \dfrac{\tfrac{1}{2}(1.50)\ \text{moles Ne}}{1.50\ \text{moles Ne}} = 4.00\,L\ (3\,SFs)$

b. $n_2 = 1.50$ moles Ne $+ 3.50$ moles Ne $= 5.00$ moles of Ne
$V_2 = V_1 \times \dfrac{n_2}{n_1} = 8.00\,L \times \dfrac{5.00\ \text{moles Ne}}{1.50\ \text{moles Ne}} = 26.7\,L\ (3\,SFs)$

c. $25.0\ \text{g Ne} \times \dfrac{1\ \text{mole Ne}}{20.18\ \text{g Ne}} = 1.24$ moles of Ne added
$n_2 = 1.50$ moles Ne $+ 1.24$ moles Ne $= 2.74$ moles of Ne
$V_2 = V_1 \times \dfrac{n_2}{n_1} = 8.00\,L \times \dfrac{2.74\ \text{moles Ne}}{1.50\ \text{moles Ne}} = 14.6\,L\ (3\,SFs)$

8.51 At STP, 1 mole of any gas occupies a volume of 22.4 L.

a. $44.8\ \text{L}\,O_2\ (STP) \times \dfrac{1\ \text{mole}\ O_2}{22.4\ \text{L}\,O_2\ (STP)} = 2.00$ moles of $O_2\ (3\,SFs)$

b. $2.50\ \text{moles}\ N_2 \times \dfrac{22.4\ \text{L}\ N_2\ (STP)}{1\ \text{mole}\ N_2} = 56.0\,L$ at STP $(3\,SFs)$

c. $50.0\ \text{g Ar} \times \dfrac{1\ \text{mole Ar}}{39.95\ \text{g Ar}} \times \dfrac{22.4\ \text{L Ar}\ (STP)}{1\ \text{mole Ar}} = 28.0\,L$ at STP $(3\,SFs)$

d. $1620\ \text{mL}\,H_2\ (STP) \times \dfrac{1\ \text{L}\,H_2}{1000\ \text{mL}\,H_2} \times \dfrac{1\ \text{mole}\,H_2}{22.4\ \text{L}\,H_2\ (STP)} \times \dfrac{2.016\ \text{g}\,H_2}{1\ \text{mole}\,H_2}$
$= 0.146\,g$ of $H_2\ (3\,SFs)$

8.53 In a gas mixture, the pressure that each gas exerts as part of the total pressure is called the partial pressure of that gas. Because the air sample is a mixture of gases, the total pressure is the sum of the partial pressures of each gas in the sample.

8.55 To obtain the total pressure of a gas mixture, add up all of the partial pressures using the same pressure unit.
$P_{\text{total}} = P_{\text{nitrogen}} + P_{\text{oxygen}} + P_{\text{helium}} = 425\ \text{Torr} + 115\ \text{Torr} + 225\ \text{Torr} = 765\ \text{Torr}\ (3\,SFs)$

8.57 Because the total pressure of a gas mixture is the sum of the partial pressures using the same pressure unit, addition and subtraction are used to obtain the "missing" partial pressure.
$P_{\text{nitrogen}} = P_{\text{total}} - (P_{\text{oxygen}} + P_{\text{helium}}) = 925\ \text{Torr} - (425\ \text{Torr} + 75\ \text{Torr}) = 425\ \text{Torr}\ (3\,SFs)$

8.59 a. If oxygen cannot readily cross from the lungs into the bloodstream, then the partial pressure of oxygen will be lower in the blood of an emphysema patient.
b. Breathing a higher concentration of oxygen will help to increase the supply of oxygen in the lungs and blood and raise the partial pressure of oxygen in the blood.

Gases

8.61 $T_1 = 37\,°C + 273 = 310\,K;$ $\quad P_1 = 745\,mmHg;$ $\quad V_1 = 3.2\,L$
$T_2 = 0\,°C + 273 = 273\,K;$ $\quad P_2 = 760\,mmHg;$ $\quad V_2 = ?\,L$
$V_2 = V_1 \times \dfrac{P_1}{P_2} \times \dfrac{T_2}{T_1} = 3.2\,L \times \dfrac{745\,mmHg}{760\,mmHg} \times \dfrac{273\,K}{310\,K} = 2.8\,L\,(2\,SFs)$

8.63 **a.** True. The flask containing helium has more moles of helium and thus more atoms of helium to collide with the walls of the container to give a higher pressure.
b. True. The mass and volume of each are the same, meaning that the mass/volume ratio or density is the same in both flasks.

8.65 **a.** 2; The fewest number of gas particles will exert the lowest pressure.
b. 1; The greatest number of gas particles will exert the highest pressure.

8.67 **a.** A; Volume decreases when temperature decreases (when P and n do not change).
b. C; Volume increases when pressure decreases (when n and T do not change).
c. A; Volume decreases when the number of moles of gas decreases (when T and P do not change).
d. B; Doubling the Kelvin temperature would double the volume, but when half of the gas escapes, the volume would decrease by half. These two opposing effects cancel each other, and there is no overall change in the volume (when P does not change).
e. C; Increasing the moles of gas causes an increase in the volume (when T and P do not change)

8.69 **a.** The volume of the chest and lungs will decrease when compressed during the Heimlich maneuver.
b. A decrease in volume causes the pressure to increase. A piece of food would be dislodged with a sufficiently high pressure.

8.71 $31\,000\,L\,H_2\,(STP) \times \dfrac{1\,mole\,H_2}{22.4\,L\,H_2\,(STP)} \times \dfrac{2.016\,g\,H_2}{1\,mole\,H_2} \times \dfrac{1\,kg\,H_2}{1000\,g\,H_2} = 2.8\,kg\,of\,H_2\,(2\,SFs)$

8.73 $T_1 = 127\,°C + 273 = 400.\,K;$ $\quad P_1 = 2.00\,atm$
$T_2 = ?\,°C;$ $\quad P_2 = 0.25\,atm$
$T_2 = T_1 \times \dfrac{P_2}{P_1} = 400.\,K \times \dfrac{0.25\,atm}{2.00\,atm} = 50.\,K \quad 50.\,K - 273 = -223\,°C$

8.75 $T_1 = 8\,°C + 273 = 281\,K;$ $\quad P_1 = 380\,Torr \times \dfrac{1\,atm}{760\,Torr} = 0.50\,atm;$ $\quad V_1 = 750\,L$
$T_2 = -45\,°C + 273 = 228\,K;$ $\quad P_2 = 0.20\,atm;$ $\quad V_2 = ?\,L$
$V_2 = V_1 \times \dfrac{P_1}{P_2} \times \dfrac{T_2}{T_1} = 750\,L \times \dfrac{0.50\,atm}{0.20\,atm} \times \dfrac{228\,K}{281\,K} = 1500\,L\,(2\,SFs)$

8.77 **a.** $25.0\,L\,He\,(STP) \times \dfrac{1\,mole\,He}{22.4\,L\,He\,(STP)} \times \dfrac{4.003\,g\,He}{1\,mole\,He} = 4.47\,g\,of\,He\,(3\,SFs)$
b. $T_1 = 0\,°C + 273 = 273\,K;$ $\quad V_1 = 25.0\,L;$ $\quad P_1 = 1.00\,atm = 760\,mmHg$
$T_2 = -35\,°C + 273 = 238\,K;$ $\quad V_2 = 2460\,L;$ $\quad P_2 = ?\,mmHg$
$P_2 = P_1 \times \dfrac{V_1}{V_2} \times \dfrac{T_2}{T_1} = 760\,mmHg \times \dfrac{25.0\,L}{2460\,L} \times \dfrac{238\,K}{273\,K}$
$= 6.73\,mmHg\,(3\,SFs)$

8.79 Because the partial pressure of nitrogen is to be reported in torr, the atm and mmHg units (for oxygen and argon, respectively) must be converted to torr, as follows:
$P_{oxygen} = 0.60\,atm \times \dfrac{760\,Torr}{1\,atm} = 460\,Torr \quad$ and
$P_{argon} = 425\,mmHg \times \dfrac{1\,Torr}{1\,mmHg} = 425\,Torr$

Chapter 8

$$\therefore P_{\text{nitrogen}} = P_{\text{total}} - (P_{\text{oxygen}} + P_{\text{argon}})$$
$$= 1250 \text{ Torr} - (460 \text{ Torr} + 425 \text{ Torr}) = 370 \text{ Torr (2 SFs)}$$

8.81 $T_1 = 24\,°\text{C} + 273 = 297 \text{ K};\quad P_1 = 745 \text{ mmHg};\quad V_1 = 425 \text{ mL}$

$T_2 = -95\,°\text{C} + 273 = 178 \text{ K};\quad P_2 = 0.115 \text{ atm} \times \dfrac{760 \text{ mmHg}}{1 \text{ atm}} = 87.4 \text{ mmHg};\quad V_2 = ? \text{ mL}$

$V_2 = V_1 \times \dfrac{P_1}{P_2} \times \dfrac{T_2}{T_1} = 425 \text{ mL} \times \dfrac{745 \text{ mmHg}}{87.4 \text{ mmHg}} \times \dfrac{178 \text{ K}}{297 \text{ K}} = 2170 \text{ mL (3 SFs)}$

8.83 $T_1 = 15\,°\text{C} + 273 = 288 \text{ K};\quad P_1 = 745 \text{ mmHg};\quad V_1 = 4250 \text{ mL}$

$T_2 = ?\,°\text{C};\quad P_2 = 1.20 \text{ atm} \times \dfrac{760 \text{ mmHg}}{1 \text{ atm}} = 912 \text{ mmHg};$

$V_2 = 2.50 \text{ L} \times \dfrac{1000 \text{ mL}}{1 \text{ L}} = 2.50 \times 10^3 \text{ mL}$

$T_2 = T_1 \times \dfrac{V_2}{V_1} \times \dfrac{P_2}{P_1} = 288 \text{ K} \times \dfrac{2.50 \times 10^3 \text{ mL}}{4250 \text{ mL}} \times \dfrac{912 \text{ mmHg}}{745 \text{ mmHg}} = 207 \text{ K} - 273$
$= -66\,°\text{C (2 SFs)}$

9 Solutions

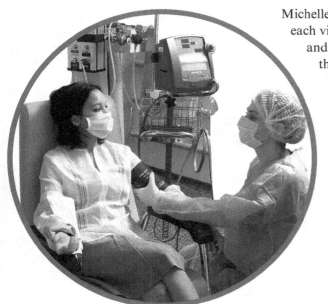

Michelle has hemodialysis three times a week for about 4 h each visit. The treatment removes harmful waste products and extra fluid. Her blood pressure is maintained because the electrolytes potassium and sodium are adjusted each time. Michelle measures the amount of fluid she takes in each day. Her diet is low in protein and includes fruits and vegetables that are low in potassium. She and her husband are now learning about performing hemodialysis at home. Soon she will have her own hemodialysis machine. A typical potassium level in blood plasma is 0.091 g in 500. mL of plasma. What is the molarity of the potassium ion in plasma?

Credit: AJPhoto/Science Source

LOOKING AHEAD

9.1 Solutions
9.2 Electrolytes and Nonelectrolytes
9.3 Solubility
9.4 Solution Concentrations
9.5 Dilution of Solutions
9.6 Properties of Solutions

 The Health icon indicates a question that is related to health and medicine.

9.1 Solutions

Learning Goal: Identify the solute and solvent in a solution; describe the formation of a solution.

> **REVIEW**
> Identifying Polarity of Molecules and Intermolecular Forces (6.9)

- A solution is a homogeneous mixture that forms when a solute dissolves in a solvent. The solvent is the substance that is present in a greater amount.
- Solvents and solutes can be solids, liquids, or gases. The solution is the same physical state as the solvent.
- A polar solute is soluble in a polar solvent; a nonpolar solute is soluble in a nonpolar solvent.
- Water molecules form hydrogen bonds because the partial positive charge of the hydrogen in one water molecule is attracted to the partial negative charge of oxygen in another water molecule.
- An ionic solute dissolves in water, a polar solvent, because the polar water molecules attract and pull the positive and negative ions into solution. In solution, water molecules surround the ions in a process called hydration.

Chapter 9

♦ **Learning Exercise 9.1A**

Indicate the solute and solvent in each of the following:

	Solute	Solvent
a. 10 g of KCl dissolved in 100 g of water	_____	_____
b. soda water: $CO_2(g)$ dissolved in water	_____	_____
c. an alloy composed of 80% Zn and 20% Cu	_____	_____
d. a mixture of O_2 (200 mmHg) and He (500 mmHg)	_____	_____
e. a solution of 40 mL of CCl_4 and 2 mL of Br_2	_____	_____

Answers   a. KCl; water   b. CO_2; water   c. Cu; Zn
            d. O_2; He   e. Br_2; CCl_4

Solute: The substance present in lesser amount

Solvent: The substance present in greater amount

A solution consists of at least one solute dispersed in a solvent.

Credit: Pearson Education / Pearson Science

♦ **Learning Exercise 9.1B**

Water is polar, and hexane is nonpolar. In which solvent is each of the following solutes soluble?

a. bromine, Br_2, nonpolar _____   b. HCl, polar _____

c. cholesterol, nonpolar _____   d. vitamin D, nonpolar _____

e. vitamin C, polar _____

Answers   a. hexane   b. water   c. hexane
            d. hexane   e. water

9.2 Electrolytes and Nonelectrolytes

Learning Goal: Identify solutes as electrolytes or nonelectrolytes.

- Electrolytes conduct an electrical current because they produce ions in aqueous solutions.
- Strong electrolytes are completely dissociated into ions, whereas weak electrolytes are partially dissociated into ions.
- Nonelectrolytes do not form ions in solution but dissolve as molecules.
- Ions are measured as equivalents (Eq), which is the amount of the ion that is equal to 1 mole of positive or negative electrical charge.

REVIEW
Writing Conversion Factors from Equalities (2.5)
Using Conversion Factors (2.6)
Writing Positive and Negative Ions (6.1)

Key Terms for Sections 9.1 and 9.2

Match each of the following key terms with the correct description:

a. solution   b. nonelectrolyte   c. equivalent
d. solute   e. strong electrolyte   f. solvent

1. _____ a substance that dissociates as molecules, not ions, when it dissolves in water

2. _____ the substance that comprises the lesser amount in a solution

Solutions

3. ____ a solute that dissociates 100% into ions in solution
4. ____ the substance that comprises the greater amount in a solution
5. ____ a homogeneous mixture of at least two components called a solute and a solvent
6. ____ the amount of positive or negative ion that supplies 1 mole of electrical charge

Answers   **1.** b   **2.** d   **3.** e   **4.** f   **5.** a   **6.** c

♦ **Learning Exercise 9.2A**

Write an equation for the formation of an aqueous solution of each of the following strong electrolytes:

a. LiCl

b. $Zn(NO_3)_2$

c. Na_3PO_4

d. K_2SO_4

e. $MgCl_2$

Answers
     a. $LiCl(s) \xrightarrow{H_2O} Li^+(aq) + Cl^-(aq)$
     b. $Zn(NO_3)_2(s) \xrightarrow{H_2O} Zn^{2+}(aq) + 2NO_3^-(aq)$
     c. $Na_3PO_4(s) \xrightarrow{H_2O} 3Na^+(aq) + PO_4^{3-}(aq)$
     d. $K_2SO_4(s) \xrightarrow{H_2O} 2K^+(aq) + SO_4^{2-}(aq)$
     e. $MgCl_2(s) \xrightarrow{H_2O} Mg^{2+}(aq) + 2Cl^-(aq)$

♦ **Learning Exercise 9.2B**

Indicate whether an aqueous solution of each of the following contains only ions, only molecules, or mostly molecules and a few ions. Write an equation for the formation of the solution.

a. glucose, $C_6H_{12}O_6$, a nonelectrolyte

b. NaOH, a strong electrolyte

c. $CaCl_2$, a strong electrolyte

d. HF, a weak electrolyte

Chapter 9

Answers
a. $C_6H_{12}O_6(s) \xrightarrow{H_2O} C_6H_{12}O_6(aq)$ only molecules
b. $NaOH(s) \xrightarrow{H_2O} Na^+(aq) + OH^-(aq)$ only ions
c. $CaCl_2(s) \xrightarrow{H_2O} Ca^{2+}(aq) + 2Cl^-(aq)$ only ions
d. $HF(aq) \xrightleftharpoons{H_2O} H^+(aq) + F^-(aq)$ mostly molecules and a few ions

♦ Learning Exercise 9.2C

Calculate the following:

a. the number of equivalents in 1 mole of Zn^{2+}

b. the number of equivalents in 2.5 moles of Cl^-

c. the number of equivalents in 2.0 moles of Ca^{2+}

Answers   a. 2 Eq   b. 2.5 Eq   c. 4.0 Eq

9.3 Solubility

Learning Goal: Define solubility; distinguish between an unsaturated and a saturated solution. Identify an ionic compound as soluble or insoluble.

REVIEW
Interpreting Graphs (1.4)

- The amount of solute that dissolves depends on the nature of the solute and solvent.
- Solubility describes the maximum amount of a solute that dissolves in exactly 100. g of solvent at a given temperature.
- A saturated solution contains the maximum amount of dissolved solute at a certain temperature whereas an unsaturated solution contains less than this amount.
- An increase in temperature increases the solubility of most solids, but decreases the solubility of gases in water.
- Henry's law states that the solubility of gases in liquids is directly related to the pressure of that gas above the liquid.
- The solubility rules describe the kinds of ionic combinations that are soluble and insoluble in water.

CdS

PbI_2

If an ionic compound contains a combination of a cation and an anion such as Cd^{2+} and S^{2-} or Pb^{2+} and I^-, that ionic compound is insoluble.

Credit: Pearson Education/Pearson Science

Solutions

♦ **Learning Exercise 9.3A**

Identify each of the following solutions as saturated or unsaturated:

a. _____ A sugar cube dissolves when added to a cup of hot tea.
b. _____ A KCl crystal added to a KCl solution does not change in size.
c. _____ A layer of sugar forms in the bottom of a glass of iced tea.
d. _____ The rate of crystal formation is equal to the rate of dissolving.
e. _____ Upon heating, all the sugar in a solution dissolves.

Answers   a. unsaturated   b. saturated   c. saturated
            d. saturated   e. unsaturated

♦ **Learning Exercise 9.3B**

Use the solubility data for $NaNO_3$ listed below to answer each of the following problems:

Temperature (°C)	Solubility (g of $NaNO_3$/100. g of H_2O)
40	100. g
60	120. g
80	150. g
100	200. g

a. How many grams of $NaNO_3$ will dissolve in 100. g of water at 40 °C?

b. How many grams of $NaNO_3$ will dissolve in 300. g of water at 60 °C?

c. A solution is prepared using 200. g of water and 350. g of $NaNO_3$ at 80 °C. Will any solute remain undissolved? If so, how much?

d. Will 150. g of $NaNO_3$ all dissolve when added to 100. g of water at 100 °C?

Answers   a. 100. g   b. 360. g
            c. 300. g of $NaNO_3$ will dissolve leaving 50. g of $NaNO_3$ that will not dissolve.
            d. Yes, all 150. g of $NaNO_3$ will dissolve.

Solubility Rules for Ionic Compounds in Water

An ionic compound is soluble in water if it contains one of the following:

Positive Ions: Li^+, Na^+, K^+, Rb^+, Cs^+, NH_4^+
Negative Ions: NO_3^-, $C_2H_3O_2^-$
                 Cl^-, Br^-, I^- except when combined with Ag^+, Pb^{2+}, or Hg_2^{2+}
                 SO_4^{2-} except when combined with Ba^{2+}, Pb^{2+}, Ca^{2+}, Sr^{2+}, or Hg_2^{2+}

Ionic compounds that do not contain at least one of these ions are usually insoluble.

Chapter 9

♦ **Learning Exercise 9.3C**

> **CORE CHEMISTRY SKILL**
> Using Solubility Rules

Predict whether the following ionic compounds are soluble or insoluble in water:

a. _____ NaCl b. _____ $LiNO_3$ c. _____ $SrSO_4$

d. _____ Ag_2S e. _____ $BaSO_4$ f. _____ Na_2CO_3

g. _____ Na_2S h. _____ $MgCl_2$ i. _____ BaS

Answers a. soluble b. soluble c. insoluble d. insoluble e. insoluble
 f. soluble g. soluble h. soluble i. insoluble

9.4 Solution Concentrations

Learning Goal: Calculate the concentration of a solute in a solution; use concentration as a conversion factor to calculate the amount of solute or solution.

> **REVIEW**
> Writing Conversion Factors from Equalities (2.5)
> Using Conversion Factors (2.6)
> Using Molar Mass as a Conversion Factor (7.3)

- The concentration of a solution is the relationship between the amount of solute, in grams, moles, or milliliters, and the amount of solution, in grams, milliliters, or liters.

- A mass percent (m/m) expresses the ratio of the mass of solute to the mass of solution multiplied by 100%. Mass units must be the same such as grams of solute/grams of solution × 100%.

$$\text{Mass percent (m/m)} = \frac{\text{mass of solute}}{\text{mass of solution}} \times 100\%$$

- A volume percent (v/v) expresses the ratio of the volume of solute to volume of solution multiplied by 100%. Volume units must be the same such as milliliters of solute/milliliters of solution × 100%.

$$\text{Volume percent (v/v)} = \frac{\text{volume of solute}}{\text{volume of solution}} \times 100\%$$

- A mass/volume percent (m/v) expresses the ratio of the mass of solute to the volume of solution multiplied by 100%. Mass units for the solute must be in grams and the volume units for the solution must be in milliliters.

$$\text{Mass/volume percent (m/v)} = \frac{\text{grams of solute}}{\text{milliliters of solution}} \times 100\%$$

- Molarity is a concentration term that indicates the number of moles of solute dissolved in 1 L (1000 mL) of solution.

$$\text{Molarity (M)} = \frac{\text{moles of solute}}{\text{liters of solution}}$$

- Concentrations can be used as conversion factors to relate the amount of solute and the volume or mass of the solution.

$$\frac{\text{amount of solute}}{\text{amount of solution}} \quad \text{and} \quad \frac{\text{amount of solution}}{\text{amount of solute}}$$

Key Terms for Sections 9.3 and 9.4

Match each of the following key terms with the correct description:

a. mass percent (m/m)   b. concentration   c. solubility
d. unsaturated   e. molarity   f. saturated

1. _____ the number of moles of solute in 1 L of solution
2. _____ the amount of solute dissolved in a certain amount of solution
3. _____ a solution containing the maximum amount of solute that can dissolve at a given temperature
4. _____ a solution with less than the maximum amount of solute that can dissolve
5. _____ the concentration of a solution in terms of mass of solute in 100. g of solution
6. _____ the maximum amount of solute that can dissolve in 100. g of solvent at a given temperature

Answers   1. e   2. b   3. f   4. d   5. a   6. c

SAMPLE PROBLEM Calculating Percent Concentration

What is the mass percent (m/m) when 2.4 g of $NaHCO_3$ dissolves in water to make 120. g of a solution of $NaHCO_3$?

Solution Guide:

STEP 1 State the given and needed quantities.

Analyze the Problem	Given	Need	Connect
	2.4 g of $NaHCO_3$ solute, 120. g of $NaHCO_3$ solution	mass percent (m/m)	$\dfrac{\text{mass of solute}}{\text{mass of solution}} \times 100\%$

STEP 2 Write the concentration expression.

$$\text{Mass percent (m/m)} = \dfrac{\text{grams of solute}}{\text{grams of solution}} \times 100\%$$

STEP 3 Substitute solute and solution quantities into the expression and calculate.

$$\text{Mass percent (m/m)} = \dfrac{2.4 \text{ g NaHCO}_3}{120. \text{ g NaHCO}_3 \text{ solution}} \times 100\% = 2.0\% \text{ (m/m)}$$

◆ Learning Exercise 9.4A

Calculate the percent concentration for each of the following solutions:

CORE CHEMISTRY SKILL
Calculating Concentration

a. mass percent (m/m) for 18.0 g of NaCl in 90.0 g of a solution of NaCl

b. mass/volume percent (m/v) for 4.0 g of KOH in 50.0 mL of a solution of KOH

Chapter 9

c. mass/volume percent (m/v) for 5.0 g of KCl in 2.0 L of a solution of KCl

d. volume percent (v/v) for 18 mL of ethanol in 350 mL of a mouthwash solution

Answers   a. 20.0% (m/m)   b. 8.0% (m/v)   c. 0.25% (m/v)   d. 5.1% (v/v)

SAMPLE PROBLEM   Using Concentration to Calculate Mass of Solute

How many grams of KI are needed to prepare 225 g of a 4.0% (m/m) KI solution?

Solution Guide:

STEP 1 State the given and needed quantities.

	Given	Need	Connect
Analyze the Problem	225 g of 4.0% (m/m) KI solution	grams of KI	mass percent factor $\dfrac{\text{g of solute}}{100.\text{ g of solution}}$

STEP 2 Write a plan to calculate the mass.

grams of KI solution $\xrightarrow{\%\text{ (m/m) factor}}$ grams of KI

STEP 3 Write equalities and conversion factors.

4.0 g of KI = 100.0 g of KI solution

$$\dfrac{4.0 \text{ g KI}}{100. \text{ g KI solution}} \quad \text{and} \quad \dfrac{100. \text{ g KI solution}}{4.0 \text{ g KI}}$$

STEP 4 Set up the problem to calculate the mass.

$$225 \text{ g KI solution} \times \dfrac{4.0 \text{ g KI}}{100. \text{ g KI solution}} = 9.0 \text{ g of KI}$$

♦ **Learning Exercise 9.4B**

Calculate the number of grams of solute in each of the following solutions:

CORE CHEMISTRY SKILL
Using Concentration as a Conversion Factor

a. grams of glucose ($C_6H_{12}O_6$) in 480 g of a 5.0% (m/m) glucose solution

b. grams of lidocaine hydrochloride in 50.0 g of a 2.0% (m/m) lidocaine hydrochloride solution

c. grams of KCl in 1.2 L of a 4.0% (m/v) KCl solution

d. grams of NaCl in 1.50 L of a 2.00% (m/v) NaCl solution

Answers   a. 24 g   b. 1.0 g   c. 48 g   d. 30.0 g

◆ **Learning Exercise 9.4C**

Use percent concentration factors to calculate each of the following:

a. How many milliliters of a 1.00% (m/v) NaCl solution contain 2.00 g of NaCl?

b. How many milliliters of a 5.0% (m/v) glucose ($C_6H_{12}O_6$) solution contain 24 g of glucose?

c. How many milliliters of a 2.5% (v/v) methanol (CH_4O) solution contain 7.0 mL of methanol?

d. How many grams of a 15% (m/m) NaOH solution contain 7.5 g of NaOH?

Answers   a. 200. mL   b. 480 mL   c. 280 mL   d. 50. g

SAMPLE PROBLEM   Calculating Molarity

A physiological glucose solution contains 25 g of glucose ($C_6H_{12}O_6$) in 0.50 L of solution. What is the molarity (M) of the glucose solution?

Solution Guide:

STEP 1 State the given and needed quantities.

	Given	Need	Connect
Analyze the Problem	25 g of glucose, 0.50 L of solution	molarity (mole/L)	molar mass of glucose, moles of solute liters of solution

To calculate the moles of glucose, we need to write the equality and conversion factors for the molar mass of glucose.

1 mole of glucose = 180.16 g of glucose

$$\frac{180.16 \text{ g glucose}}{1 \text{ mole glucose}} \text{ and } \frac{1 \text{ mole glucose}}{180.16 \text{ g glucose}}$$

moles of glucose = 25 g glucose × $\frac{1 \text{ mole glucose}}{180.16 \text{ g glucose}}$ = 0.14 mole of glucose

Chapter 9

STEP 2 Write the concentration expression.

$$\text{Molarity (M)} = \frac{\text{moles of solute}}{\text{liters of solution}}$$

STEP 3 Substitute solute and solution quantities into the expression and calculate.

$$\text{Molarity (M)} = \frac{0.14 \text{ mole glucose}}{0.50 \text{ L solution}} = 0.28 \text{ M glucose solution}$$

♦ **Learning Exercise 9.4D**

Calculate the molarity (M) of the following solutions:

a. 1.45 moles of HCl in 0.250 L of a solution of HCl

b. 10.0 moles of glucose ($C_6H_{12}O_6$) in 2.50 L of a glucose solution

c. 80.0 g of NaOH in 1.60 L of a solution of NaOH

d. 38.8 g of NaBr in 175 mL of a solution of NaBr

Answers   a. 5.80 M   b. 4.00 M   c. 1.25 M   d. 2.15 M

SAMPLE PROBLEM   Using Molarity

How many grams of NaOH are in 0.250 L of a 5.00 M NaOH solution?

Solution Guide:

STEP 1 State the given and needed quantities.

Analyze the Problem	Given	Need	Connect
	0.250 L of a 5.00 M NaOH solution	grams of NaOH	molar mass of NaOH, molarity

STEP 2 Write a plan to calculate the mass.

liters of solution $\xrightarrow{\text{Molarity}}$ moles of solute $\xrightarrow{\text{Molar mass}}$ grams of solute

STEP 3 Write equalities and conversion factors.

1 L of solution = 5.00 moles of NaOH   1 mole of NaOH = 40.00 g of NaOH

$$\frac{5.00 \text{ moles NaOH}}{1 \text{ L solution}} \text{ and } \frac{1 \text{ L solution}}{5.00 \text{ moles NaOH}} \qquad \frac{40.00 \text{ g NaOH}}{1 \text{ mole NaOH}} \text{ and } \frac{1 \text{ mole NaOH}}{40.00 \text{ g NaOH}}$$

Solutions

STEP 4 Set up the problem to calculate the mass.

$$0.250 \text{ L solution} \times \frac{5.00 \text{ moles NaOH}}{1 \text{ L solution}} \times \frac{40.00 \text{ g NaOH}}{1 \text{ mole NaOH}} = 50.0 \text{ g of NaOH}$$

♦ **Learning Exercise 9.4E**

Calculate the quantity of solute in each of the following solutions:

a. moles of HCl in 1.50 L of a 6.00 M HCl solution

b. moles of KOH in 0.750 L of a 10.0 M KOH solution

c. grams of NaOH in 0.500 L of a 4.40 M NaOH solution

d. grams of NaCl in 285 mL of a 1.75 M NaCl solution

Answers a. 9.00 moles   b. 7.50 moles   c. 88.0 g   d. 29.1 g

♦ **Learning Exercise 9.4F**

Calculate the volume, in milliliters, needed of each solution to obtain the given quantity of solute:

a. 1.50 moles of $Mg(OH)_2$ from a 2.00 M $Mg(OH)_2$ solution

b. 0.150 mole of glucose from a 2.20 M glucose solution

c. 18.5 g of KI from a 3.00 M KI solution

d. 18.0 g of NaOH from a 6.00 M NaOH solution

Answers a. 750 mL   b. 68.2 mL   c. 37.1 mL   d. 75.0 mL

Chapter 9

9.5 Dilution of Solutions

REVIEW
Solving Equations (1.4)

Learning Goal: Describe the dilution of a solution; calculate the unknown concentration or volume when a solution is diluted.

- Dilution is the process of mixing a solution with solvent to obtain a solution with a lower concentration.
- For dilutions, use the expression $C_1 V_1 = C_2 V_2$ and solve for the unknown value.

SAMPLE PROBLEM Concentration of a Diluted Solution

What is the molarity when 150. mL of a 2.00 M NaCl solution is diluted to a volume of 400. mL?

Solution Guide:

STEP 1 Prepare a table of the concentrations and volumes of the solutions.

	Given	Need	Connect
Analyze the Problem	$C_1 = 2.00$ M $V_1 = 150.$ mL $V_2 = 400.$ mL	C_2	$C_1 V_1 = C_2 V_2$ V_2 increases, C_2 decreases

STEP 2 Rearrange the dilution expression to solve for the unknown quantity.

$$C_1 V_1 = C_2 V_2 \qquad C_2 = C_1 \times \frac{V_1}{V_2}$$

STEP 3 Substitute the known quantities into the dilution expression and calculate.

$$C_2 = 2.00 \text{ M} \times \frac{150. \text{ mL}}{400. \text{ mL}} = 0.750 \text{ M}$$

♦ **Learning Exercise 9.5**

Solve each of the following dilution problems:

a. What is the molarity when 100. mL of a 5.0 M KCl solution is diluted with water to give a final volume of 200. mL?

b. What is the molarity of the diluted solution when 5.0 mL of a 1.5 M KCl solution is diluted to a total volume of 25 mL?

c. What is the molarity when 250 mL of an 8.0 M NaOH solution is diluted with 750 mL of water?

d. What is the final volume when 42 mL of a 1.0 M NaCl solution is diluted to give a 0.20 M NaCl solution?

e. What volume of 6.0 M HCl is needed to prepare 300. mL of a 1.0 M HCl solution? How much water must be added? Assume that the volumes add.

Answers   **a.** 2.5 M   **b.** 0.30 M   **c.** 2.0 M   **d.** 210 mL
e. $V_1 = 50.$ mL; add 250. mL of water

9.6 Properties of Solutions

Learning Goal: Identify a mixture as a solution, a colloid, or a suspension. Describe how the number of particles in a solution affects the osmotic pressure.

- Colloids contain particles that do not settle out and pass through filters but not through semipermeable membranes.
- Suspensions are composed of large particles that settle out of solution.
- In the process of osmosis, water (solvent) moves through a semipermeable membrane from the solution that has a lower solute concentration to a solution where the solute concentration is higher.
- Osmotic pressure is the pressure that prevents the flow of water into a more concentrated solution.
- Isotonic solutions have osmotic pressures equal to that of body fluids. A hypotonic solution has a lower osmotic pressure than body fluids; a hypertonic solution has a higher osmotic pressure.
- A red blood cell maintains its volume in an isotonic solution, but it swells (hemolysis) in a hypotonic solution and shrinks (crenation) in a hypertonic solution.
- In dialysis, water and small solute particles can pass through a dialyzing membrane, while larger particles such as blood cells cannot.

Key Terms for Sections 9.5 and 9.6

Match each of the following key terms with the correct description:

    a. hypertonic   **b.** dilution   **c.** suspension
    d. osmosis   **e.** semipermeable membrane   **f.** colloid

1. ____ the particles that pass through filters but are too large to pass through semipermeable membranes
2. ____ has large particles that settle out on standing
3. ____ a solution that has a higher osmotic pressure than the red blood cells of the body
4. ____ a process by which water is added to a solution and decreases its concentration
5. ____ a membrane that permits the passage of small particles while blocking larger particles
6. ____ the flow of solvent through a semipermeable membrane into a solution of higher solute concentration

Answers   **1.** f   **2.** c   **3.** a   **4.** b   **5.** e   **6.** d

Chapter 9

♦ **Learning Exercise 9.6A**

Identify each of the following as characteristic of a solution, colloid, or suspension:

a. _____ the solute consists of single atoms, ions, or small molecules

b. _____ settles out with gravity

c. _____ retained by filters

d. _____ cannot diffuse through a cellular membrane

e. _____ contains large particles that are visible

Answers   a. solution   b. suspension   c. suspension
            d. colloid, suspension   e. suspension

♦ **Learning Exercise 9.6B**

Fill in the blanks:

In osmosis, the direction of solvent flow is from the solution with (**a.**) [higher/lower] solute concentration to the solution with (**b.**) [higher/lower] solute concentration. A semipermeable membrane separates 5% (m/v) and 10% (m/v) sucrose solutions. The (**c.**) _____ % (m/v) solution has the greater osmotic pressure. Water will move from the (**d.**) _____ % (m/v) solution into the (**e.**) _____ % (m/v) solution. The compartment that contains the (**f.**) _____ % (m/v) solution increases in volume.

Answers   a. lower   b. higher   c. 10   d. 5   e. 10   f. 10

♦ **Learning Exercise 9.6C**

A semipermeable membrane separates a 2% starch solution from a 10% starch solution. Complete each of the following with 2% or 10%:

a. Water will flow from the _____ starch solution into the _____ starch solution.

b. The volume of the _____ starch solution will increase and the volume of the _____ starch solution will decrease.

Answers   a. 2%, 10%   b. 10%, 2%

♦ **Learning Exercise 9.6D**

Fill in the blanks:

A (**a.**) _____ % (m/v) NaCl solution and a (**b.**) _____ % (m/v) glucose solution are isotonic to the body fluids. A red blood cell placed in these solutions does not change in volume because these solutions are (**c.**) _____ tonic. When a red blood cell is placed in water, it undergoes (**d.**) _____ because water is (**e.**) _____ tonic. A 20% (m/v) glucose solution will cause a red blood cell to undergo (**f.**) _____ because the 20% (m/v) glucose solution is (**g.**) _____ tonic.

Answers   a. 0.9   b. 5   c. iso   d. hemolysis
            e. hypo   f. crenation   g. hyper

Solutions

♦ **Learning Exercise 9.6E**

Are the following solutions **a.** hypotonic, **b.** hypertonic, or **c.** isotonic, compared with a red blood cell?
1. ____ 5% (m/v) glucose
2. ____ 3% (m/v) NaCl
3. ____ 2% (m/v) glucose
4. ____ water
5. ____ 0.9% (m/v) NaCl
6. ____ 10% (m/v) glucose

Answers   **1.** c   **2.** b   **3.** a   **4.** a   **5.** c   **6.** b

♦ **Learning Exercise 9.6F**

Indicate whether the following will cause a red blood cell to undergo

   a. crenation   **b.** hemolysis   **c.** no change

1. ____ 10% (m/v) NaCl
2. ____ 1% (m/v) glucose
3. ____ 5% (m/v) glucose
4. ____ 0.5% (m/v) NaCl
5. ____ 10% (m/v) glucose
6. ____ water

Answers   **1.** a   **2.** b   **3.** c   **4.** b   **5.** a   **6.** b

♦ **Learning Exercise 9.6G**

A dialysis bag contains starch, glucose, NaCl, protein, and urea.

a. When the dialysis bag is placed in water, what components would you expect to dialyze through the bag? Why?

b. Which components will stay inside the dialysis bag? Why?

Answers   **a.** Glucose, Na^+, Cl^- ions, and urea are solution particles. Solution particles will pass through semipermeable membranes.
b. Starch and protein form colloidal particles; because of their size, colloidal particles are retained by semipermeable membranes.

Checklist for Chapter 9

You are ready to take the Practice Test for Chapter 9. Be sure you have accomplished the following learning goals for this chapter. If not, review the Section listed at the end of the goal. Then apply your new skills and understanding to the Practice Test.

After studying Chapter 9, I can successfully:

____ Identify the solute and solvent in a solution and describe the process of dissolving an ionic solute in water. (9.1)

____ Identify the components in solutions of electrolytes and nonelectrolytes. (9.2)

____ Identify a saturated and an unsaturated solution. (9.3)

Chapter 9

_____ Describe the effects of temperature and nature of a solute on its solubility in a solvent. (9.3)

_____ Determine the solubility of an ionic compound in water. (9.3)

_____ Calculate the percent concentration, m/m, v/v, and m/v, of a solute in a solution, and use percent concentration to calculate the amount of solute or solution. (9.4)

_____ Calculate the molarity of a solution. (9.4)

_____ Use molarity as a conversion factor to calculate the moles (or grams) of a solute or the volume of the solution. (9.4)

_____ Calculate the unknown concentration or volume when a solution is diluted. (9.5)

_____ Identify a mixture as a solution, a colloid, or a suspension. (9.6)

_____ Identify a solution as isotonic, hypotonic, or hypertonic. (9.6)

_____ Explain the processes of osmosis and dialysis. (9.6)

Practice Test for Chapter 9

The chapter Sections to review are shown in parentheses at the end of each question.

For questions 1 through 4, indicate if each of the following is more soluble in (W) *water, a polar solvent, or* (B) *benzene, a nonpolar solvent*: (9.1)

1. $I_2(s)$, nonpolar
2. $NaBr(s)$, polar
3. $KI(s)$, polar
4. C_6H_{12}, nonpolar

5. When dissolved in water, $Ca(NO_3)_2(s)$ dissociates into (9.2)
   A. $Ca^{2+}(aq) + (NO_3)_2^{2-}(aq)$
   B. $Ca^+(aq) + NO_3^-(aq)$
   C. $Ca^{2+}(aq) + 2NO_3^-(aq)$
   D. $Ca^{2+}(aq) + 2N^{5+}(aq) + 2O_3^{6-}(aq)$
   E. $CaNO_3^+(aq) + NO_3^-(aq)$

6. Ethanol (C_2H_6O) is a nonelectrolyte. When placed in water it (9.2)
   A. dissociates completely.   B. dissociates partially.
   C. does not dissociate.   D. makes the solution acidic.
   E. makes the solution basic.

7. The solubility of NH_4Cl is 46 g in 100. g of water at 40 °C. How much NH_4Cl can dissolve in 500. g of water at 40 °C? (9.3)
   A. 9.2 g   B. 46 g   C. 100 g   D. 184 g   E. 230 g

For questions 8 through 11, indicate if each of the following is soluble (S) *or insoluble* (I) *in water*: (9.3)

8. NaCl
9. AgCl
10. $BaSO_4$
11. FeO

12. Which of the following ionic compounds is soluble in water? (9.3)
    A. FeS   B. $BaCO_3$ C. K_2SO_4 D. PbS   E. MgO

13. Which of the following ionic compounds is insoluble in water? (9.3)
    A. $CuCl_2$ B. $Pb(NO_3)_2$ C. K_2CO_3 D. $(NH_4)_2SO_4$ E. $CaCO_3$

14. A solution made by dissolving 48 g of KCl in 1200 mL of solution has a percent (m/v) concentration of (9.4)
    A. 0.040%   B. 0.40%   C. 4.0%   D. 25%   E. 40.%

Solutions

15. A solution containing 1.20 g of sucrose in 50.0 g of solution has a percent (m/m) concentration of (9.4)
 A. 0.024%   B. 1.20%   C. 2.40%   D. 30.0%   E. 41.7%

16. A solution containing 6.0 g of NaCl in 1500 g of solution has a percent (m/m) concentration of (9.4)
 A. 0.40%   B. 0.25%   C. 4.0%   D. 0.90%   E. 2.5%

17. The grams of lactose in 250 g of a 3.0% (m/m) lactose solution for infant formula are (9.4)
 A. 0.15 g   B. 1.2 g   C. 6.0 g   D. 7.5 g   E. 30 g

18. The mass of solution needed to obtain 0.40 g of glucose from a 5.0% (m/m) glucose solution is (9.4)
 A. 1.0 g   B. 2.0 g   C. 4.0 g   D. 5.0 g   E. 8.0 g

19. The amount of NaCl needed to prepare 50.0 g of a 4.00% (m/m) NaCl solution is (9.4)
 A. 20.0 g   B. 15.0 g   C. 10.0 g   D. 4.00 g   E. 2.00 g

20. The number of moles of KOH needed to prepare 2400 mL of a 2.0 M KOH solution is (9.4)
 A. 1.2 moles   B. 2.4 moles   C. 4.8 moles   D. 12 moles   E. 48 moles

21. The number of grams of NaOH needed to prepare 7.5 mL of a 5.0 M NaOH solution is (9.4)
 A. 1.5 g   B. 3.8 g   C. 6.7 g   D. 15 g   E. 38 g

For questions 22 through 24, consider a 20.0-g sample of a solution that contains 2.0 g of NaOH: (9.4)

22. The percent (m/m) concentration of the solution is
 A. 1.0%   B. 4.0%   C. 5.0%   D. 10.%   E. 20.%

23. The number of moles of NaOH in the sample is
 A. 0.050 mole   B. 0.40 mole   C. 1.0 mole   D. 2.5 moles   E. 4.0 moles

24. If the solution has a volume of 0.025 L, what is the molarity of the sample?
 A. 0.10 M   B. 0.5 M   C. 1.0 M   D. 1.5 M   E. 2.0 M

25. A 20.-mL sample of 5.0 M HCl solution is diluted with water to give 100. mL of solution. The final concentration of the HCl solution is (9.5)
 A. 10 M   B. 5.0 M   C. 2.0 M   D. 1.0 M   E. 0.50 M

26. Water is added to 200. mL of a 4.00 M KNO_3 solution to give 400. mL of solution. The final concentration of the diluted solution is (9.5)
 A. 1.00 M   B. 2.00 M   C. 4.00 M   D. 0.500 M   E. 0.100 M

27. 5.0 mL of a 2.0 M KOH solution is diluted with water to give a 0.40 M solution. The final volume of the KOH solution is (9.5)
 A. 10. mL   B. 25 mL   C. 40. mL   D. 250 mL   E. 400 mL

For questions 28 through 32, indicate whether each statement describes a (9.6)
 A. solution   B. colloid   C. suspension

28. _____ can be separated by filtering

29. _____ can be separated by semipermeable membranes

30. _____ passes through semipermeable membranes

31. _____ contains single atoms, ions, or small molecules of solute

32. _____ settles out upon standing

33. Two solutions that have identical osmotic pressures are (9.6)
 A. hypotonic   B. hypertonic   C. isotonic   D. isotopic   E. hyperactive

Chapter 9

34. In osmosis, the net flow of water is (9.6)
    A. between solutions of equal concentrations
    B. from higher solute concentration to lower solute concentration
    C. from lower solute concentration to higher solute concentration
    D. from a colloid to a solution of equal concentration
    E. from lower solvent concentration to higher solvent concentration

35. A red blood cell undergoes hemolysis when placed in a solution that is (9.6)
    A. isotonic   B. hypotonic   C. hypertonic   D. colloidal   E. semitonic

36. A solution that has the same osmotic pressure as body fluids is (9.6)
    A. 0.1% (m/v) NaCl   B. 0.9% (m/v) NaCl   C. 5% (m/v) NaCl
    D. 10% (m/v) glucose   E. 15% (m/v) glucose

Answers to the Practice Test

1. B	2. W	3. W	4. B	5. C
6. C	7. E	8. S	9. I	10. I
11. I	12. C	13. E	14. C	15. C
16. A	17. D	18. E	19. E	20. C
21. A	22. D	23. A	24. E	25. D
26. B	27. B	28. C	29. B	30. A
31. A	32. C	33. C	34. C	35. B
36. B				

Selected Answers and Solutions to Text Problems

9.1 The component present in the smaller amount is the solute; the component present in the larger amount is the solvent.
  a. NaCl, solute; water, solvent
  b. water, solute; ethanol, solvent
  c. oxygen, solute; nitrogen, solvent

9.3 The K^+ and I^- ions at the surface of the solid are pulled into solution by the polar water molecules, where the hydration process surrounds separate ions with water molecules.

9.5
  a. $CaCO_3$ (an ionic solute) would be soluble in water (a polar solvent).
  b. Retinol (a nonpolar solute) would be soluble in CCl_4 (a nonpolar solvent).
  c. Sucrose (a polar solute) would be soluble in water (a polar solvent).
  d. Cholesterol (a nonpolar solute) would be soluble in CCl_4 (a nonpolar solvent).

9.7 The strong electrolyte KF completely dissociates into K^+ and F^- ions when it dissolves in water. When the weak electrolyte HF dissolves in water, there are a few ions of H^+ and F^- present, but mostly dissolved HF molecules.

9.9 Strong electrolytes dissociate completely into ions.
  a. $KCl(s) \xrightarrow{H_2O} K^+(aq) + Cl^-(aq)$
  b. $CaCl_2(s) \xrightarrow{H_2O} Ca^{2+}(aq) + 2Cl^-(aq)$
  c. $K_3PO_4(s) \xrightarrow{H_2O} 3K^+(aq) + PO_4^{3-}(aq)$
  d. $Fe(NO_3)_3(s) \xrightarrow{H_2O} Fe^{3+}(aq) + 3NO_3^-(aq)$

9.11
  a. $HC_2H_3O_2(l) \rightleftharpoons H^+(aq) + C_2H_3O_2^-(aq)$ mostly molecules, a few ions
  An aqueous solution of a weak electrolyte like acetic acid will contain mostly $HC_2H_3O_2$ molecules, with a few H^+ ions and a few $C_2H_3O_2^-$ ions.
  b. $NaBr(s) \xrightarrow{H_2O} Na^+(aq) + Br^-(aq)$ only ions
  An aqueous solution of a strong electrolyte like NaBr will contain only the ions Na^+ and Br^-.
  c. $C_6H_{12}O_6(s) \xrightarrow{H_2O} C_6H_{12}O_6(aq)$ only molecules
  An aqueous solution of a nonelectrolyte like fructose will contain only $C_6H_{12}O_6$ molecules.

9.13
  a. K_2SO_4 is a strong electrolyte because only ions are present in the K_2SO_4 solution.
  b. NH_3 is a weak electrolyte because only a few NH_4^+ and OH^- ions are present in the solution.
  c. $C_6H_{12}O_6$ is a nonelectrolyte because only $C_6H_{12}O_6$ molecules are present in the solution.

9.15
  a. $1 \text{ mole } K^+ \times \dfrac{1 \text{ Eq } K^+}{1 \text{ mole } K^+} = 1 \text{ Eq of } K^+$

  b. $2 \text{ moles } OH^- \times \dfrac{1 \text{ Eq } OH^-}{1 \text{ mole } OH^-} = 2 \text{ Eq of } OH^-$

  c. $1 \text{ mole } Ca^{2+} \times \dfrac{2 \text{ Eq } Ca^{2+}}{1 \text{ mole } Ca^{2+}} = 2 \text{ Eq of } Ca^{2+}$

  d. $3 \text{ moles } CO_3^{2-} \times \dfrac{2 \text{ Eq } CO_3^{2-}}{1 \text{ mole } CO_3^{2-}} = 6 \text{ Eq of } CO_3^{2-}$

9.17
$1.00 \text{ L} \times \dfrac{154 \text{ mEq } Na^+}{1 \text{ L}} \times \dfrac{1 \text{ Eq } Na^+}{1000 \text{ mEq } Na^+} \times \dfrac{1 \text{ mole } Na^+}{1 \text{ Eq } Na^+} = 0.154 \text{ mole of } Na^+ \text{ (3 SFs)}$

$1.00 \text{ L} \times \dfrac{154 \text{ mEq } Cl^-}{1 \text{ L}} \times \dfrac{1 \text{ Eq } Cl^-}{1000 \text{ mEq } Cl^-} \times \dfrac{1 \text{ mole } Cl^-}{1 \text{ Eq } Cl^-} = 0.154 \text{ mole of } Cl^- \text{ (3 SFs)}$

Chapter 9

9.19 In any solution, the total equivalents of anions must be equal to the equivalents of cations. mEq of anions = 40. mEq/L Cl⁻ + 15 mEq/L HPO₄²⁻ = 55 mEq/L of anions
mEq of Na⁺ = mEq of anions = 55 mEq/L of Na⁺

9.21 **a.** Given 0.45 g/dL Cl⁻   Need mEq/L

Plan g → mole → Eq → mEq; dL → L $\dfrac{1 \text{ mole Cl}^-}{35.45 \text{ g Cl}^-}$ $\dfrac{1 \text{ Eq}}{1 \text{ mole Cl}^-}$ $\dfrac{1000 \text{ mEq}}{1 \text{ Eq}}$ $\dfrac{10 \text{ dL}}{1 \text{ L}}$

Set-up $\dfrac{0.45 \text{ g Cl}^-}{1 \text{ dL}} \times \dfrac{1 \text{ mole Cl}^-}{35.45 \text{ g Cl}^-} \times \dfrac{1 \text{ Eq}}{1 \text{ mole Cl}^-} \times \dfrac{1000 \text{ mEq}}{1 \text{ Eq}} \times \dfrac{10 \text{ dL}}{1 \text{ L}}$
= 130 mEq/L (2 SFs)

b. The normal range for Cl⁻ is 95 to 105 mEq/L of blood plasma, so a concentration of 130 mEq/L is above the normal range.

9.23 **a.** The solution must be saturated because no additional solute dissolves.
b. The solution was unsaturated because the sugar cube dissolves completely.
c. The solution is unsaturated because the uric acid concentration does not exceed its solubility of 7.6 mg/100 mL at 37 °C.

9.25 **a.** At 20 °C, KCl has a solubility of 34 g of KCl in 100. g of H₂O. Because 25 g of KCl is less than the maximum amount that can dissolve in 100. g of H₂O at 20 °C, the KCl solution is unsaturated.
b. At 20 °C, NaNO₃ has a solubility of 88 g of NaNO₃ in 100. g of H₂O. Using the solubility as a conversion factor, we can calculate the maximum amount of NaNO₃ that can dissolve in 25 g of H₂O:

$25 \text{ g H}_2\text{O} \times \dfrac{88 \text{ g NaNO}_3}{100. \text{ g H}_2\text{O}} = 22 \text{ g of NaNO}_3$ (2 SFs)

Because 11 g of NaNO₃ is less than the maximum amount that can dissolve in 25 g of H₂O at 20 °C, the NaNO₃ solution is unsaturated.

c. At 20 °C, sugar has a solubility of 204 g of C₁₂H₂₂O₁₁ in 100. g of H₂O. Using the solubility as a conversion factor, we can calculate the maximum amount of sugar that can dissolve in 125 g of H₂O:

$125 \text{ g H}_2\text{O} \times \dfrac{204 \text{ g sugar}}{100. \text{ g H}_2\text{O}} = 255 \text{ g of sugar}$ (3 SFs)

Because 400. g of C₁₂H₂₂O₁₁ exceeds the maximum amount that can dissolve in 125 g of H₂O at 20 °C, the sugar solution is saturated, and excess undissolved sugar will be present on the bottom of the container.

9.27 **a.** At 20 °C, KCl has a solubility of 34 g of KCl in 100. g of H₂O.
∴ 200. g of H₂O will dissolve:

$200. \text{ g H}_2\text{O} \times \dfrac{34 \text{ g KCl}}{100. \text{ g H}_2\text{O}} = 68 \text{ g of KCl}$ (2 SFs)

At 20 °C, 68 g of KCl will remain in solution.

b. Since 80. g of KCl dissolves at 50 °C and 68 g remains in solution at 20 °C, the mass of solid KCl that crystallizes after cooling is (80. g KCl − 68 g KCl =) 12 g of KCl. (2 SFs)

9.29 **a.** In general, the solubility of solid solutes (like sugar) increases as temperature is increased.
b. The solubility of a gaseous solute (CO₂) is less at a higher temperature.
c. The solubility of a gaseous solute is less at a higher temperature, and the CO₂ pressure in the can is increased. When the can of warm soda is opened, more CO₂ is released, producing more spray.

9.31 **a.** Salts containing Li⁺ ions are soluble.
b. Salts containing S²⁻ ions are usually insoluble.

Solutions

    c. Salts containing CO_3^{2-} ions are usually insoluble.
    d. Salts containing K^+ ions are soluble.
    e. Salts containing NO_3^- ions are soluble.

9.33 Mass percent $(m/m) = \dfrac{\text{mass of solute (g)}}{\text{mass of solution (g)}} \times 100\%$

    a. mass of solution = 25 g KCl + 125 g H_2O = 150. g of solution

$$\dfrac{25 \text{ g KCl}}{150. \text{ g solution}} \times 100\% = 17\% \text{ (m/m) KCl solution (2 SFs)}$$

    b. $\dfrac{12 \text{ g sucrose}}{225 \text{ g solution}} \times 100\% = 5.3\%$ (m/m) sucrose solution (2 SFs)

    c. $\dfrac{8.0 \text{ g } CaCl_2}{80.0 \text{ g solution}} \times 100\% = 10.\%$ (m/m) $CaCl_2$ solution (2 SFs)

9.35 355 mL solution × $\dfrac{22.5 \text{ mL alcohol}}{100. \text{ mL solution}} = 79.9$ mL of alcohol (3 SFs)

9.37 A 5.00% (m/m) glucose solution can be made by adding 5.00 g of glucose to 95.00 g of water, while a 5.00% (m/v) glucose solution can be made by adding 5.00 g of glucose to enough water to make 100.0 mL of solution.

9.39 Mass/volume percent $(m/v) = \dfrac{\text{mass of solute (g)}}{\text{volume of solution (mL)}} \times 100\%$

    a. $\dfrac{75 \text{ g } Na_2SO_4}{250 \text{ mL solution}} \times 100\% = 30.\%$ (m/v) Na_2SO_4 solution (2 SFs)

    b. $\dfrac{39 \text{ g sucrose}}{355 \text{ mL solution}} \times 100\% = 11\%$ (m/v) sucrose solution (2 SFs)

9.41   **a.** 50. g solution × $\dfrac{5.0 \text{ g KCl}}{100. \text{ g solution}} = 2.5$ g of KCl (2 SFs)

    b. 1250 mL solution × $\dfrac{4.0 \text{ g } NH_4Cl}{100. \text{ mL solution}} = 50.$ g of NH_4Cl (2 SFs)

    c. 250. mL solution × $\dfrac{10.0 \text{ mL acetic acid}}{100. \text{ mL solution}} = 25.0$ mL of acetic acid (3 SFs)

9.43   **a.** 5.0 g $LiNO_3$ × $\dfrac{100. \text{ g solution}}{25 \text{ g } LiNO_3} = 20.$ g of $LiNO_3$ solution (2 SFs)

    b. 40.0 g KOH × $\dfrac{100. \text{ mL solution}}{10.0 \text{ g KOH}} = 400.$ mL of KOH solution (3 SFs)

    c. 2.0 mL formic acid × $\dfrac{100. \text{ mL solution}}{10.0 \text{ mL formic acid}} = 20.$ mL of formic acid solution (2 SFs)

9.45 Molarity $(M) = \dfrac{\text{moles of solute}}{\text{liters of solution}}$

    a. $\dfrac{2.00 \text{ moles glucose}}{4.00 \text{ L solution}} = 0.500$ M glucose solution (3 SFs)

    b. $\dfrac{4.00 \text{ g KOH}}{2.00 \text{ L solution}} \times \dfrac{1 \text{ mole KOH}}{56.11 \text{ g KOH}} = 0.0356$ M KOH solution (3 SFs)

    c. $\dfrac{5.85 \text{ g NaCl}}{400. \text{ mL solution}} \times \dfrac{1 \text{ mole NaCl}}{58.44 \text{ g NaCl}} \times \dfrac{1000 \text{ mL solution}}{1 \text{ L solution}} = 0.250$ M NaCl solution (3 SFs)

9.47 a. $2.00 \text{ L solution} \times \dfrac{1.50 \text{ moles NaOH}}{1 \text{ L solution}} \times \dfrac{40.00 \text{ g NaOH}}{1 \text{ mole NaOH}} = 120. \text{ g of NaOH (3 SFs)}$

  b. $4.00 \text{ L solution} \times \dfrac{0.200 \text{ mole KCl}}{1 \text{ L solution}} \times \dfrac{74.55 \text{ g KCl}}{1 \text{ mole KCl}} = 59.6 \text{ g of KCl (3 SFs)}$

  c. $25.0 \text{ mL solution} \times \dfrac{1 \text{ L solution}}{1000 \text{ mL solution}} \times \dfrac{6.00 \text{ moles HCl}}{1 \text{ L solution}} \times \dfrac{36.46 \text{ g HCl}}{1 \text{ mole HCl}}$
  $= 5.47 \text{ g of HCl (3 SFs)}$

9.49 a. $3.00 \text{ moles KBr} \times \dfrac{1 \text{ L solution}}{2.00 \text{ moles KBr}} = 1.50 \text{ L of KBr solution (3 SFs)}$

  b. $15.0 \text{ moles NaCl} \times \dfrac{1 \text{ L solution}}{1.50 \text{ moles NaCl}} = 10.0 \text{ L of NaCl solution (3 SFs)}$

  c. $0.0500 \text{ mole Ca(NO}_3)_2 \times \dfrac{1 \text{ L solution}}{0.800 \text{ mole Ca(NO}_3)_2} \times \dfrac{1000 \text{ mL solution}}{1 \text{ L solution}}$
  $= 62.5 \text{ mL of Ca(NO}_3)_2 \text{ solution (3 SFs)}$

9.51 a. $1 \text{ h} \times \dfrac{100. \text{ mL solution}}{1 \text{ h}} \times \dfrac{20. \text{ g mannitol}}{100. \text{ mL solution}} = 20. \text{ g of mannitol (2 SFs)}$

  b. $12 \text{ h} \times \dfrac{100. \text{ mL solution}}{1 \text{ h}} \times \dfrac{20. \text{ g mannitol}}{100. \text{ mL solution}} = 240 \text{ g of mannitol (2 SFs)}$

9.53 $100. \text{ g glucose} \times \dfrac{100. \text{ mL solution}}{5 \text{ g glucose}} \times \dfrac{1 \text{ L}}{1000 \text{ mL}} = 2 \text{ L of glucose solution (1 SF)}$

9.55 $C_1 V_1 = C_2 V_2$

  a. $C_2 = C_1 \times \dfrac{V_1}{V_2} = 6.0 \text{ M} \times \dfrac{2.0 \text{ L}}{6.0 \text{ L}} = 2.0 \text{ M HCl solution (2 SFs)}$

  b. $C_2 = C_1 \times \dfrac{V_1}{V_2} = 12 \text{ M} \times \dfrac{0.50 \text{ L}}{3.0 \text{ L}} = 2.0 \text{ M NaOH solution (2 SFs)}$

  c. $C_2 = C_1 \times \dfrac{V_1}{V_2} = 25\% \times \dfrac{10.0 \text{ mL}}{100.0 \text{ mL}} = 2.5\% \text{ (m/v) KOH solution (2 SFs)}$

  d. $C_2 = C_1 \times \dfrac{V_1}{V_2} = 15\% \times \dfrac{50.0 \text{ mL}}{250 \text{ mL}} = 3.0\% \text{ (m/v) H}_2\text{SO}_4 \text{ solution (2 SFs)}$

9.57 $C_1 V_1 = C_2 V_2$

  a. $V_2 = V_1 \times \dfrac{C_1}{C_2} = 20.0 \text{ mL} \times \dfrac{6.0 \text{ M}}{1.5 \text{ M}} = 80. \text{ mL of diluted HCl solution (2 SFs)}$

  b. $V_2 = V_1 \times \dfrac{C_1}{C_2} = 50.0 \text{ mL} \times \dfrac{10.0\%}{2.0\%} = 250 \text{ mL of diluted LiCl solution (2 SFs)}$

  c. $V_2 = V_1 \times \dfrac{C_1}{C_2} = 50.0 \text{ mL} \times \dfrac{6.00 \text{ M}}{0.500 \text{ M}} = 600. \text{ mL of diluted H}_3\text{PO}_4 \text{ solution (3 SFs)}$

  d. $V_2 = V_1 \times \dfrac{C_1}{C_2} = 75 \text{ mL} \times \dfrac{12\%}{5.0\%} = 180 \text{ mL of diluted glucose solution (2 SFs)}$

9.59 $C_1 V_1 = C_2 V_2$

  a. $V_1 = V_2 \times \dfrac{C_2}{C_1} = 255 \text{ mL} \times \dfrac{0.200 \text{ M}}{4.00 \text{ M}} = 12.8 \text{ mL of the HNO}_3 \text{ solution (3 SFs)}$

  b. $V_1 = V_2 \times \dfrac{C_2}{C_1} = 715 \text{ mL} \times \dfrac{0.100 \text{ M}}{6.00 \text{ M}} = 11.9 \text{ mL of the MgCl}_2 \text{ solution (3 SFs)}$

c. $V_2 = 0.100 \,\cancel{L} \times \dfrac{1000 \text{ mL}}{1 \,\cancel{L}} = 100. \text{ mL}$

$V_1 = V_2 \times \dfrac{C_2}{C_1} = 100. \text{ mL} \times \dfrac{0.150 \,\cancel{M}}{8.00 \,\cancel{M}} = 1.88 \text{ mL of the KCl solution (3 SFs)}$

9.61 $C_1 V_1 = C_2 V_2$

$V_1 = V_2 \times \dfrac{C_2}{C_1} = 500. \text{ mL} \times \dfrac{5.0\%}{25\%} = 1.0 \times 10^2 \text{ mL of the glucose solution (2 SFs)}$

9.63 **a.** A solution cannot be separated by a semipermeable membrane.
b. A suspension settles out upon standing.

9.65 **a.** The 10% (m/v) starch solution has the higher solute concentration, more solute particles, and therefore the higher osmotic pressure.
b. Initially, water will flow out of the 1% (m/v) starch solution into the more concentrated 10% (m/v) starch solution.
c. The volume of the 10% (m/v) starch solution will increase due to inflow of water.

9.67 Water will flow from a region of higher solvent concentration (which corresponds to a lower solute concentration) to a region of lower solvent concentration (which corresponds to a higher solute concentration).
a. B; The volume level will rise as water flows into compartment B, which contains the 10% (m/v) sucrose solution.
b. A; The volume level will rise as water flows into compartment A, which contains the 8% (m/v) albumin solution.
c. B; The volume level will rise as water flows into compartment B, which contains the 10% (m/v) starch solution.

9.69 A red blood cell has the same osmotic pressure as a 5% (m/v) glucose solution or a 0.9% (m/v) NaCl solution; these are isotonic solutions. Solutions with higher concentrations than these are hypertonic, and solutions with lower concentrations than these are hypotonic.
a. Distilled water is a hypotonic solution when compared with a red blood cell's contents.
b. A 1% (m/v) glucose solution is a hypotonic solution.
c. A 0.9% (m/v) NaCl solution is isotonic with a red blood cell's contents.
d. A 15% (m/v) glucose solution is a hypertonic solution.

9.71 Colloids cannot pass through the semipermeable dialysis membrane; water and solutions freely pass through semipermeable membranes.
a. Sodium and chloride ions will both pass through the membrane into the distilled water.
b. The amino acid alanine will pass through the semipermeable membrane into the distilled water; the colloid starch will not.
c. Sodium and chloride ions will both be present in the water surrounding the dialysis bag; the colloid starch will not.
d. Urea will diffuse through the dialysis bag into the surrounding water.

9.73 $0.075 \text{ g chlorpromazine} \times \dfrac{100. \text{ mL solution}}{2.5 \text{ g chlorpromazine}}$
$= 3.0 \text{ mL of chlorpromazine solution (2 SFs)}$

9.75 $5.0 \text{ mL solution} \times \dfrac{10. \text{ g CaCl}_2}{100. \text{ mL solution}} = 0.50 \text{ g of CaCl}_2 \text{ (2 SFs)}$

9.77 **a.** 1; A solution will form because both the solute and the solvent are polar.
b. 2; Two layers will form because one component is nonpolar and the other is polar.
c. 1; A solution will form because both the solute and the solvent are nonpolar.

9.79 **a.** 3; A nonelectrolyte will show no dissociation.
  b. 1; A weak electrolyte will show some dissociation, producing a few ions, but mostly remaining as molecules.
  c. 2; A strong electrolyte will be completely dissociated into ions.

9.81 A "brine" salt-water solution has a high concentration of Na^+ and Cl^- ions, which is hypertonic to the cucumber. The skin of the cucumber acts like a semipermeable membrane; therefore, water flows from the more dilute solution inside the cucumber into the more concentrated brine solution that surrounds it. The loss of water causes the cucumber to become a wrinkled pickle.

9.83 **a.** 2; Water will flow into the B (8% starch solution) side.
  b. 1; Water will continue to flow equally in both directions; no change in volumes.
  c. 3; Water will flow into the A (5% sucrose solution) side.
  d. 2; Water will flow into the B (1% sucrose solution) side.

9.85 Because iodine is a nonpolar molecule, it will dissolve in hexane, a nonpolar solvent. Iodine does not dissolve in water because water is a polar solvent.

9.87 At 20 °C, KNO_3 has a solubility of 32 g of KNO_3 in 100. g of H_2O.
  a. 200. g of H_2O will dissolve:

  $$200.\ \text{g}\ H_2O \times \frac{32\ \text{g}\ KNO_3}{100.\ \text{g}\ H_2O} = 64\ \text{g of}\ KNO_3\ (2\ \text{SFs})$$

  Because 32 g of KNO_3 is less than the maximum amount that can dissolve in 200. g of H_2O at 20 °C, the KNO_3 solution is unsaturated.

  b. 50. g of H_2O will dissolve:

  $$50.\ \text{g}\ H_2O \times \frac{32\ \text{g}\ KNO_3}{100.\ \text{g}\ H_2O} = 16\ \text{g of}\ KNO_3\ (2\ \text{SFs})$$

  Because 19 g of KNO_3 exceeds the maximum amount that can dissolve in 50. g of H_2O at 20 °C, the KNO_3 solution is saturated, and excess undissolved KNO_3 will be present on the bottom of the container.

  c. 150. g of H_2O will dissolve:

  $$150.\ \text{g}\ H_2O \times \frac{32\ \text{g}\ KNO_3}{100.\ \text{g}\ H_2O} = 48\ \text{g of}\ KNO_3\ (2\ \text{SFs})$$

  Because 68 g of KNO_3 exceeds the maximum amount that can dissolve in 150. g of H_2O at 20 °C, the KNO_3 solution is saturated, and excess undissolved KNO_3 will be present on the bottom of the container.

9.89 **a.** K^+ salts are soluble.
  b. The SO_4^{2-} salt of Mg^{2+} is soluble.
  c. Salts containing S^{2-} ions are usually insoluble.
  d. Salts containing NO_3^- ions are soluble.
  e. Salts containing OH^- ions are usually insoluble.

9.91 $80.0\ \text{g}\ NaCl \times \dfrac{100.\ \text{g water}}{36.0\ \text{g}\ NaCl} = 222\ \text{g of water needed (3 SFs)}$

9.93 mass of solution = $15.5\ \text{g}\ Na_2SO_4 + 75.5\ \text{g}\ H_2O = 91.0\ \text{g of solution}$

  $\dfrac{15.5\ \text{g}\ Na_2SO_4}{91.0\ \text{g solution}} \times 100\% = 17.0\%\ (m/m)\ Na_2SO_4\ \text{solution (3 SFs)}$

9.95 $4.5\ \text{mL propyl alcohol} \times \dfrac{100.\ \text{mL solution}}{12\ \text{mL propyl alcohol}} = 38\ \text{mL of propyl alcohol solution (2 SFs)}$

9.97 $86.0\ \text{g}\ KOH \times \dfrac{100.\ \text{mL solution}}{12\ \text{g}\ KOH} \times \dfrac{1\ \text{L solution}}{1000\ \text{mL solution}} = 0.72\ \text{L of KOH solution (2 SFs)}$

9.99 $\dfrac{8.0 \text{ g NaOH}}{400. \text{ mL solution}} \times \dfrac{1 \text{ mole NaOH}}{40.00 \text{ g NaOH}} \times \dfrac{1000 \text{ mL solution}}{1 \text{ L solution}} = 0.50 \text{ M NaOH solution (2 SFs)}$

9.101 $15.2 \text{ g LiCl} \times \dfrac{1 \text{ mole LiCl}}{42.39 \text{ g LiCl}} \times \dfrac{1 \text{ L solution}}{1.75 \text{ moles LiCl}} \times \dfrac{1000 \text{ mL solution}}{1 \text{ L solution}}$
= 205 mL of LiCl solution (3 SFs)

9.103 $60.0 \text{ g KNO}_3 \times \dfrac{1 \text{ mole KNO}_3}{101.11 \text{ g KNO}_3} \times \dfrac{1 \text{ L solution}}{2.50 \text{ moles KNO}_3}$
= 0.237 L of KNO_3 solution (3 SFs)

9.105 a. $2.5 \text{ L solution} \times \dfrac{3.0 \text{ moles Al(NO}_3)_3}{1 \text{ L solution}} \times \dfrac{213.0 \text{ g Al(NO}_3)_3}{1 \text{ mole Al(NO}_3)_3}$
= 1600 g of $Al(NO_3)_3$ (2 SFs)

b. $75 \text{ mL solution} \times \dfrac{1 \text{ L solution}}{1000 \text{ mL solution}} \times \dfrac{0.50 \text{ mole C}_6\text{H}_{12}\text{O}_6}{1 \text{ L solution}} \times \dfrac{180.16 \text{ g C}_6\text{H}_{12}\text{O}_6}{1 \text{ mole C}_6\text{H}_{12}\text{O}_6}$
= 6.8 g of $C_6H_{12}O_6$ (2 SFs)

c. $235 \text{ mL solution} \times \dfrac{1 \text{ L solution}}{1000 \text{ mL solution}} \times \dfrac{1.80 \text{ moles LiCl}}{1 \text{ L solution}} \times \dfrac{42.39 \text{ g LiCl}}{1 \text{ mole LiCl}}$
= 17.9 g of LiCl (3 SFs)

9.107 $C_1 V_1 = C_2 V_2$

a. $C_2 = C_1 \times \dfrac{V_1}{V_2} = 0.200 \text{ M} \times \dfrac{25.0 \text{ mL}}{50.0 \text{ mL}} = 0.100 \text{ M NaBr solution (3 SFs)}$

b. $C_2 = C_1 \times \dfrac{V_1}{V_2} = 12.0\% \times \dfrac{15.0 \text{ mL}}{40.0 \text{ mL}} = 4.50\% \text{ (m/v) } K_2SO_4 \text{ solution (3 SFs)}$

c. $C_2 = C_1 \times \dfrac{V_1}{V_2} = 6.00 \text{ M} \times \dfrac{75.0 \text{ mL}}{255 \text{ mL}} = 1.76 \text{ M NaOH solution (3 SFs)}$

9.109 $C_1 V_1 = C_2 V_2$

a. $V_1 = V_2 \times \dfrac{C_2}{C_1} = 250 \text{ mL} \times \dfrac{3.0\%}{10.0\%} = 75 \text{ mL of the HCl solution (2 SFs)}$

b. $V_1 = V_2 \times \dfrac{C_2}{C_1} = 500. \text{ mL} \times \dfrac{0.90\%}{5.0\%} = 90. \text{ mL of the NaCl solution (2 SFs)}$

c. $V_1 = V_2 \times \dfrac{C_2}{C_1} = 350. \text{ mL} \times \dfrac{2.00 \text{ M}}{6.00 \text{ M}} = 117 \text{ mL of the NaOH solution (3 SFs)}$

9.111 $C_1 V_1 = C_2 V_2$

a. $V_2 = V_1 \times \dfrac{C_1}{C_2} = 25.0 \text{ mL} \times \dfrac{10.0\%}{2.50\%} = 100. \text{ mL of diluted HCl solution (3 SFs)}$

b. $V_2 = V_1 \times \dfrac{C_1}{C_2} = 25.0 \text{ mL} \times \dfrac{5.00 \text{ M}}{1.00 \text{ M}} = 125 \text{ mL of diluted HCl solution (3 SFs)}$

c. $V_2 = V_1 \times \dfrac{C_1}{C_2} = 25.0 \text{ mL} \times \dfrac{6.00 \text{ M}}{0.500 \text{ M}} = 300. \text{ mL of diluted HCl solution (3 SFs)}$

9.113 a. mass of NaCl = 25.50 g − 24.10 g = 1.40 g of NaCl
mass of solution = 36.15 g − 24.10 g = 12.05 g of solution

Mass percent (m/m) = $\dfrac{1.40 \text{ g NaCl}}{12.05 \text{ g solution}} \times 100\% = 11.6\% \text{ (m/m) NaCl solution (3 SFs)}$

Chapter 9

b. Molarity (M) = $\dfrac{1.40 \text{ g NaCl}}{10.0 \text{ mL solution}} \times \dfrac{1 \text{ mole NaCl}}{58.44 \text{ g NaCl}} \times \dfrac{1000 \text{ mL solution}}{1 \text{ L solution}}$

= 2.40 M NaCl solution (3 SFs)

c. $C_1 V_1 = C_2 V_2 \quad C_2 = C_1 \times \dfrac{V_1}{V_2} = 2.40 \text{ M} \times \dfrac{10.0 \text{ mL}}{60.0 \text{ mL}} = 0.400$ M NaCl solution (3 SFs)

9.115 At 18 °C, KF has a solubility of 92 g of KF in 100. g of H_2O.

a. 25 g of H_2O will dissolve:

$25 \text{ g } H_2O \times \dfrac{92 \text{ g KF}}{100. \text{ g } H_2O} = 23$ g of KF (2 SFs)

Because 35 g of KF exceeds the maximum amount that can dissolve in 25 g of H_2O at 18 °C, the KF solution is saturated, and excess undissolved KF will be present on the bottom of the container.

b. 50. g of H_2O will dissolve:

$50. \text{ g } H_2O \times \dfrac{92 \text{ g KF}}{100. \text{ g } H_2O} = 46$ g of KF (2 SFs)

Because 42 g of KF is less than the maximum amount that can dissolve in 50. g of H_2O at 18 °C, the solution is unsaturated.

c. 150. g of H_2O will dissolve:

$150. \text{ g } H_2O \times \dfrac{92 \text{ g KF}}{100. \text{ g } H_2O} = 140$ g of KF (2 SFs)

Because 145 g of KF exceeds the maximum amount that can dissolve in 150. g of H_2O at 18 °C, the KF solution is saturated, and excess undissolved KF will be present on the bottom of the container.

9.117 mass of solution: 70.0 g HNO_3 + 130.0 g H_2O = 200.0 g of solution

a. $\dfrac{70.0 \text{ g } HNO_3}{200.0 \text{ g solution}} \times 100\% = 35.0\%$ (m/m) HNO_3 solution (3 SFs)

b. $200.0 \text{ g solution} \times \dfrac{1 \text{ mL solution}}{1.21 \text{ g solution}} = 165$ mL of solution (3 SFs)

c. $\dfrac{70.0 \text{ g } HNO_3}{165 \text{ mL solution}} \times 100\% = 42.4\%$ (m/v) HNO_3 solution (3 SFs)

d. $\dfrac{70.0 \text{ g } HNO_3}{165 \text{ mL solution}} \times \dfrac{1 \text{ mole } HNO_3}{63.02 \text{ g } HNO_3} \times \dfrac{1000 \text{ mL solution}}{1 \text{ L solution}}$

= 6.73 M HNO_3 solution (3 SFs)

Selected Answers to Combining Ideas from Chapters 7 to 9

CI.13
a. Reactants: A and B_2; Products: AB_3
b. $2A + 3B_2 \longrightarrow 2AB_3$
c. combination

CI.15
a. Sodium hypochlorite ($NaClO$) is an ionic compound, composed of sodium (Na^+) and hypochlorite (ClO^-) ions.

b. Volume of solution = $1.42 \text{ gal bleach} \times \dfrac{4 \text{ qt solution}}{1 \text{ gal solution}} \times \dfrac{946 \text{ mL solution}}{1 \text{ qt solution}}$
$= 5.37 \times 10^3$ mL of solution (3 SFs)

Mass/volume % = $\dfrac{\text{mass of solute (g)}}{\text{volume of solution (mL)}} \times 100\%$

$= \dfrac{282 \text{ g NaClO}}{5.37 \times 10^3 \text{ mL solution}} \times 100\% = 5.25\%$ (m/v) NaClO solution (3 SFs)

c. $2NaOH(aq) + Cl_2(g) \longrightarrow NaClO(aq) + NaCl(aq) + H_2O(l)$

d. **Given** 1 bottle of bleach   **Need** liters of Cl_2 gas at STP
    Plan bottle → gallons of bleach → grams of NaClO → moles of NaClO
    → moles of Cl_2 → liters of Cl_2 gas (STP)

$1 \text{ bottle bleach} \times \dfrac{1.42 \text{ gal bleach}}{1 \text{ bottle bleach}} \times \dfrac{282 \text{ g NaClO}}{1.42 \text{ gal bleach}} \times \dfrac{1 \text{ mole NaClO}}{74.44 \text{ g NaClO}}$

$\times \dfrac{1 \text{ mole } Cl_2}{1 \text{ mole NaClO}} \times \dfrac{22.4 \text{ L } Cl_2 \text{ (STP)}}{1 \text{ mole } Cl_2}$

$= 84.9$ L of Cl_2 at STP (3 SFs)

CI.17
a. CH_4 H:C:H with H above and below   or   H—C—H with H above and below

b. $7.0 \times 10^6 \text{ gal} \times \dfrac{4 \text{ qt}}{1 \text{ gal}} \times \dfrac{946 \text{ mL}}{1 \text{ qt}} \times \dfrac{0.45 \text{ g}}{1 \text{ mL}} \times \dfrac{1 \text{ kg}}{1000 \text{ g}}$
$= 1.2 \times 10^7$ kg of LNG (methane) (2 SFs)

c. Molar mass of methane (CH_4) = $12.01 \text{ g} + 4(1.008 \text{ g}) = 16.04$ g

$7.0 \times 10^6 \text{ gal} \times \dfrac{4 \text{ qt}}{1 \text{ gal}} \times \dfrac{946 \text{ mL}}{1 \text{ qt}} \times \dfrac{0.45 \text{ g}}{1 \text{ mL}} \times \dfrac{1 \text{ mole } CH_4}{16.04 \text{ g}} \times \dfrac{22.4 \text{ L } CH_4 \text{ (STP)}}{1 \text{ mole } CH_4}$

$= 1.7 \times 10^{10}$ L of LNG (methane) at STP (2 SFs)

d. $CH_4(g) + 2O_2(g) \xrightarrow{\Delta} CO_2(g) + 2H_2O(g) + 883 \text{ kJ}$

e. $7.0 \times 10^6 \text{ gal} \times \dfrac{4 \text{ qt}}{1 \text{ gal}} \times \dfrac{946 \text{ mL}}{1 \text{ qt}} \times \dfrac{0.45 \text{ g}}{1 \text{ mL}} \times \dfrac{1 \text{ mole } CH_4}{16.04 \text{ g}} \times \dfrac{2 \text{ moles } O_2}{1 \text{ mole } CH_4}$

$\times \dfrac{32.00 \text{ g } O_2}{1 \text{ mole } O_2} \times \dfrac{1 \text{ kg } O_2}{1000 \text{ g } O_2}$

$= 4.8 \times 10^7$ kg of O_2 (2 SFs)

f. $7.0 \times 10^6 \text{ gal} \times \dfrac{4 \text{ qt}}{1 \text{ gal}} \times \dfrac{946 \text{ mL}}{1 \text{ qt}} \times \dfrac{0.45 \text{ g}}{1 \text{ mL}} \times \dfrac{1 \text{ mole } CH_4}{16.04 \text{ g}} \times \dfrac{883 \text{ kJ}}{1 \text{ mole } CH_4}$

$= 6.6 \times 10^{11}$ kJ (2 SFs)

CI.19 **a.** The formula of shikimic acid is $C_7H_{10}O_5$.

b. Molar mass of shikimic acid $(C_7H_{10}O_5)$ = $7(12.01 \text{ g}) + 10(1.008 \text{ g}) + 5(16.00 \text{ g})$
= 174.15 g (5 SFs)

c. 130 g shikimic acid $\times \dfrac{1 \text{ mole shikimic acid}}{174.15 \text{ g shikimic acid}}$ = 0.75 mole of shikimic acid (2 SFs)

d. 155 g star anise $\times \dfrac{0.13 \text{ g shikimic acid}}{2.6 \text{ g star anise}} \times \dfrac{1 \text{ capsule Tamiflu}}{0.13 \text{ g shikimic acid}}$
= 59 capsules of Tamiflu (2 SFs)

e. Molar mass of Tamiflu $(C_{16}H_{28}N_2O_4)$
= $16(12.01 \text{ g}) + 28(1.008 \text{ g}) + 2(14.01 \text{ g}) + 4(16.00 \text{ g})$ = 312.4 g (4 SFs)

f. 500 000 people $\times \dfrac{2 \text{ capsules}}{1 \text{ day } 1 \text{ person}} \times 5 \text{ days} \times \dfrac{75 \text{ mg Tamiflu}}{1 \text{ capsule}} \times \dfrac{1 \text{ g Tamiflu}}{1000 \text{ mg Tamiflu}}$
$\times \dfrac{1 \text{ kg Tamiflu}}{1000 \text{ g Tamiflu}}$ = 400 kg of Tamiflu (1 SF)

10 Acids and Bases and Equilibrium

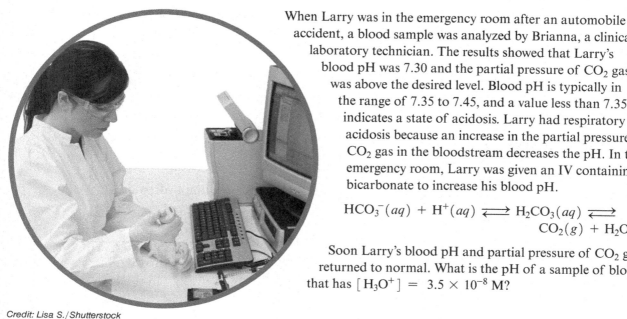

When Larry was in the emergency room after an automobile accident, a blood sample was analyzed by Brianna, a clinical laboratory technician. The results showed that Larry's blood pH was 7.30 and the partial pressure of CO_2 gas was above the desired level. Blood pH is typically in the range of 7.35 to 7.45, and a value less than 7.35 indicates a state of acidosis. Larry had respiratory acidosis because an increase in the partial pressure of CO_2 gas in the bloodstream decreases the pH. In the emergency room, Larry was given an IV containing bicarbonate to increase his blood pH.

$$HCO_3^-(aq) + H^+(aq) \rightleftharpoons H_2CO_3(aq) \rightleftharpoons CO_2(g) + H_2O(l)$$

Soon Larry's blood pH and partial pressure of CO_2 gas returned to normal. What is the pH of a sample of blood that has $[H_3O^+] = 3.5 \times 10^{-8}$ M?

Credit: Lisa S./Shutterstock

LOOKING AHEAD

10.1 Acids and Bases
10.2 Brønsted–Lowry Acids and Bases
10.3 Strengths of Acids and Bases
10.4 Acid–Base Equilibrium
10.5 Dissociation of Water
10.6 The pH Scale
10.7 Reactions of Acids and Bases
10.8 Buffers

 The Health icon indicates a question that is related to health and medicine.

10.1 Acids and Bases

Learning Goal: Describe and name acids and bases.

> **REVIEW**
> Writing Ionic Formulas (6.2)

- In water, an Arrhenius acid produces hydrogen ions, H^+, and an Arrhenius base produces hydroxide ions, OH^-.
- An acid with a simple nonmetal anion is named by placing the prefix *hydro* in front of the name of the anion, and changing its *ide* ending to *ic acid*.
- An acid with a polyatomic anion is named as an *ic acid* when its anion ends in *ate* and as an *ous acid* when its anion ends in *ite*.
- Typical Arrhenius bases are named as hydroxides.

Chapter 10

♦ **Learning Exercise 10.1A**

Indicate if each of the following characteristics describes an acid or a base:

a. _____ turns litmus red
b. _____ tastes sour
c. _____ contains more OH^- than H_3O^+
d. _____ neutralizes bases
e. _____ tastes bitter
f. _____ turns litmus blue
g. _____ contains more H_3O^+ than OH^-
h. _____ neutralizes acids

Answers a. acid   b. acid   c. base   d. acid
          e. base   f. base   g. acid   h. base

♦ **Learning Exercise 10.1B**

Fill in the blanks with the formula or name of an acid or base.

a. KOH _____
b. _____ sodium hydroxide
c. _____ sulfurous acid
d. _____ chlorous acid
e. $Ba(OH)_2$ _____
f. H_2CO_3 _____
g. $Zn(OH)_2$ _____
h. _____ lithium hydroxide
i. $HBrO_3$ _____
j. H_3PO_3 _____

Answers a. potassium hydroxide   b. NaOH   c. H_2SO_3
          d. $HClO_2$ e. barium hydroxide   f. carbonic acid
          g. zinc hydroxide   h. LiOH   i. bromic acid
          j. phosphorous acid

10.2 Brønsted–Lowry Acids and Bases

Learning Goal: Identify conjugate acid–base pairs for Brønsted–Lowry acids and bases.

- According to the Brønsted–Lowry theory, acids are hydrogen ion, H^+, donors, and bases are hydrogen ion, H^+, acceptors.
- In solution, hydrogen ions, H^+, from acids bond to polar water molecules to form hydronium ions, H_3O^+.
- Conjugate acid–base pairs are molecules or ions linked by the loss and gain of one H^+.
- Every hydrogen ion transfer reaction involves two acid–base conjugate pairs.

Solution Guide to Writing Conjugate Acid–Base Pairs	
STEP 1	Identify the reactant that loses H^+ as the acid.
STEP 2	Identify the reactant that gains H^+ as the base.
STEP 3	Write the conjugate acid–base pairs.

Acids and Bases and Equilibrium

♦ Learning Exercise 10.2A

Complete the following:

Conjugate acid–base pair
HF — Donates H⁺ → F⁻

Conjugate acid–base pair
H$_2$O — Accepts H⁺ → H$_3$O⁺

Acid	Conjugate Base
a. H$_2$O	_____
b. H$_2$SO$_4$	_____
c. _____	Cl⁻
d. HCO$_3^-$	_____
e. _____	NO$_3^-$
f. NH$_4^+$	_____
g. _____	HS⁻
h. _____	H$_2$PO$_4^-$

Answers a. OH⁻   b. HSO$_4^-$ c. HCl   d. CO$_3^{2-}$
e. HNO$_3$ f. NH$_3$ g. H$_2$S   h. H$_3$PO$_4$

♦ Learning Exercise 10.2B

Identify the reactant that is a Brønsted–Lowry acid and the reactant that is a Brønsted–Lowry base in each of the following:

a. HBr(aq) + CO$_3^{2-}$(aq) ⟶ Br⁻(aq) + HCO$_3^-$(aq)

b. HSO$_4^-$(aq) + OH⁻(aq) ⇌ SO$_4^{2-}$(aq) + H$_2$O(l)

c. NH$_4^+$(aq) + H$_2$O(l) ⇌ NH$_3$(aq) + H$_3$O⁺(aq)

d. SO$_4^{2-}$(aq) + HCl(aq) ⟶ HSO$_4^-$(aq) + Cl⁻(aq)

Answers a. Brønsted–Lowry acid: HBr; Brønsted–Lowry base CO$_3^{2-}$
b. Brønsted–Lowry acid: HSO$_4^-$; Brønsted–Lowry base OH⁻
c. Brønsted–Lowry acid: NH$_4^+$; Brønsted–Lowry base H$_2$O
d. Brønsted–Lowry acid: HCl; Brønsted–Lowry base SO$_4^{2-}$

Chapter 10

♦ **Learning Exercise 10.2C**

> **CORE CHEMISTRY SKILL**
> Identifying Conjugate Acid–Base Pairs

Identify the Brønsted–Lowry acid–base pairs in each of the following reactions:

a. $NH_3(aq) + H_2O(l) \rightleftharpoons NH_4^+(aq) + OH^-(aq)$

b. $NH_4^+(aq) + SO_4^{2-}(aq) \rightleftharpoons NH_3(aq) + HSO_4^-(aq)$

c. $HCO_3^-(aq) + H_2O(l) \rightleftharpoons CO_3^{2-}(aq) + H_3O^+(aq)$

d. $HNO_3(aq) + OH^-(aq) \longrightarrow NO_3^-(aq) + H_2O(l)$

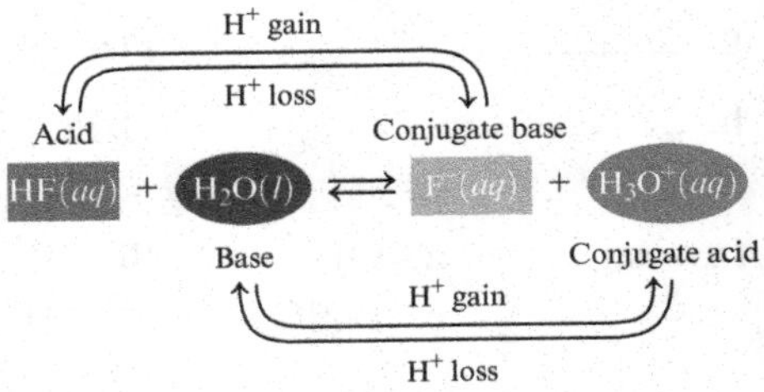

HF, an acid, loses one H^+ to form its conjugate base F^-. Water acts as a base by gaining one H^+ to form its conjugate acid H_3O^+.

Answers
a. H_2O/OH^- and NH_4^+/NH_3
b. NH_4^+/NH_3 and HSO_4^-/SO_4^{2-}
c. HCO_3^-/CO_3^{2-} and H_3O^+/H_2O
d. HNO_3/NO_3^- and H_2O/OH^-

10.3 Strengths of Acids and Bases

Learning Goal: Write equations for the dissociation of strong and weak acids and bases.

- Strong acids dissociate completely in water, and the H^+ is accepted by H_2O acting as a base.
- A weak acid dissociates slightly in water, producing only small amounts of H_3O^+.
- All hydroxides of Group 1A (1) and most hydroxides of Group 2A (2) are strong bases, which dissociate completely in water.
- In an aqueous ammonia solution, a weak base, NH_3, accepts only a small percentage of hydrogen ions to form the conjugate acid, NH_4^+.

Study Note
Only six common acids are strong acids; other acids are weak acids.
HI   HBr
HClO₄   HCl
H₂SO₄   HNO₃
Example: Is H_2S a strong or a weak acid?
Solution: H_2S is a weak acid because it is not one of the six strong acids.

Acids and Bases and Equilibrium

♦ **Learning Exercise 10.3A**

Identify each of the following as a strong acid, a weak acid, a strong base, or a weak base:

a. HNO_3 _____   b. H_2CO_3 _____

c. $H_2PO_4^-$ _____   d. NH_3 _____

e. LiOH _____   f. H_3BO_3 _____

g. $Ca(OH)_2$ _____   h. H_2SO_4 _____

Answers   a. strong acid   b. weak acid   c. weak acid
d. weak base   e. strong base   f. weak acid
g. strong base   h. strong acid

♦ **Learning Exercise 10.3B**

Identify the stronger acid in each of the following pairs of acids:

a. HCl or H_2CO_3 _____   b. HNO_2 or HCN _____

c. H_2S or HBr _____   d. H_2SO_4 or HSO_4^- _____

e. HF or H_3PO_4 _____

Answers   a. HCl   b. HNO_2 c. HBr   d. H_2SO_4 e. H_3PO_4

Acetic acid ($HC_2H_3O_2$) is a weak acid because it dissociates slightly in water, producing only a small percentage of H_3O^+.

Credit: Pearson Education / Pearson Science

10.4 Acid–Base Equilibrium

Learning Goal: Use the concept of reversible reactions to explain acid–base equilibrium. Use Le Châtelier's principle to determine the effect on equilibrium concentrations when reaction conditions change.

- A reversible reaction proceeds in both the forward and reverse directions.
- Chemical equilibrium is achieved when the rate of the forward reaction becomes equal to the rate of the reverse reaction.
- In a system at equilibrium, there is no change in the concentrations of reactants and products.
- Le Châtelier's principle states when a stress (change in conditions) is placed on a reaction at equilibrium, the equilibrium will shift in a direction (forward or reverse) that relieves the stress.
- A change in the concentration of a reactant or product or the temperature of any reaction will cause the reaction to shift to relieve the stress.

♦ **Learning Exercise 10.4A**

Indicate if each of the following represents a system at equilibrium (E) or not at equilibrium (NE):

a. _____ The rate of the forward reaction is faster than the rate of the reverse reaction.

b. _____ There is no change in the concentrations of reactants and products.

c. ___ The rate of the forward reaction is equal to the rate of the reverse reaction.
d. ___ The concentrations of reactants are decreasing.
e. ___ The concentrations of products are increasing.

Answers   a. NE   b. E   c. E   d. NE   e. NE

♦ **Learning Exercise 10.4B**

CORE CHEMISTRY SKILL
Using Le Châtelier's Principle

Identify the effect of each of the following on the reaction at equilibrium:

$$HClO(aq) + H_2O(l) \rightleftharpoons H_3O^+(aq) + ClO^-(aq)$$

a. shift in the direction of products   b. shift in the direction of reactants   c. no change

1. ___ adding more $HClO(aq)$
2. ___ removing some $HClO(aq)$
3. ___ removing some $H_3O^+(aq)$
4. ___ adding more $ClO^-(aq)$

Answers   1. a   2. b   3. a   4. b

10.5 Dissociation of Water

REVIEW
Solving Equations (1.4)

Learning Goal: Use the water dissociation expression to calculate the $[H_3O^+]$ and $[OH^-]$ in an aqueous solution.

- In pure water, a few water molecules transfer hydrogen ions to other water molecules, producing small, but equal, amounts of each ion such that $[H_3O^+]$ and $[OH^-]$ each $= 1.0 \times 10^{-7}$ M at 25 °C.

$$H_2O(l) + H_2O(l) \rightleftharpoons H_3O^+(aq) + OH^-(aq)$$

- K_w, the water dissociation expression $[H_3O^+][OH^-] = [1.0 \times 10^{-7}][1.0 \times 10^{-7}] = 1.0 \times 10^{-14}$ at 25 °C, applies to all aqueous solutions.
- In acidic solutions, the $[H_3O^+]$ is greater than the $[OH^-]$. In basic solutions, the $[OH^-]$ is greater than the $[H_3O^+]$.

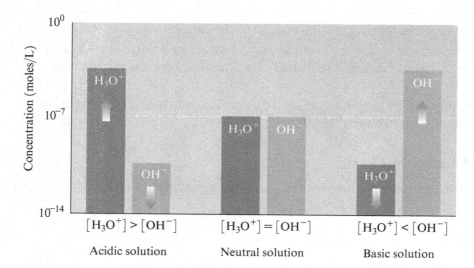

Acids and Bases and Equilibrium

♦ **Learning Exercise 10.5A**

Indicate whether each of the following solutions is acidic, basic, or neutral:

a. $[H_3O^+] = 2.5 \times 10^{-9}$ M _____ b. $[OH^-] = 1.6 \times 10^{-2}$ M _____

c. $[H_3O^+] = 7.9 \times 10^{-3}$ M _____ d. $[OH^-] = 2.9 \times 10^{-12}$ M _____

Answers   a. basic   b. basic   c. acidic   d. acidic

SAMPLE PROBLEM Calculating $[H_3O^+]$ in Aqueous Solutions

What is the $[H_3O^+]$ in a urine sample that has $[OH^-] = 4.0 \times 10^{-10}$ M?

Solution Guide:

STEP 1 State the given and needed quantities.

Analyze the Problem	Given	Need	Connect
	$[OH^-] = 4.0 \times 10^{-10}$ M	$[H_3O^+]$	$K_w = [H_3O^+][OH^-]$

STEP 2 Write the K_w for water and solve for the unknown $[H_3O^+]$.

$$K_w = [H_3O^+][OH^-] = 1.0 \times 10^{-14}$$

$$[H_3O^+] = \frac{1.0 \times 10^{-14}}{[OH^-]}$$

STEP 3 Substitute the known $[OH^-]$ into the equation and calculate.

$$[H_3O^+] = \frac{1.0 \times 10^{-14}}{[4.0 \times 10^{-10}]} = 2.5 \times 10^{-5} \text{ M}$$

A dipstick is used to measure the acidity of a urine sample.
Credit: Alexander Gospodinov/Fotolia

♦ **Learning Exercise 10.5B**

Use the K_w to calculate the $[OH^-]$ of each aqueous solution with the following $[H_3O^+]$:

a. $[H_3O^+] = 1.0 \times 10^{-3}$ M $[OH^-] =$ _____

b. $[H_3O^+] = 3.0 \times 10^{-10}$ M $[OH^-] =$ _____

c. $[H_3O^+] = 4.0 \times 10^{-6}$ M $[OH^-] =$ _____

d. $[H_3O^+] = 2.8 \times 10^{-13}$ M $[OH^-] =$ _____

e. $[H_3O^+] = 8.6 \times 10^{-7}$ M $[OH^-] =$ _____

Answers   a. 1.0×10^{-11} M   b. 3.3×10^{-5} M   c. 2.5×10^{-9} M
             d. 3.6×10^{-2} M   e. 1.2×10^{-8} M

Chapter 10

♦ **Learning Exercise 10.5C**

CORE CHEMISTRY SKILL
Calculating $[H_3O^+]$ and $[OH^-]$ in Solutions

Use the K_w to calculate the $[H_3O^+]$ of each aqueous solution with the following $[OH^-]$:

a. $[OH^-] = 1.0 \times 10^{-10}$ M $[H_3O^+] =$ _____

b. $[OH^-] = 2.0 \times 10^{-5}$ M $[H_3O^+] =$ _____

c. $[OH^-] = 4.5 \times 10^{-7}$ M $[H_3O^+] =$ _____

d. $[OH^-] = 8.0 \times 10^{-4}$ M $[H_3O^+] =$ _____

e. $[OH^-] = 5.5 \times 10^{-8}$ M $[H_3O^+] =$ _____

Answers   a. 1.0×10^{-4} M   b. 5.0×10^{-10} M   c. 2.2×10^{-8} M
           d. 1.3×10^{-11} M   e. 1.8×10^{-7} M

10.6 The pH Scale

Learning Goal: Calculate the pH of a solution from $[H_3O^+]$; given the pH, calculate $[H_3O^+]$.

- The pH scale is a range of numbers from 0 to 14 related to the $[H_3O^+]$.
- A neutral solution has a pH of 7.0. In an acidic solution, the pH is below 7.0, and in a basic solution, the pH is above 7.0.
- Mathematically, pH is the negative logarithm of the hydronium ion concentration: $pH = -\log[H_3O^+]$.

♦ **Learning Exercise 10.6A**

State whether each of the following pH values is acidic, basic, or neutral:

a. _____ blood plasma, pH = 7.4   b. _____ soft drink, pH = 2.8
c. _____ maple syrup, pH = 6.8   d. _____ beans, pH = 5.0
e. _____ tomatoes, pH = 4.2   f. _____ lemon juice, pH = 2.2
g. _____ saliva, pH = 7.0   h. _____ egg white, pH = 7.8
i. _____ lye, pH = 12.4   j. _____ strawberries, pH = 3.0

Answers   a. basic   b. acidic   c. acidic   d. acidic   e. acidic
           f. acidic   g. neutral   h. basic   i. basic   j. acidic

Solution Guide to Calculating pH of an Aqueous Solution	
STEP 1	State the given and needed quantities.
STEP 2	Enter the $[H_3O^+]$ into the pH equation and calculate.
STEP 3	Adjust the number of SFs on the *right* of the decimal point.

Learning Exercise 10.6B

KEY MATH SKILL: Calculating pH from $[H_3O^+]$

Calculate the pH of each of the following solutions:

a. $[H_3O^+] = 1.0 \times 10^{-9}$ M _____

b. $[OH^-] = 1 \times 10^{-12}$ M _____

c. $[H_3O^+] = 1 \times 10^{-3}$ M _____

d. $[OH^-] = 1 \times 10^{-10}$ M _____

e. $[H_3O^+] = 3.4 \times 10^{-8}$ M _____

f. $[OH^-] = 7.8 \times 10^{-2}$ M _____

Answers   a. 9.00   b. 2.0   c. 3.0   d. 4.0   e. 7.47   f. 12.89

Solution Guide to Calculating $[H_3O^+]$ from pH	
STEP 1	State the given and needed quantities.
STEP 2	Enter the pH value into the inverse log equation and calculate.
STEP 3	Adjust the SFs for the coefficient.

Learning Exercise 10.6C

Complete the following table:

KEY MATH SKILL: Calculating $[H_3O^+]$ from pH

	$[H_3O^+]$	$[OH^-]$	pH
a.	_____	1×10^{-12} M	_____
b.	_____	_____	8.0
c.	5.0×10^{-11} M	_____	_____
d.	_____	_____	7.80
e.	_____	_____	4.25
f.	2.0×10^{-10} M	_____	_____

Answers

	$[H_3O^+]$	$[OH^-]$	pH
a.	1×10^{-2} M	1×10^{-12} M	2.0
b.	1×10^{-8} M	1×10^{-6} M	8.0
c.	5.0×10^{-11} M	2.0×10^{-4} M	10.30
d.	1.6×10^{-8} M	6.3×10^{-7} M	7.80
e.	5.6×10^{-5} M	1.8×10^{-10} M	4.25
f.	2.0×10^{-10} M	5.0×10^{-5} M	9.70

10.7 Reactions of Acids and Bases

REVIEW: Balancing a Chemical Equation (7.4); Using Concentration as a Conversion Factor (9.4)

Learning Goal: Write balanced equations for reactions of acids with metals, carbonates or bicarbonates, and bases; calculate the molarity or volume of an acid from titration information.

- Acids react with many metals to yield hydrogen gas, H_2, and a salt.
- Acids react with carbonates and bicarbonates to yield CO_2, H_2O, and a salt.

Chapter 10

- Acids neutralize bases in a reaction that produces water and a salt.
- The net ionic equation for any strong acid–strong base neutralization is
  $H^+(aq) + OH^-(aq) \longrightarrow H_2O(l)$.
- In a balanced neutralization equation, an equal number of moles of H^+ and OH^- must react.
- In a laboratory procedure called titration, an acid or base sample is neutralized.
- A titration is used to determine the volume or concentration of an acid or a base from the laboratory data.

♦ **Learning Exercise 10.7A**

CORE CHEMISTRY SKILL
Writing Equations for Reactions of Acids and Bases

Complete and balance each of the following reactions of acids:

a. ____ Fe(s) + ____ HCl(aq) ⟶ ____ + ____ $FeCl_2$(aq)
b. ____ HCl(aq) + ____ Li_2CO_3(aq) ⟶ ____ + ____ + ____
c. ____ HBr(aq) + ____ $KHCO_3$(s) ⟶ ____ CO_2(g) + ____ H_2O(l) + ____
d. ____ Al(s) + ____ H_2SO_4(aq) ⟶ ____ + ____ $Al_2(SO_4)_3$(aq)

Answers
a. Fe(s) + 2HCl(aq) ⟶ H_2(g) + $FeCl_2$(aq)
b. 2HCl(aq) + Li_2CO_3(aq) ⟶ CO_2(g) + H_2O(l) + 2LiCl(aq)
c. HBr(aq) + $KHCO_3$(s) ⟶ CO_2(g) + H_2O(l) + KBr(aq)
d. 2Al(s) + 3H_2SO_4(aq) ⟶ 3H_2(g) + $Al_2(SO_4)_3$(aq)

	Solution Guide to Balancing Equations for Neutralization of Acids and Bases
STEP 1	Write the reactants and products.
STEP 2	Balance the H^+ in the acid with the OH^- in the base.
STEP 3	Balance the H_2O with the H^+ and the OH^-.
STEP 4	Write the formula of the salt from the remaining ions.

♦ **Learning Exercise 10.7B**

Complete and balance each of the following neutralization reactions:

a. ____ H_2SO_4(aq) + ____ NaOH(aq) ⟶ ____ + ____
b. ____ HCl(aq) + ____ $Mg(OH)_2$(s) ⟶ ____ + ____
c. ____ HNO_3(aq) + ____ $Al(OH)_3$(s) ⟶ ____ + ____
d. ____ H_3PO_4(aq) + ____ $Ca(OH)_2$(s) ⟶ ____ + ____

Answers
a. H_2SO_4(aq) + 2NaOH(aq) ⟶ 2H_2O(l) + Na_2SO_4(aq)
b. 2HCl(aq) + $Mg(OH)_2$(s) ⟶ 2H_2O(l) + $MgCl_2$(aq)
c. 3HNO_3(aq) + $Al(OH)_3$(s) ⟶ 3H_2O(l) + $Al(NO_3)_3$(aq)
d. 2H_3PO_4(aq) + 3$Ca(OH)_2$(s) ⟶ 6H_2O(l) + $Ca_3(PO_4)_2$(s)

Acids and Bases and Equilibrium

♦ **Learning Exercise 10.7C**

Complete and balance each of the following neutralization reactions:

a. ___ $H_3PO_4(aq)$ + ___ $KOH(aq)$ ⟶ ___ $H_2O(l)$ + _____

b. _____ + ___ $NaOH(aq)$ ⟶ _____ + ___ $Na_2CO_3(aq)$

c. _____ + _____ ⟶ _____ + ___ $AlCl_3(aq)$

d. _____ + _____ ⟶ _____ + ___ $Fe_2(SO_4)_3(aq)$

Answers

a. $H_3PO_4(aq) + 3KOH(aq) \longrightarrow 3H_2O(l) + K_3PO_4(aq)$

b. $H_2CO_3(aq) + 2NaOH(aq) \longrightarrow 2H_2O(l) + Na_2CO_3(aq)$

c. $3HCl(aq) + Al(OH)_3(s) \longrightarrow 3H_2O(l) + AlCl_3(aq)$

d. $3H_2SO_4(aq) + 2Fe(OH)_3(s) \longrightarrow 6H_2O(l) + Fe_2(SO_4)_3(aq)$

SAMPLE PROBLEM   Titration of an Acid for Molarity

A 32.0-mL (0.0320 L) sample of an HCl solution is placed in a flask with a few drops of indicator. If 46.2 mL (0.0462 L) of a 0.214 M NaOH solution is needed to reach the endpoint, what is the molarity of the HCl solution?

$$NaOH(aq) + HCl(aq) \longrightarrow H_2O(l) + NaCl(aq)$$

Solution Guide:

STEP 1 State the given and needed quantities and concentrations.

	Given	Need	Connect
Analyze the Problem	32.0 mL (0.0320 L) of HCl solution, 46.2 mL (0.0462 L) of 0.214 M NaOH solution	molarity of HCl solution	molarity, mole–mole factor
	Neutralization Equation		
	$NaOH(aq) + HCl(aq) \longrightarrow H_2O(l) + NaCl(aq)$		

STEP 2 Write a plan to calculate the molarity.

liters of NaOH solution $\xrightarrow{\text{Molarity of NaOH}}$ moles of NaOH $\xrightarrow{\text{Mole–mole factor}}$ moles of HCl $\xrightarrow{\text{divide by liters HCl}}$ molarity of HCl

STEP 3 State equalities and conversion factors, including concentrations.

1 L of NaOH solution = 0.214 mole of NaOH   1 mole of NaOH = 1 mole of HCl

$$\frac{0.214 \text{ mole NaOH}}{1 \text{ L NaOH solution}} \text{ and } \frac{1 \text{ L NaOH solution}}{0.214 \text{ mole NaOH}} \qquad \frac{1 \text{ mole NaOH}}{1 \text{ mole HCl}} \text{ and } \frac{1 \text{ mole HCl}}{1 \text{ mole NaOH}}$$

STEP 4 Set up the problem to calculate the needed quantity.

$$0.0462 \text{ L NaOH solution} \times \frac{0.214 \text{ mole NaOH}}{1 \text{ L NaOH solution}} \times \frac{1 \text{ mole HCl}}{1 \text{ mole NaOH}} = 0.009\ 89 \text{ mole of HCl}$$

$$\text{molarity of HCl} = \frac{0.009\ 89 \text{ mole HCl}}{0.0320 \text{ L soution}} = 0.309 \text{ M HCl solution}$$

Chapter 10

SAMPLE PROBLEM Titration of an Acid for Volume

> **CORE CHEMISTRY SKILL**
> Calculating Molarity or Volume of an Acid or Base in a Titration

A 15.7-mL (0.0157 L) sample of a 0.165 M H_2SO_4 solution reacts completely with a 0.187 M NaOH solution. How many liters of the NaOH solution are needed?

$$H_2SO_4(aq) + 2NaOH(aq) \longrightarrow 2H_2O(l) + Na_2SO_4(aq)$$

Solution Guide:

STEP 1 State the given and needed quantities and concentrations.

	Given	Need	Connect
Analyze the Problem	0.0157 L of 0.165 M H_2SO_4 solution, 0.187 M NaOH solution	liters of NaOH solution	molarity, mole–mole factor
	Neutralization Equation		
	$H_2SO_4(aq) + 2NaOH(aq) \longrightarrow 2H_2O(l) + Na_2SO_4(aq)$		

STEP 2 Write a plan to calculate the volume.

liters of H_2SO_4 solution $\xrightarrow{\text{Molarity of } H_2SO_4}$ moles of H_2SO_4 $\xrightarrow{\text{Mole–mole factor}}$ moles of NaOH $\xrightarrow{\text{Molarity of NaOH}}$ liters of NaOH solution

STEP 3 State equalities and conversion factors, including concentrations.

1 L of H_2SO_4 solution = 0.165 mole of H_2SO_4 2 moles of NaOH = 1 mole of H_2SO_4

$$\frac{0.165 \text{ mole } H_2SO_4}{1 \text{ L } H_2SO_4 \text{ solution}} \text{ and } \frac{1 \text{ L } H_2SO_4 \text{ solution}}{0.165 \text{ mole } H_2SO_4} \qquad \frac{2 \text{ moles NaOH}}{1 \text{ mole } H_2SO_4} \text{ and } \frac{1 \text{ mole } H_2SO_4}{2 \text{ moles NaOH}}$$

1 L of NaOH solution = 0.187 mole of NaOH

$$\frac{0.187 \text{ mole NaOH}}{1 \text{ L NaOH solution}} \text{ and } \frac{1 \text{ L NaOH solution}}{0.187 \text{ mole NaOH}}$$

STEP 4 Set up the problem to calculate the needed quantity.

$$0.0157 \text{ L } H_2SO_4 \text{ solution} \times \frac{0.165 \text{ mole } H_2SO_4}{1 \text{ L } H_2SO_4 \text{ solution}} \times \frac{2 \text{ moles NaOH}}{1 \text{ mole } H_2SO_4} \times \frac{1 \text{ L NaOH solution}}{0.187 \text{ mole NaOH}}$$

= 0.0277 L of NaOH solution

♦ **Learning Exercise 10.7D**

Solve the following problems using the titration data given:

a. A 5.00-mL sample of HCl solution is placed in a flask. If 15.0 mL of a 0.200 M NaOH solution is required for neutralization, what is the molarity of the HCl solution?

$$HCl(aq) + NaOH(aq) \longrightarrow H_2O(l) + NaCl(aq)$$

b. How many milliliters of a 0.200 M KOH solution are required to neutralize completely 8.50 mL of a 0.500 M H_2SO_4 solution?

$$H_2SO_4(aq) + 2KOH(aq) \longrightarrow 2H_2O(l) + K_2SO_4(aq)$$

c. A 10.0-mL sample of H_3PO_4 solution is placed in a flask. If titration requires 42.0 mL of a 0.100 M NaOH solution for complete neutralization, what is the molarity of the H_3PO_4 solution?

$$H_3PO_4(aq) + 3NaOH(aq) \longrightarrow 3H_2O(l) + Na_3PO_4(aq)$$

d. A 24.6-mL sample of HCl solution reacts completely with 33.0 mL of a 0.222 M NaOH solution. What is the molarity of the HCl solution (see reaction in part **a**)?

Answers **a.** 0.600 M HCl solution **b.** 42.5 mL
c. 0.140 M H_3PO_4 solution **d.** 0.298 M HCl solution

10.8 Buffers

Learning Goal: Describe the role of buffers in maintaining the pH of a solution.

- A buffer solution resists a change in pH when small amounts of acid or base are added.
- A buffer contains either (1) a weak acid and its salt or (2) a weak base and its salt.
- The weak acid reacts with added OH^-, and the anion of the salt reacts with added H_3O^+, which maintains the pH of the solution.
- The weak base reacts with added H_3O^+, and the cation of the salt reacts with added OH^-, which maintains the pH of the solution.

Key Terms for Sections 10.1 to 10.8

Match each of the following key terms with the correct description:

   a. acid   **b.** base   **c.** pH   **d.** neutralization
   e. buffer   **f.** titration   **g.** dissociation

1. ____ a substance that forms OH^- in water and/or accepts H^+
2. ____ a reaction between an acid and a base to form water and a salt
3. ____ a substance that forms H^+ in water
4. ____ a mixture of a weak acid (or base) and its salt that maintains the pH of a solution
5. ____ a measure of the acidity of a solution
6. ____ the separation of an acid or base into ions in water
7. ____ the addition of base to an acid sample to determine the concentration of the acid

Answers **1.** b **2.** d **3.** a **4.** e **5.** c **6.** g **7.** f

Chapter 10

♦ **Learning Exercise 10.8A**

State whether each of the following represents a buffer system and explain why:

a. HCl + NaCl

b. K_2SO_4

c. H_2CO_3

d. H_2CO_3 + $KHCO_3$

Answers
    a. No. A strong acid and its salt do not make a buffer.
    b. No. A salt alone cannot act as a buffer.
    c. No. A weak acid alone cannot act as a buffer.
    d. Yes. A weak acid and its salt act as a buffer system.

♦ **Learning Exercise 10.8B**

A buffer system that functions in the fluid of the cells contains dihydrogen phosphate, $H_2PO_4^-$, and its conjugate base, hydrogen phosphate, HPO_4^{2-}.

$$H_2PO_4^-(aq) + H_2O(l) \rightleftharpoons H_3O^+(aq) + HPO_4^{2-}(aq)$$

a. The purpose of this buffer system is to _____
    1. maintain $H_2PO_4^-$ **2.** maintain HPO_4^{2-} **3.** maintain pH

b. The weak acid is needed to _____
    1. provide the conjugate base   **2.** neutralize added OH^- **3.** provide the conjugate acid

c. If H_3O^+ is added, it is neutralized by _____
    1. HPO_4^{2-} **2.** H_2O **3.** OH^-

d. When OH^- is added, the equilibrium shifts in the direction of _____
    1. the reactants   **2.** the products   **3.** does not change

Answers   **a.** 3   **b.** 2   **c.** 1   **d.** 2

Checklist for Chapter 10

You are ready to take the Practice Test for Chapter 10. Be sure you have accomplished the following learning goals for this chapter. If not, review the Section listed at the end of the goal. Then apply your new skills and understanding to the Practice Test.

After studying Chapter 10, I can successfully:

_____ Describe the properties of Arrhenius acids and bases and write their names. (10.1)

_____ Describe the Brønsted–Lowry concept of acids and bases. (10.2)

_____ Write conjugate acid–base pairs for an acid–base reaction. (10.2)

_____ Write equations for the dissociation of strong and weak acids and bases. (10.3)

___ Use Le Châtelier's principle to determine the effect on equilibrium concentrations when reaction conditions change. (10.4)

___ Use the water dissociation expression to calculate $[H_3O^+]$ and $[OH^-]$. (10.5)

___ Calculate pH from the $[H_3O^+]$ or $[OH^-]$ of a solution. (10.6)

___ Write a balanced equation for the reactions of acids with metals, carbonates or bicarbonates, and bases. (10.7)

___ Calculate the molarity or volume of an acid or base from titration information. (10.7)

___ Describe the role of buffers in maintaining the pH of a solution. (10.8)

Practice Test for Chapter 10

The chapter Sections to review are shown in parentheses at the end of each question.

1. An acid is a compound that when placed in water yields this characteristic ion. (10.1)
   A. H_3O^+ B. OH^- C. Na^+ D. Cl^- E. CO_3^{2-}

2. $MgCl_2$ would be classified as a(an) (10.1)
   A. acid B. base C. salt D. buffer E. nonelectrolyte

3. $Mg(OH)_2$ would be classified as a (10.1)
   A. weak acid B. strong base C. salt D. buffer E. nonelectrolyte

4. Which of the following would turn litmus blue? (10.1)
   A. HCl B. NH_4Cl C. Na_2SO_4 D. KOH E. $NaNO_3$

5. What is the name of $HClO_3$? (10.1)
   A. hypochlorous acid
   B. chloric acid
   C. chlorous acid
   D. chloric trioxide acid
   E. perchloric acid

6. What is the name of NH_4OH? (10.1)
   A. ammonium oxide
   B. nitrogen tetrahydride hydroxide
   C. ammonium hydroxide
   D. perammonium hydroxide
   E. amine oxide hydride

7. Which of the following is a conjugate acid–base pair? (10.2)
   A. HCl/HNO_3 B. HNO_2/NO_2^- C. $NaOH/KOH$ D. HSO_4^-/HCO_3^- E. H_2S/S^{2-}

8. The conjugate base of HSO_4^- is (10.2)
   A. SO_4^{2-} B. H_2SO_4 C. HS^- D. H_2S E. SO_3^{2-}

9. In which of the following reactions does H_2O act as an acid? (10.2)
   A. $H_3PO_4(aq) + H_2O(l) \longrightarrow H_3O^+(aq) + H_2PO_4^-(aq)$
   B. $H_2SO_4(aq) + H_2O(l) \longrightarrow H_3O^+(aq) + HSO_4^-(aq)$
   C. $H_2O(l) + HS^-(aq) \longrightarrow H_3O^+(aq) + S^{2-}(aq)$
   D. $NaOH(aq) + HCl(aq) \longrightarrow H_2O(l) + NaCl(aq)$
   E. $NH_3(g) + H_2O(l) \longrightarrow NH_4^+(aq) + OH^-(aq)$

10. Acetic acid is a weak acid because (10.3)
    A. it forms a dilute acid solution
    B. it is isotonic
    C. it is slightly dissociated in water
    D. it is a nonpolar molecule
    E. it can form a buffer

11. A weak base when added to water (10.3)
   A. makes the solution slightly basic
   B. does not affect the pH
   C. dissociates completely
   D. does not dissociate
   E. makes the solution slightly acidic

For questions 12 to 14, predict whether each of the following changes causes the system to shift in the direction of products (P) or reactants (R): (10.4)

$$HF(aq) + H_2O(l) \rightleftharpoons F^-(aq) + H_3O^+(aq)$$

12. removing some $HF(aq)$

13. adding more $F^-(aq)$

14. removing some $H_3O^+(aq)$

15. In the K_w expression for H_2O at 25 °C, the $[H_3O^+]$ has the value (10.5)
   A. 1.0×10^{-7} M
   B. 1.0×10^{-1} M
   C. 1.0×10^{-14} M
   D. 1.0×10^{-6} M
   E. 1.0×10^{12} M

For questions 16 and 17, consider a solution with $[H_3O^+] = 1 \times 10^{-11}$ M. (10.5)

16. The hydroxide ion concentration is
   A. 1×10^{-1} M   B. 1×10^{-3} M   C. 1×10^{-4} M   D. 1×10^{-7} M   E. 1×10^{-11} M

17. The solution is
   A. acidic   B. basic   C. neutral   D. a buffer   E. neutralized

For questions 18 and 19, consider a solution with $[OH^-] = 1 \times 10^{-5}$ M. (10.5)

18. The hydrogen ion concentration of the solution is
   A. 1×10^{-5} M   B. 1×10^{-7} M   C. 1×10^{-9} M   D. 1×10^{-10} M   E. 1×10^{-14} M

19. The solution is
   A. acidic   B. basic   C. neutral   D. a buffer   E. neutralized

20. What is the $[H_3O^+]$ of a solution that has a pH = 8.7? (10.6)
   A. 7×10^{-8} B. 2×10^{-9} C. 2×10^9 D. 8×10^{-7} E. 2×10^{-8}

21. Of the following pH values, which is the most acidic? (10.6)
   A. 8.0   B. 5.5   C. 1.5   D. 3.2   E. 9.0

22. Of the following pH values, which is the most basic? (10.6)
   A. 10.0   B. 4.0   C. 2.2   D. 11.5   E. 9.0

23. Which is an equation for neutralization of an acid and a base? (10.7)
   A. $CaCO_3(s) \longrightarrow CaO(s) + CO_2(g)$
   B. $Na_2SO_4(s) \longrightarrow 2Na^+(aq) + SO_4^{2-}(aq)$
   C. $H_2SO_4(aq) + 2NaOH(aq) \longrightarrow 2H_2O(l) + Na_2SO_4(aq)$
   D. $Na_2O(s) + SO_3(g) \longrightarrow Na_2SO_4(aq)$
   E. $H_2CO_3(aq) \longrightarrow CO_2(g) + H_2O(l)$

24. What is the molarity of a 10.0-mL sample of HCl solution that is neutralized by 15.0 mL of a 2.0 M NaOH solution? (10.7)
   A. 0.50 M HCl   B. 1.0 M HCl   C. 1.5 M HCl   D. 2.0 M HCl   E. 3.0 M HCl

25. In a titration, 6.0 moles of NaOH will completely neutralize _____ mole(s) of H_2SO_4. (10.7)
   A. 1.0   B. 2.0   C. 3.0   D. 6.0   E. 12

26. What is the name given to components in the body that keep blood pH within its normal 7.35 to 7.45 range? (10.8)
    A. nutrients   B. buffers   C. metabolites   D. regufluids   E. neutralizers
27. A buffer system (10.8)
    A. maintains a pH of 7.0
    B. contains only a weak base
    C. contains only a salt
    D. contains a strong acid and its salt
    E. maintains the pH of a solution
28. Which of the following would act as a buffer system? (10.8)
    A. HCl
    B. Na_2CO_3
    C. NaOH + $NaNO_3$
    D. NH_4OH
    E. $NaHCO_3$ + H_2CO_3

Answers to the Practice Test

1. A	2. C	3. B	4. D	5. B
6. C	7. B	8. A	9. E	10. C
11. A	12. R	13. R	14. P	15. A
16. B	17. B	18. C	19. B	20. B
21. C	22. D	23. C	24. E	25. C
26. B	27. E	28. E		

Chapter 10

Selected Answers and Solutions to Text Problems

10.1
  a. Acids taste sour.
  b. Acids neutralize bases.
  c. Acids produce H^+ ions in water.
  d. Barium hydroxide is the name of a base.
  e. Both acids and bases are electrolytes.

10.3 Acids containing a simple nonmetal anion use the prefix *hydro*, followed by the name of the anion with its *ide* ending changed to *ic acid*. When the anion is an oxygen-containing polyatomic ion, the *ate* ending of the polyatomic anion is replaced with *ic acid*. Acids with one oxygen less than the common *ic acid* name are named as *ous acids*. Bases are named as ionic compounds containing hydroxide anions.
  a. hydrochloric acid
  b. calcium hydroxide
  c. perchloric acid
  d. nitric acid
  e. sulfurous acid
  f. bromous acid

10.5
  a. RbOH
  b. HF
  c. H_3PO_4
  d. LiOH
  e. NH_4OH
  f. HIO_4

10.7 A Brønsted–Lowry acid donates a hydrogen ion (H^+), whereas a Brønsted–Lowry base accepts a hydrogen ion.
  a. HI is the acid (H^+ donor); H_2O is the base (H^+ acceptor).
  b. H_2O is the acid (H^+ donor); F^- is the base (H^+ acceptor).

10.9 To form the conjugate base, remove a hydrogen ion (H^+) from the acid.
  a. F^-
  b. OH^-
  c. HPO_3^{2-}
  d. SO_4^{2-}

10.11 To form the conjugate acid, add a hydrogen ion (H^+) to the base.
  a. HCO_3^-
  b. H_3O^+
  c. H_3PO_4
  d. HBr

10.13 The conjugate acid is an H^+ donor, and the conjugate base is an H^+ acceptor.
  a. In the reaction, the acid H_2CO_3 donates an H^+ to the base H_2O. The conjugate acid–base pairs are H_2CO_3/HCO_3^- and H_3O^+/H_2O.
  b. In the reaction, the acid NH_4^+ donates an H^+ to the base H_2O. The conjugate acid–base pairs are NH_4^+/NH_3 and H_3O^+/H_2O.
  c. In the reaction, the acid HCN donates an H^+ to the base NO_2^-. The conjugate acid–base pairs are HCN/CN^- and HNO_2/NO_2^-.

10.15 Use Table 10.3 to answer (the stronger acid will be closer to the top of the table).
  a. HBr is the stronger acid.
  b. HSO_4^- is the stronger acid.
  c. H_2CO_3 is the stronger acid.

10.17 Use Table 10.3 to answer (the weaker acid will be closer to the bottom of the table).
  a. HSO_4^- is the weaker acid.
  b. HF is the weaker acid.
  c. HCO_3^- is the weaker acid.

10.19 A reversible reaction is one in which a forward reaction converts reactants to products, while a reverse reaction converts products to reactants.

10.21
  a. When the rate of the forward reaction is faster than the rate of the reverse reaction, the process is not at equilibrium.
  b. When the concentrations of the reactants and the products do not change, the process is at equilibrium.
  c. When the rate of either the forward or reverse reaction does not change, the process is at equilibrium.

Acids and Bases and Equilibrium

10.23 **a.** Adding more product shifts the system in the direction of the reactants.
b. Removing reactant shifts the system in the direction of the reactants.
c. Removing product shifts the system in the direction of the products.
d. Adding more reactant shifts the system in the direction of the products.

10.25 In pure water, $[H_3O^+] = [OH^-]$ because one of each is produced every time an H^+ transfers from one water molecule to another.

10.27 In an acidic solution, the $[H_3O^+]$ is greater than the $[OH^-]$, which means that at 25 °C, the $[H_3O^+]$ is greater than 1×10^{-7} M and the $[OH^-]$ is less than 1×10^{-7} M.

10.29 The value of $K_w = [H_3O^+][OH^-] = 1.0 \times 10^{-14}$ at 25 °C.
If $[H_3O^+]$ needs to be calculated from $[OH^-]$, then rearranging the K_w for $[H_3O^+]$ gives
$[H_3O^+] = \dfrac{1.0 \times 10^{-14}}{[OH^-]}$.
If $[OH^-]$ needs to be calculated from $[H_3O^+]$, then rearranging the K_w for $[OH^-]$ gives
$[OH^-] = \dfrac{1.0 \times 10^{-14}}{[H_3O^+]}$.
A neutral solution has $[OH^-] = [H_3O^+]$. If the $[OH^-] > [H_3O^+]$, the solution is basic; if the $[H_3O^+] > [OH^-]$, the solution is acidic.

a. $[OH^-] = \dfrac{1.0 \times 10^{-14}}{[H_3O^+]} = \dfrac{1.0 \times 10^{-14}}{[2.0 \times 10^{-5}]} = 5.0 \times 10^{-10}$ M;
since $[H_3O^+] > [OH^-]$, the solution is acidic.

b. $[OH^-] = \dfrac{1.0 \times 10^{-14}}{[H_3O^+]} = \dfrac{1.0 \times 10^{-14}}{[1.4 \times 10^{-9}]} = 7.1 \times 10^{-6}$ M;
since $[OH^-] > [H_3O^+]$, the solution is basic.

c. $[H_3O^+] = \dfrac{1.0 \times 10^{-14}}{[OH^-]} = \dfrac{1.0 \times 10^{-14}}{[8.0 \times 10^{-3}]} = 1.3 \times 10^{-12}$ M;
since $[OH^-] > [H_3O^+]$, the solution is basic.

d. $[H_3O^+] = \dfrac{1.0 \times 10^{-14}}{[OH^-]} = \dfrac{1.0 \times 10^{-14}}{[3.5 \times 10^{-10}]} = 2.9 \times 10^{-5}$ M;
since $[H_3O^+] > [OH^-]$, the solution is acidic.

10.31 The value of $K_w = [H_3O^+][OH^-] = 1.0 \times 10^{-14}$ at 25 °C.
When $[H_3O^+]$ is known, the $[OH^-]$ can be calculated by rearranging the K_w for $[OH^-]$:
$[OH^-] = \dfrac{1.0 \times 10^{-14}}{[H_3O^+]}$

a. $[OH^-] = \dfrac{1.0 \times 10^{-14}}{[H_3O^+]} = \dfrac{1.0 \times 10^{-14}}{[1.0 \times 10^{-5}]} = 1.0 \times 10^{-9}$ M (2 SFs)

b. $[OH^-] = \dfrac{1.0 \times 10^{-14}}{[H_3O^+]} = \dfrac{1.0 \times 10^{-14}}{[1.0 \times 10^{-8}]} = 1.0 \times 10^{-6}$ M (2 SFs)

c. $[OH^-] = \dfrac{1.0 \times 10^{-14}}{[H_3O^+]} = \dfrac{1.0 \times 10^{-14}}{[5.0 \times 10^{-10}]} = 2.0 \times 10^{-5}$ M (2 SFs)

d. $[OH^-] = \dfrac{1.0 \times 10^{-14}}{[H_3O^+]} = \dfrac{1.0 \times 10^{-14}}{[2.5 \times 10^{-2}]} = 4.0 \times 10^{-13}$ M (2 SFs)

10.33 The value of $K_w = [H_3O^+][OH^-] = 1.0 \times 10^{-14}$ at 25 °C.

When $[OH^-]$ is known, the $[H_3O^+]$ can be calculated by rearranging the K_w for $[H_3O^+]$:

$$[H_3O^+] = \frac{1.0 \times 10^{-14}}{[OH^-]}$$

a. $[H_3O^+] = \dfrac{1.0 \times 10^{-14}}{[OH^-]} = \dfrac{1.0 \times 10^{-14}}{[2.5 \times 10^{-13}]} = 4.0 \times 10^{-2}$ M (2 SFs)

b. $[H_3O^+] = \dfrac{1.0 \times 10^{-14}}{[OH^-]} = \dfrac{1.0 \times 10^{-14}}{[2.0 \times 10^{-9}]} = 5.0 \times 10^{-6}$ M (2 SFs)

c. $[H_3O^+] = \dfrac{1.0 \times 10^{-14}}{[OH^-]} = \dfrac{1.0 \times 10^{-14}}{[5.0 \times 10^{-11}]} = 2.0 \times 10^{-4}$ M (2 SFs)

d. $[H_3O^+] = \dfrac{1.0 \times 10^{-14}}{[OH^-]} = \dfrac{1.0 \times 10^{-14}}{[2.5 \times 10^{-6}]} = 4.0 \times 10^{-9}$ M (2 SFs)

10.35 An acidic solution has a pH less than 7.0. A basic solution has a pH greater than 7.0. A neutral solution has a pH equal to 7.0.
a. basic (pH 7.38 > 7.0)   **b.** acidic (pH 2.8 < 7.0)
c. basic (pH 11.2 > 7.0)   **d.** acidic (pH 5.52 < 7.0)
e. acidic (pH 4.2 < 7.0)   **f.** basic (pH 7.6 > 7.0)

10.37 In a neutral solution, the $[H_3O^+] = 1 \times 10^{-7}$ M at 25 °C.
pH $= -\log[H_3O^+] = -\log[1 \times 10^{-7}] = 7.0$. The pH value contains one *decimal place*, which represent the one significant figure in the coefficient 1.

10.39 pH $= -\log[H_3O^+]$

Since the value of $K_w = [H_3O^+][OH^-] = 1.0 \times 10^{-14}$ at 25 °C, if $[H_3O^+]$ needs to be calculated from $[OH^-]$, rearranging the K_w for $[H_3O^+]$ gives $[H_3O^+] = \dfrac{1.0 \times 10^{-14}}{[OH^-]}$.

a. pH $= -\log[H_3O^+] = -\log[1 \times 10^{-4}] = 4.0$ (1 SF on the right of the decimal point)
b. pH $= -\log[H_3O^+] = -\log[3 \times 10^{-9}] = 8.5$ (1 SF on the right of the decimal point)
c. $[H_3O^+] = \dfrac{1.0 \times 10^{-14}}{[1 \times 10^{-5}]} = 1 \times 10^{-9}$ M

pH $= -\log[1 \times 10^{-9}] = 9.0$ (1 SF on the right of the decimal point)

d. $[H_3O^+] = \dfrac{1.0 \times 10^{-14}}{[2.5 \times 10^{-11}]} = 4.0 \times 10^{-4}$ M

pH $= -\log[4.0 \times 10^{-4}] = 3.40$ (2 SFs on the right of the decimal point)

e. pH $= -\log[H_3O^+] = -\log[6.7 \times 10^{-8}] = 7.17$ (2 SFs on the right of the decimal point)

f. $[H_3O^+] = \dfrac{1.0 \times 10^{-14}}{[8.2 \times 10^{-4}]} = 1.2 \times 10^{-11}$ M

pH $= -\log[1.2 \times 10^{-11}] = 10.92$ (2 SFs on the right of the decimal point)

10.41 On a calculator, pH is calculated by entering $-log$, followed by the coefficient *EE* (*EXP*) key and the power of 10 followed by the change sign (+/−) key. On some calculators, the concentration is entered first (coefficient *EXP* –power) followed by *log* and +/− key.

$[H_3O^+] = \dfrac{1.0 \times 10^{-14}}{[OH^-]}$; $[OH^-] = \dfrac{1.0 \times 10^{-14}}{[H_3O^+]}$; pH $= -\log[H_3O^+]$

[H_3O^+]	[OH^-]	pH	Acidic, Basic, or Neutral?
1.0×10^{-8} M	1.0×10^{-6} M	8.00	Basic
3.2×10^{-4} M	3.1×10^{-11} M	3.49	Acidic
2.8×10^{-5} M	3.6×10^{-10} M	4.55	Acidic

10.43 [H_3O^+] = 10^{-pH} = $10^{-6.92}$ = 1.2×10^{-7} M (2 SFs)

10.45 Acids react with active metals to form $H_2(g)$ and a salt of the metal. The reaction of acids with carbonates yields CO_2, H_2O, and a salt. In a neutralization reaction, an acid and a base react to form H_2O and a salt.
   a. $ZnCO_3(s) + 2HBr(aq) \longrightarrow CO_2(g) + H_2O(l) + ZnBr_2(aq)$
   b. $Zn(s) + 2HCl(aq) \longrightarrow H_2(g) + ZnCl_2(aq)$
   c. $HCl(aq) + NaHCO_3(s) \longrightarrow CO_2(g) + H_2O(l) + NaCl(aq)$
   d. $H_2SO_4(aq) + Mg(OH)_2(s) \longrightarrow 2H_2O(l) + MgSO_4(aq)$

10.47 In balancing a neutralization equation, the number of H^+ and OH^- must be equalized by placing coefficients in front of the formulas for the acid and base.
   a. $2HCl(aq) + Mg(OH)_2(s) \longrightarrow 2H_2O(l) + MgCl_2(aq)$
   b. $H_3PO_4(aq) + 3LiOH(aq) \longrightarrow 3H_2O(l) + Li_3PO_4(aq)$

10.49 The products of a neutralization are water and a salt. In balancing a neutralization equation, the number of H^+ and OH^- must be equalized by placing coefficients in front of the formulas for the acid and base.
   a. $H_2SO_4(aq) + 2NaOH(aq) \longrightarrow 2H_2O(l) + Na_2SO_4(aq)$
   b. $3HCl(aq) + Fe(OH)_3(s) \longrightarrow 3H_2O(l) + FeCl_3(aq)$
   c. $H_2CO_3(aq) + Mg(OH)_2(s) \longrightarrow 2H_2O(l) + MgCO_3(s)$

10.51 In the titration equation, 1 mole of HCl reacts with 1 mole of NaOH.

$$28.6 \text{ mL NaOH solution} \times \frac{1 \text{ L solution}}{1000 \text{ mL solution}} \times \frac{0.145 \text{ mole NaOH}}{1 \text{ L solution}} \times \frac{1 \text{ mole HCl}}{1 \text{ mole NaOH}}$$
$$= 0.004\ 15 \text{ mole of HCl}$$

$$5.00 \text{ mL HCl solution} \times \frac{1 \text{ L solution}}{1000 \text{ mL solution}} = 0.005\ 00 \text{ L of HCl solution}$$

$$\text{molarity (M) of HCl} = \frac{\text{moles of solute}}{\text{liters of solution}} = \frac{0.004\ 15 \text{ mole HCl}}{0.005\ 00 \text{ L solution}}$$
$$= 0.830 \text{ M HCl solution (3 SFs)}$$

10.53 In the titration equation, 1 mole of H_2SO_4 reacts with 2 moles of NaOH.

$$32.8 \text{ mL NaOH solution} \times \frac{1 \text{ L solution}}{1000 \text{ mL solution}} \times \frac{0.162 \text{ mole NaOH}}{1 \text{ L solution}} \times \frac{1 \text{ mole } H_2SO_4}{2 \text{ moles NaOH}}$$
$$= 0.002\ 66 \text{ mole of } H_2SO_4$$

$$25.0 \text{ mL } H_2SO_4 \text{ solution} \times \frac{1 \text{ L solution}}{1000 \text{ mL solution}} = 0.0250 \text{ L of } H_2SO_4 \text{ solution}$$

$$\text{molarity (M) of } H_2SO_4 = \frac{\text{moles of solute}}{\text{liters of solution}} = \frac{0.002\ 66 \text{ mole } H_2SO_4}{0.0250 \text{ L solution}}$$
$$= 0.106 \text{ M } H_2SO_4 \text{ solution (3 SFs)}$$

10.55 A buffer system contains a weak acid and a salt containing its conjugate base (or a weak base and a salt containing its conjugate acid).
  a. This is not a buffer system because it only contains the strong base NaOH and the neutral salt NaCl.
  b. This is a buffer system; it contains the weak acid H_2CO_3 and a salt containing its conjugate base HCO_3^-.
  c. This is a buffer system; it contains the weak acid HF and a salt containing its conjugate base F^-.
  d. This is not a buffer system because it only contains the neutral salts KCl and NaCl.

10.57 a. The purpose of this buffer system is to (3) maintain pH.
  b. The salt of the weak acid is needed to (1) provide the conjugate base and (2) neutralize added H_3O^+.
  c. If OH^- is added, it is neutralized by (3) H_3O^+.
  d. When H_3O^+ is added, the equilibrium shifts in the direction of the (1) reactants.

10.59 If you breathe fast, CO_2 is expelled and the equilibrium shifts to lower $[H_3O^+]$, which raises the pH.

10.61 If large amounts of HCO_3^- are lost, the equilibrium shifts to higher $[H_3O^+]$, which lowers the pH.

10.63 $pH = -\log[H_3O^+] = -\log[2.0 \times 10^{-4}] = 3.70$ (2 SFs on the right of the decimal point)

10.65 $[H_3O^+] = 10^{-pH} = 10^{-3.60} = 2.5 \times 10^{-4}$ M (2 SFs)

10.67 $CaCO_3(s) + 2HCl(aq) \longrightarrow CO_2(g) + H_2O(l) + CaCl_2(aq)$

10.69 From the neutralization equation in problem 10.67, 1 mole of $CaCO_3$ reacts with 2 moles of HCl.

$$100. \text{ mL HCl solution} \times \frac{1 \text{ L HCl solution}}{1000 \text{ mL HCl solution}} \times \frac{0.0400 \text{ mole HCl}}{1 \text{ L HCl solution}} \times \frac{1 \text{ mole CaCO}_3}{2 \text{ moles HCl}}$$

$$\times \frac{100.09 \text{ g CaCO}_3}{1 \text{ mole CaCO}_3} = 0.200 \text{ g of CaCO}_3 \text{ (3 SFs)}$$

10.71 a. H_2SO_4 is an acid.
  c. $Ca(OH)_2$ is a base.
  b. RbOH is a base.
  d. HI is an acid.

10.73

Acid	Conjugate Base
H_2O	OH^-
HCN	CN^-
HNO_2	NO_2^-
H_3PO_4	$H_2PO_4^-$

10.75 An acidic solution has a pH less than 7.0. A neutral solution has a pH equal to 7.0. A basic solution has a pH greater than 7.0.
  a. acidic (pH 5.2 < 7.0)
  b. basic (pH 7.5 > 7.0)
  c. acidic (pH 3.8 < 7.0)
  d. acidic (pH 2.5 < 7.0)
  e. basic (pH 12.0 > 7.0)

10.77 a. This diagram represents a weak acid; only a few HX molecules dissociate into H_3O^+ and X^- ions.
  b. This diagram represents a strong acid; all of the HX molecules dissociate into H_3O^+ and X^- ions.

10.79 a. Hyperventilation will lower the CO_2 concentration in the blood, which lowers the $[H_2CO_3]$, which decreases the $[H_3O^+]$ and increases the blood pH.
   b. Breathing into a paper bag will increase the CO_2 concentration in the blood, increase the $[H_2CO_3]$, increase $[H_3O^+]$, and lower the blood pH back toward the normal range.

10.81 a. base; lithium hydroxide
   b. salt; calcium nitrate
   c. acid; hydrobromic acid
   d. base; barium hydroxide
   e. acid; carbonic acid

10.83 Use Table 10.3 to answer (the stronger acid will be closer to the top of the table).
   a. HF is the stronger acid.
   b. H_3O^+ is the stronger acid.
   c. HNO_2 is the stronger acid.
   d. HCO_3^- is the stronger acid.

10.85 a. Adding more reactant shifts the system in the direction of the products.
   b. Removing product shifts the system in the direction of the products.
   c. Adding more product shifts the system in the direction of the reactants.
   d. Removing reactant shifts the system in the direction of the reactants.

10.87 $[H_3O^+] = \dfrac{1.0 \times 10^{-14}}{[OH^-]}$; $pH = -\log[H_3O^+]$

   a. $pH = -\log[H_3O^+] = -\log[1.0 \times 10^{-11}] = 11.00$ (2 SFs on the right of the decimal point)
   b. $pH = -\log[5.0 \times 10^{-2}] = 1.30$ (2 SFs on the right of the decimal point)
   c. $[H_3O^+] = \dfrac{1.0 \times 10^{-14}}{[3.5 \times 10^{-4}]} = 2.9 \times 10^{-11}$ M

   $pH = -\log[2.9 \times 10^{-11}] = 10.54$ (2 SFs on the right of the decimal point)

   d. $[H_3O^+] = \dfrac{1.0 \times 10^{-14}}{[0.005]} = 2 \times 10^{-12}$ M

   $pH = -\log[2 \times 10^{-12}] = 11.7$ (1 SF on the right of the decimal point)

10.89 a. basic (pH > 7.0)
   b. acidic (pH < 7.0)
   c. basic (pH > 7.0)
   d. basic (pH > 7.0)

10.91 If the pH is given, the $[H_3O^+]$ can be found by using the relationship $[H_3O^+] = 10^{-pH}$. The $[OH^-]$ can be found by rearranging $K_w = [H_3O^+][OH^-] = 1 \times 10^{-14}$.

   a. pH = 3.40; $[H_3O^+] = 10^{-pH} = 10^{-3.40} = 4.0 \times 10^{-4}$ M (2 SFs)

   $[OH^-] = \dfrac{1.0 \times 10^{-14}}{[H_3O^+]} = \dfrac{1.0 \times 10^{-14}}{[4.0 \times 10^{-4}]} = 2.5 \times 10^{-11}$ M (2 SFs)

   b. pH = 6.00; $[H_3O^+] = 10^{-pH} = 10^{-6.00} = 1.0 \times 10^{-6}$ M (2 SFs)

   $[OH^-] = \dfrac{1.0 \times 10^{-14}}{[H_3O^+]} = \dfrac{1.0 \times 10^{-14}}{[1.0 \times 10^{-6}]} = 1.0 \times 10^{-8}$ M (2 SFs)

   c. pH = 8.0; $[H_3O^+] = 10^{-pH} = 10^{-8.0} = 1 \times 10^{-8}$ M (1 SF)

   $[OH^-] = \dfrac{1.0 \times 10^{-14}}{[H_3O^+]} = \dfrac{1.0 \times 10^{-14}}{[1 \times 10^{-8}]} = 1 \times 10^{-6}$ M (1 SF)

   d. pH = 11.0; $[H_3O^+] = 10^{-pH} = 10^{-11.0} = 1 \times 10^{-11}$ M (1 SF)

   $[OH^-] = \dfrac{1.0 \times 10^{-14}}{[H_3O^+]} = \dfrac{1.0 \times 10^{-14}}{[1 \times 10^{-11}]} = 1 \times 10^{-3}$ M (1 SF)

   e. pH = 9.20; $[H_3O^+] = 10^{-pH} = 10^{-9.20} = 6.3 \times 10^{-10}$ M (2 SFs)

   $[OH^-] = \dfrac{1.0 \times 10^{-14}}{[H_3O^+]} = \dfrac{1.0 \times 10^{-14}}{[6.3 \times 10^{-10}]} = 1.6 \times 10^{-5}$ M (2 SFs)

10.93 a. Solution A, with a pH of 4.0, is more acidic than solution B.

b. In solution A, the $[H_3O^+] = 10^{-pH} = 10^{-4.0} = 1 \times 10^{-4}$ M (1 SF)
In solution B, the $[H_3O^+] = 10^{-pH} = 10^{-6.0} = 1 \times 10^{-6}$ M (1 SF)

c. In solution A, the $[OH^-] = \dfrac{1.0 \times 10^{-14}}{[H_3O^+]} = \dfrac{1.0 \times 10^{-14}}{[1 \times 10^{-4}]} = 1 \times 10^{-10}$ M (1 SF)

In solution B, the $[OH^-] = \dfrac{1.0 \times 10^{-14}}{[H_3O^+]} = \dfrac{1.0 \times 10^{-14}}{[1 \times 10^{-6}]} = 1 \times 10^{-8}$ M (1 SF)

10.95 In the titration equation, 1 mole of H_2SO_4 reacts with 2 moles of NaOH.

45.6 mL NaOH solution × $\dfrac{1 \text{ L solution}}{1000 \text{ mL solution}}$ × $\dfrac{0.205 \text{ mole NaOH}}{1 \text{ L solution}}$ × $\dfrac{1 \text{ mole } H_2SO_4}{2 \text{ moles NaOH}}$

= 0.004 67 mole of H_2SO_4

20.0 mL H_2SO_4 solution × $\dfrac{1 \text{ L solution}}{1000 \text{ mL solution}}$ = 0.0200 L of H_2SO_4 solution

molarity (M) of H_2SO_4 = $\dfrac{\text{moles of solute}}{\text{liters of solution}}$ = $\dfrac{0.004\ 67 \text{ mole } H_2SO_4}{0.0200 \text{ L solution}}$

= 0.234 M H_2SO_4 solution (3 SFs)

10.97 a. In the titration equation, 1 mole of HCl reacts with 1 mole of NaOH.
$HCl(aq) + NaOH(aq) \longrightarrow H_2O(l) + NaCl(aq)$

25.0 mL HCl solution × $\dfrac{1 \text{ L HCl solution}}{1000 \text{ mL HCl solution}}$ × $\dfrac{0.288 \text{ mole HCl}}{1 \text{ L HCl solution}}$ × $\dfrac{1 \text{ mole NaOH}}{1 \text{ mole HCl}}$

× $\dfrac{1 \text{ L NaOH solution}}{0.150 \text{ mole NaOH}}$ × $\dfrac{1000 \text{ mL NaOH solution}}{1 \text{ L NaOH solution}}$ = 48.0 mL of NaOH solution (3 SFs)

b. In the titration equation, 1 mole of H_2SO_4 reacts with 2 moles of NaOH.
$H_2SO_4(aq) + 2NaOH(aq) \longrightarrow 2H_2O(l) + Na_2SO_4(aq)$

10.0 mL H_2SO_4 solution × $\dfrac{1 \text{ L } H_2SO_4 \text{ solution}}{1000 \text{ mL } H_2SO_4 \text{ solution}}$ × $\dfrac{0.560 \text{ mole } H_2SO_4}{1 \text{ L } H_2SO_4 \text{ solution}}$ × $\dfrac{2 \text{ moles NaOH}}{1 \text{ mole } H_2SO_4}$

× $\dfrac{1 \text{ L NaOH solution}}{0.150 \text{ mole NaOH}}$ × $\dfrac{1000 \text{ mL NaOH solution}}{1 \text{ L NaOH solution}}$ = 74.7 mL of NaOH solution (3 SFs)

c. In the titration equation, 1 mole of HBr reacts with 1 mole of NaOH.
$HBr(aq) + NaOH(aq) \longrightarrow H_2O(l) + NaBr(aq)$

5.00 mL HBr solution × $\dfrac{1 \text{ L HBr solution}}{1000 \text{ mL HBr solution}}$ × $\dfrac{0.618 \text{ mole HBr}}{1 \text{ L HBr solution}}$ × $\dfrac{1 \text{ mole NaOH}}{1 \text{ mole HBr}}$

× $\dfrac{1 \text{ L NaOH solution}}{0.150 \text{ mole NaOH}}$ × $\dfrac{1000 \text{ mL NaOH solution}}{1 \text{ L NaOH solution}}$ = 20.6 mL of NaOH solution (3 SFs)

10.99 This buffer solution is made from the weak acid H_3PO_4 and a salt containing its conjugate base $H_2PO_4^-$.
  a. Acid added: $H_2PO_4^-(aq) + H_3O^+(aq) \longrightarrow H_2O(l) + H_3PO_4(aq)$
  b. Base added: $H_3PO_4(aq) + OH^-(aq) \longrightarrow H_2O(l) + H_2PO_4^-(aq)$

10.101 a. To form the conjugate base, remove a hydrogen ion (H^+) from the acid.
  1. HS^- **2.** $H_2PO_4^-$
  b. H_2S (see Table 10.3; the weaker acid will be closer to the bottom of the table)

10.103 a. In the reaction, the acid HNO_3 donates an H^+ to the base NH_3. The conjugate acid–base pairs are HNO_3/NO_3^- and NH_4^+/NH_3.
  b. In the reaction, the acid HBr donates an H^+ to the base H_2O. The conjugate acid–base pairs are HBr/Br^- and H_3O^+/H_2O.

10.105 a. $ZnCO_3(s) + H_2SO_4(aq) \longrightarrow CO_2(g) + H_2O(l) + ZnSO_4(aq)$
  b. $2Al(s) + 6HBr(aq) \longrightarrow 3H_2(g) + 2AlBr_3(aq)$

10.107 $KOH \text{ (strong base)} \longrightarrow K^+(aq) + OH^-(aq)$ (100% dissociation)
  $[OH^-] = 0.050 \text{ M} = 5.0 \times 10^{-2} \text{ M}$

  a. $[H_3O^+] = \dfrac{1.0 \times 10^{-14}}{[OH^-]} = \dfrac{1.0 \times 10^{-14}}{[5.0 \times 10^{-2}]} = 2.0 \times 10^{-13}$ M (2 SFs)

  b. $pH = -\log[H_3O^+] = -\log[2.0 \times 10^{-13}] = 12.70$ (2 SFs on the right of the decimal point)

  c. $H_2SO_4(aq) + 2KOH(aq) \longrightarrow 2H_2O(l) + K_2SO_4(aq)$

  d. In the titration equation, 1 mole of H_2SO_4 reacts with 2 moles of KOH.

  $40.0 \text{ mL } H_2SO_4 \text{ solution} \times \dfrac{1 \text{ L } H_2SO_4 \text{ solution}}{1000 \text{ mL } H_2SO_4 \text{ solution}} \times \dfrac{0.035 \text{ mole } H_2SO_4}{1 \text{ L } H_2SO_4 \text{ solution}} \times \dfrac{2 \text{ moles KOH}}{1 \text{ mole } H_2SO_4}$

  $\times \dfrac{1 \text{ L KOH solution}}{0.050 \text{ mole KOH}} \times \dfrac{1000 \text{ mL KOH solution}}{1 \text{ L KOH solution}} = 56$ mL of KOH solution (2 SFs)

10.109 a. $H_3PO_4(aq) + 3NaOH(aq) \longrightarrow 3H_2O(l) + Na_3PO_4(aq)$
  b. In the titration equation, 1 mole of H_3PO_4 reacts with 3 moles of $NaOH$.

  $16.4 \text{ mL NaOH solution} \times \dfrac{1 \text{ L NaOH solution}}{1000 \text{ mL NaOH solution}} \times \dfrac{0.204 \text{ mole NaOH}}{1 \text{ L NaOH solution}} \times \dfrac{1 \text{ mole } H_3PO_4}{3 \text{ moles NaOH}}$

  $= 1.12 \times 10^{-3}$ mole of H_3PO_4 (3 SFs)

  $50.0 \text{ mL } H_3PO_4 \text{ solution} \times \dfrac{1 \text{ L solution}}{1000 \text{ mL solution}} = 0.0500$ L of H_3PO_4 solution

  molarity (M) of $H_3PO_4 = \dfrac{\text{moles of solute}}{\text{liters of solution}} = \dfrac{1.12 \times 10^{-3} \text{ mole } H_3PO_4}{0.0500 \text{ L solution}}$

  $= 0.0224$ M H_3PO_4 solution (3 SFs)

10.111 a. The $[H_3O^+]$ can be found by using the relationship $[H_3O^+] = 10^{-pH}$.
  $[H_3O^+] = 10^{-pH} = 10^{-4.2} = 6 \times 10^{-5}$ M (1 SF)

  $[OH^-] = \dfrac{1.0 \times 10^{-14}}{[H_3O^+]} = \dfrac{1.0 \times 10^{-14}}{[6 \times 10^{-5}]} = 2 \times 10^{-10}$ M (1 SF)

  b. $[H_3O^+] = 10^{-pH} = 10^{-6.5} = 3 \times 10^{-7}$ M (1 SF)

  $[OH^-] = \dfrac{1.0 \times 10^{-14}}{[H_3O^+]} = \dfrac{1.0 \times 10^{-14}}{[3 \times 10^{-7}]} = 3 \times 10^{-8}$ M (1 SF)

  c. In the titration equation, 1 mole of $CaCO_3$ reacts with 1 mole of H_2SO_4.

  $1.0 \text{ kL solution} \times \dfrac{1000 \text{ L solution}}{1 \text{ kL solution}} \times \dfrac{6 \times 10^{-5} \text{ mole } H_3O^+}{1 \text{ L solution}}$

  $\times \dfrac{1 \text{ mole } H_2SO_4}{2 \text{ moles } H_3O^+} \times \dfrac{1 \text{ mole } CaCO_3}{1 \text{ mole } H_2SO_4} \times \dfrac{100.09 \text{ g } CaCO_3}{1 \text{ mole } CaCO_3} = 3$ g of $CaCO_3$ (1 SF)

11
Introduction to Organic Chemistry: Hydrocarbons

Credit: Andrew Poertner/AP Images

For three months after her house fire, Diane remained in the hospital's burn unit, where she continued to have treatment for burns, including removing damaged tissue, skin grafts, and preventing infection. For some of the large burn areas, some of Diane's own skin cells were removed and grown for three weeks to give a layer of skin that covered the burn areas. By covering the burn areas, the incidence of dehydration and infection were decreased. After three months, she was discharged.

The arson investigators determined that gasoline was the primary accelerant used to start the fire at Diane's house. However, they later determined there was also kerosene and some diesel fuel. About 70% of kerosene consists of alkanes and cycloalkanes; the rest is aromatic hydrocarbons. Write a balanced chemical equation for the combustion of $C_{11}H_{24}$, which is a compound in kerosene.

LOOKING AHEAD

11.1 Organic Compounds
11.2 Alkanes
11.3 Alkanes with Substituents
11.4 Properties of Alkanes
11.5 Alkenes and Alkynes
11.6 Cis–Trans Isomers
11.7 Addition Reactions for Alkenes
11.8 Aromatic Compounds

 The Health icon indicates a question that is related to health and medicine.

11.1 Organic Compounds

Learning Goal: Identify properties characteristic of organic or inorganic compounds.

> **REVIEW**
> Drawing Lewis Structures (6.6)
> Predicting Shape (6.8)

- Organic compounds are compounds of carbon and hydrogen that have covalent bonds, have low melting and boiling points, burn vigorously, are mostly nonpolar molecules, and are usually more soluble in nonpolar solvents than in water.
- An expanded structural formula shows all of the atoms and the bonds connected to each atom.
- A condensed structural formula depicts each carbon atom and its attached hydrogen atoms as a group.
- A molecular formula gives the total number of each kind of atom.

Introduction to Organic Chemistry: Hydrocarbons

Ball-and-Stick Model	Expanded Structural Formula	Condensed Structural Formula	Molecular Formula
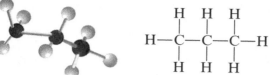	H H H H—C—C—C—H H H H	CH$_3$—CH$_2$—CH$_3$	C$_3$H$_8$

♦ **Learning Exercise 11.1A**

Identify each of the following as characteristic of organic or inorganic compounds:

1. _____ have covalent bonds
2. _____ have low boiling points
3. _____ burn in air
4. _____ are soluble in water
5. _____ have high melting points
6. _____ are soluble in nonpolar solvents
7. _____ have ionic bonds
8. _____ form long carbon chains
9. _____ contain carbon and hydrogen
10. _____ do not burn in air

Answers 1. organic 2. organic 3. organic 4. inorganic
5. inorganic 6. organic 7. inorganic 8. organic
9. organic 10. inorganic

♦ **Learning Exercise 11.1B**

1. What type of three-dimensional representation is shown here?

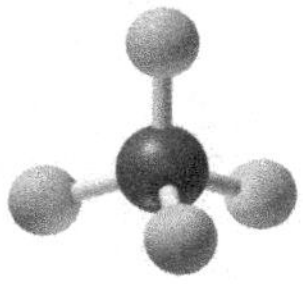

2. Draw the expanded structural formula for methane.

Vegetable oil, an organic compound, is not soluble in water.
Credit: Pearson Education / Pearson Science

Answers 1. ball-and-stick model   2.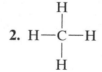

11.2 Alkanes

Learning Goal: Write the IUPAC names and draw the condensed structural and line-angle formulas for alkanes and cycloalkanes.

- Alkanes are hydrocarbons that have only single bonds, C—C and C—H.
- Each carbon in an alkane has four bonds arranged so that the bonded atoms are in the corners of a tetrahedron.
- A line-angle formula shows the carbon atoms as the corners or ends of a zigzag line.

Chapter 11

- In an IUPAC (International Union of Pure and Applied Chemistry) name, the prefix indicates the number of carbon atoms, and the suffix describes the family of the compound. For example, in the name *propane*, the prefix *prop* indicates a chain of three carbon atoms and the ending *ane* indicates single bonds (alkane). The names of the first six alkanes follow:

Number of Carbon Atoms	Name	Condensed Structural Formula	Line-Angle Formula
1	**Meth**ane	CH_4	
2	**Eth**ane	CH_3-CH_3	
3	**Prop**ane	$CH_3-CH_2-CH_3$	
4	**But**ane	$CH_3-CH_2-CH_2-CH_3$	
5	**Pent**ane	$CH_3-CH_2-CH_2-CH_2-CH_3$	
6	**Hex**ane	$CH_3-CH_2-CH_2-CH_2-CH_2-CH_3$	

♦ **Learning Exercise 11.2A**

CORE CHEMISTRY SKILL
Naming and Drawing Alkanes

Indicate if each of the following is an expanded structural formula (E), a condensed structural formula (C), a line-angle formula (L), or a molecular formula (M):

1. ____ CH_3-CH_3

2. ____ C_5H_{12}

3. ____ $CH_3-CH_2-CH_3$

4. ____ H—C—C—C—H (with H's on each C)

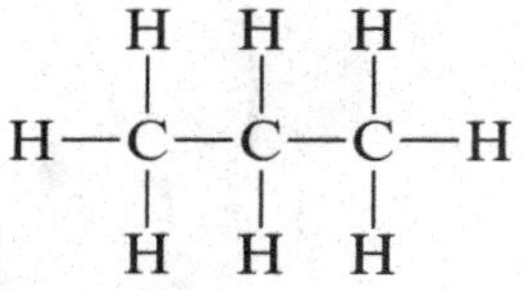

5. ____ (line-angle zigzag)

Answers   1. C   2. M   3. C   4. E   5. L

	Solution Guide to Drawing Formulas for an Alkane
STEP 1	Draw the carbon chain.
STEP 2	Draw the expanded structural formula by adding the hydrogen atoms using single bonds to each of the carbon atoms.
STEP 3	Draw the condensed structural formula by combining the H atoms with each C atom.
STEP 4	Draw the line-angle formula as a zigzag line in which the ends and corners represent C atoms.

♦ **Learning Exercise 11.2B**

Draw the condensed structural formula for each of the following expanded structural or line-angle formulas:

1. 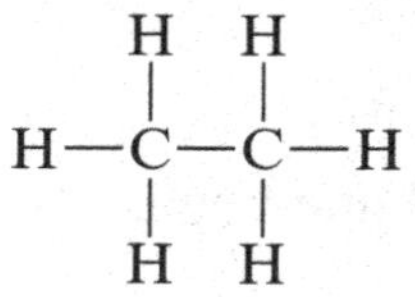 H—C—C—H (with H's, ethane expanded)

2. 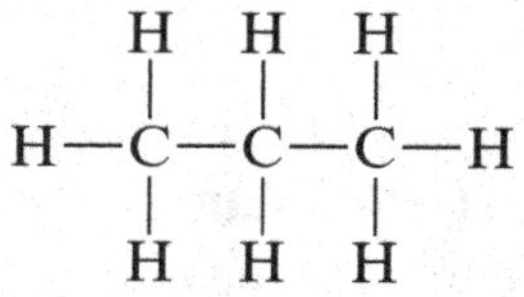 H—C—C—C—H (with H's, propane expanded)

3. 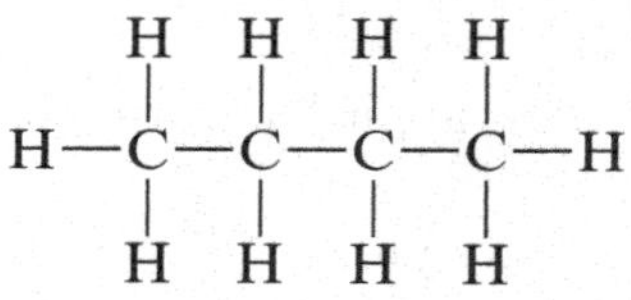 H—C—C—C—C—H (with H's, butane expanded)

4.  (line-angle, propane)

256

Introduction to Organic Chemistry: Hydrocarbons

Answers
1. CH_3-CH_3
2. $CH_3-CH_2-CH_3$
3. $CH_3-CH_2-CH_2-CH_3$
4. $CH_3-CH_2-CH_2-CH_2-CH_3$

♦ **Learning Exercise 11.2C**

Draw the condensed structural formula and give the name for the continuous-chain alkane with each of the following molecular formulas:

1. C_2H_6 _____
2. C_3H_8 _____
3. C_4H_{10} _____
4. C_5H_{12} _____
5. C_6H_{14} _____

Answers
1. CH_3-CH_3, ethane
2. $CH_3-CH_2-CH_3$, propane
3. $CH_3-CH_2-CH_2-CH_3$, butane
4. $CH_3-CH_2-CH_2-CH_2-CH_3$, pentane
5. $CH_3-CH_2-CH_2-CH_2-CH_2-CH_3$, hexane

♦ **Learning Exercise 11.2D**

Draw the line-angle formula for each of the following:

1. butane _____
2. pentane _____
3. hexane _____
4. octane _____
5. cyclohexane _____

Answers
1.  ∧
2.  ∧∧
3.  ∧∧∧
4.  ∧∧∧∧
5.  ⬡

Study Note

A cycloalkane is named by adding the prefix *cyclo* to the name of the corresponding alkane. For example, the name *cyclopropane* indicates a cyclic structure of three carbon atoms, usually represented by the geometric shape of a triangle.

♦ **Learning Exercise 11.2E**

Give the IUPAC name for each of the following cycloalkanes:

1.  ⬡
2. ☐
3. 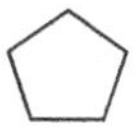 ⬠
4.  △

Answers
1. cyclohexane
2. cyclobutane
3. cyclopentane
4. cyclopropane

Chapter 11

11.3 Alkanes with Substituents

Learning Goal: Write the IUPAC names for alkanes with substituents and draw their condensed structural or line-angle formulas.

- In the IUPAC system, each substituent is numbered and listed alphabetically in front of the name of the longest chain.
- Carbon groups that are substituents are named as alkyl groups or alkyl substituents. An alkyl group is named by replacing the *ane* of the alkane name with *yl*. For example, CH_3- is methyl (from CH_4 methane), and CH_3-CH_2- is ethyl (from CH_3-CH_3 ethane).
- In a haloalkane, a halogen atom, $-F$, $-Cl$, $-Br$, or $-I$ replaces a hydrogen atom in an alkane.
- A halogen atom is named as a substituent (*fluoro, chloro, bromo, iodo*) attached to the alkane chain.

SAMPLE PROBLEM   Naming Alkanes with Substituents

Write the IUPAC name for the following compound:

$$CH_3-CH_2-\underset{\underset{CH_3}{|}}{CH}-CH_3$$

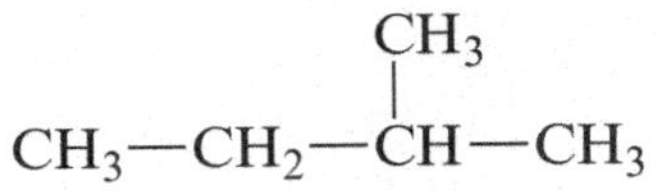

Solution Guide:

STEP 1 Write the alkane name for the longest chain of carbon atoms. In this compound, the longest chain has four carbon atoms, which is named butane.

STEP 2 Number the carbon atoms from the end nearer a substituent. In this compound, the methyl group (substituent) is on carbon 2.

STEP 3 Give the location and name for each substituent (alphabetical order) as a prefix to the name of the main chain. The compound with a methyl group on carbon 2 of a four-carbon chain butane is named 2-methylbutane.

When there are two or more of the same substituent, a prefix (*di, tri, tetra*) is used in front of the name. Commas are used to separate the numbers for the location of the substituents.

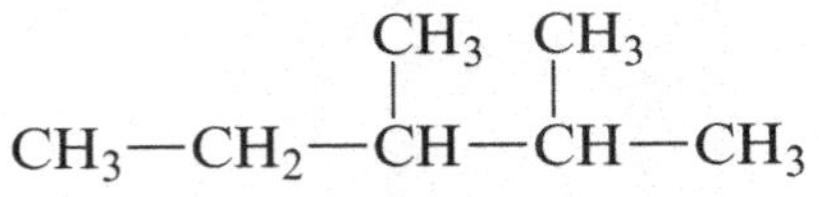

$$CH_3-CH_2-\underset{\underset{CH_3}{|}}{CH}-\underset{\underset{CH_3}{|}}{CH}-CH_3$$
2,3-Dimethylpentane

♦ Learning Exercise 11.3A

Give the IUPAC name for each of the following compounds:

1. $CH_3-\underset{\underset{CH_3}{|}}{CH}-CH_3$

2. $CH_3-\underset{\underset{CH_3}{|}}{CH}-CH_2-\underset{\underset{CH_3}{|}}{CH}-CH_2-CH_3$

3. 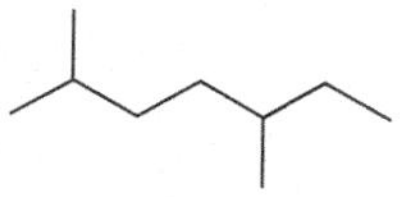 [structure: 2,5-dimethylheptane skeleton]   4. 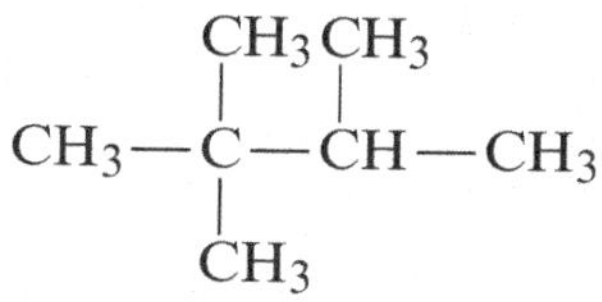 $CH_3-\underset{\underset{CH_3}{|}}{\overset{\overset{CH_3}{|}}{C}}-\overset{\overset{CH_3}{|}}{CH}-CH_3$

Answers   1. methylpropane   2. 2,4-dimethylhexane   3. 2,5-dimethylheptane
4. 2,2,3-trimethylbutane

♦ **Learning Exercise 11.3B**

Give the IUPAC name for each of the following:

1. CH_3-CH_2-Br

2. 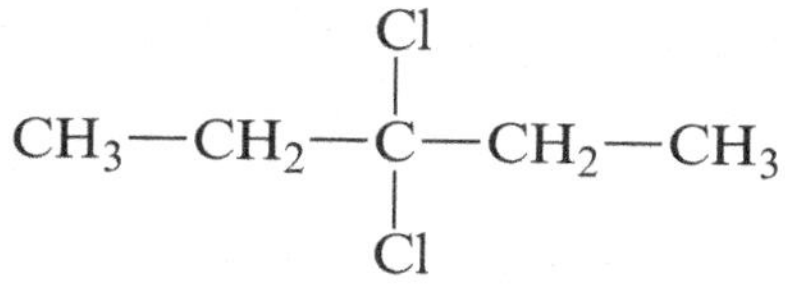 $CH_3-CH_2-\underset{\underset{Cl}{|}}{\overset{\overset{Cl}{|}}{C}}-CH_2-CH_3$

3. [line-angle structure with Br and Cl]

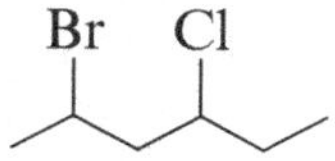

4. 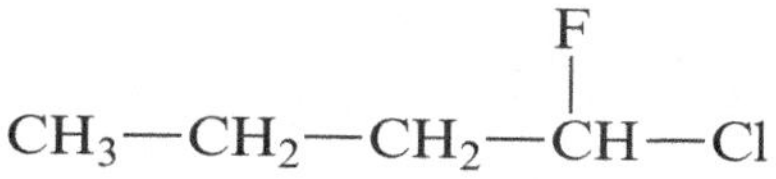 $CH_3-CH_2-CH_2-\overset{\overset{F}{|}}{CH}-Cl$

5.  [cyclopropane with Br]

Answers   1. bromoethane   2. 3,3-dichloropentane
3. 2-bromo-4-chlorohexane   4. 1-chloro-1-fluorobutane
5. bromocyclopropane

Solution Guide to Drawing Condensed Structural and Line-Angle Formulas from IUPAC Names	
STEP 1	Draw the main chain of carbon atoms.
STEP 2	Number the chain and place the substituents on the carbons indicated by the numbers.
STEP 3	For the condensed structural formula, add the correct number of hydrogen atoms to give four bonds to each C atom.

♦ **Learning Exercise 11.3C**

Draw the condensed structural formulas for **1, 2,** and **3** and the line-angle formulas for **4, 5,** and **6**.

1. 2-methylbutane

2. 2-methylpentane

Chapter 11

3. 4-ethyl-2-methylhexane

4. 2,2,4-trimethylhexane

5. 3-ethylpentane

6. methylcyclohexane

Answers

1. CH₃—CH(CH₃)—CH₂—CH₃

2. CH₃—CH(CH₃)—CH₂—CH₂—CH₃

3. CH₃—CH(CH₃)—CH₂—CH(CH₂—CH₃)—CH₂—CH₃

4. 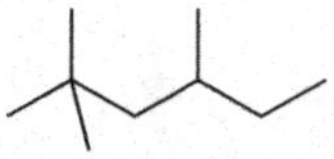 (line-angle structure)

5. 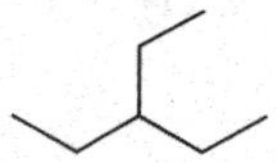 (line-angle structure)

6. 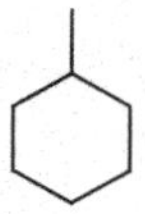 (line-angle structure of methylcyclohexane)

♦ **Learning Exercise 11.3D**

Draw the condensed structural formulas for **1**, **2**, and **3** and the line-angle formulas for **4**, **5**, and **6**.

1. chloroethane

2. bromomethane

3. 3-bromo-1-chloropentane

4. 1,1-dichlorohexane

5. 2,2,3-trichlorobutane

6. 2,4-dibromo-2,4-dichloropentane

Answers

1. CH₃—CH₂—Cl

2. CH₃—Br

3. Cl—CH₂—CH₂—CH(Br)—CH₂—CH₃

4. 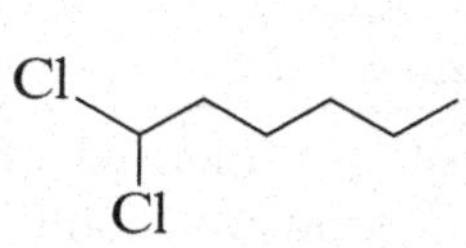 (line-angle structure with two Cl on C1)

5. 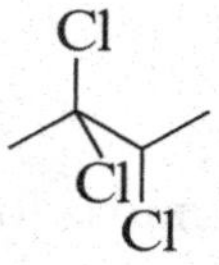 (line-angle structure with Cl, Cl, Cl)

6. 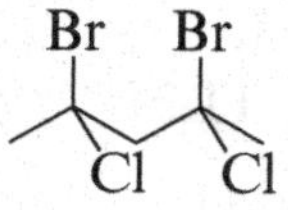 (line-angle structure with Br, Br, Cl, Cl)

260

Introduction to Organic Chemistry: Hydrocarbons

11.4 Properties of Alkanes

Learning Goal: Identify the properties of alkanes and write a balanced chemical equation for combustion.

- Alkanes are less dense than water, and mostly unreactive, except that they burn vigorously.
- Alkanes are found in natural gas, gasoline, and diesel fuels.
- Alkanes are nonpolar and insoluble in water, and they have low boiling points.
- In combustion, an alkane reacts with oxygen at a high temperature to produce carbon dioxide, water, and energy.

Key Terms for Sections 11.1 to 11.4

Match each of the following key terms with the correct description:

    a. line-angle formula   b. expanded structural formula   c. alkane
    d. condensed structural formula   e. combustion   f. cycloalkane
    g. substituent

1. _____ groups of atoms such as an alkyl group or a halogen bonded to a carbon chain
2. _____ a type of formula for carbon compounds that indicates the carbon atoms as corners or ends of a zigzag line
3. _____ the type of formula that shows all of the atoms and the bonds connected to each atom
4. _____ a straight-chain hydrocarbon that contains only carbon–carbon and carbon–hydrogen single bonds
5. _____ an alkane that exists as a cyclic structure
6. _____ the chemical reaction of an alkane and oxygen that yields CO_2, H_2O, and energy
7. _____ the type of formula that shows the arrangement of the carbon atoms grouped with their attached H atoms

Answers   **1.** g   **2.** a   **3.** b   **4.** c   **5.** f   **6.** e   **7.** d

SAMPLE PROBLEM   Combustion of Alkanes

Write the balanced chemical equation for the combustion of methane.

Solution: Write the molecular formulas for the reactants: methane (CH_4) and oxygen (O_2)
Write the products CO_2 and H_2O and balance the equation.

$$CH_4 + O_2 \xrightarrow{\Delta} CO_2 + H_2O + \text{energy}$$

$$CH_4 + 2O_2 \xrightarrow{\Delta} CO_2 + 2H_2O + \text{energy} \quad \textit{Balanced}$$

◆ Learning Exercise 11.4A

Write a balanced chemical equation for the complete combustion of each of the following:

1. propane _____
2. hexane _____
3. pentane _____
4. cyclobutane _____

Chapter 11

Answers
1. $C_3H_8 + 5O_2 \xrightarrow{\Delta} 3CO_2 + 4H_2O + \text{energy}$
2. $2C_6H_{14} + 19O_2 \xrightarrow{\Delta} 12CO_2 + 14H_2O + \text{energy}$
3. $C_5H_{12} + 8O_2 \xrightarrow{\Delta} 5CO_2 + 6H_2O + \text{energy}$
4. $C_4H_8 + 6O_2 \xrightarrow{\Delta} 4CO_2 + 4H_2O + \text{energy}$

♦ Learning Exercise 11.4B

Indicate which property in each of the following pairs is more likely to be a property of hexane:

a. density is 0.66 g/mL or 1.66 g/mL   _____

b. solid or liquid at room temperature   _____

c. soluble or insoluble in water   _____

d. flammable or nonflammable   _____

e. melting point 95 °C or −95 °C   _____

Answers a. 0.66 g/mL   b. liquid   c. insoluble
  d. flammable   e. −95 °C

11.5 Alkenes and Alkynes

Learning Goal: Write the IUPAC names or draw the condensed structural or line-angle formulas for alkenes and alkynes.

- Alkenes are unsaturated hydrocarbons that contain one or more carbon–carbon double bonds.
- In alkenes, each carbon in the double bond is connected to two other groups in addition to the other carbon of the double bond, all being in the same plane and having bond angles of 120°.
- Alkynes are unsaturated hydrocarbons that contain at least one carbon–carbon triple bond.
- The IUPAC names of alkenes are derived by changing the *ane* ending of the parent alkane to *ene*. For example, the IUPAC name of $H_2C=CH_2$ is ethene. It has a common name of ethylene. In alkenes, the longest carbon chain containing the double bond is numbered from the end nearer the double bond. In cycloalkenes with substituents, the double-bond carbons are given positions of 1 and 2, and the ring is numbered to give the next lower numbers to the substituents.

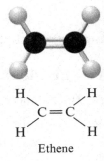
Ethene

$CH_3-CH=CH_2$	$H_2C=CH-CH_2-CH_3$	$CH_3-CH=CH-CH_3$
Propene (propylene)	1-Butene	2-Butene

Ethyne

- The IUPAC names of alkynes are derived by changing the *ane* ending of the parent alkane to *yne*. In alkynes, the longest carbon chain containing the triple bond is numbered from the end nearer the triple bond.

$HC\equiv CH$	$CH_3-C\equiv CH$	$CH_3-CH_2-C\equiv CH$
Ethyne	Propyne	1-Butyne

Ball-and-stick models show the double bond of an alkene and the triple bond of an alkyne.

Introduction to Organic Chemistry: Hydrocarbons

♦ **Learning Exercise 11.5A**

Classify each of the following as an alkane, alkene, cycloalkene, or alkyne:

1. _____ $CH_3-CH_2-CH_3$

2. _____ (CH₃)₂C=CH₂

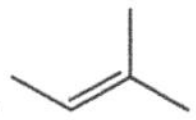

3. _____ $CH_3-C \equiv C-CH_3$

4. _____ $CH_3-CH_2-CH=\underset{\underset{CH_3}{|}}{C}-CH_2-CH_3$

Answers   1. alkane   2. alkene   3. alkyne   4. alkene

Solution Guide to Naming Alkenes and Alkynes	
STEP 1	Name the longest carbon chain that contains the double or triple bond.
STEP 2	Number the carbon chain starting from the end nearer the double or triple bond.
STEP 3	Give the location and name for each substituent (alphabetical order) as a prefix to the alkene or alkyne name.

♦ **Learning Exercise 11.5B**

Write the IUPAC name for each of the following alkenes:

1. $CH_3-CH=CH_2$

2. $CH_3-\underset{\underset{CH_3}{|}}{C}=CH-CH_3$

3.  (cyclohexene structure)

4. $H_2C=CH-\underset{\underset{Cl}{|}}{CH}-CH_2-\underset{\underset{CH_3}{|}}{CH}-CH_3$

5. $CH_3-CH=\underset{\underset{CH_2-CH_3}{|}}{C}-CH_2-CH_3$

6. 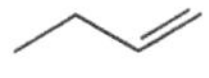 (1-butene structure)

Answers   1. propene   2. 2-methyl-2-butene   3. cyclohexene
              4. 3-chloro-5-methyl-1-hexene   5. 3-ethyl-2-pentene   6. 1-butene

Chapter 11

♦ **Learning Exercise 11.5C**

Write the IUPAC name for each of the following alkynes:

1. HC≡CH

2. $CH_3-C\equiv CH$

3. $CH_3-CH_2-C\equiv CH$

4. $CH_3-\underset{\underset{CH_3}{|}}{CH}-C\equiv C-CH_3$

Answers   1. ethyne   2. propyne
            3. 1-butyne   4. 4-methyl-2-pentyne

♦ **Learning Exercise 11.5D**

Draw the condensed structural formulas for **1** and **2** and the line-angle formulas for **3** and **4**.

1. 2-pentyne

2. 2-chloro-2-butene

3. 3-bromo-2-methyl-2-pentene

4. 4-methylcyclohexene

Answers   1. $CH_3-C\equiv C-CH_2-CH_3$ 2. $CH_3-\underset{\underset{Cl}{|}}{C}=CH-CH_3$

3. [line-angle structure with Br]   4. [cyclohexene with methyl group]

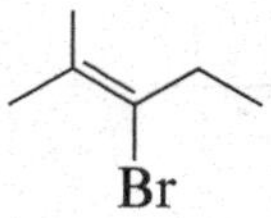

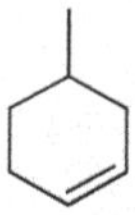

11.6 Cis–Trans Isomers

Learning Goal: Draw the condensed structural formulas and give the names for the cis–trans isomers of alkenes.

- Cis–trans isomers are possible for alkenes because there is no rotation around the rigid double bond.
- In the cis isomer, carbon groups or halogen atoms are attached on the same side of the double bond, whereas in the trans isomer, they are attached on opposite sides of the double bond.

Introduction to Organic Chemistry: Hydrocarbons

♦ **Learning Exercise 11.6A**

Draw the condensed structural formulas for the cis and trans isomers of 1,2-dibromoethene.

cis-2-Butene

Answers In the cis isomer, the bromine (halogen) atoms are attached on the same side of the double bond; in the trans isomer, they are on opposite sides.

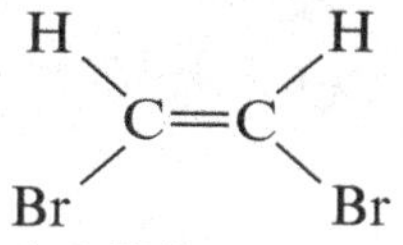

```
    H       H              H       Br
     \     /                \     /
      C = C                  C = C
     /     \                /     \
    Br      Br             Br      H
  cis-1,2-Dibromoethene   trans-1,2-Dibromoethene
```

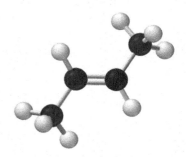
trans-2-Butene

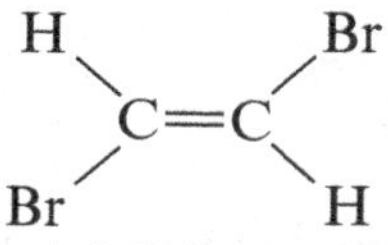

Ball-and-stick models show the cis and trans isomers of 2-butene.

♦ **Learning Exercise 11.6B**

Name the following alkenes using the cis–trans prefix where isomers are possible:

1. 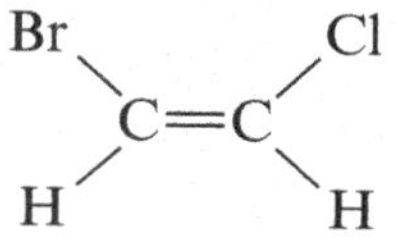
```
    Br      Cl
     \     /
      C = C
     /     \
    H       H
```

2. 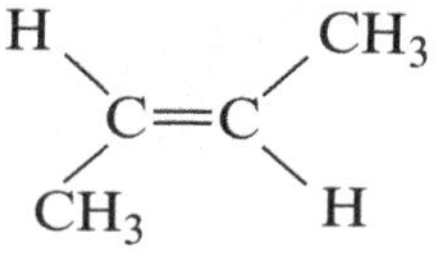
```
    H       CH₃
     \     /
      C = C
     /     \
    CH₃     H
```

_____   _____

3. 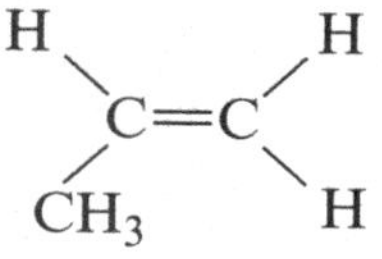
```
    H       H
     \     /
      C = C
     /     \
    CH₃     H
```

4. 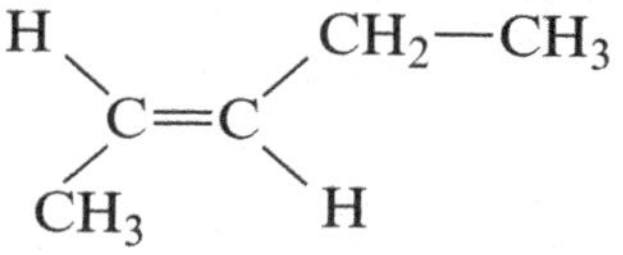
```
    H       CH₂—CH₃
     \     /
      C = C
     /     \
    CH₃     H
```

_____   _____

Answers 1. *cis*-1-bromo-2-chloroethene   2. *trans*-2-butene
3. propene (not a cis–trans isomer)   4. *trans*-2-pentene

11.7 Addition Reactions for Alkenes

Learning Goal: Draw the condensed structural and line-angle formulas and give the names for the organic products of addition reactions of alkenes.

- The addition of small molecules to the double bond is a characteristic reaction of alkenes.
- Hydrogenation adds hydrogen atoms to the double bond of an alkene to yield an alkane.

$$H_2C=CH_2 + H_2 \xrightarrow{Pt} CH_3-CH_3$$

Chapter 11

Table 11.1 Summary of Addition Reactions

Name of Addition Reaction	Reactants	Conditions/ Catalysts	Product
Hydrogenation	Alkene + H_2	Pt, Ni, or Pd	Alkane
Hydration	Alkene + H_2O	H^+ (strong acid)	Alcohol

- Hydration, in the presence of a strong acid, adds water to a double bond. For an asymmetrical alkene, the H— from H—OH bonds to the carbon in the double bond that has the greater number of hydrogen atoms.

$$CH_3-CH=CH_2 + H_2O \xrightarrow{H^+} CH_3-\underset{\underset{\textstyle OH}{|}}{CH}-CH_3$$

Hydrogenation is used to convert unsaturated fats in vegetable oils to saturated fats that make a more solid product.
Credit: Pearson Education Inc./Pearson Science

♦ Learning Exercise 11.7

Draw the condensed structural or line-angle formulas for the products of the following reactions:

1. $CH_3-CH_2-CH=CH_2 + H_2 \xrightarrow{Pt}$

2.   + $H_2 \xrightarrow{Pt}$

3. $CH_3-CH=CH_2 + H_2 \xrightarrow{Pt}$

4. 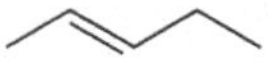 ∧∧ + $H_2 \xrightarrow{Ni}$

5. $CH_3-CH=CH_2 + H_2O \xrightarrow{H^+}$

6.  + $H_2O \xrightarrow{H^+}$

CORE CHEMISTRY SKILL
Writing Equations for Hydrogenation and Hydration

Answers
1. $CH_3-CH_2-CH_2-CH_3$
2. 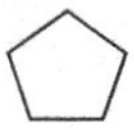 ⬠
3. $CH_3-CH_2-CH_3$
4.  ∧∧∧
5. $CH_3-\underset{\underset{\textstyle OH}{|}}{CH}-CH_3$
6. 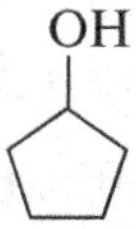 cyclopentanol (⬠ with OH)

11.8 Aromatic Compounds

Learning Goal: Describe the bonding in benzene; name aromatic compounds and draw their line-angle formulas.

- Most aromatic compounds contain benzene, a cyclic structure containing six CH units. The structure of benzene can be represented as a hexagon with alternating single and double bonds or with a circle in the center.

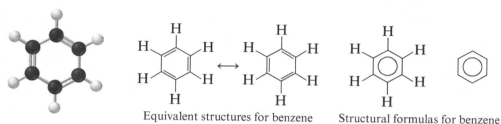

Equivalent structures for benzene   Structural formulas for benzene

- When benzene has only one substituent, the ring is not numbered. The names of many aromatic compounds use the parent name benzene, although many common names were retained as IUPAC names, such as toluene, aniline, and phenol.

- When there are two or more substituents, the benzene ring is numbered to give the lowest numbers to the substituents.

Key Terms for Sections 11.5 to 11.8

Match each of the following key terms with the correct description:

    a. alkene   **b.** hydrogenation   **c.** alkyne
    d. hydration   **e.** benzene

1. _____ contains a cyclic structure with six CH units
2. _____ the addition of H_2 to a carbon–carbon double bond
3. _____ a compound that contains a carbon–carbon double bond
4. _____ the addition of H_2O to a carbon–carbon double bond
5. _____ a compound that contains a carbon–carbon triple bond

Answers   **1.** e   **2.** b   **3.** a   **4.** d   **5.** c

Chapter 11

♦ **Learning Exercise 11.8**

Write the IUPAC name for each of the following:

1. [structure: benzene with Cl at position 1 and Cl at position 3]

2. [structure: benzene with CH₃ at top and Br at para position]

3. [structure: benzene with Br, Cl adjacent and Br at another position]

4. [structure: benzene with OH, Br, and Cl substituents]

Answers 1. 1,3-dichlorobenzene 2. 4-bromotoluene
3. 1,3-dibromo-4-chlorobenzene 4. 3-bromo-4-chlorophenol

Checklist for Chapter 11

You are ready to take the Practice Test for Chapter 11. Be sure you have accomplished the following learning goals for this chapter. If not, review the Section listed at the end of the goal. Then apply your new skills and understanding to the Practice Test.

After studying Chapter 11, I can successfully:

_____ Identify properties as characteristic of organic or inorganic compounds. (11.1)

_____ Identify the number of bonds to carbon. (11.1)

_____ Describe the shape around carbon in organic compounds. (11.1)

_____ Draw the expanded structural, condensed structural, and line-angle formulas for an alkane. (11.2)

_____ Use the IUPAC system to write the names for alkanes and cycloalkanes. (11.2)

_____ Use the IUPAC system to write the names for alkanes with substituents. (11.3)

_____ Draw the condensed structural and line-angle formulas for alkanes with substituents. (11.3)

_____ Describe some physical properties of alkanes. (11.4)

_____ Write and balance equations for the combustion of alkanes. (11.4)

_____ Identify the structural features of alkenes and alkynes. (11.5)

_____ Name alkenes and alkynes using IUPAC rules and draw their condensed structural and line-angle formulas. (11.5)

_____ Identify alkenes that exist as cis–trans isomers; draw their condensed structural and line-angle formulas and give their names. (11.6)

_____ Draw the condensed structural and line-angle formulas for the products of the addition of hydrogen and water to alkenes. (11.7)

_____ Give the names and draw the line-angle formulas for compounds that contain a benzene ring. (11.8)

Practice Test for Chapter 11

The chapter Sections to review are shown in parentheses at the end of each question.

For questions 1 through 8, indicate whether the following characteristics are typical of organic (O) or inorganic (I) compounds: (11.1)

1. ____ high melting points
2. ____ fewer compounds
3. ____ covalent bonds
4. ____ soluble in water
5. ____ ionic bonds
6. ____ flammable
7. ____ low boiling points
8. ____ soluble in nonpolar solvents

For questions 9 through 13, match the formula with the correct name: (11.2)

    A. methane   **B.** ethane   **C.** propane   **D.** pentane   **E.** heptane

9. ____ $CH_3-CH_2-CH_3$
10. ____ $CH_3-CH_2-CH_2-CH_2-CH_2-CH_2-CH_3$
11. ____ CH_4
12. ____ (zigzag structure)
13. ____ CH_3-CH_3

For questions 14 through 17, match the compounds with one of the following families: (11.2, 11.5)

    A. alkane   **B.** alkene   **C.** alkyne   **D.** cycloalkene

14. $CH_3-CH_2-CH=CH-CH_3$

15.  (cyclohexene ring)

16. $CH_3-CH_2-\underset{\underset{CH_3}{|}}{CH}-CH_2-CH_3$

17. $CH_3-CH_2-C{\equiv}CH$

For questions 18 through 21, match the formula with the correct name: (11.2, 11.3)

    A. butane   **B.** 1,1-dichlorobutane   **C.** 1,2-dichloropropane
    D. 3,5-dimethylhexane   **E.** 2,4-dimethylhexane   **F.** 1,2-dichlorobutane

18. $CH_3-CH_2-CH_2-CH_3$

19. $CH_3-\underset{\underset{CH_3}{|}}{CH}-CH_2-\underset{\underset{CH_3}{|}}{CH}-CH_2-CH_3$

20. 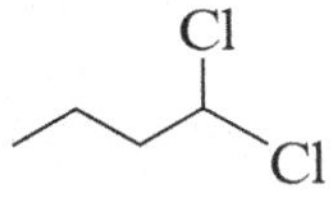 (zigzag with Cl, Cl on same carbon)

21. 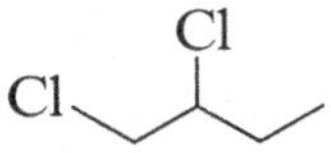 (zigzag with Cl, Cl)

For questions 22 through 24, match the formula with the correct name: (11.2, 11.3)

    A. methylcyclopentane   **B.** cyclobutane
    C. cyclohexane   **D.** ethylcyclopentane

22. ____ (cyclohexane)
23. ____ (methylcyclopentane)
24. ____ (ethylcyclopentane)

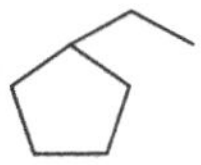

Chapter 11

For questions 25 through 27, match the formula with the correct name: (11.3)

    A. 2,4-dichloropentane   **B.** chlorocyclopentane
    C. 1,2-dichloropentane   **D.** 4,5-dichloropentane

25. ____ $CH_3-CH_2-CH_2-CH(Cl)-CH_2-Cl$

26. ____ (chlorocyclopentane structure)

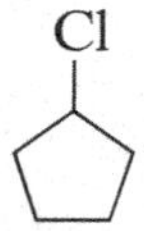

27. ____ $CH_3-CH(Cl)-CH_2-CH(Cl)-CH_3$

For questions 28 through 30, refer to the following compounds (A) and (B): (11.2, 11.5)

$H_2C=CH-CH_3$ (cyclopropane structure)
   (A)   (B)

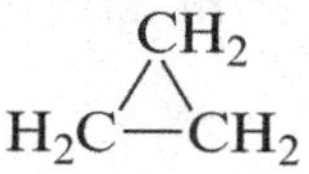

28. Compound (A) is a(an)

    A. alkane   **B.** alkene   **C.** cycloalkane   **D.** alkyne   **E.** aromatic

29. Compound (B) is named

    A. propane   **B.** propylene   **C.** cyclobutane   **D.** cyclopropane   **E.** cyclopropene

30. Compound (A) is named

    A. propane   **B.** propene   **C.** 2-propene   **D.** propyne   **E.** 1-butene

31. The correctly balanced equation for the complete combustion of ethane is (11.4)

    A. $C_2H_6 + O_2 \xrightarrow{\Delta} 2CO + 3H_2O$ + energy
    B. $C_2H_6 + O_2 \xrightarrow{\Delta} CO_2 + H_2O$ + energy
    C. $C_2H_6 + 2O_2 \xrightarrow{\Delta} 2CO_2 + 3H_2O$ + energy
    D. $2C_2H_6 + 7O_2 \xrightarrow{\Delta} 4CO_2 + 6H_2O$ + energy
    E. $2C_2H_6 + 4O_2 \xrightarrow{\Delta} 4CO_2 + 6H_2O$ + energy

In questions 32 through 35, match each formula with its name: (11.5)

    A. cyclopentene   **B.** methylpropene   **C.** cyclohexene
    D. ethene   **E.** 3-methylcyclopentene

32. $H_2C=CH_2$

33. $CH_3-C(CH_3)=CH_2$

34. 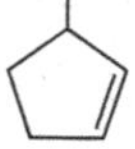 (3-methylcyclopentene structure)

35.  (cyclohexene structure)

Introduction to Organic Chemistry: Hydrocarbons

36. What is the IUPAC name of $CH_3-CH_2-C\equiv CH$? (11.5)
  A. methylacetylene
  B. propyne
  C. propylene
  D. 4-butyne
  E. 1-butyne

37. The cis isomer of 2-butene is (11.6)

  A. $H_2C=CH-CH_2-CH_3$
  B. $CH_3-CH=CH-CH_3$
  C. 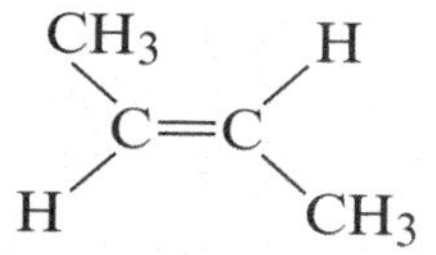 $\begin{array}{c}CH_3H\\ C=C\\ HCH_3\end{array}$
  D. 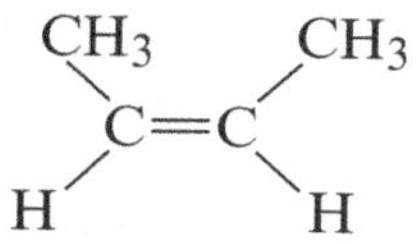 $\begin{array}{c}CH_3CH_3\\ C=C\\ HH\end{array}$
  E. 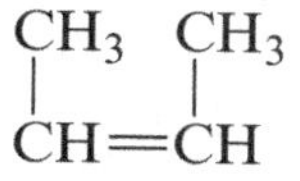 $\begin{array}{c}CH_3CH_3\\ ||\\ CH=CH\end{array}$

38. The name of this compound is (11.6)

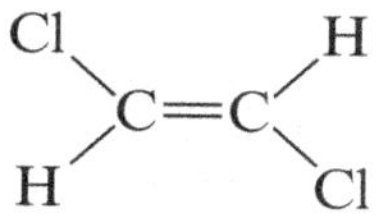

$\begin{array}{c}ClH\\ C=C\\ HCl\end{array}$

  A. dichloroethene
  B. *cis*-1,2-dichloroethene
  C. *trans*-1,2-dichloroethene
  D. *cis*-chloroethene
  E. *trans*-chloroethene

39. Hydrogenation of $CH_3-CH=CH_2$ gives (11.7)
  A. $3CO_2 + 6H_2$
  B. $CH_3-CH_2-CH_3$
  C. $H_2C=CH-CH_3$
  D. no reaction
  E. $CH_3-CH_2-CH_2-CH_3$

For problems 40 and 41, use the reaction $CH_3-CH=CH_2 + H_2O \xrightarrow{H^+}$ (11.7)

40. Choose the product of the reaction.
  A. $CH_3-CH_2-CH_2-OH$
  B. no reaction
  C. $CH_3-CH_2-CH_3$
  D. 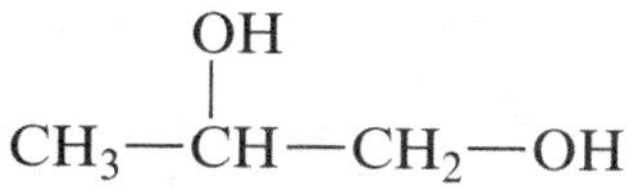 $CH_3-\underset{\underset{OH}{|}}{CH}-CH_2-OH$
  E. 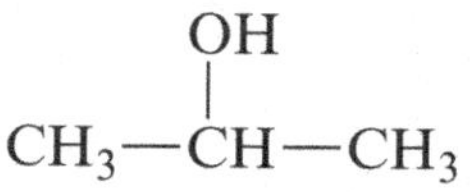 $CH_3-\underset{\underset{OH}{|}}{CH}-CH_3$

41. The reaction is called
  A. hydrogenation
  B. decomposition
  C. reduction
  D. hydration
  E. combustion

For questions 42 through 45, match the name of each of the following aromatic compounds with the correct structure: (11.8)

A. 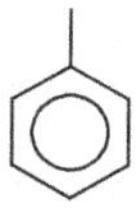   B. 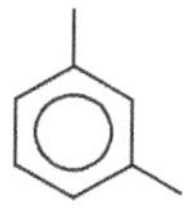   C. 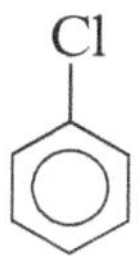   D.

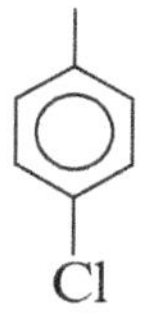

42. ____ chlorobenzene

43. ____ toluene

44. ____ 4-chlorotoluene

45. ____ 1,3-dimethylbenzene

Answers to the Practice Test

1. I	2. I	3. O	4. I	5. I
6. O	7. O	8. O	9. C	10. E
11. A	12. D	13. B	14. B	15. D
16. A	17. C	18. A	19. E	20. B
21. F	22. C	23. A	24. D	25. C
26. B	27. A	28. B	29. D	30. B
31. D	32. D	33. B	34. E	35. C
36. E	37. D	38. C	39. B	40. E
41. D	42. C	43. A	44. D	45. B

Selected Answers and Solutions to Text Problems

11.1 Organic compounds always contain C and H and sometimes O, S, N, P, or a halogen atom. Inorganic compounds usually contain elements other than C and H.
  a. inorganic
  b. organic, condensed structural formula
  c. organic, expanded structural formula
  d. inorganic
  e. inorganic
  f. organic, molecular formula

11.3 **a.** Inorganic compounds are usually soluble in water.
  b. Organic compounds have lower boiling points than most inorganic compounds.
  c. Organic compounds contain carbon and hydrogen.
  d. Inorganic compounds contain ionic bonds.

11.5 **a.** Ethane boils at −89 °C.
  b. Ethane burns vigorously in air.
  c. NaBr is a solid at 250 °C.
  d. NaBr dissolves in water.

11.7 **a.** Pentane has a chain of five carbon atoms.
  b. Ethane has a chain of two carbon atoms.
  c. Hexane has a chain of six carbon atoms.
  d. Cycloheptane has a ring of seven carbon atoms.

11.9 **a.** CH_4 **b.** CH_3-CH_3
  c. $CH_3-CH_2-CH_2-CH_3$ **d.** △

11.11 Two structures are isomers if they have the same molecular formula, but different arrangements of atoms.
  a. These condensed structural formulas represent the same molecule; the only difference is due to rotation of the structure. Each has a CH_3- group attached to the middle carbon in a three-carbon chain.
  b. The molecular formula of both these condensed structural formulas is C_5H_{12}. However, they represent structural isomers because the C atoms are bonded in a different order; they have different arrangements. In the first, there is a CH_3- group attached to carbon 2 of a four-carbon chain, and in the other, there is a five-carbon chain.
  c. The molecular formula of both these line-angle formulas is C_6H_{14}. However, they represent structural isomers because the C atoms are bonded in a different order; they have different arrangements. In the first, there is a CH_3- group attached to carbon 3 of a five-carbon chain, and in the other, there is a CH_3- group on carbon 2 and carbon 3 of a four-carbon chain.

11.13 **a.** 2,2-dimethylbutane
  b. 2-chloro-3-methylpentane
  c. 4-ethyl-2,2-dimethylhexane
  d. methylcyclobutane

11.15 Draw the main chain with the number of carbon atoms in the ending of the name. For example, butane has a main chain of four carbon atoms, and hexane has a main chain of six carbon atoms. Attach substituents on the carbon atoms indicated. For example, in 3-methylpentane, a CH_3- group is bonded to carbon 3 of a five-carbon chain.

```
           CH3
            |
a. CH3—CH2—C—CH2—CH3
            |
           CH3

      CH3  CH3       CH3
       |    |         |
b. CH3—CH—CH—CH2—CH—CH3
```

Chapter 11

c.
$$CH_3-\underset{\underset{CH_3}{|}}{CH}-\underset{\underset{CH_2-CH_3}{|}}{CH}-CH_2-\underset{\underset{CH_3}{|}}{CH}-CH_2-CH_2-CH_3$$

d. $Br-CH_2-CH_2-Cl$

11.17 a. (line-angle structure)   **b.** (cyclopentane with ethyl group)

c. (cyclobutane with Br)   **d.** (structure with two Cl substituents)

11.19 a. $CH_3-CH_2-CH_2-CH_2-CH_2-CH_2-CH_3$ (line-angle structure)
  b. Heptane is a liquid at room temperature since it has a boiling point of 98 °C.
  c. Heptane contains only nonpolar C—H bonds, which makes it insoluble in water.
  d. Since the density of heptane (0.68 g/mL) is less than that of water, heptane will float on water.
  e. $C_7H_{16} + 11O_2 \xrightarrow{\Delta} 7CO_2 + 8H_2O + \text{energy}$

11.21 In combustion, a hydrocarbon reacts with oxygen to yield CO_2 and H_2O.
  a. $2C_2H_6 + 7O_2 \xrightarrow{\Delta} 4CO_2 + 6H_2O + \text{energy}$
  b. $2C_3H_6 + 9O_2 \xrightarrow{\Delta} 6CO_2 + 6H_2O + \text{energy}$
  c. $2C_8H_{18} + 25O_2 \xrightarrow{\Delta} 16CO_2 + 18H_2O + \text{energy}$

11.23 a. A condensed structural formula with a carbon–carbon double bond is an alkene.
  b. A condensed structural formula with a carbon–carbon triple bond is an alkyne.
  c. A line-angle formula with a carbon–carbon double bond is an alkene.
  d. A line-angle formula with a carbon–carbon double bond in a ring is a cycloalkene.

11.25 a. The two-carbon compound with a double bond is named ethene.
  b. This alkene has a three-carbon chain with a methyl substituent. The name is methylpropene.
  c. This alkyne has a five-carbon chain with the triple bond between carbon 2 and carbon 3. The name is 2-pentyne.
  d. This is a four-carbon cyclic structure with a double bond and a methyl group on carbon 3 of the ring. The name is 3-methylcyclobutene.

11.27 a. 1-Pentene is the five-carbon compound with a double bond between carbon 1 and carbon 2.
  $H_2C=CH-CH_2-CH_2-CH_3$
  b. 2-Methyl-1-butene has a four-carbon chain with a double bond between carbon 1 and carbon 2 and a methyl group attached to carbon 2.
  $$H_2C=\underset{\underset{CH_3}{|}}{C}-CH_2-CH_3$$
  c. 3-Methylcyclohexene is a six-carbon cyclic compound with a double bond and a methyl group attached to carbon 3 of the ring.
  (cyclohexene with methyl group structure)
  d. 3-Chloro-1-butyne is a four-carbon compound with a triple bond between carbon 1 and carbon 2 and a chlorine atom bonded to carbon 3.
  $$HC\equiv C-\underset{\underset{Cl}{|}}{CH}-CH_3$$

11.29 a. *cis*-3-Heptene; this is a seven-carbon compound with a double bond between carbon 3 and carbon 4. Both alkyl groups are on the same side of the double bond; it is the cis isomer.
  b. *trans*-3-Octene; this compound has eight carbons with a double bond between carbon 3 and carbon 4. The alkyl groups are on the opposite sides of the double bond; it is the trans isomer.
  c. Propene; this is a three-carbon compound with a double bond between carbon 1 and carbon 2. This compound does not have cis–trans isomers since there are 2 Hs attached to carbon 1 in the double bond.

11.31 a. *trans*-1-Bromo-2-chloroethene has a two-carbon chain with a double bond. The trans isomer has the attached groups on the opposite sides of the double bond.

$$\begin{array}{c} Br \\ \\ H \end{array} C=C \begin{array}{c} H \\ \\ Cl \end{array}$$

  b. *cis*-2-Hexene has a six-carbon chain with a double bond between carbon 2 and carbon 3. The cis isomer has the alkyl groups on the same side of the double bond.

$$\begin{array}{c} CH_3 \\ \\ H \end{array} C=C \begin{array}{c} CH_2-CH_2-CH_3 \\ \\ H \end{array}$$

  c. *cis*-4-Octene has an eight-carbon chain with a double bond between carbon 4 and carbon 5. The cis isomer has alkyl groups on the same side of the double bond.

$$\begin{array}{c} CH_3-CH_2-CH_2 \\ \\ H \end{array} C=C \begin{array}{c} CH_2-CH_2-CH_3 \\ \\ H \end{array}$$

11.33 a. Hydrogenation of an alkene gives the saturated compound, the alkane.
    $CH_3-CH_2-CH_2-CH_2-CH_3$
  b. In a hydration reaction, the H— and —OH from water add to the carbon atoms in the double bond to form an alcohol. The H— adds to the carbon with more hydrogens, and the —OH bonds to the carbon with fewer hydrogen atoms.

$$CH_3-\underset{\underset{OH}{|}}{\overset{\overset{CH_3}{|}}{C}}-CH_2-CH_3$$

  c. Hydrogenation of an alkene gives the saturated compound, the alkane.

  d. The H— and —OH from water add to the carbon atoms in the double bond to form an alcohol.

11.35 Aromatic compounds that contain a benzene ring with a single substituent are usually named as benzene derivatives. A benzene ring with a methyl substituent is named toluene. The methyl group is attached to carbon 1 and the ring is numbered to give the lower numbers to other substituents.
  a. 2-chlorotoluene
  b. ethylbenzene
  c. 1,3,5-trichlorobenzene

11.37 a. [phenol structure: benzene with OH] **b.** [1,3-dichlorobenzene] **c.** [1,4-dimethylbenzene]

11.39 a. $2C_8H_{18} + 25O_2 \xrightarrow{\Delta} 16CO_2 + 18H_2O + \text{energy}$
   b. $C_5H_{12} + 8O_2 \xrightarrow{\Delta} 5CO_2 + 6H_2O + \text{energy}$
   c. $C_9H_{12} + 12O_2 \xrightarrow{\Delta} 9CO_2 + 6H_2O + \text{energy}$

11.41 a. Butane melts at −138 °C.
   b. Butane burns vigorously in air.
   c. Potassium chloride melts at 770 °C.
   d. Potassium chloride contains ionic bonds.
   e. Butane is a gas at room temperature.

11.43 Two structures are isomers if they have the same molecular formula, but different arrangements of atoms.
   a. The molecular formula of both these line-angle formulas is C_6H_{12}. However, they represent structural isomers because the C atoms are bonded in a different order; they have different arrangements. In the first, there is a methyl group attached to a five-carbon ring, and in the other, there is a six-carbon ring.
   b. The molecular formula of the first line-angle structure is C_5H_{12}; the molecular formula of the second is C_6H_{14}. They are not structural isomers.

11.45 a. $CH_3-CH_2-CH_2-CH(CH_3)-CH(CH_3)-CH_3$ 2,3-Dimethylhexane

   b. $CH_3-CH(CH_3)-CH(CH_3)-CH(CH_3)-CH_3$ 2,3,4-Trimethylpentane

11.47 a. 2,2-dimethylbutane
   b. chloroethane
   c. 2-bromo-4-ethylhexane
   d. methylcyclopentane

11.49 a. This alkene has a five-carbon chain with a double bond between carbon 2 and carbon 3. The alkyl groups are on the opposite sides of the double bond. The IUPAC name is *trans*-2-pentene.
   b. This alkene has a six-carbon chain with the double bond between carbon 1 and carbon 2. The substituents are a bromine atom attached to carbon 4 and a methyl group on carbon 5. The IUPAC name is 4-bromo-5-methyl-1-hexene.
   c. This is a five-carbon cyclic structure with a double bond and a methyl group attached to carbon 1 of the ring. The name is 1-methylcyclopentene.

11.51 a. These structures represent a pair of structural isomers. In one isomer, the chlorine is attached to one of the carbons in the double bond; in the other isomer, the carbon bonded to the chlorine is not part of the double bond.
   b. These structures are cis–trans isomers. In the cis isomer, the two methyl groups are on the same side of the double bond. In the trans isomer, the methyl groups are on the opposite sides of the double bond.

11.53 Aromatic compounds that contain a benzene ring with a single substituent are usually named as benzene derivatives. A benzene ring with a methyl or amino substituent is named toluene or aniline, respectively. The methyl or amino group is attached to carbon 1 and the ring is numbered to give the lower numbers to other substituents.
   a. 3-chlorotoluene
   b. 4-bromoaniline

Introduction to Organic Chemistry: Hydrocarbons

11.55 a. Bromocyclopropane has a three-carbon cyclic structure with a bromine atom attached to one of the carbons of the ring.

b. 1,1-Dibromo-2-pentyne has a five-carbon chain with a triple bond between carbon 2 and carbon 3 and two bromine atoms attached to carbon 1.

$$\begin{array}{c} \text{Br} \\ | \\ \text{Br}-\text{CH}-\text{C}\equiv\text{C}-\text{CH}_2-\text{CH}_3 \end{array}$$

c. *cis*-2-Heptene has a seven-carbon chain with a double bond between carbon 2 and carbon 3. In the cis isomer, both alkyl groups are on the same side of the double bond.

$$\begin{array}{c} \text{CH}_3 \\ \diagdown \\ \text{C}=\text{C} \\ \diagup \\ \text{H} \end{array} \begin{array}{c} \text{CH}_2-\text{CH}_2-\text{CH}_2-\text{CH}_3 \\ \diagup \\ \\ \diagdown \\ \text{H} \end{array}$$

11.57 a.

cis-2-Pentene; both alkyl groups are on the same side of the double bond.

trans-2-Pentene; both alkyl groups are on the opposite sides of the double bond.

b.

cis-3-Hexene; both alkyl groups are on the same side of the double bond.

trans-3-Hexene; both alkyl groups are on the opposite sides of the double bond.

11.59 a. (ethylbenzene) **b.** (2,4-dibromophenol) **c.** (3-chloroaniline)

11.61 a. $C_5H_{12} + 8O_2 \xrightarrow{\Delta} 5CO_2 + 6H_2O + \text{energy}$
b. $C_4H_8 + 6O_2 \xrightarrow{\Delta} 4CO_2 + 4H_2O + \text{energy}$
c. $C_6H_{12} + 9O_2 \xrightarrow{\Delta} 6CO_2 + 6H_2O + \text{energy}$

11.63 During hydrogenation, the multiple bonds are converted to single bonds and two H atoms are added for each bond converted.

a. 3-methylpentane **b.** cyclohexane **c.** propane

11.65 a. Hydrogenation of a cycloalkene gives the saturated compound, the cycloalkane.

b. The H— and —OH from water add to the carbon atoms in the double bond to form an alcohol. The H— adds to the carbon with more hydrogens (carbon 1), and the —OH bonds to the carbon with fewer hydrogen atoms (carbon 2).

11.67 a. $C_5H_{12} + 8O_2 \xrightarrow{\Delta} 5CO_2 + 6H_2O + \text{energy}$

b. Molar mass of pentane (C_5H_{12}) = $5(12.01 \text{ g}) + 12(1.008 \text{ g}) = 72.15$ g/mole

c. $1 \text{ gal} \times \dfrac{4 \text{ qt}}{1 \text{ gal}} \times \dfrac{946 \text{ mL}}{1 \text{ qt}} \times \dfrac{0.63 \text{ g}}{1 \text{ mL}} \times \dfrac{1 \text{ mole } C_5H_{12}}{72.15 \text{ g } C_5H_{12}} \times \dfrac{845 \text{ kcal}}{1 \text{ mole } C_5H_{12}}$
$= 2.8 \times 10^4$ kcal (2 SFs)

d. $1 \text{ gal} \times \dfrac{4 \text{ qt}}{1 \text{ gal}} \times \dfrac{946 \text{ mL}}{1 \text{ qt}} \times \dfrac{0.63 \text{ g}}{1 \text{ mL}} \times \dfrac{1 \text{ mole } C_5H_{12}}{72.15 \text{ g } C_5H_{12}} \times \dfrac{5 \text{ moles } CO_2}{1 \text{ mole } C_5H_{12}} \times \dfrac{22.4 \text{ L } CO_2 \text{ (STP)}}{1 \text{ mole } CO_2}$
$= 3.7 \times 10^3$ L of CO_2 at STP (2 SFs)

11.69
$$\text{CH}_3-\underset{\underset{\text{CH}_3}{|}}{\text{CH}}-\underset{\underset{\text{CH}_3}{|}}{\text{CH}}-\text{CH}_3 \qquad \text{CH}_3-\underset{\underset{\text{CH}_3}{|}}{\overset{\overset{\text{CH}_3}{|}}{\text{C}}}-\text{CH}_2-\text{CH}_3$$

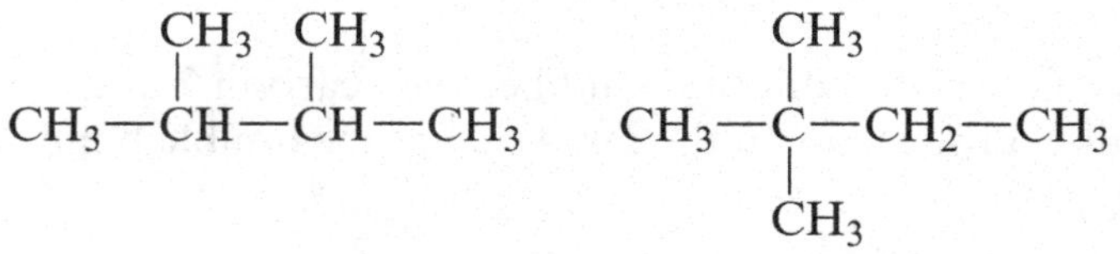

11.71 a. $\text{CH}_3-\text{CH}_2-\text{CH}_3$

b. $C_3H_8 + 5O_2 \xrightarrow{\Delta} 3CO_2 + 4H_2O + \text{energy}$

c. $12.0 \text{ L } C_3H_8 \times \dfrac{1 \text{ mole } C_3H_8}{22.4 \text{ L } C_3H_8 \text{ (STP)}} \times \dfrac{5 \text{ moles } O_2}{1 \text{ mole } C_3H_8} \times \dfrac{32.00 \text{ g } O_2}{1 \text{ mole } O_2} = 85.7$ g of O_2 (3 SFs)

d. $12.0 \text{ L } C_3H_8 \times \dfrac{1 \text{ mole } C_3H_8}{22.4 \text{ L } C_3H_8 \text{ (STP)}} \times \dfrac{3 \text{ moles } CO_2}{1 \text{ mole } C_3H_8} \times \dfrac{44.01 \text{ g } CO_2}{1 \text{ mole } CO_2} = 70.7$ g of CO_2 (3 SFs)

11.73 a. 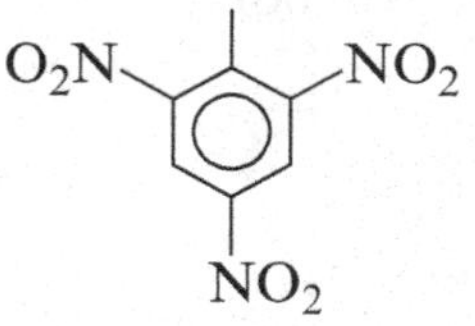 2,4,6-trinitrotoluene structure (methylbenzene with NO_2 groups at 2, 4, and 6 positions)

b. 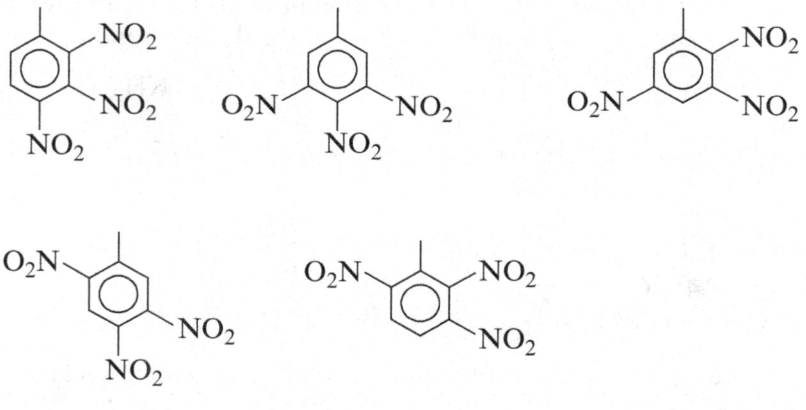 Five isomeric structures of trinitrotoluene with NO_2 groups at various positions on the methylbenzene ring.

12
Alcohols, Thiols, Ethers, Aldehydes, and Ketones

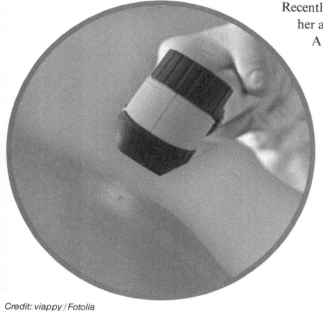

Credit: viappy/Fotolia

Recently, Diana noticed a change in color of a small mole on her arm, which her dermatologist determined was melanoma. A surgeon excised the mole including subcutaneous fat. Skin cancer begins in the cells that make up the outer layer (epidermis) of the skin. Limiting exposure to UV (ultraviolet) radiation from the Sun or from tanning salons can help to reduce the risk of developing skin cancer. Sunscreen absorbs UV light radiation when the skin is exposed to sunlight and thus helps protect against sunburn. The sun protection factor (SPF) number gives the amount of time for protected skin to sunburn compared to the time for unprotected skin to sunburn. One of the principal ingredients in sunscreens is oxybenzone. Identify the functional groups in oxybenzone.

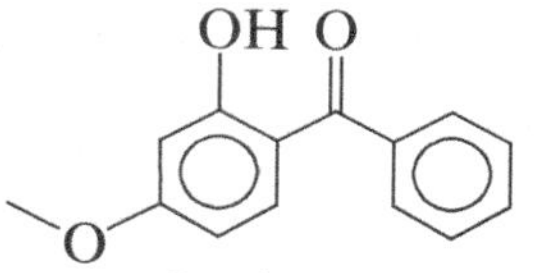

Oxybenzone

LOOKING AHEAD

12.1 Alcohols, Phenols, Thiols, and Ethers

12.2 Properties of Alcohols

12.3 Aldehydes and Ketones

12.4 Reactions of Alcohols, Thiols, Aldehydes, and Ketones

 The Health icon indicates a question that is related to health and medicine.

12.1 Alcohols, Phenols, Thiols, and Ethers

Learning Goal: Write the IUPAC and common names for alcohols, phenols, and thiols and the common names for ethers. Draw their condensed structural or line-angle formulas.

REVIEW
Naming and Drawing Alkanes (11.2)

- An alcohol contains the hydroxyl group (—OH) attached to a carbon chain.
- In the IUPAC system, alcohols are named by replacing the *e* of the alkane name with *ol*. The location of the —OH group is given by numbering the carbon chain.

Chapter 12

- The common name of a simple alcohol uses the name of the alkyl group followed by *alcohol*. For example, CH_3-OH is methyl alcohol, and CH_3-CH_2-OH is ethyl alcohol.

CORE CHEMISTRY SKILL
Identifying Functional Groups

$$CH_3-OH \qquad CH_3-CH_2-OH \qquad CH_3-CH_2-CH_2-OH$$
Methanol · · · · · · · · · Ethanol · · · · · · · · · 1-Propanol
(methyl alcohol) · · · · (ethyl alcohol) · · · (propyl alcohol)

- When a hydroxyl group is attached to a benzene ring, the compound is a phenol.
- Thiols are similar to alcohols, except they have an $-SH$ group in place of the $-OH$ group.

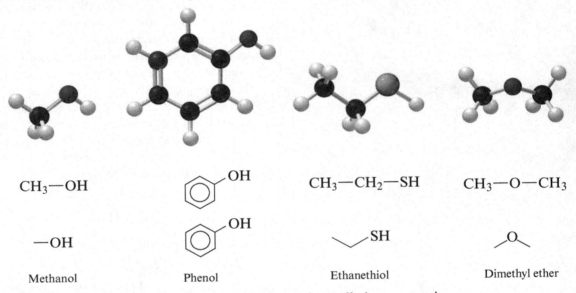

- In ethers, an oxygen atom is connected by single bonds to alkyl or aromatic groups.
- In the common names of ethers, the alkyl groups are listed alphabetically followed by the name *ether*.

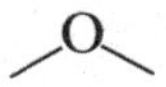

	Solution Guide to Naming Alcohols
STEP 1	Name the longest carbon chain attached to the $-OH$ group by replacing the *e* in the corresponding alkane name with *ol*. Name an aromatic alcohol as a *phenol*.
STEP 2	Number the chain starting at the end nearer to the $-OH$ group.
STEP 3	Give the location and name for each substituent relative to the $-OH$ group.

♦ **Learning Exercise 12.1A**

Write the correct IUPAC and common name (if any) for each of the following compounds:

CORE CHEMISTRY SKILL
Naming Alcohols and Phenols

1. $CH_3-CH_2-CH_2-OH$

2. $CH_3-\underset{\underset{OH}{|}}{CH}-CH_2-CH_2-CH_3$

3. 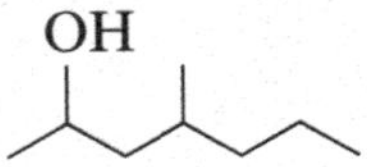 [structure: hexane chain with OH on C2 and methyl on C3]

4. 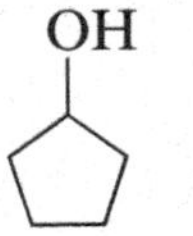 [cyclopentane with OH]

5.  [phenol structure]

280

Alcohols, Thiols, Ethers, Aldehydes, and Ketones

Answers
1. 1-propanol (propyl alcohol)
2. 2-pentanol
3. 4-methyl-2-heptanol
4. cyclopentanol
5. phenol

♦ **Learning Exercise 12.1B**

Draw the condensed structural formulas for **1, 2,** and **3** and the line-angle formulas for **4, 5,** and **6**.

1. 2-butanol

2. 2-chloro-1-propanol

3. 2,4-dimethyl-1-pentanol

4. 2-bromoethanol

5. cyclohexanol

6. 2,3-dichlorophenol

Answers

1. CH$_3$—CH(OH)—CH$_2$—CH$_3$

2. CH$_3$—CH(Cl)—CH$_2$—OH

3. CH$_3$—CH(CH$_3$)—CH$_2$—CH(CH$_3$)—CH$_2$—OH

4. 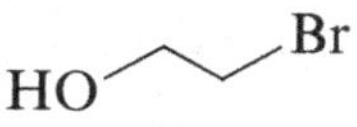 HO—CH$_2$—CH$_2$—Br

5. 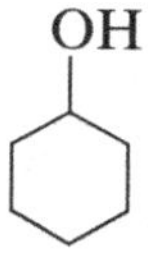 (cyclohexane with OH)

6. 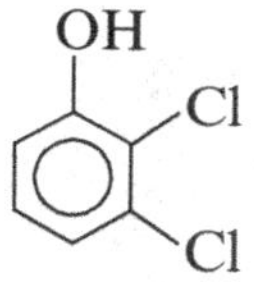 (phenol with Cl at 2 and 3 positions)

Study Note

Most ethers have common names. The name of each alkyl or aromatic group attached to the oxygen atom is written in alphabetical order, followed by the word *ether*.

Example: Write the common name for CH$_3$—CH$_2$—O—CH$_3$.
Solution: Ethyl group   Methyl group
               CH$_3$—CH$_2$—O—CH$_3$
               Common: ethyl methyl ether

♦ **Learning Exercise 12.1C**

Write the common name for each of the following ethers:

1. CH$_3$—O—CH$_3$

2. 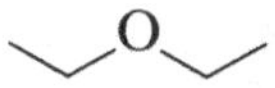 (line-angle: ethyl-O-ethyl)

Chapter 12

3. $CH_3-CH_2-CH_2-CH_2-O-CH_3$ 4. 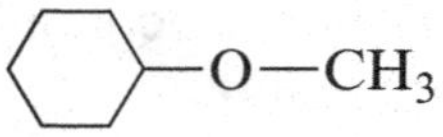 [cyclohexyl]$-O-CH_3$

Answers   1. dimethyl ether   2. diethyl ether
3. butyl methyl ether   4. methyl cyclohexyl ether

♦ **Learning Exercise 12.1D**

Draw the condensed structural formula for **1** and the line-angle formula for **2**.

1. ethyl propyl ether   2. dibutyl ether

Answers   1. $CH_3-CH_2-O-CH_2-CH_2-CH_3$ 2.  [line-angle formula of dibutyl ether]

♦ **Learning Exercise 12.1E**

Identify the functional groups of alcohol, phenol, thiol, and/or ether in each of the following compounds:

1. 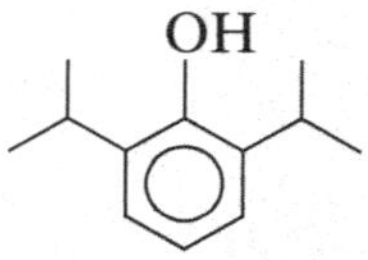 [2,6-diisopropylphenol structure with OH]

   an anesthetic

2. $CH_3-\overset{\overset{\displaystyle CH_3}{|}}{CH}-CH_2-CH_2-SH$

   a component of skunk odor

3. $CH_3-O-CH_2-CH_2-OH$

   a solvent

4. $CH_3-O-CH_2-CH_2-O-CH_3$

   used in lithium batteries

5. 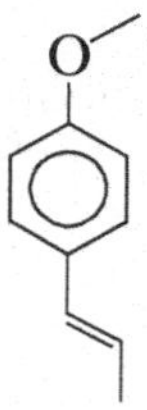 [4-methoxystyrene structure]

   the odor of fennel

6. 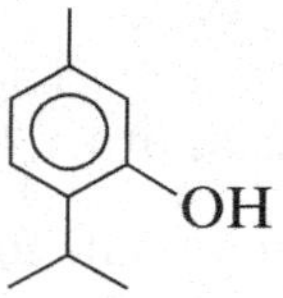 [thymol structure with OH and isopropyl]

   the odor of thyme

Answers   1. phenol   2. thiol   3. alcohol, ether
4. ether   5. ether   6. phenol

282

Alcohols, Thiols, Ethers, Aldehydes, and Ketones

12.2 Properties of Alcohols

Learning Goal: Describe the classification of alcohols; describe the solubility of alcohols in water.

- Alcohols are classified according to the number of alkyl groups attached to the carbon atom bonded to the —OH group.
- In a primary alcohol, there is one alkyl group attached to the carbon atom bonded to the —OH. In a secondary alcohol, there are two alkyl groups, and in a tertiary alcohol, there are three alkyl groups attached to the carbon atom with the —OH group.

$$CH_3-CH_2-OH \qquad CH_3-\underset{\underset{}{}}{\overset{\overset{CH_3}{|}}{CH}}-OH \qquad CH_3-\underset{\underset{CH_3}{|}}{\overset{\overset{CH_3}{|}}{C}}-OH$$

Primary (1°)　　　Secondary (2°) 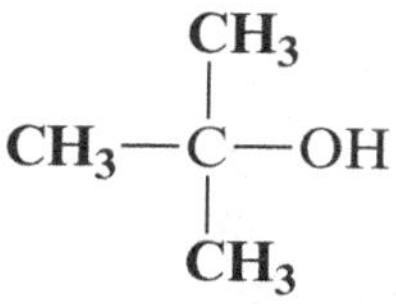　　　Tertiary (3°)

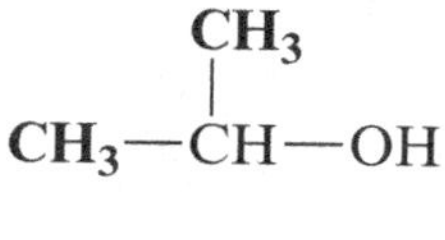

- Alcohols with one to three carbons are completely soluble in water because the —OH group forms hydrogen bonds with water molecules.
- Phenol is slightly soluble in water and acts as a weak acid.

♦ Learning Exercise 12.2A

Classify each of the following alcohols as primary (1°), secondary (2°), or tertiary (3°):

1. _____ CH_3-CH_2-OH

2. _____ (structure with OH on secondary carbon)

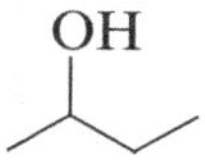

3. _____ $CH_3-\underset{\underset{CH_3}{|}}{\overset{\overset{OH}{|}}{C}}-CH_3$

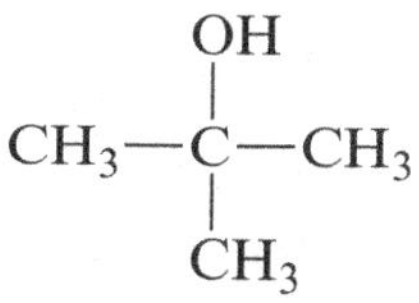

4. _____ (structure with OH on tertiary carbon)

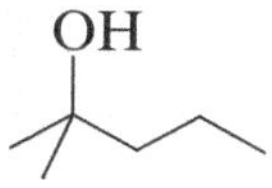

5. _____ $CH_3-\underset{\underset{CH_3}{|}}{\overset{\overset{CH_3}{|}}{C}}-CH_2-OH$

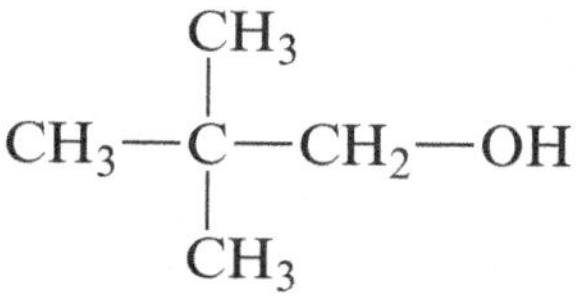

6. _____ (cyclopentanol structure)

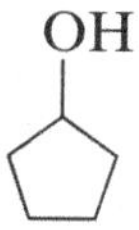

Answers
1. primary (1°)　　2. secondary (2°)　　3. tertiary (3°)
4. tertiary (3°)　　5. primary (1°)　　6. secondary (2°)

♦ Learning Exercise 12.2B

Select the compound in each pair that is more soluble in water.

1. CH_3-CH_3 or CH_3-CH_2-OH

2. $CH_3-CH_2-CH_2-OH$ or $CH_3-CH_2-CH_2-CH_2-CH_2-OH$

3. ~~~ or ~~~OH

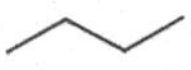

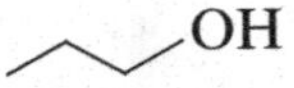

4. benzene or phenol

Answers   1. CH_3-CH_2-OH 2. $CH_3-CH_2-CH_2-OH$
          3. 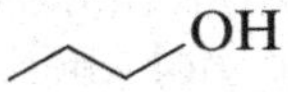 ~~~OH   4. phenol

12.3 Aldehydes and Ketones

Learning Goal: Write the IUPAC and common names for aldehydes and ketones; draw their condensed structural or line-angle formulas. Describe the solubility of aldehydes and ketones in water.

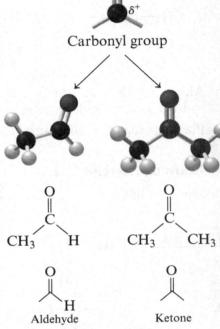

Carbonyl group

- In an aldehyde, the carbonyl group appears at the end of a carbon chain attached to at least one hydrogen atom.
- In a ketone, the carbonyl group occurs between carbon groups.
- In the IUPAC system, aldehydes and ketones are named by replacing the *e* in the longest chain containing the carbonyl group with *al* for aldehydes, and *one* for ketones. The location of the carbonyl group in a ketone is given if there are more than four carbon atoms in the chain.

$$CH_3-\overset{O}{\underset{\|}{C}}-H \qquad CH_3-\overset{O}{\underset{\|}{C}}-CH_3$$
Ethanal   Propanone
(acetaldehyde)   (dimethyl ketone)

Aldehyde   Ketone
The carbonyl group is found in aldehydes and ketones.

♦ **Learning Exercise 12.3A**
Classify each of the following compounds:

   a. alcohol   b. aldehyde   c. ketone   d. ether   e. thiol

1. ___ $CH_3-CH_2-CH_2-\overset{O}{\underset{\|}{C}}-H$ 2. ___ ~~~OH

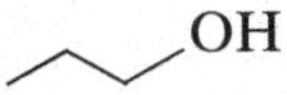

3. ___ $CH_3-CH_2-\overset{O}{\underset{\|}{C}}-CH_2-CH_3$ 4. ___ $CH_3-CH_2-O-CH_3$

5. ___ (cyclohexanone)   6. ___ $CH_3-\overset{O}{\underset{\|}{C}}-H$

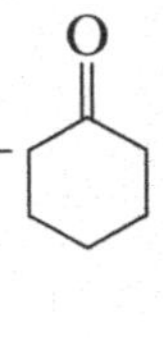

7. ___ $CH_3-CH_2-\overset{SH}{\underset{|}{CH}}-CH_3$ 8. ___ (2-propanol with OH)

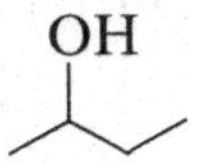

Answers   1. b   2. a   3. c   4. d
           5. c   6. b   7. e   8. a

Alcohols, Thiols, Ethers, Aldehydes, and Ketones

	Solution Guide to Naming Aldehydes
STEP 1	Name the longest carbon chain by replacing the *e* in the alkane name with *al*.
STEP 2	Name and number any substituents by counting the carbonyl group as carbon 1.

♦ **Learning Exercise 12.3B**

CORE CHEMISTRY SKILL
Naming Aldehydes and Ketones

Write the correct IUPAC name and common name, if any, for the following aldehydes:

1. 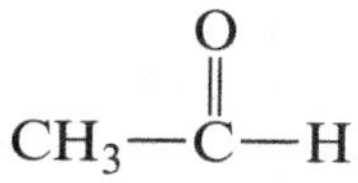 $CH_3-\overset{\overset{O}{\|}}{C}-H$

2. $CH_3-CH_2-CH_2-CH_2-\overset{\overset{O}{\|}}{C}-H$

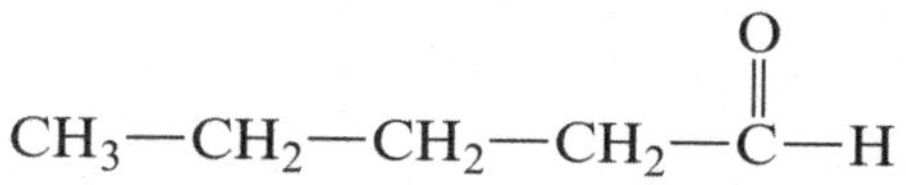

3. $CH_3-CH_2-\overset{\overset{CH_3}{|}}{CH}-CH_2-CH_2-\overset{\overset{O}{\|}}{C}-H$

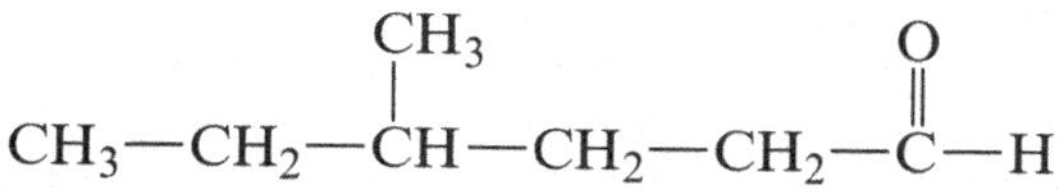

4. 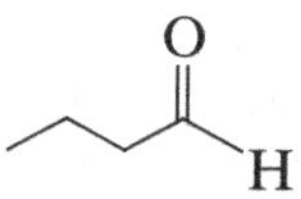 (propanal structure with CHO)

5. 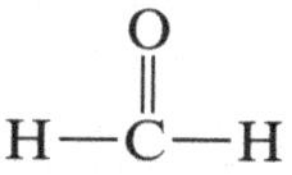 $H-\overset{\overset{O}{\|}}{C}-H$

Answers
1. ethanal; acetaldehyde
2. pentanal
3. 4-methylhexanal
4. butanal; butyraldehyde
5. methanal; formaldehyde

	Solution Guide to Naming Ketones
STEP 1	Name the longest carbon chain by replacing the *e* in the alkane name with *one*.
STEP 2	Number the carbon chain starting from the end nearer the carbonyl group and indicate its position.
STEP 3	Name and number any substituents on the carbon chain.

♦ **Learning Exercise 12.3C**

Write the IUPAC name and common name, if any, for the following ketones:

1. 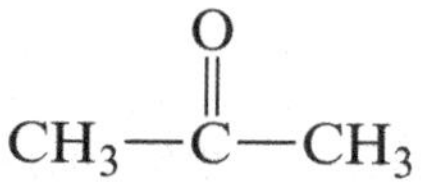 $CH_3-\overset{\overset{O}{\|}}{C}-CH_3$

2. 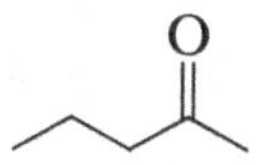 (2-pentanone structure)

3. $CH_3-CH_2-\overset{\overset{O}{\|}}{C}-CH_2-CH_3$

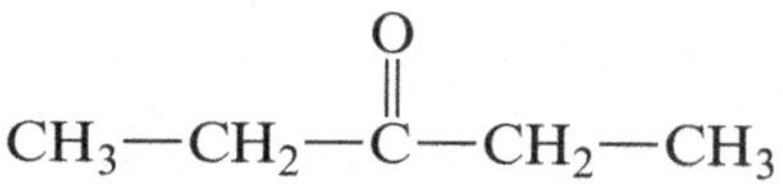

4. 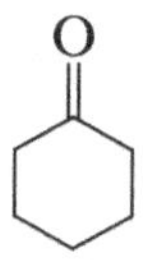 (cyclohexanone structure)

Answers
1. propanone; dimethyl ketone, acetone
2. 2-pentanone; methyl propyl ketone
3. 3-pentanone; diethyl ketone
4. cyclohexanone

♦ Learning Exercise 12.3D

Draw the condensed structural formulas for **1** to **4** and the line-angle formulas for **5** and **6**.

1. ethanal
2. 2-methylbutanal
3. 2-chloropropanal
4. ethyl methyl ketone
5. 2-hexanone
6. benzaldehyde

Answers

1. 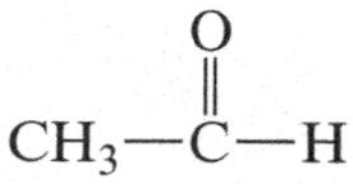 $CH_3-\underset{\underset{O}{\|}}{C}-H$

2. $CH_3-CH_2-\underset{\underset{CH_3}{|}}{CH}-\underset{\underset{O}{\|}}{C}-H$

3. $CH_3-\underset{\underset{Cl}{|}}{CH}-\underset{\underset{O}{\|}}{C}-H$

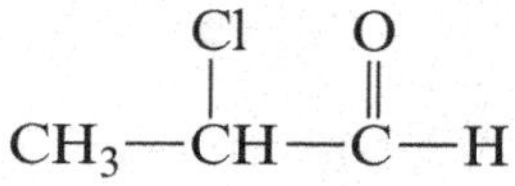

4. 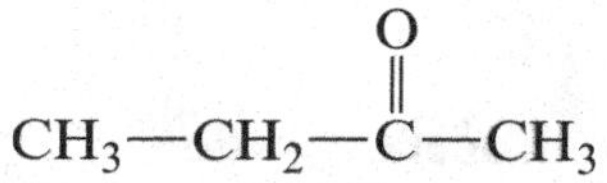 $CH_3-CH_2-\underset{\underset{O}{\|}}{C}-CH_3$

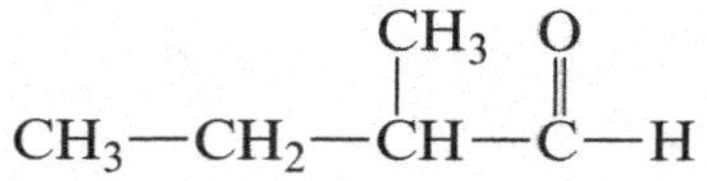

5. 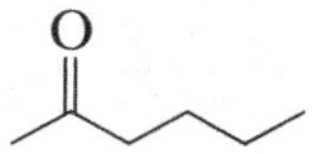 (line-angle structure of 2-hexanone)

6. 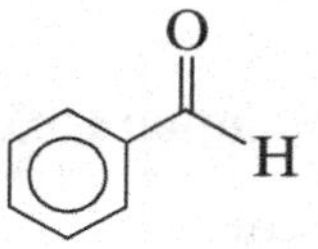 (line-angle structure of benzaldehyde)

12.4 Reactions of Alcohols, Thiols, Aldehydes, and Ketones

Learning Goal: Write balanced chemical equations for the combustion, dehydration, and oxidation of alcohols. Write balanced chemical equations for the reduction of aldehydes and ketones.

- Alcohols undergo combustion with O_2 to form CO_2, H_2O, and energy.
- At high temperatures, an alcohol dehydrates in the presence of an acid to yield an alkene and water.

$$CH_3-CH_2-OH \xrightarrow{H^+, \text{ heat}} H_2C=CH_2 + H_2O$$

- Using an oxidizing agent [O], primary alcohols oxidize to aldehydes, which usually oxidize further to carboxylic acids. Secondary alcohols are oxidized to ketones, but tertiary alcohols do not oxidize.

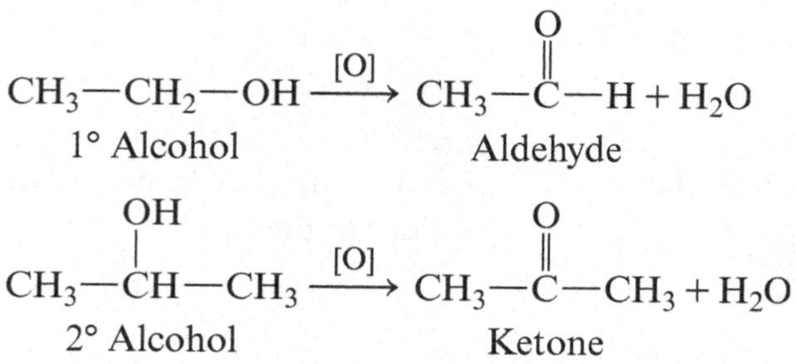

$$CH_3-CH_2-OH \xrightarrow{[O]} CH_3-\underset{\underset{O}{\|}}{C}-H + H_2O$$
1° Alcohol → Aldehyde

$$CH_3-\underset{\underset{OH}{|}}{CH}-CH_3 \xrightarrow{[O]} CH_3-\underset{\underset{O}{\|}}{C}-CH_3 + H_2O$$
2° Alcohol → Ketone

- Aldehydes and ketones can be reduced to alcohols with $NaBH_4$ or H_2 along with a catalyst.

Alcohols, Thiols, Ethers, Aldehydes, and Ketones

Key Terms for Sections 12.1 to 12.4

Match each of the following key terms with the correct description:

    a. primary alcohol   **b.** thiol   **c.** ether   **d.** phenol
    e. tertiary alcohol   **f.** aldehyde   **g.** ketone

1. ____ an organic compound with one alkyl group bonded to the carbon attached to the —OH group
2. ____ an organic compound that contains an —SH group
3. ____ an organic compound that contains an oxygen atom —O— attached to two alkyl groups
4. ____ an organic compound with three alkyl groups bonded to the carbon attached to the —OH group
5. ____ an organic compound that contains a benzene ring bonded to a hydroxyl group
6. ____ an organic compound with a carbonyl group attached to two alkyl groups
7. ____ an organic compound that contains a carbonyl group bonded to a hydrogen atom at the end of the carbon chain

Answers   **1.** a   **2.** b   **3.** c   **4.** e   **5.** d   **6.** g   **7.** f

♦ **Learning Exercise 12.4A**

Write the balanced chemical equation for the complete combustion of the following:
1. ethanol

2. 1-hexanol

Answers
1. $C_2H_6O + 3O_2 \longrightarrow 2CO_2 + 3H_2O + \text{energy}$
2. $C_6H_{14}O + 9O_2 \longrightarrow 6CO_2 + 7H_2O + \text{energy}$

♦ **Learning Exercise 12.4B**

Draw the condensed structural or line-angle formulas for the products expected from dehydration of the following:

CORE CHEMISTRY SKILL
Writing Equations for the Dehydration of Alcohols

1. $CH_3-\underset{\underset{CH_3}{|}}{CH}-CH_2-CH_2-OH \xrightarrow{H^+,\ \text{heat}}$

2. 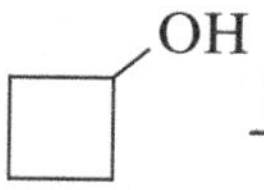 (cyclobutanol)—OH $\xrightarrow{H^+,\ \text{heat}}$

3.  (isopropanol with OH) $\xrightarrow{H^+,\ \text{heat}}$

4. $CH_3-CH_2-\underset{\underset{OH}{|}}{CH}-CH_2-CH_3 \xrightarrow{H^+,\ \text{heat}}$

Answers
1. $CH_3-\underset{\underset{CH_3}{|}}{CH}-CH=CH_2$

2. 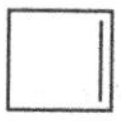 (cyclobutene)

3. 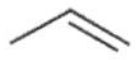 (propene line-angle)

4. $CH_3-CH_2-CH=CH-CH_3$

Copyright © 2018 Pearson Education, Inc.

Chapter 12

♦ **Learning Exercise 12.4C**

CORE CHEMISTRY SKILL
Writing Equations for the Oxidation of Alcohols

Draw the condensed structural or line-angle formula for the aldehyde or ketone expected in the oxidation of each of the following:

1. $CH_3-CH_2-CH_2-CH_2-OH \xrightarrow{[O]}$

2. 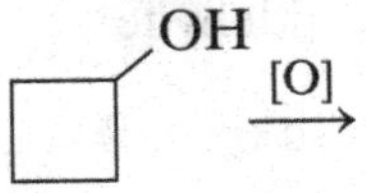 (cyclobutanol) $\xrightarrow{[O]}$

3. 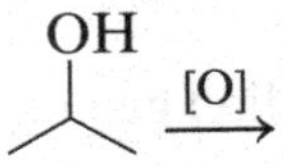 (2-propanol, line-angle with OH) $\xrightarrow{[O]}$

4. $CH_3-CH_2-\underset{\underset{OH}{|}}{CH}-CH_2-CH_3 \xrightarrow{[O]}$

Answers

1. $CH_3-CH_2-CH_2-\underset{\underset{H}{}}{\overset{\overset{O}{\|}}{C}}-H$

2. 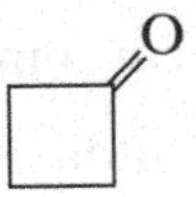 (cyclobutanone)

3.  (acetone line-angle)

4. $CH_3-CH_2-\overset{\overset{O}{\|}}{C}-CH_2-CH_3$

♦ **Learning Exercise 12.4D**

Draw the condensed structural formula for the reduction product from each of the following:

1. $CH_3-\overset{\overset{O}{\|}}{C}-CH_3 + H_2 \xrightarrow{Pt}$

2. $CH_3-CH_2-\overset{\overset{O}{\|}}{C}-H + H_2 \xrightarrow{Pt}$

3. $CH_3-\overset{\overset{O}{\|}}{C}-H + H_2 \xrightarrow{Pt}$

4. 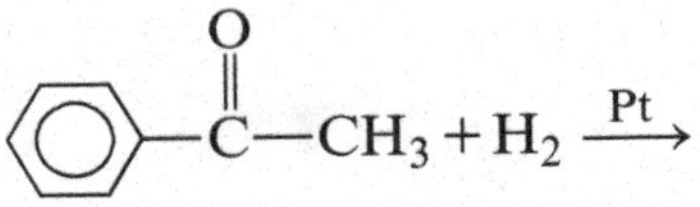 $\text{C}_6\text{H}_5-\overset{\overset{O}{\|}}{C}-CH_3 + H_2 \xrightarrow{Pt}$

Answers

1. $CH_3-\underset{\underset{OH}{|}}{CH}-CH_3$

2. $CH_3-CH_2-CH_2-OH$

3. CH_3-CH_2-OH

4. 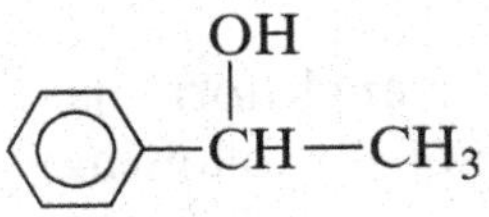 $\text{C}_6\text{H}_5-\underset{\underset{OH}{|}}{CH}-CH_3$

Checklist for Chapter 12

You are ready to take the Practice Test for Chapter 12. Be sure you have accomplished the following learning goals for this chapter. If not, review the Section listed at the end of the goal. Then apply your new skills and understanding to the Practice Test.

After studying Chapter 12, I can successfully:

____ Write the IUPAC or common name for an alcohol. (12.1)

____ Write the common name for an ether and draw the condensed structural or line-angle formula. (12.1)

____ Classify an alcohol as primary, secondary, or tertiary. (12.2)

____ Describe the solubility of alcohols in water. (12.2)

_____ Identify condensed structural or line-angle formulas as aldehydes and ketones. (12.3)

_____ Write the IUPAC and common names for an aldehyde or ketone; draw the condensed structural or line-angle formula from the name. (12.3)

_____ Describe the solubility of aldehydes and ketones in water. (12.3)

_____ Draw the condensed structural or line-angle formulas for the reactants and products of the oxidation or dehydration of alcohols. (12.4)

_____ Draw the condensed structural or line-angle formulas for the products of the reduction of aldehydes and ketones. (12.4)

Practice Test for Chapter 12

The chapter Sections to review are shown in parentheses at the end of each question.

For questions 1 through 4, identify each of the following condensed structural formulas as: (12.1)
    A. alcohol   B. ether

1. $CH_3-CH_2-\underset{\underset{\displaystyle OH}{|}}{CH}-CH_3$

2. $CH_3-\underset{\underset{\displaystyle CH_3}{|}}{\overset{\overset{\displaystyle O-CH_3}{|}}{C}}-CH_3$

3. $CH_3-CH_2-\underset{\underset{\displaystyle CH_3}{|}}{\overset{\overset{\displaystyle CH_2-OH}{|}}{CH}}$

4. $CH_3-CH_2-O-CH_3$

For questions 5 through 9, match the names with the correct structures. (12.1, 12.3)
    A. ethyl methyl ether   B. acetaldehyde   C. methanal   D. dimethyl ketone   E. propanal

5. _____ $H-\overset{\overset{\displaystyle O}{\|}}{C}-H$

6. _____ $CH_3-\overset{\overset{\displaystyle O}{\|}}{C}-H$

7. _____

8. _____ $CH_3-CH_2-\overset{\overset{\displaystyle O}{\|}}{C}-H$

9. _____ $CH_3-\overset{\overset{\displaystyle O}{\|}}{C}-CH_3$

For questions 10 through 14, match the names of the following compounds with their correct structure: (12.1, 12.2)
    A. 1-propanol   B. cyclobutanol   C. 2-propanol
    D. ethyl methyl ether   E. diethyl ether

10. (structure with OH on branched carbon)

11. $CH_3-CH_2-CH_2-OH$

12. $CH_3-O-CH_2-CH_3$

13.

14. $CH_3-CH_2-O-CH_2-CH_3$

15. Phenol is (12.1)
    A. the alcohol of benzene   B. the aldehyde of benzene   C. the phenyl group of benzene
    D. the ketone of benzene   E. another name for cyclohexanol

16. The condensed structural formula for ethanethiol is (12.1)
  A. CH₃—SH
  B. CH₃—CH₂—OH
  C. CH₃—CH₂—SH
  D. CH₃—CH₂—S—CH₃
  E. CH₃—S—OH

17. Why are short-chain alcohols water-soluble? (12.2)
  A. They are nonpolar.
  B. They can hydrogen bond.
  C. They are organic.
  D. They are bases.
  E. They are acids.

For questions 18 through 22, classify each alcohol as (12.2)
  A. primary (1°)
  B. secondary (2°)
  C. tertiary (3°)

18. CH₃—CH₂—CH₂—OH

19.

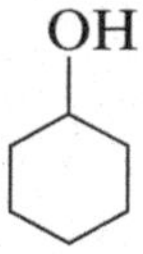

20. 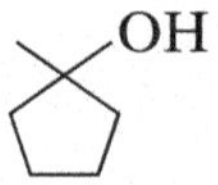 (cyclopentane with OH and CH₃ on same carbon)

21. CH₃—C(OH)(CH₃)—CH₂—CH₂—CH₃

22. CH₃—CH(OH)—CH₂—CH₂—CH₂—CH₃

23. When (CH₃—CH(OH)—CH₃) undergoes oxidation, the product is (12.4)
  A. an alkane
  B. an aldehyde
  C. a ketone
  D. an ether
  E. a phenol

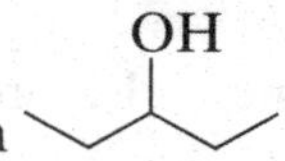

24. The compound CH₃—C(=O)—CH₃ is formed by the oxidation of (12.4)
  A. 2-propanol
  B. propane
  C. 1-propanol
  D. dimethyl ether
  E. ethyl methyl ketone

25. The dehydration of cyclohexanol gives (12.4)
  A. cyclohexane
  B. cyclohexene
  C. cyclohexyne
  D. benzene
  E. phenol

Complete questions 26 through 28 by indicating the product (**A** *to* **E**) *formed in each of the following reactions:* (12.4)

  A. primary alcohol   B. secondary alcohol   C. aldehyde   D. ketone   E. alkene

26. _____ oxidation of a primary alcohol

27. _____ oxidation of a secondary alcohol

28. _____ dehydration of 1-propanol

29. The major product from the dehydration of 3-methylcyclobutanol is: (12.4)
   A. cyclobutene   **B.** 1-methylcyclobutene   **C.** 2-methylcyclobutene
   D. 3-methylcyclobutene   **E.** 1-methylcyclobutane

30. When butanal is reduced with H_2 and a nickel catalyst, the product is: (12.4)
   A. 1-butanol   **B.** 2-butanol   **C.** butanone
   D. butyraldehyde   **E.** 1-butene

Answers to the Practice Test

1. A	**2.** B	**3.** A	**4.** B	**5.** C
6. B	**7.** A	**8.** E	**9.** D	**10.** C
11. A	**12.** D	**13.** B	**14.** E	**15.** A
16. C	**17.** B	**18.** A	**19.** B	**20.** C
21. C	**22.** B	**23.** C	**24.** A	**25.** B
26. C	**27.** D	**28.** E	**29.** D	**30.** A

Chapter 12

Selected Answers and Solutions to Text Problems

12.1 **a.** This compound has a two-carbon chain. The final *e* from ethane is dropped, and *ol* added to indicate an alcohol. The IUPAC name is ethanol.
  b. This compound has a four-carbon chain with a hydroxyl group attached to carbon 2. The IUPAC name is 2-butanol.
  c. This is the line-angle formula of a five-carbon chain with a hydroxyl group attached to carbon 2. The IUPAC name is 2-pentanol.
  d. This compound is a phenol because the —OH group is attached to a benzene ring. For a phenol, the carbon atom attached to the —OH is understood to be carbon 1. No number is needed to give the location of the —OH group. This compound also has a bromine atom attached to carbon 3 of the ring; the IUPAC name is 3-bromophenol.

12.3 **a.** Propyl alcohol (1-propanol) has a three-carbon chain with a hydroxyl group attached to carbon 1.
  CH_3—CH_2—CH_2—OH
  b. 3-Pentanethiol has a five-carbon chain with a thiol group attached to carbon 3.

  $$CH_3-CH_2-\underset{\underset{SH}{|}}{CH}-CH_2-CH_3$$

  c. 2-Methyl-2-butanol has a four-carbon chain with a methyl group and a hydroxyl group attached to carbon 2.

  $$CH_3-\underset{\underset{CH_3}{|}}{\overset{\overset{OH}{|}}{C}}-CH_2-CH_3$$

  d. 4-Bromophenol has a hydroxyl group attached to a benzene ring and a bromine atom attached to carbon 4 of the ring.

  [Structure: benzene ring with OH at top and Br at bottom]

12.5 **a.** The common name of the ether with two three-carbon alkyl groups attached to an oxygen atom is dipropyl ether.
  b. The common name of the ether with a one-carbon alkyl group and a six-carbon cycloalkyl group attached to an oxygen atom is cyclohexyl methyl ether.
  c. The common name of the ether with a four-carbon and a three-carbon alkyl group attached to an oxygen atom is butyl propyl ether.

12.7 **a.** Ethyl methyl ether has a two-carbon group and a one-carbon group attached to oxygen by single bonds.
  CH_3—CH_2—O—CH_3
  b. Cyclopropyl ethyl ether has a two-carbon group and a three-carbon cycloalkyl group attached to oxygen by single bonds.

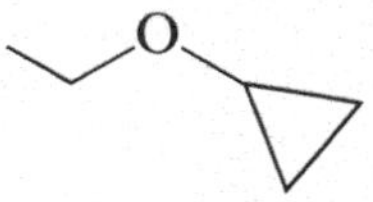

  [Line-angle structure of cyclopropyl ethyl ether]

Alcohols, Thiols, Ethers, Aldehydes, and Ketones

12.9 The carbon bonded to the hydroxyl group ($-$OH) is attached to one alkyl group in a primary (1°) alcohol, except for methanol; to two alkyl groups in a secondary alcohol (2°); and to three alkyl groups in a tertiary alcohol (3°).
   a. primary alcohol (1°)
   b. primary alcohol (1°)
   c. tertiary alcohol (3°)
   d. secondary alcohol (2°)

12.11 a. Soluble; ethanol with a short carbon chain is soluble because the hydroxyl group forms hydrogen bonds with water.
   b. Insoluble; an alcohol with a carbon chain of five or more carbon atoms is not soluble in water.

12.13 a. Methanol has a polar $-$OH group that can form hydrogen bonds with water, but the alkane ethane does not.
   b. 1-Propanol is more soluble because it can form more hydrogen bonds with water than the ether can.

12.15 a. A carbonyl group (C=O) attached to two carbon atoms within the carbon chain makes this compound a ketone.
   b. A carbonyl group (C=O) attached to a hydrogen atom at the end of the carbon chain makes this compound an aldehyde.
   c. A carbonyl group (C=O) attached to two carbon atoms within the carbon chain makes this compound a ketone.
   d. A carbonyl group (C=O) attached to a hydrogen atom at the end of the carbon chain makes this compound an aldehyde.

12.17 a. acetaldehyde   b. methyl propyl ketone   c. formaldehyde

12.19 a. propanal   b. 2-methyl-3-pentanone
   c. 3,4-dimethylcyclohexanone   d. 2-bromobenzaldehyde

12.21 a. $CH_3-\overset{\overset{O}{\|}}{C}-H$

b. $CH_3-\overset{\overset{O}{\|}}{C}-CH_2-\overset{\overset{Cl}{|}}{CH}-CH_3$

c. $CH_3-\overset{\overset{O}{\|}}{C}-CH_2-CH_2-CH_2-CH_3$

d. $CH_3-CH_2-\overset{\overset{CH_3}{|}}{CH}-CH_2-\overset{\overset{O}{\|}}{C}-H$

12.23 a. $CH_3-\overset{\overset{O}{\|}}{C}-\overset{\overset{O}{\|}}{C}-CH_2-CH_3$ is more soluble in water because it has two polar carbonyl groups to hydrogen bond with water.
   b. Acetone is more soluble in water because it has a shorter carbon chain than 2-pentanone.
   c. Propanal is more soluble in water because it has a shorter carbon chain than pentanal.

12.25 a. $2CH_4O + 3O_2 \longrightarrow 2CO_2 + 4H_2O + energy$
   b. $C_4H_{10}O + 6O_2 \longrightarrow 4CO_2 + 5H_2O + energy$

12.27 Dehydration is the removal of an H$-$ and an $-$OH from adjacent carbon atoms of an alcohol to form a water molecule and the corresponding alkene.
   a. $CH_3-CH_2-CH=CH_2$ b. (cyclopentene)   c. (2-butene)

12.29 A primary alcohol oxidizes to an aldehyde and a secondary alcohol oxidizes to a ketone.

a. $CH_3-CH_2-\overset{\overset{O}{\|}}{C}-H$

b. $CH_3-\overset{\overset{O}{\|}}{C}-CH_2-CH_2-CH_2-CH_3$

c. (cyclohexanone)

d. (4-methylpentanal structure with H)

Chapter 12

12.31 A primary alcohol will be oxidized to an aldehyde and then further oxidized to a carboxylic acid.

    a. $CH_3-CH_2-CH_2-CH_2-\overset{\overset{O}{\|}}{C}-H$; $CH_3-CH_2-CH_2-CH_2-\overset{\overset{O}{\|}}{C}-OH$

    b. $CH_3-\overset{\overset{CH_3}{|}}{CH}-CH_2-\overset{\overset{O}{\|}}{C}-H$; $CH_3-\overset{\overset{CH_3}{|}}{CH}-CH_2-\overset{\overset{O}{\|}}{C}-OH$

    c. $CH_3-CH_2-CH_2-\overset{\overset{O}{\|}}{C}-H$; $CH_3-CH_2-CH_2-\overset{\overset{O}{\|}}{C}-OH$

12.33 In reduction, an aldehyde will give a primary alcohol and a ketone will give a secondary alcohol.
    a. Butyraldehyde is the four-carbon aldehyde; it will be reduced to a four-carbon primary alcohol.
    $CH_3-CH_2-CH_2-CH_2-OH$
    b. Acetone is a three-carbon ketone; it will be reduced to a three-carbon secondary alcohol.
    $CH_3-\overset{\overset{OH}{|}}{CH}-CH_3$
    c. Hexanal is a six-carbon aldehyde; it reduces to a six-carbon primary alcohol.
    $CH_3-CH_2-CH_2-CH_2-CH_2-CH_2-OH$
    d. 2-Methyl-3-pentanone is a five-carbon ketone with a methyl group attached to carbon 2. It will be reduced to a five-carbon secondary alcohol with a methyl group attached to carbon 2.
    $CH_3-\overset{\overset{CH_3}{|}}{CH}-\overset{\overset{OH}{|}}{CH}-CH_2-CH_3$

12.35 a. Oxybenzone contains ether, phenol, ketone, and aromatic functional groups.
    b. Molecular formula of oxybenzone = $C_{14}H_{12}O_3$
    Molar mass of oxybenzone ($C_{14}H_{12}O_3$)
    = 14(12.01 g) + 12(1.008 g) + 3(16.00 g) = 228.2 g/mole (4 SFs)
    c. 178 mL sunscreen × $\dfrac{6.0 \text{ g oxybenzone}}{100 \text{ mL sunscreen}}$ = 11 g of oxybenzone (2 SFs)

12.37 phenol, ether, ketone, alcohol

12.39 phenol, alkene

12.41 Compound **a** will give a positive Tollens' test because it is an aldehyde.

12.43 The carbon bonded to the hydroxyl group (—OH) is attached to one alkyl group in a primary (1°) alcohol, except for methanol; to two alkyl groups in a secondary alcohol (2°); and to three alkyl groups in a tertiary alcohol (3°).
    a. secondary (2°) alcohol   **b.** primary (1°) alcohol
    c. secondary (2°) alcohol   **d.** tertiary (3°) alcohol

12.45 a. 2-methyl-2-propanol   **b.** 4-bromo-2-pentanol   **c.** 3-methylphenol

12.47 a. 4-chlorophenol (OH on benzene ring, Cl para)

    b. $CH_3-\overset{\overset{CH_3}{|}}{CH}-\overset{\overset{OH}{|}}{CH}-CH_2-CH_3$

    c. $CH_3-\overset{\overset{CH_3}{|}}{CH}-CH_2-OH$ **d.** $CH_3-\overset{\overset{OH}{|}}{CH}-\overset{\overset{CH_3}{|}}{CH}-CH_3$

12.49 a. 1-Propanol has a polar —OH group that can form hydrogen bonds with water, but the alkane butane does not.

b. 2-Propanol is more soluble because it can form more hydrogen bonds with water than the ether can.

c. Ethanol is more soluble in water than 1-hexanol because it has a shorter carbon chain.

12.51 a. Dehydration of an alcohol produces an alkene. $CH_3-CH=CH_2$

b. Oxidation of a primary alcohol produces an aldehyde.

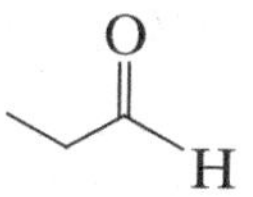

$$\underset{H}{\overset{O}{\|}}\diagup$$

c. Dehydration of an alcohol produces an alkene. $CH_3-CH=CH-CH_2-CH_3$

d. Oxidation of a secondary alcohol produces a ketone.

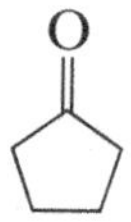

12.53 In reduction, an aldehyde will give a primary alcohol and a ketone will give a secondary alcohol.

a. $CH_3-CH_2-\underset{\underset{OH}{|}}{CH}-CH_3$

b. 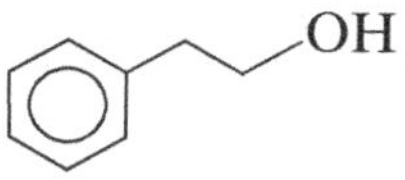 (phenethyl alcohol: benzene ring—CH$_2$CH$_2$—OH)

c.  (cyclopropyl with OH)

12.55 a. 3-bromo-4-chlorobenzaldehyde
b. 3-chloropropanal
c. 2-chloro-3-pentanone

12.57 a. 4-Chlorobenzaldehyde is the aldehyde of benzene with a chlorine atom attached to carbon 4 of the ring.

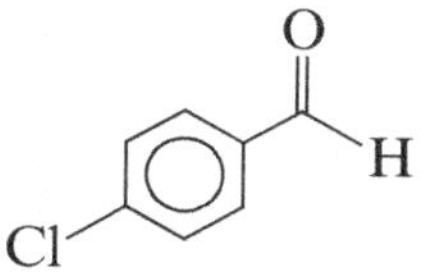

(structure: Cl-benzene-CHO)

b. 3-Chloropropionaldehyde is a three-carbon aldehyde with a chlorine atom located two carbons from the carbonyl group.

$$Cl-CH_2-CH_2-\overset{\overset{O}{\|}}{C}-H$$

c. Ethyl methyl ketone (butanone) is a four-carbon ketone.

$$CH_3-\overset{\overset{O}{\|}}{C}-CH_2-CH_3$$

d. 3-Methylhexanal is a six-carbon aldehyde with a methyl group attached two carbons from the carbonyl group.

$$CH_3-CH_2-CH_2-\underset{\underset{CH_3}{|}}{CH}-CH_2-\overset{\overset{O}{\|}}{C}-H$$

12.59 Compounds **a** and **b** are soluble in water because they each have a polar group with an oxygen atom that hydrogen bonds with water and fewer than five carbon atoms.

12.60 Compounds **a** and **b** are soluble in water because they each have a polar group with an oxygen atom that hydrogen bonds with water and fewer than five carbon atoms.

12.61 Primary alcohols oxidize to aldehydes and then to carboxylic acids. Secondary alcohols oxidize to ketones.

a. $CH_3-CH_2-\overset{\overset{O}{\|}}{C}-OH$

b. 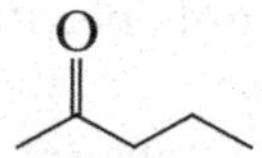 (butan-2-one structure: $CH_3-\overset{\overset{O}{\|}}{C}-CH_2-CH_3$)

c. $CH_3-CH_2-CH_2-\overset{\overset{O}{\|}}{C}-OH$

d. 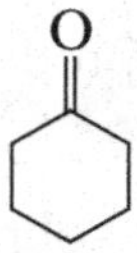 (cyclohexanone)

12.63

$CH_3-CH_2-CH_2-CH_2-CH_2-OH$ 1-Pentanol

$CH_3-\underset{\underset{}{}}{\overset{\overset{OH}{|}}{CH}}-CH_2-CH_2-CH_3$ 2-Pentanol

$CH_3-CH_2-\overset{\overset{OH}{|}}{CH}-CH_2-CH_3$ 3-Pentanol

$HO-CH_2-\overset{\overset{CH_3}{|}}{CH}-CH_2-CH_3$ 2-Methyl-1-butanol

$HO-CH_2-CH_2-\overset{\overset{CH_3}{|}}{CH}-CH_3$ 3-Methyl-1-butanol

$CH_3-\overset{\overset{CH_3}{|}}{\underset{\underset{OH}{|}}{C}}-CH_2-CH_3$ 2-Methyl-2-butanol

$CH_3-\overset{\overset{OH}{|}}{CH}-\overset{\overset{CH_3}{|}}{CH}-CH_3$ 3-Methyl-2-butanol

$CH_3-\overset{\overset{CH_3}{|}}{\underset{\underset{CH_3}{|}}{C}}-CH_2-OH$ 2,2-Dimethyl-1-propanol

12.65 Since the compound is synthesized from a primary alcohol and oxidizes to give a carboxylic acid, it must be an aldehyde.

$CH_3-\overset{\overset{CH_3}{|}}{CH}-\overset{\overset{O}{\|}}{C}-H$ 2-Methylpropanal

12.67 A $CH_3-CH_2-CH_2-OH$ 1-Propanol
      B $CH_3-CH=CH_2$ Propene
      C $CH_3-CH_2-\overset{\overset{O}{\|}}{C}-H$ Propanal

Selected Answers to Combining Ideas from Chapters 10 to 12

CI.21 **a.** $2M(s) + 6HCl(aq) \longrightarrow 3H_2(g) + 2MCl_3(aq)$

b. $34.8 \text{ mL solution} \times \dfrac{1 \text{ L solution}}{1000 \text{ mL solution}} \times \dfrac{0.520 \text{ mole HCl}}{1 \text{ L solution}} \times \dfrac{3 \text{ moles } H_2}{6 \text{ moles HCl}}$

$\times \dfrac{22.4 \text{ L } H_2 \text{(STP)}}{1 \text{ mole } H_2} \times \dfrac{1000 \text{ mL } H_2}{1 \text{ L } H_2}$

$= 203 \text{ mL of } H_2 \text{ at STP (3 SFs)}$

c. $34.8 \text{ mL solution} \times \dfrac{1 \text{ L solution}}{1000 \text{ mL solution}} \times \dfrac{0.520 \text{ mole HCl}}{1 \text{ L solution}} \times \dfrac{2 \text{ moles M}}{6 \text{ moles HCl}}$

$= 6.03 \times 10^{-3} \text{ mole of M (3 SFs)}$

d. $\dfrac{0.420 \text{ g M}}{6.03 \times 10^{-3} \text{ mole M}} = 69.7 \text{ g/mole of M (3 SFs)}$

e. The metal is gallium, Ga.

f. $2Ga(s) + 6HCl(aq) \longrightarrow 3H_2(g) + 2GaCl_3(aq)$

CI.23 **a.** The molecular formula of acetone is C_3H_6O.

b. Molar mass of acetone $(C_3H_6O) = 3(12.01 \text{ g}) + 6(1.008 \text{ g}) + 1(16.00 \text{ g})$
$= 58.08 \text{ g (4 SFs)}$

c. Polar covalent bonds in C_3H_6O: C—O
Nonpolar covalent bonds in C_3H_6O: C—C, C—H

d. $C_3H_6O + 4O_2 \xrightarrow{\Delta} 3CO_2 + 3H_2O + \text{energy}$

e. $15.0 \text{ mL } C_3H_6O \times \dfrac{0.786 \text{ g } C_3H_6O}{1 \text{ mL } C_3H_6O} \times \dfrac{1 \text{ mole } C_3H_6O}{58.08 \text{ g } C_3H_6O} \times \dfrac{4 \text{ moles } O_2}{1 \text{ mole } C_3H_6O} \times \dfrac{32.00 \text{ g } O_2}{1 \text{ mole } O_2}$

$= 26.0 \text{ g of } O_2 \text{ (3 SFs)}$

CI.25 **a.** $CH_3-CH_2-CH_2-\underset{\underset{\text{H}}{\|}}{\overset{\overset{\text{O}}{\|}}{C}}-H$

b. (structural formula of butanal)

c. The IUPAC name of butyraldehyde is butanal.

d. $CH_3-CH_2-CH_2-CH_2-OH$

e. $2C_4H_8O + 11O_2 \xrightarrow{\Delta} 8CO_2 + 8H_2O + \text{energy}$

f. $15.0 \text{ mL } C_4H_8O \times \dfrac{0.802 \text{ g } C_4H_8O}{1 \text{ mL } C_4H_8O} \times \dfrac{1 \text{ mole } C_4H_8O}{72.10 \text{ g } C_4H_8O} \times \dfrac{11 \text{ moles } O_2}{2 \text{ moles } C_4H_8O} \times \dfrac{32.00 \text{ g } O_2}{1 \text{ mole } O_2}$

$= 29.4 \text{ g of } O_2 \text{ (3 SFs)}$

g. $15.0 \text{ mL } C_4H_8O \times \dfrac{0.802 \text{ g } C_4H_8O}{1 \text{ mL } C_4H_8O} \times \dfrac{1 \text{ mole } C_4H_8O}{72.10 \text{ g } C_4H_8O} \times \dfrac{8 \text{ moles } O_2}{2 \text{ moles } C_4H_8O} \times \dfrac{22.4 \text{ L } CO_2 \text{(STP)}}{1 \text{ mole } O_2}$

$= 14.9 \text{ L of } CO_2 \text{ at STP (3 SFs)}$

13
Carbohydrates

Kate completed her diabetes educational class taught by Paula, a diabetes nurse. Kate now follows a healthy meal plan, which includes 45 to 60 g of carbohydrates in the form of fruits and vegetables. She is walking for 30 minutes twice a day and has lost 10 lb. Kate measures her blood glucose level before and after a meal using a blood glucose meter. She writes the results in her log, which helps determine if she is maintaining a consistent blood sugar level or if her blood glucose is high or low. Her blood sugar level should be 70 to 110 mg/dL before a meal, and less than 180 mg/dL two hours after a meal. Draw the Fischer projection for D-glucose.

Credit: Rolf Bruderer/Alamy

LOOKING AHEAD

13.1 Carbohydrates
13.2 Chiral Molecules
13.3 Fischer Projections of Monosaccharides
13.4 Haworth Structures of Monosaccharides
13.5 Chemical Properties of Monosaccharides
13.6 Disaccharides
13.7 Polysaccharides

The Health icon indicates a question that is related to health and medicine.

13.1 Carbohydrates

Learning Goal: Classify a monosaccharide as an aldose or a ketose, and indicate the number of carbon atoms.

- Photosynthesis is the process of using water, carbon dioxide, and the energy from the Sun to form monosaccharides and oxygen.
- Carbohydrates are classified as monosaccharides (simple sugars), disaccharides (two monosaccharide units), and polysaccharides (many monosaccharide units).
- Monosaccharides are polyhydroxy aldehydes (aldoses) or ketones (ketoses).
- Monosaccharides are classified as *aldo* for an aldehyde or *keto* for a ketone and by the number of carbon atoms as *trioses*, *tetroses*, *pentoses*, or *hexoses*.

Carbohydrates

♦ Learning Exercise 13.1A

Complete and balance the equations for the photosynthesis of

REVIEW
Identifying Functional Groups (12.1)

1. Glucose: _____ + _____ ⟶ $C_6H_{12}O_6$ + _____

2. Ribose: _____ + _____ ⟶ $C_5H_{10}O_5$ + _____

Answers   1. $6CO_2 + 6H_2O \longrightarrow C_6H_{12}O_6 + 6O_2$ 2. $5CO_2 + 5H_2O \longrightarrow C_5H_{10}O_5 + 5O_2$

♦ Learning Exercise 13.1B

Indicate the number of monosaccharide units (one, two, or many) in each of the following carbohydrates:

1. sucrose, a disaccharide _____
2. cellulose, a polysaccharide _____
3. glucose, a monosaccharide _____
4. amylose, a polysaccharide _____
5. maltose, a disaccharide _____

Answers   1. two   2. many   3. one   4. many   5. two

♦ Learning Exercise 13.1C

Identify each of the following monosaccharides as an aldotriose, a ketotriose, an aldotetrose, a ketotetrose, an aldopentose, a ketopentose, an aldohexose, or a ketohexose:

1. 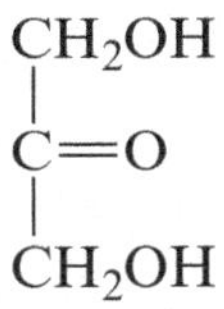 Dihydroxyacetone
2. 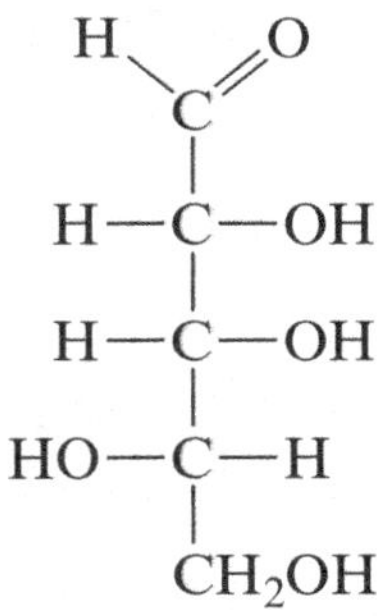 Lyxose
3. 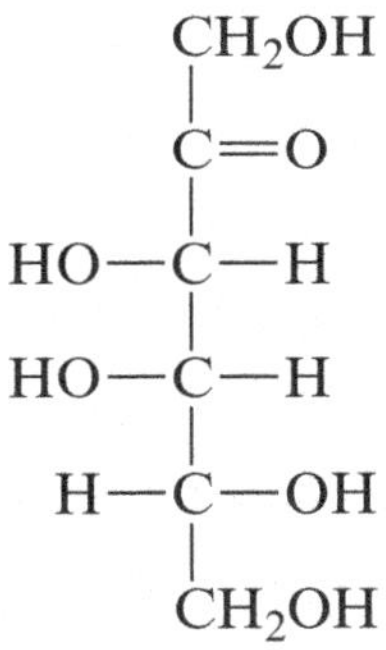 Tagatose
4. 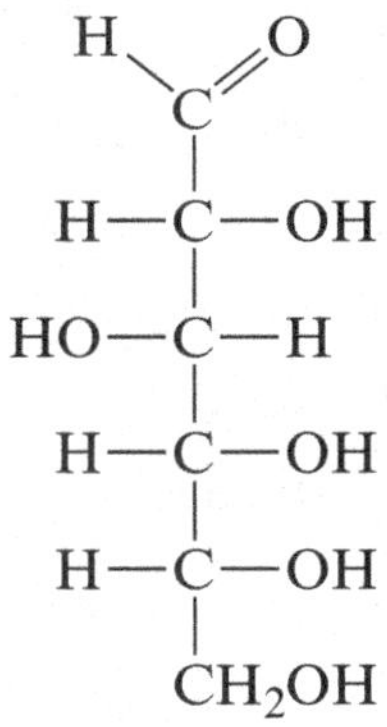 Glucose
5. 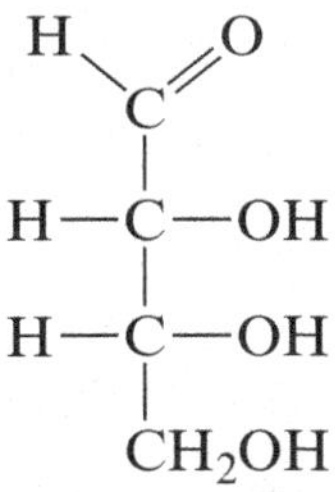 Erythrose

1. _____   2. _____   3. _____
4. _____   5. _____

Answers   1. ketotriose   2. aldopentose   3. ketohexose
4. aldohexose   5. aldotetrose

Chapter 13

13.2 Chiral Molecules

Learning Goal: Identify chiral and achiral carbon atoms in an organic molecule. Identify D and L enantiomers.

- Chiral molecules have mirror images that cannot be superimposed.
- When the mirror image of an object is identical and can be superimposed on the original, it is achiral.
- In a chiral molecule, one or more carbon atoms are attached to four different atoms or groups.
- The mirror images of a chiral molecule represent two different molecules called *enantiomers*.
- In a Fischer projection, the prefixes D and L are used to distinguish between the mirror images.
- In D-glyceraldehyde, the —OH group is on the right of the chiral carbon; it is on the left in L-glyceraldehyde.

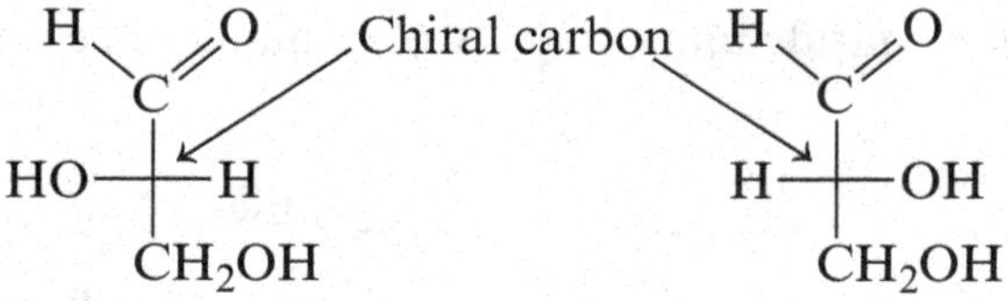

$$
\begin{array}{cc}
\text{L-Glyceraldehyde} & \text{D-Glyceraldehyde}
\end{array}
$$

- For compounds with two or more chiral carbons, the designation as a D and L isomer is determined by the position of the —OH group attached to the chiral carbon *farthest from the carbonyl group*.

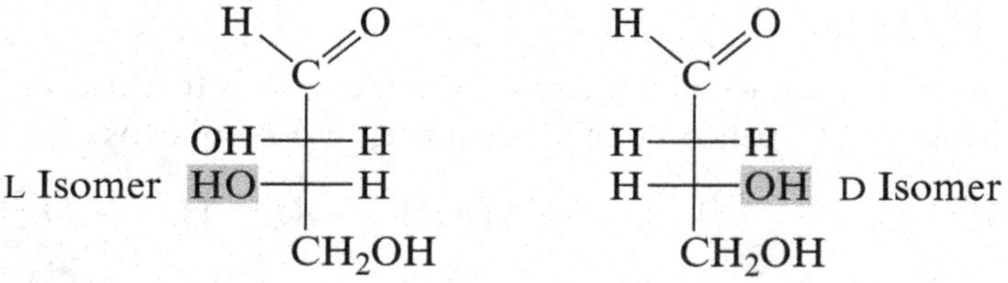

L Isomer — L-Erythrose   D-Erythrose — D Isomer

♦ **Learning Exercise 13.2A**

CORE CHEMISTRY SKILL
Identifying Chiral Molecules

Indicate whether each of the following objects is chiral or achiral:

1. a piece of blank computer paper _____   2. a glove _____
3. a plain baseball cap _____   4. a volleyball net _____
5. a left foot _____

Answers   1. achiral   2. chiral   3. achiral   4. achiral   5. chiral

Carbohydrates

♦ **Learning Exercise 13.2B**

State whether each of the following molecules is chiral or achiral:

1. 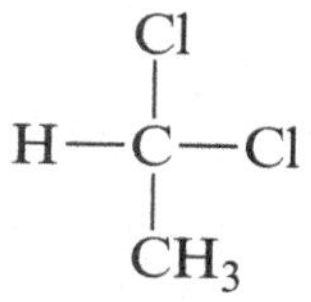 H—C(Cl)(Cl)—CH₃ with H
2. 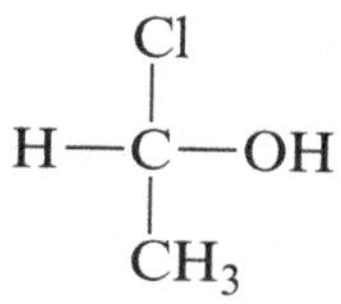 H—C(Cl)(OH)—CH₃
3. 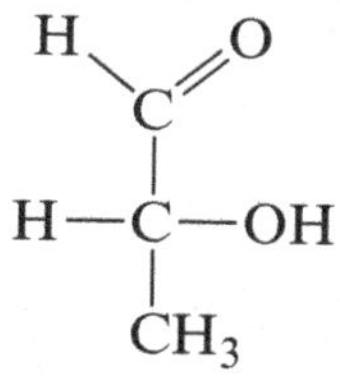 H—C(CHO)(OH)—CH₃

_____   _____   _____

Answers   1. achiral   2. chiral   3. chiral

♦ **Learning Exercise 13.2C**

Identify each of the following features as characteristic of a chiral compound or a compound that is achiral:

1. central atom attached to two identical groups   _____
2. contains a carbon attached to four different groups   _____
3. has identical mirror images   _____

Answers   1. achiral   2. chiral   3. achiral

♦ **Learning Exercise 13.2D**

Indicate whether each pair of Fischer projections represents enantiomers (E) or identical structures (I).

1. 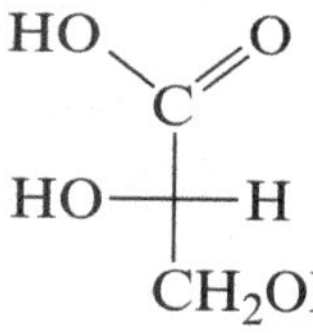 HO—C(=O), HO—|—H, CH₂OH   and   HO—C(=O), H—|—OH, CH₂OH

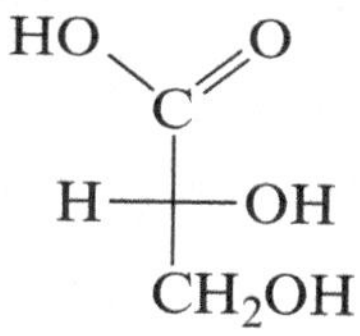

2. 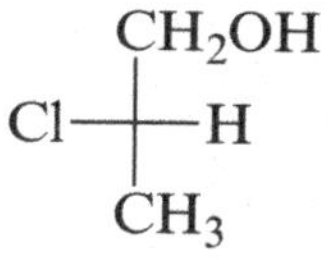 CH₂OH, Cl—|—H, CH₃   and   CH₂OH, Cl—|—H, CH₃

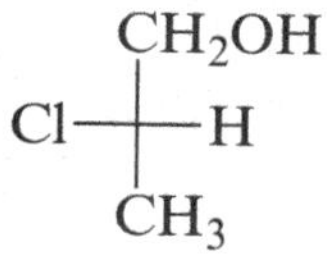

3. 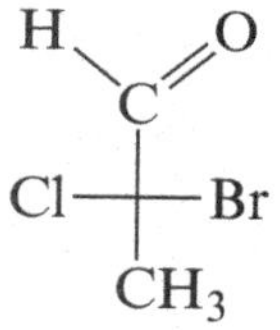 H—C(=O), Cl—|—Br, CH₃   and   H—C(=O), Br—|—Cl, CH₃

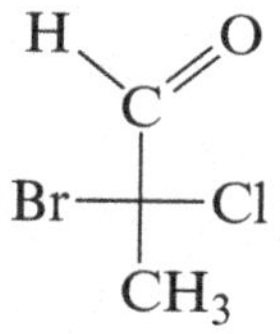

4. 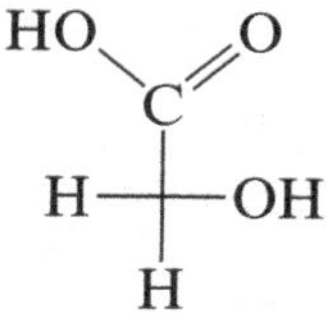 HO—C(=O), H—|—OH, H   and   HO—C(=O), HO—|—H, H

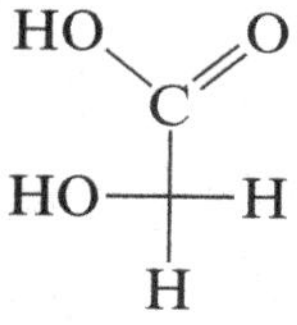

Answers   1. E   2. I   3. E   4. I

13.3 Fischer Projections of Monosaccharides

Learning Goal: Identify or draw the D and L configurations of the Fischer projections for common monosaccharides.

- In a Fischer projection, the carbon chain is drawn vertically, with the most oxidized carbon (usually an aldehyde or ketone) at the top.
- In the Fischer projection for a monosaccharide, the chiral —OH group *farthest* from the carbonyl group (C=O) is drawn on the left side in the L isomer and on the right side in the D isomer.

Copyright © 2018 Pearson Education, Inc.

Chapter 13

- The carbon atom in the —CH$_2$OH group at the bottom of the Fischer projection is not chiral because it does not have four different groups bonded to it.
- Important monosaccharides are the aldopentose ribose, the aldohexoses glucose and galactose, and the ketohexose fructose.

♦ Learning Exercise 13.3A

CORE CHEMISTRY SKILL
Identifying D and L Fischer Projections for Carbohydrates

Identify each of the following Fischer projections of monosaccharides as the D or L isomer:

1. 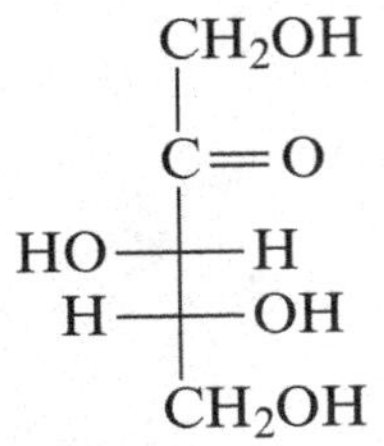
```
    CH₂OH
    |
    C=O
HO—|—H
 H—|—OH
    |
    CH₂OH
```

2. 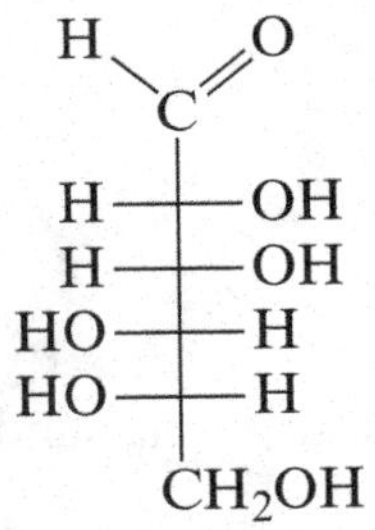
```
     H   O
      \\//
       C
    H—|—OH
    H—|—OH
   HO—|—H
   HO—|—H
       |
      CH₂OH
```

3. 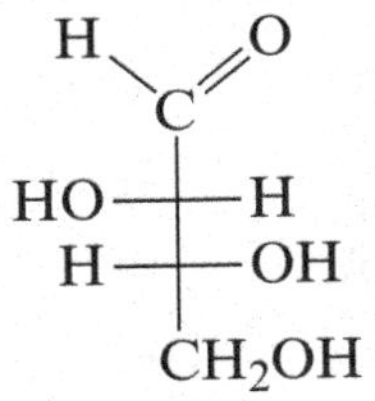
```
     H   O
      \\//
       C
   HO—|—H
    H—|—OH
       |
      CH₂OH
```

4. 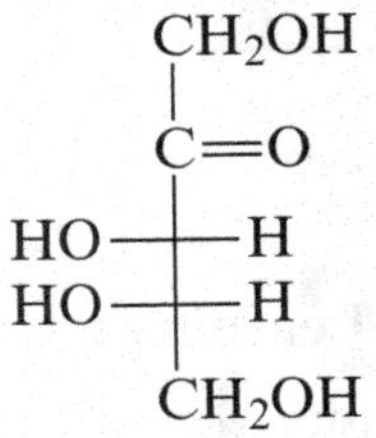
```
    CH₂OH
    |
    C=O
   HO—|—H
   HO—|—H
    |
    CH₂OH
```

_____ Xylulose   _____ Mannose   _____ Threose   _____ Ribulose

Answers   1. D-Xylulose   2. L-Mannose   3. D-Threose   4. L-Ribulose

♦ Learning Exercise 13.3B

Draw the mirror image of each of the monosaccharides in Learning Exercise 13.3A and give the D or L name.

1.   2.   3.   4.

Answers

1. 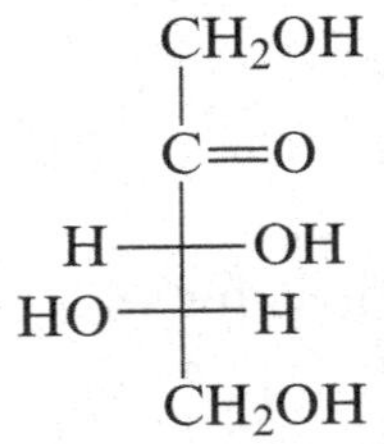
```
    CH₂OH
    |
    C=O
    H—|—OH
   HO—|—H
    |
    CH₂OH
```
L-Xylulose

2. 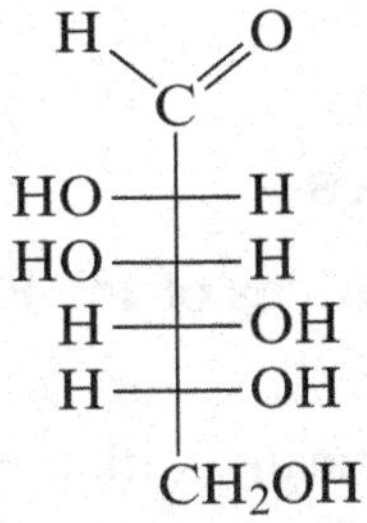
```
     H   O
      \\//
       C
   HO—|—H
   HO—|—H
    H—|—OH
    H—|—OH
       |
      CH₂OH
```
D-Mannose

3. 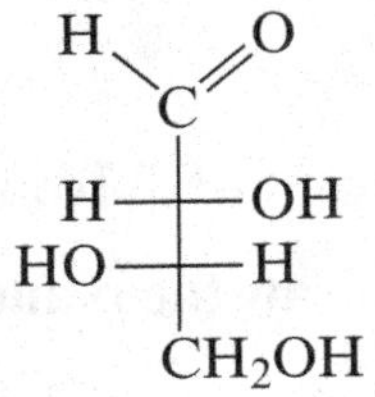
```
     H   O
      \\//
       C
    H—|—OH
   HO—|—H
       |
      CH₂OH
```
L-Threose

4. 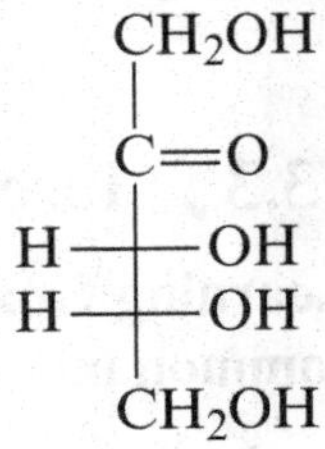
```
    CH₂OH
    |
    C=O
    H—|—OH
    H—|—OH
    |
    CH₂OH
```
D-Ribulose

Carbohydrates

♦ Learning Exercise 13.3C

Identify the monosaccharide (D-glucose, D-fructose, or D-galactose) that fits each of the following descriptions:

1. a building block in cellulose   _____

2. also known as fruit sugar   _____

3. accumulates in the disease known as *galactosemia*   _____

4. the most common monosaccharide   _____

5. the sweetest monosaccharide   _____

Answers   1. D-glucose   2. D-fructose   3. D-galactose   4. D-glucose   5. D-fructose

♦ Learning Exercise 13.3D

Draw the Fischer projection for each of the following monosaccharides:

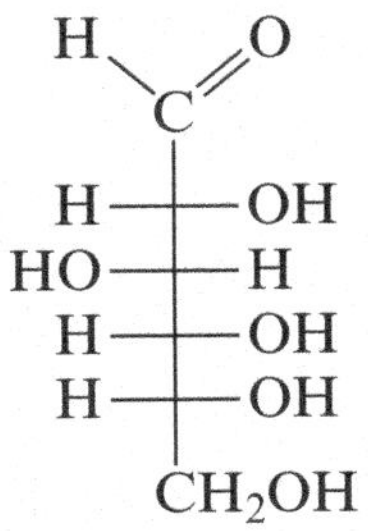

D-Glucose   D-Galactose   D-Fructose

Answers

D-Glucose:
```
    H   O
     \\//
      C
  H ──┼── OH
  HO ─┼── H
  H ──┼── OH
  H ──┼── OH
      CH₂OH
```

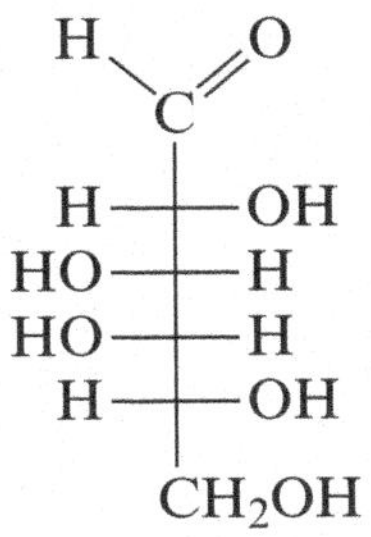

D-Galactose:
```
    H   O
     \\//
      C
  H ──┼── OH
  HO ─┼── H
  HO ─┼── H
  H ──┼── OH
      CH₂OH
```

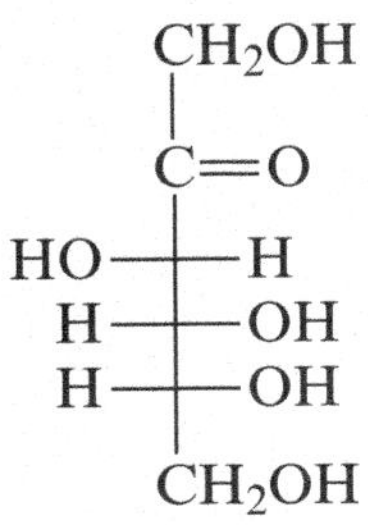

D-Fructose:
```
      CH₂OH
       │
       C═O
  HO ─┼── H
  H ──┼── OH
  H ──┼── OH
      CH₂OH
```

Chapter 13

13.4 Haworth Structures of Monosaccharides

Learning Goal: Draw and identify the Haworth structures for monosaccharides.

- The Haworth structure is a representation of the cyclic, stable form of monosaccharides, which are rings of five or six atoms.
- The Haworth structure forms by a reaction between the —OH group on carbon 5 of hexoses and the carbonyl group on carbon 1 or 2 of the same molecule.
- The formation of a new —OH group on carbon 1 (or 2 in fructose) gives α and β isomers of the cyclic monosaccharide. Because the molecule opens and closes continuously in solution, both α and β isomers are present.

	Solution Guide to Drawing Haworth Structures
Step 1	Turn the Fischer projection clockwise by 90°.
Step 2	Fold the horizontal chain into a hexagon, rotate the groups on carbon 5, and bond the O on carbon 5 to carbon 1.
Step 3	Draw the new —OH group on carbon 1 below the ring to give the α isomer or above the ring to give the β isomer.

♦ Learning Exercise 13.4

CORE CHEMISTRY SKILL
Drawing Haworth Structures

Draw the Haworth structure for the α isomer of each of the following:

1. D-Glucose
2. D-Galactose
3. D-Fructose

Answers

1. [Haworth structure of α-D-Glucose with CH₂OH, O, H, OH, HO, H positions]
2. [Haworth structure of α-D-Galactose with CH₂OH, O, HO, H, OH positions]
3. [Haworth structure of α-D-Fructose with HOCH₂, O, CH₂OH, H, OH positions]

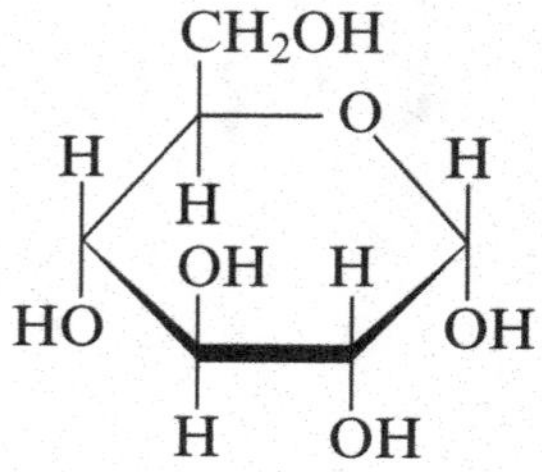

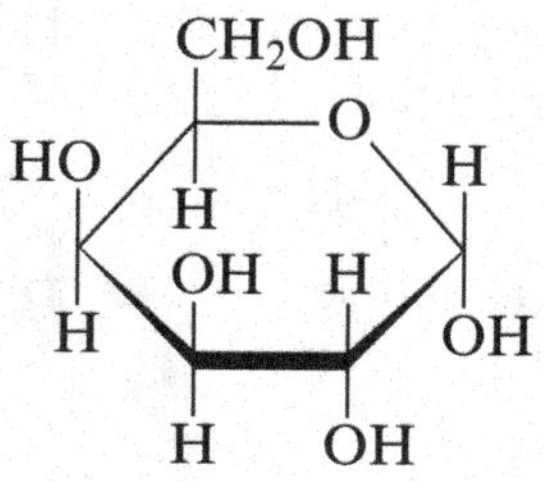

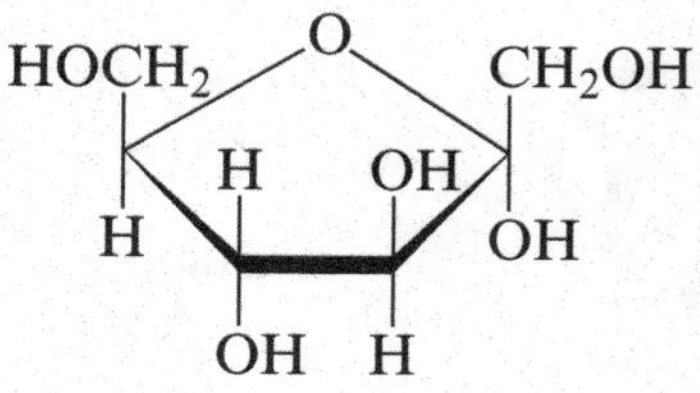

13.5 Chemical Properties of Monosaccharides

Learning Goal: Identify the products of oxidation or reduction of monosaccharides; determine whether a carbohydrate is a reducing sugar.

- Monosaccharides are called *reducing sugars* because an aldehyde group is oxidized by Cu^{2+} in Benedict's solution, which is reduced.
- Monosaccharides are also reduced to give sugar alcohols.

♦ **Learning Exercise 13.5A**

What changes occur when a reducing sugar reacts with Benedict's reagent?

> **REVIEW**
> Writing Equations for the Oxidation of Alcohols (12.4)

Answer   The carbonyl group of the reducing sugar is oxidized to a carboxylic acid group; the Cu^{2+} ion in Benedict's reagent is reduced to Cu^+, which forms a brick-red solid Cu_2O.

♦ **Learning Exercise 13.5B**

Draw the Fischer projection for D-lyxitol that is produced when D-lyxose is reduced.

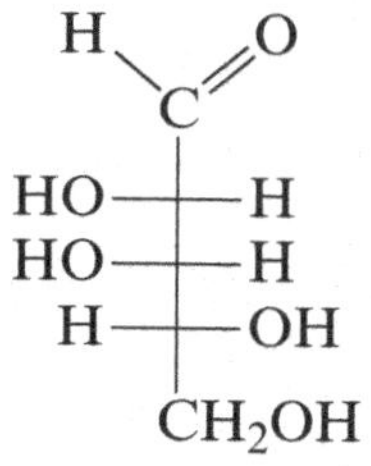

```
       H   O
        \ //
         C
   HO ──┼── H
   HO ──┼── H
    H ──┼── OH
        CH₂OH
      D-Lyxose
```

Answer

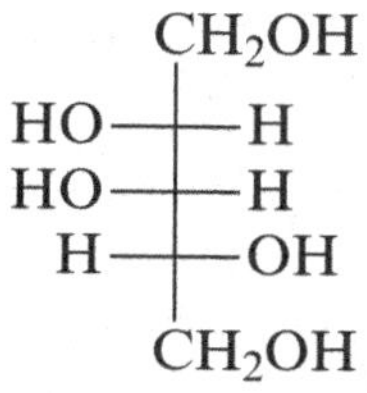

```
        CH₂OH
   HO ──┼── H
   HO ──┼── H
    H ──┼── OH
        CH₂OH
      D-Lyxitol
```

13.6 Disaccharides

Learning Goal: Describe the monosaccharide units and linkages in disaccharides.

- Disaccharides have two monosaccharide units joined together by a glycosidic bond.

    Monosaccharide (1) + monosaccharide (2) ⟶ disaccharide + H_2O

- In the most common disaccharides, maltose, lactose, and sucrose, there is at least one glucose unit.
- In the disaccharide maltose, two glucose units are linked by an $\alpha(1\rightarrow4)$ bond. The $\alpha(1\rightarrow4)$ indicates that the —OH group on carbon 1 of α-D-glucose is bonded to carbon 4 of the other glucose molecule.
- In lactose there is a $\beta(1\rightarrow4)$-glycosidic bond because the —OH group on carbon 1 of β-D-galactose forms a glycosidic bond with the —OH group on carbon 4 of a D-glucose molecule.
- Sucrose has an α-D-glucose and a β-D-fructose molecule joined by an $\alpha,\beta(1\rightarrow2)$-glycosidic bond.

    Glucose + glucose ⟶ maltose + H_2O

    Glucose + galactose ⟶ lactose + H_2O

    Glucose + fructose ⟶ sucrose + H_2O

Lactose is a disaccharide found in milk and milk products.
Credit: Pearson Education, Inc.

◆ **Learning Exercise 13.6**

For the following disaccharides, state (**a**) the monosaccharide units, (**b**) the type of glycosidic bond, and (**c**) the name of the disaccharide:

Carbohydrates

	a. Monosaccharide(s)	b. Type of Glycosidic Bond	c. Name of Disaccharide
1.			
2.			
3.			
4.			

Answers
1. a. two glucose units   b. α(1→4)-glycosidic bond   c. β-maltose
2. a. galactose + glucose   b. β(1→4)-glycosidic bond   c. α-lactose
3. a. fructose + glucose   b. α,β(1→2)-glycosidic bond   c. sucrose
4. a. two glucose units   b. α(1→4)-glycosidic bond   c. α-maltose

13.7 Polysaccharides

Learning Goal: Describe the structural features of amylose, amylopectin, glycogen, and cellulose.

The polysaccharide cellulose is the structural material in plants such as cotton.
Credit: Danny E Hooks / Shutterstock

- Polysaccharides are polymers of monosaccharide units.
- Starches consist of amylose and amylopectin. Amylose is an unbranched chain of glucose connected by α(1→4)-bonds.
- Amylopectin is a branched-chain polysaccharide; the glucose molecules are connected by α(1→4)-glycosidic bonds. However, at about every 25 glucose units, there is a branch of glucose molecules attached by an α(1→6)-glycosidic bond between carbon 1 of the branch and carbon 6 in the main chain.
- Glycogen, the storage form of glucose in animals, is similar to amylopectin, but has more branching.
- Cellulose is also a polymer of glucose, but in cellulose the glycosidic bonds are β(1→4)-bonds rather than α bonds as in the starches. Humans can digest starches, but not cellulose, to obtain energy. However, cellulose is important as a source of fiber in our diets.

Key Terms for Sections 13.1 to 13.7

Match the following key terms with the correct description:

    a. carbohydrate   **b.** glucose   **c.** chiral carbon   **d.** Haworth structure
    e. disaccharide   **f.** cellulose   **g.** Fischer projection

1. ____ a simple or complex sugar composed of carbon, hydrogen, and oxygen
2. ____ a carbon that is bonded to four different groups
3. ____ a cyclic structure that represents the closed-chain form of a monosaccharide
4. ____ an unbranched polysaccharide that cannot be digested by humans
5. ____ an aldohexose that is the most prevalent monosaccharide in the diet
6. ____ a carbohydrate that contains two monosaccharides linked by a glycosidic bond
7. ____ a system for drawing chiral molecules that uses horizontal lines for bonds coming forward and vertical lines for bonds going back, with the chiral atom at the center

Answers   1. a   2. c   3. d   4. f   5. b   6. e   7. g

Chapter 13

♦ **Learning Exercise 13.7**

List the monosaccharides and describe the glycosidic bonds in each of the following carbohydrates:

	Monosaccharides	Type(s) of glycosidic bonds
a. amylose	_____	_____
b. amylopectin	_____	_____
c. glycogen	_____	_____
d. cellulose	_____	_____

Answers
a. glucose; $\alpha(1\rightarrow 4)$-glycosidic bonds
b. glucose; $\alpha(1\rightarrow 4)$- and $\alpha(1\rightarrow 6)$-glycosidic bonds
c. glucose; $\alpha(1\rightarrow 4)$- and $\alpha(1\rightarrow 6)$-glycosidic bonds
d. glucose; $\beta(1\rightarrow 4)$-glycosidic bonds

Checklist for Chapter 13

You are ready to take the Practice Test for Chapter 13. Be sure you have accomplished the following learning goals for this chapter. If not, review the Section listed at the end of the goal. Then apply your new skills and understanding to the Practice Test.

After studying Chapter 13, I can successfully:

_____ Classify carbohydrates as monosaccharides, disaccharides, and polysaccharides. (13.1)

_____ Classify a monosaccharide as an aldose or ketose, and indicate the number of carbon atoms. (13.1)

_____ Identify a molecule as chiral or achiral; draw the D- and L-Fischer projections. (13.2)

_____ Draw and identify D- and L-Fischer projections for carbohydrate molecules. (13.3)

_____ Draw the open-chain structures for D-glucose, D-galactose, and D-fructose. (13.3)

_____ Draw or identify the Haworth structures of monosaccharides. (13.4)

_____ Describe some chemical properties of carbohydrates. (13.5)

_____ Describe the monosaccharide units and linkages in disaccharides. (13.6)

_____ Describe the structural features of amylose, amylopectin, glycogen, and cellulose. (13.7)

Practice Test for Chapter 13

The chapter Sections to review are shown in parentheses at the end of each question.

1. The requirements for photosynthesis are (13.1)
   A. sun
   B. sun and water
   C. water and carbon dioxide
   D. sun, water, and carbon dioxide
   E. carbon dioxide and sun

2. What are the products of photosynthesis? (13.1)
   A. carbohydrates
   B. carbohydrates and oxygen
   C. carbon dioxide and oxygen
   D. carbohydrates and carbon dioxide
   E. water and oxygen

Carbohydrates

3. The name "carbohydrate" came from the fact that (13.1)
   A. carbohydrates are hydrates of water
   B. carbohydrates contain hydrogen and oxygen in a 2:1 ratio
   C. carbohydrates contain a great quantity of water
   D. all plants produce carbohydrates
   E. carbon and hydrogen atoms are abundant in carbohydrates

4. What functional groups are in the open-chain forms of monosaccharides? (13.1)
   A. hydroxyl groups
   B. aldehyde groups
   C. ketone groups
   D. hydroxyl and aldehyde or ketone groups
   E. hydroxyl and ether groups

5. What is the classification of the following monosaccharide? (13.1)

   CH_2OH
   |
   $C=O$
   |
   CH_2OH

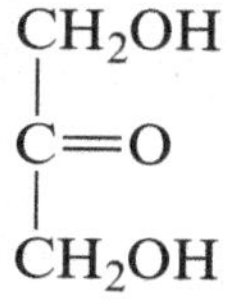

   A. aldotriose   B. ketotriose   C. aldotetrose   D. ketotetrose   E. ketopentose

For questions 6 through 9, identify each of the following pairs of Fischer projections as enantiomers (E) *or identical compounds* (I): (13.2)

6. 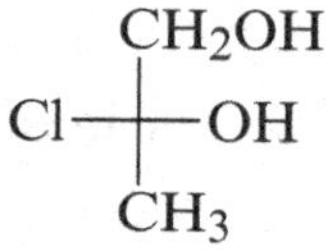 Cl—|—OH and HO—|—Cl
      CH₂OH   CH₂OH
      CH₃   CH₃

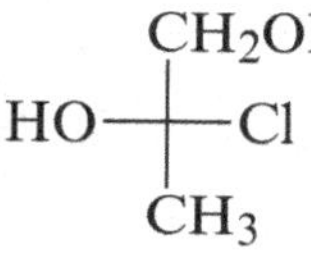

7. 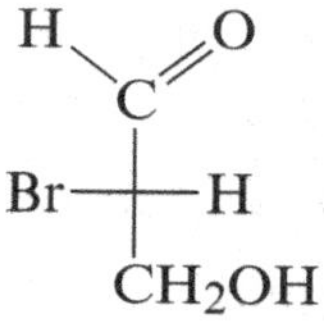 Br—|—H and H—|—Br
      CHO   CHO
      CH₂OH   CH₂OH

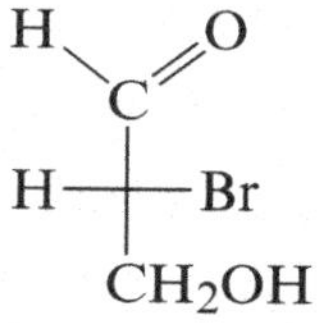

8. 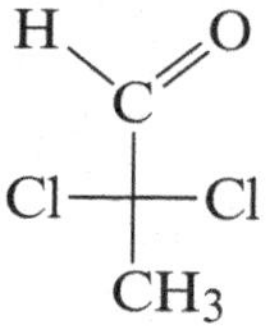 Cl—|—Cl and Cl—|—Cl
      CHO   CHO
      CH₃   CH₃

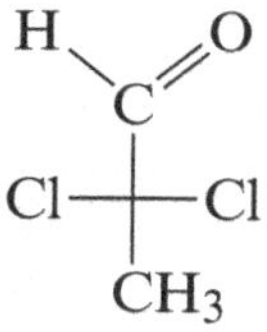

9. 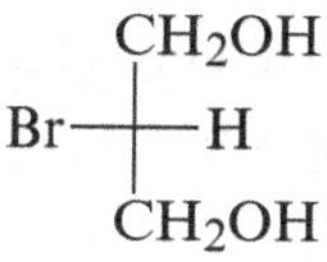 Br—|—H and H—|—Br
      CH₂OH   CH₂OH
      CH₂OH   CH₂OH

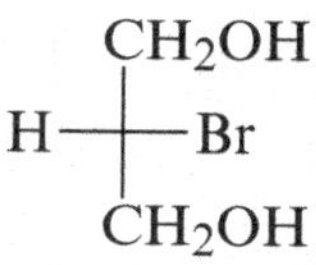

10. Identify the following as the D or L isomer: (13.3)

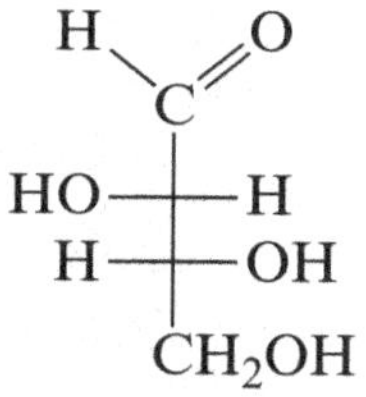

    CHO
    HO—|—H
    H—|—OH
    CH₂OH

For questions 11 through 15, refer to the following monosaccharide: (13.1, 13.4, 13.5, 13.7)

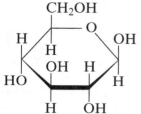

Chapter 13

11. It is the Haworth structure of a(n)
   A. aldotriose   B. ketopentose   C. aldopentose   D. aldohexose   E. aldoheptose

12. It is the Haworth structure for
   A. fructose   B. glucose   C. ribose   D. glyceraldehyde   E. galactose

13. It is one of the products of the complete hydrolysis of
   A. maltose   B. sucrose   C. lactose   D. glycogen   E. all of these

14. A Benedict's test with this sugar would
   A. be positive   B. be negative   C. produce a blue precipitate
   D. give no color change   E. produce a silver mirror

15. It is the monosaccharide unit used to build the following polymers:
   A. amylose   B. amylopectin   C. cellulose   D. glycogen   E. all of these

For questions 16 through 20, identify each carbohydrate described as one of the following: (13.6, 13.7)
   A. maltose   B. sucrose   C. cellulose   D. amylopectin   E. glycogen

16. _____ a disaccharide that is not a reducing sugar

17. _____ a disaccharide that occurs as a breakdown product of amylose

18. _____ a carbohydrate that is produced as a storage form of energy in plants

19. _____ the storage form of energy in humans

20. _____ a carbohydrate that is used for structural purposes by plants

For questions 21 through 25, identify each carbohydrate described as one of the following: (13.6, 13.7)
   A. amylose   B. cellulose   C. glycogen   D. lactose   E. sucrose

21. _____ a polysaccharide composed of many glucose units linked only by $\alpha(1\rightarrow4)$-glycosidic bonds

22. _____ a sugar containing both glucose and galactose

23. _____ a polysaccharide composed of glucose units joined by both $\alpha(1\rightarrow4)$- and $\alpha(1\rightarrow6)$-glycosidic bonds

24. _____ a disaccharide that is a reducing sugar

25. _____ a carbohydrate composed of glucose units joined by $\beta(1\rightarrow4)$-glycosidic bonds

For questions 26 through 30, identify each carbohydrate described as one of the following: (13.6)
   A. glucose   B. lactose   C. sucrose   D. maltose

26. _____ a sugar composed of glucose and fructose

27. _____ also called table sugar

28. _____ found in milk and milk products

29. _____ gives glucitol upon reduction

30. _____ gives galactose upon hydrolysis

Answers to the Practice Test

1. D	2. B	3. B	4. D	5. B
6. E	7. E	8. I	9. I	10. D
11. D	12. B	13. E	14. A	15. E
16. B	17. A	18. D	19. E	20. C
21. A	22. D	23. C	24. D	25. B
26. C	27. C	28. B	29. A	30. B

Carbohydrates

Selected Answers and Solutions to Text Problems

13.1 Photosynthesis requires CO_2, H_2O, and the energy from the Sun. Respiration requires O_2 from the air and glucose from our foods.

13.3 Monosaccharides are composed of a chain of three to eight carbon atoms—one in a carbonyl group as an aldehyde or ketone, and the rest attached to hydroxyl groups. A monosaccharide cannot be split or hydrolyzed into smaller carbohydrates. A disaccharide consists of two monosaccharide units joined together. A disaccharide can be hydrolyzed into two monosaccharide units.

13.5 Hydroxyl groups are found in all monosaccharides, along with a carbonyl on the first or second carbon that gives an aldehyde or ketone functional group.

13.7 A ketopentose contains hydroxyl and ketone functional groups and has five carbon atoms.

13.9 **a.** This six-carbon monosaccharide has a carbonyl group on carbon 2; it is a ketohexose.
  b. This five-carbon monosaccharide has a carbonyl group on carbon 1; it is an aldopentose.

13.11 **a.** Achiral; there are no carbon atoms attached to four different groups.
  b. Chiral; (Br, Chiral carbon shown)
  c. Chiral; $CH_3-CH(Br)-C(=O)-H$ (Chiral carbon)
  d. Achiral; there are no carbon atoms attached to four different groups.

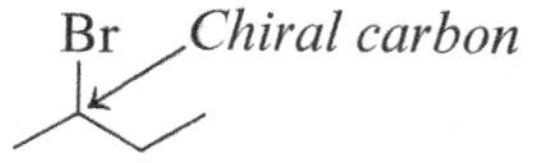

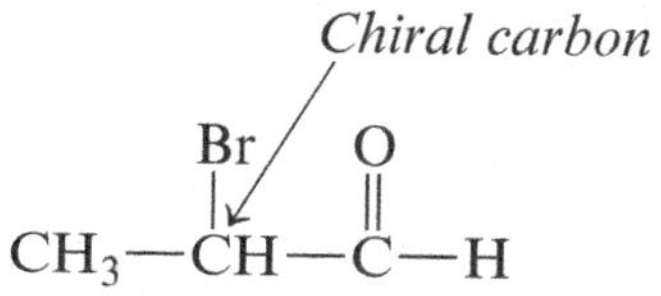

13.13 **a.** Fischer projection: CHO top, HO—C—Br, CH₃ bottom
  b. Fischer projection: CHO top, Cl—C—Br, OH bottom
  c. Fischer projection: CHO top, HO—C—H, CH₂CH₃ bottom

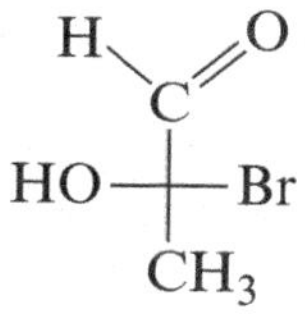

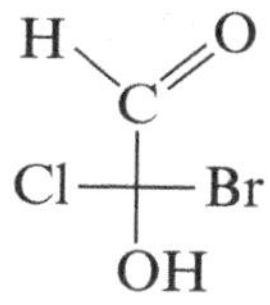

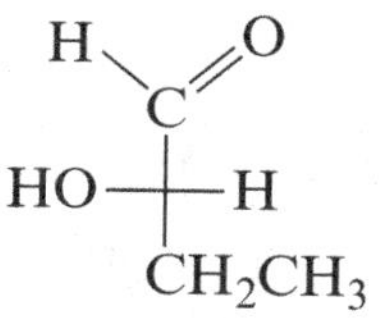

13.15 Enantiomers are nonsuperimposable mirror images.
  a. identical structures (no chiral carbons)
  b. enantiomers
  c. enantiomers
  d. enantiomers

13.17 **a.** When the —OH group is drawn to the right of the chiral carbon, it is the D isomer.
  b. When the —OH group is drawn to the right of the chiral carbon, it is the D isomer.
  c. When the —OH group is drawn to the left of the chiral carbon farthest from the top of the Fischer projection, it is the L isomer.

13.19 **a.** $CH_3-C(CH_3)=CH-CH_2-CH_2-CH(CH_3)-CH_2-CH_2-OH$ (Chiral carbon on the CH)
  b. $H_2N-CH(CH_3)-C(=O)-OH$ (Chiral carbon)

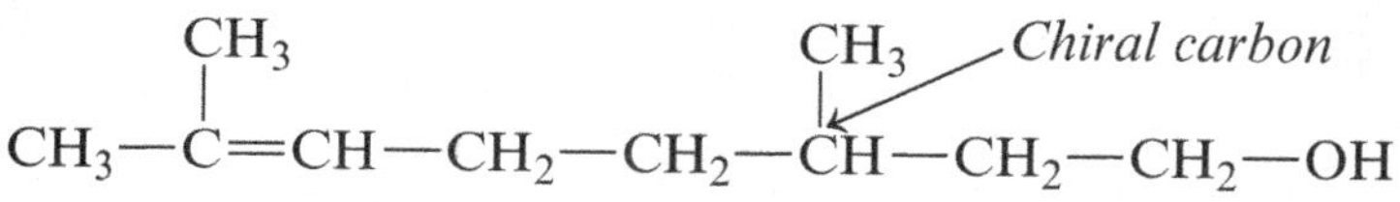

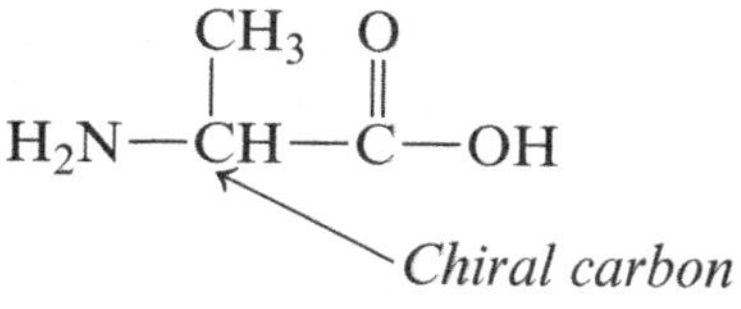

13.21 **a.** This structure is the D enantiomer since the hydroxyl group on the chiral carbon farthest from the carbonyl group is on the right.
  b. This structure is the D enantiomer since the hydroxyl group on the chiral carbon farthest from the carbonyl group is on the right.

c. This structure is the L enantiomer since the hydroxyl group on the chiral carbon farthest from the carbonyl group is on the left.

d. This structure is the D enantiomer since the hydroxyl group on the chiral carbon farthest from the carbonyl group is on the right.

13.23

a.
```
    H   O
     \ //
      C
  H——OH
 HO——H
     CH₂OH
```

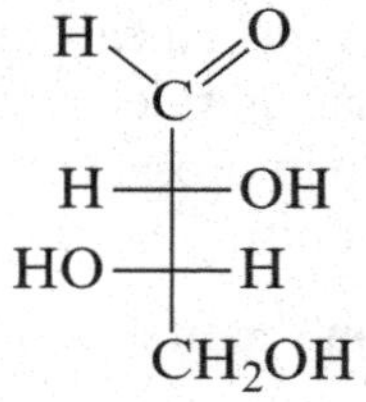

b.
```
    CH₂OH
     |
     C=O
  H——OH
 HO——H
     CH₂OH
```

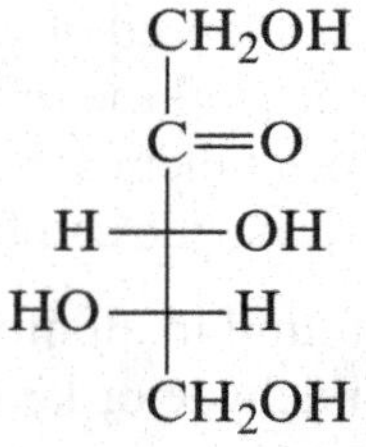

c.
```
    H   O
     \ //
      C
 HO——H
 HO——H
  H——OH
  H——OH
     CH₂OH
```

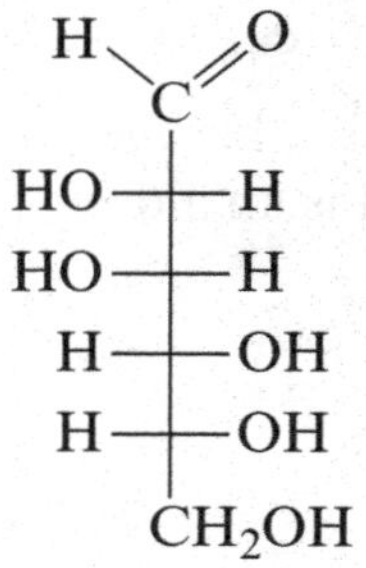

d.
```
    H   O
     \ //
      C
 HO——H
 HO——H
 HO——H
 HO——H
     CH₂OH
```

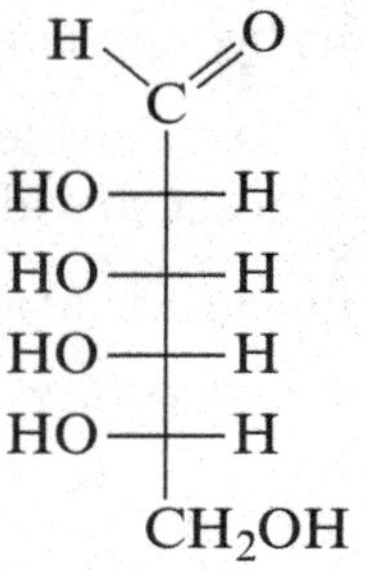

13.25 L-Glucose is the mirror image of D-glucose.

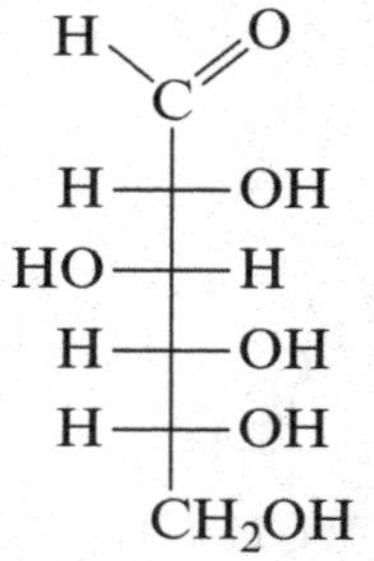

```
    H   O              H   O
     \ //               \ //
      C                  C
  H——OH             HO——H
 HO——H              H——OH
  H——OH             HO——H
  H——OH             HO——H
     CH₂OH              CH₂OH
   D-Glucose          L-Glucose
```

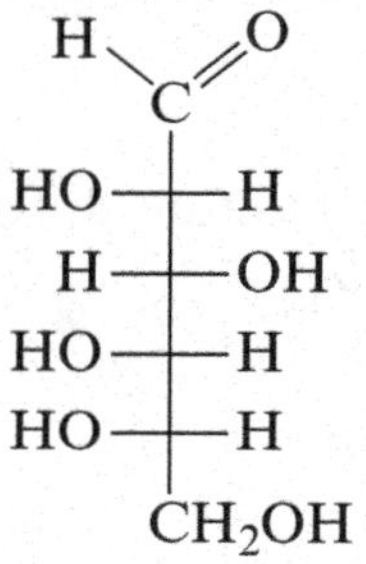

13.27 In D-galactose, the —OH group on carbon 4 extends to the left; in D-glucose, this —OH group goes to the right.

13.29
a. Glucose is also called blood sugar.
b. Galactose is not metabolized in the condition called galactosemia.

13.31 In the cyclic structure of glucose, there are five carbon atoms and an oxygen atom.

13.33 In the α isomer, the —OH group on carbon 1 is drawn down; in the β isomer, the —OH group on carbon 1 is drawn up.

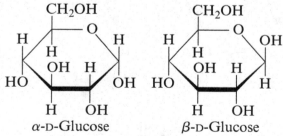
α-D-Glucose   β-D-Glucose

13.35
a. This is the α isomer because the —OH group on carbon 2 is down.
b. This is the α isomer because the —OH group on carbon 1 is down.

13.37 Oxidation product (sugar acid):

```
      O
HO   ‖
  \\ C
H ──┼── OH
HO ─┼── H
H ──┼── OH
    CH₂OH
```
D-Xylonic acid

Reduction product (sugar alcohol):

```
    CH₂OH
H ──┼── OH
HO ─┼── H
H ──┼── OH
    CH₂OH
```
D-Xylitol

13.39 Oxidation product (sugar acid):

```
      O
HO   ‖
  \\ C
HO ─┼── H
H ──┼── OH
H ──┼── OH
    CH₂OH
```
D-Arabinonic acid

Reduction product (sugar alcohol):

```
    CH₂OH
HO ─┼── H
H ──┼── OH
H ──┼── OH
    CH₂OH
```
D-Arabitol

13.41 a. When this disaccharide is hydrolyzed, galactose and glucose are produced. The glycosidic bond is a $\beta(1\rightarrow 4)$ bond since the ether bond is drawn up from carbon 1 of the galactose unit, which is on the left in the drawing, to carbon 4 of the glucose on the right. β-Lactose is the name of this disaccharide since the —OH group on carbon 1 of the glucose unit is drawn up.

b. When this disaccharide is hydrolyzed, two molecules of glucose are produced. The glycosidic bond is an $\alpha(1\rightarrow 4)$ bond since the ether bond is drawn down from carbon 1 of the glucose unit on the left to carbon 4 of the glucose on the right. α-Maltose is the name of this disaccharide since the —OH group on the rightmost glucose unit is drawn down.

13.43 a. β-Lactose is a reducing sugar; the ring on the right can open up to form an aldehyde that can undergo oxidation.

b. α-Maltose is a reducing sugar; the ring on the right can open up to form an aldehyde that can undergo oxidation.

13.45 a. Another name for table sugar is sucrose.
b. Lactose is the disaccharide found in milk and milk products.
c. Maltose is also called malt sugar.
d. When lactose is hydrolyzed, the products are the monosaccharides galactose and glucose.

13.47 a. Amylose is an unbranched polymer of glucose units joined by $\alpha(1\rightarrow 4)$-glycosidic bonds. Amylopectin is a branched polymer of glucose units joined by $\alpha(1\rightarrow 4)$- and $\alpha(1\rightarrow 6)$-glycosidic bonds.

b. Amylopectin, which is produced in plants, is a branched polymer of glucose units joined by $\alpha(1\rightarrow 4)$- and $\alpha(1\rightarrow 6)$-glycosidic bonds. The branches in amylopectin occur about every 25 glucose units. Glycogen, which is produced in animals, is a highly branched polymer of glucose units joined by $\alpha(1\rightarrow 4)$- and $\alpha(1\rightarrow 6)$-glycosidic bonds. The branches in glycogen occur about every 10 to 15 glucose units.

13.49 a. Cellulose is not digestible by humans.
b. Amylose and amylopectin are the storage forms of carbohydrates in plants.
c. Amylose is the polysaccharide that contains only $\alpha(1\rightarrow 4)$-glycosidic bonds.
d. Glycogen is the most highly branched polysaccharide.

13.51 $3.9 \text{ L blood} \times \dfrac{10 \text{ dL blood}}{1 \text{ L blood}} \times \dfrac{178 \text{ mg glucose}}{1 \text{ dL blood}} \times \dfrac{1 \text{ g glucose}}{1000 \text{ mg glucose}} = 6.9 \text{ g of glucose (2 SFs)}$

Chapter 13

13.53 a. Total carbohydrate = 23 g + 24 g + 26 g + 0 g = 73 g (2 SFs)
  Therefore, Kate has exceeded her limit of 45 to 60 g of carbohydrate.
  b. 73 g carbohydrate × $\dfrac{4 \text{ kcal}}{1 \text{ g carbohydrate}}$ = 290 kcal (rounded off to the tens place)

13.55 a. Isomaltose is a disaccharide.
  b. Isomaltose consists of two α-D-glucose units.
  c. The glycosidic link in isomaltose is an α(1→6)-glycosidic bond.
  d. The structure shown is α-isomaltose.
  e. α-Isomaltose is a reducing sugar; the ring on the right can open up to form an aldehyde that can undergo oxidation.

13.57 a. Melezitose is a trisaccharide.
  b. Melezitose contains two units of the aldohexose α-D-glucose and one unit of the ketohexose β-D-fructose.
  c. Melezitose, like sucrose, is not a reducing sugar; the rings on the right cannot open up to form an aldehyde that can undergo oxidation.

13.59 a. Sucrose is the disaccharide found in sugarcane and table sugar.
  b. Cellulose is the structural polysaccharide found in cotton and other plants.

13.61 A chiral carbon is bonded to four different groups.

  a. 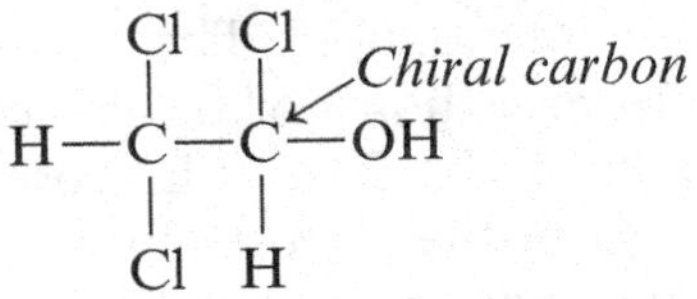 H—C(Cl)(H)—C(Cl)(Cl)—OH *Chiral carbon* (second C)
  b. none
  c. none
  d. 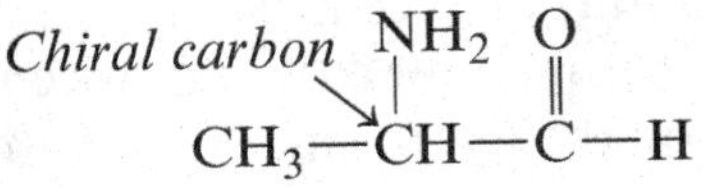 CH₃—CH(NH₂)—C(=O)—H *Chiral carbon* (CH)
  e. 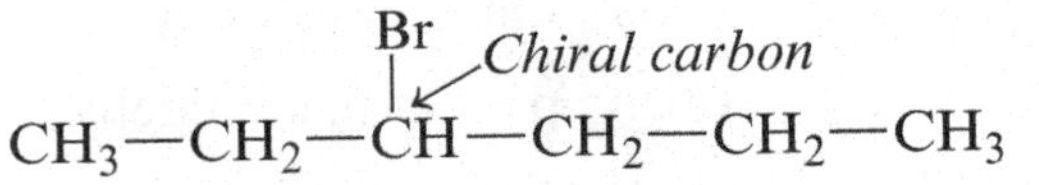 CH₃—CH₂—CH(Br)—CH₂—CH₂—CH₃ *Chiral carbon* (CH)

13.63 Enantiomers are nonsuperimposable mirror images.
  a. identical compounds (no chiral carbons) **b.** enantiomers
  c. enantiomers **d.** identical compounds (no chiral carbons)

13.65 D-Fructose is a ketohexose with the carbonyl group on carbon 2; D-galactose is an aldohexose where the carbonyl group is on carbon 1. In the Fischer projection of D-galactose, the —OH group on carbon 4 is drawn on the left; in fructose, the —OH group on carbon 4 is drawn on the right.

13.67 D-Galactose is the mirror image of L-galactose. In the Fischer projection of D-galactose, the —OH groups on carbon 2 and carbon 5 are drawn on the right side, but they are drawn on the left for carbon 3 and carbon 4. In L-galactose, the —OH groups are reversed: carbon 2 and carbon 5 have —OH groups drawn on the left, and carbon 3 and carbon 4 have —OH groups drawn on the right.

13.69 a.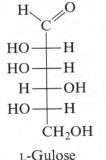

b. 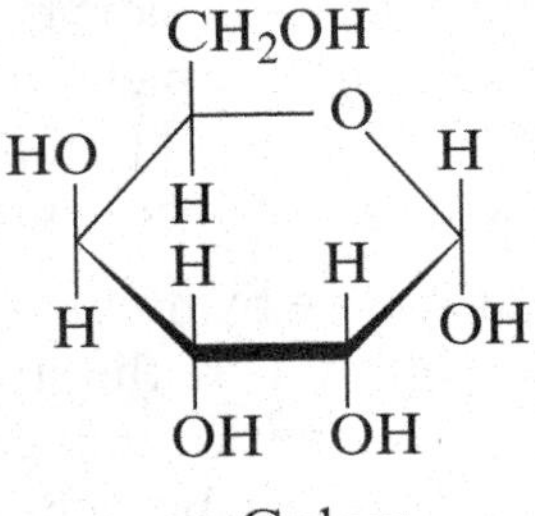 α-D-Gulose 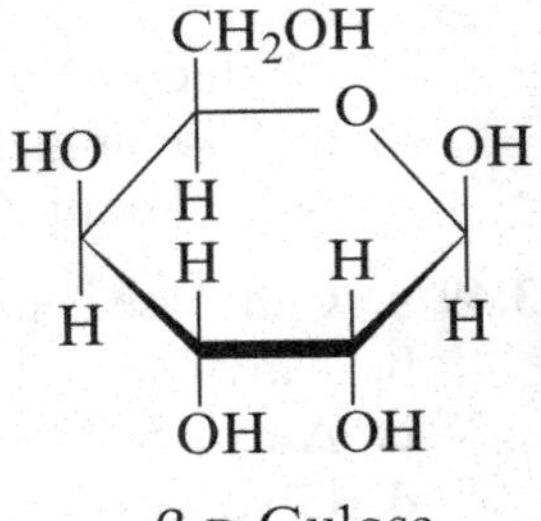   β-D-Gulose

13.71 Since D-sorbitol can be oxidized to D-glucose, it must contain the same number of carbons with the same groups attached as glucose. The difference is that sorbitol has only hydroxyl groups, while glucose has an aldehyde group. In sorbitol, the aldehyde group is changed to a hydroxyl group.

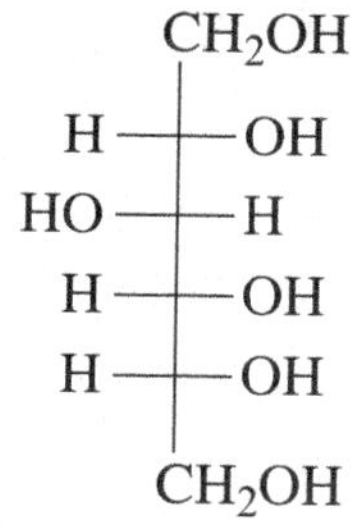

```
        CH₂OH  ←—— This hydroxyl group is an aldehyde in glucose.
   H ——|—— OH
  HO ——|—— H
   H ——|—— OH
   H ——|—— OH
        CH₂OH
      D-Sorbitol
```

13.73 When α-galactose forms an open-chain structure in water, it can close to form either α- or β-galactose.

13.75 [Structure showing a disaccharide with a β(1→4)-Glycosidic bond]

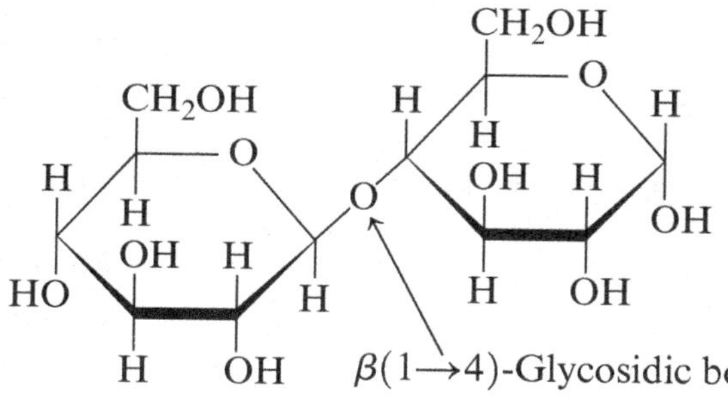

13.77 a. [Structure showing gentiobiose with a β(1→6)-Glycosidic bond]

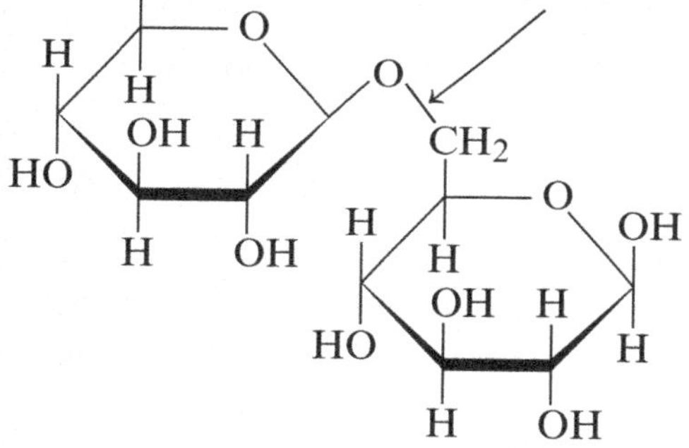

b. Yes. Gentiobiose is a reducing sugar. The ring on the right can open up to form an aldehyde that can undergo oxidation.

14
Carboxylic Acids, Esters, Amines, and Amides

Lance, an environmental health practitioner, has returned to a farm where he found small amounts of fenbendazole, a sheep dewormer, in the soil. When he is working on-site, he wears protective clothing while he obtains samples of soil and water. His recent tests indicate that the presence of fenbendazole is now within the acceptable range in both the soil and the water in the lake near the farm. Lance specializes in measuring levels of contaminants in soil and water caused by manufacturing or farming. In his work, Lance also acts as an advisor, consultant, and safety officer to ensure that soil and water are safe and to maintain protection of the environment.

Circle the amine and amide functional groups in the line-angle formula of fenbendazole.

Credit: Bart Coenders/E+/Getty Images

Fenbendazole

LOOKING AHEAD

14.1 Carboxylic Acids
14.2 Properties of Carboxylic Acids
14.3 Esters
14.4 Hydrolysis of Esters
14.5 Amines
14.6 Amides

The Health icon indicates a question that is related to health and medicine.

14.1 Carboxylic Acids

Learning Goal: Write the IUPAC and common names for carboxylic acids; draw their condensed structural or line-angle formulas.

REVIEW
Naming and Drawing Alkanes (11.2)

- The IUPAC name of a carboxylic acid replaces the *e* of the corresponding alkane with *oic acid*. Simple acids usually are named by the common system using the prefixes *form* (1C), *acet* (2C), *propion* (3C), or *butyr* (4C), followed by *ic acid*.

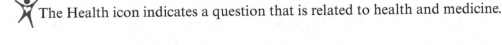

Methanoic acid (formic acid) Ethanoic acid (acetic acid) Butanoic acid (butyric acid) Pentanoic acid

316   Copyright © 2018 Pearson Education, Inc.

Carboxylic Acids, Esters, Amines, and Amides

- The name of the carboxylic acid of benzene is benzoic acid. When substituents are bonded to benzene, the ring is numbered from the carboxylic acid on carbon 1 in the direction that gives the substituents the smallest possible numbers.

Benzoic acid

2,5-Dibromobenzoic acid

The sour taste of vinegar is due to ethanoic acid (acetic acid).
Credit: Pearson Education/Pearson Science

Solution Guide to Naming Carboxylic Acids	
STEP 1	Identify the longest carbon chain and replace the *e* in the corresponding alkane name with *oic acid*.
STEP 2	Name and number any substituents by counting the carboxyl group as carbon 1.

♦ Learning Exercise 14.1A

CORE CHEMISTRY SKILL
Naming Carboxylic Acids

Write the IUPAC name (and common name, if any) for each of the following carboxylic acids:

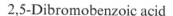

1. $CH_3-\overset{O}{\underset{\|}{C}}-OH$

2. $CH_3-\overset{Br}{\underset{|}{CH}}-\overset{O}{\underset{\|}{C}}-OH$

3. 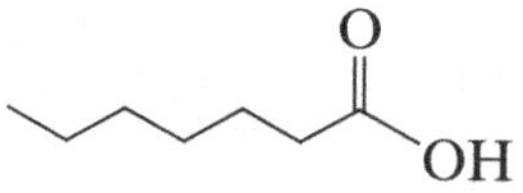 (line-angle structure of hexyl chain ending in COOH)

4. 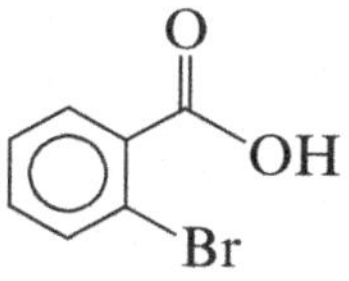 (benzene ring with COOH and Br ortho)

Answers
1. ethanoic acid (acetic acid)
2. 2-bromopropanoic acid
3. heptanoic acid
4. 2-bromobenzoic acid

♦ Learning Exercise 14.1B

Draw the condensed structural formulas for **1, 2,** and **3** and the line-angle formulas for **4, 5,** and **6**.

1. acetic acid
2. 3-chloropropanoic acid
3. formic acid

4. 2-bromobutanoic acid
5. 3-iodobenzoic acid
6. 3-methylpentanoic acid

Chapter 14

Answers

1. $CH_3-\underset{\underset{O}{\parallel}}{C}-OH$

2. $Cl-CH_2-CH_2-\underset{\underset{O}{\parallel}}{C}-OH$

3. $H-\underset{\underset{O}{\parallel}}{C}-OH$

4. 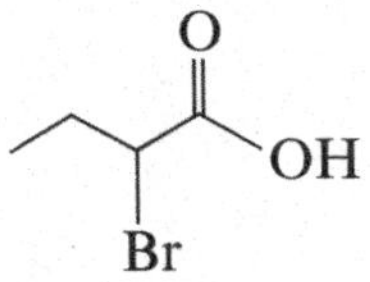 $CH_3-CH(Br)-C(=O)-OH$

5. 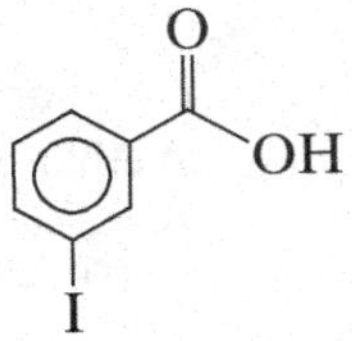 3-iodobenzoic acid

6. 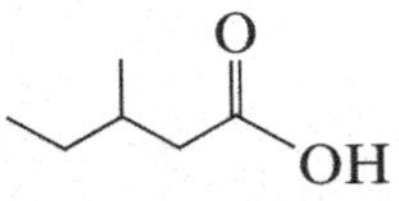 3-methylbutanoic acid

14.2 Properties of Carboxylic Acids

Learning Goal: Describe the solubility, dissociation, and neutralization of carboxylic acids.

REVIEW
Writing Equations for Reactions of Acids and Bases (10.7)

- Carboxylic acids with one to five carbon atoms are soluble in water.
- As weak acids, carboxylic acids dissociate slightly in water to form acidic solutions of H_3O^+ and carboxylate ions.
- When bases neutralize carboxylic acids, the products are carboxylate salts and water.

♦ **Learning Exercise 14.2A**

Indicate whether each of the following carboxylic acids is soluble (S) or not soluble (NS) in water:

1. _____ hexanoic acid
2. _____ acetic acid
3. _____ propanoic acid
4. _____ benzoic acid
5. _____ formic acid
6. _____ octanoic acid

Answers   1. NS   2. S   3. S   4. NS   5. S   6. NS

♦ **Learning Exercise 14.2B**

Draw the condensed structural formulas for the products from the dissociation of each of the following carboxylic acids in water:

1. $CH_3-\underset{\underset{O}{\parallel}}{C}-OH + H_2O \rightleftarrows$

2. 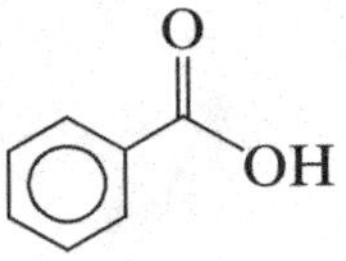 $C_6H_5-C(=O)-OH + H_2O \rightleftarrows$

Answers

1. $CH_3-\underset{\underset{O}{\parallel}}{C}-O^- + H_3O^+$

2. 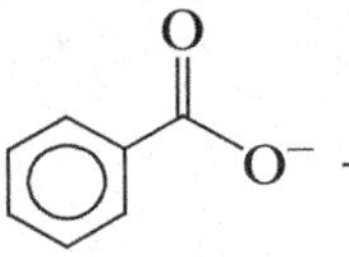 $C_6H_5-C(=O)-O^- + H_3O^+$

318

Carboxylic Acids, Esters, Amines, and Amides

♦ **Learning Exercise 14.2C**

Draw the condensed structural formulas for the products and give the name of the carboxylate salt produced in each of the following reactions:

1. $CH_3-\overset{\overset{O}{\|}}{C}-OH + NaOH \longrightarrow$

2. $H-\overset{\overset{O}{\|}}{C}-OH + KOH \longrightarrow$

Answers

1. $CH_3-\overset{\overset{O}{\|}}{C}-O^- Na^+ + H_2O$
Sodium ethanoate
(sodium acetate)

2. $H-\overset{\overset{O}{\|}}{C}-O^- K^+ + H_2O$
Potassium methanoate
(potassium formate)

14.3 Esters

Learning Goal: Write the IUPAC and common names for an ester; write the balanced chemical equation for the formation of an ester.

- In the presence of a strong acid, a carboxylic acid reacts with an alcohol to produce an ester and water.

$CH_3-\overset{\overset{O}{\|}}{C}-OH + HO-CH_2-CH_3 \underset{}{\overset{H^+, heat}{\rightleftharpoons}} CH_3-\overset{\overset{O}{\|}}{C}-O-CH_2-CH_3 + H_2O$

- The names of esters consist of two words, one from the alcohol and the other from the carboxylic acid, with the *ic acid* ending replaced by *ate*.

$CH_3-\overset{\overset{O}{\|}}{C}-O-CH_3$
Methyl ethanoate
(methyl acetate)

$CH_3-CH_2-\overset{\overset{O}{\|}}{C}-O-CH_2-CH_3$
Ethyl propanoate
(ethyl propionate)

♦ **Learning Exercise 14.3A**

Draw the condensed structural formulas for the products of each of the following esterification reactions:

1. $CH_3-\overset{\overset{O}{\|}}{C}-OH + HO-CH_3 \underset{}{\overset{H^+, heat}{\rightleftharpoons}}$

2. $H-\overset{\overset{O}{\|}}{C}-OH + HO-CH_2-CH_3 \underset{}{\overset{H^+, heat}{\rightleftharpoons}}$

3. 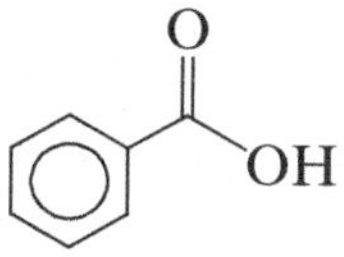 $\text{C}_6\text{H}_5-\overset{\overset{O}{\|}}{C}-OH + HO-CH_3 \underset{}{\overset{H^+, heat}{\rightleftharpoons}}$

4. propanoic acid and ethanol $\underset{}{\overset{H^+, heat}{\rightleftharpoons}}$

Answers

1. $CH_3-\overset{\overset{O}{\|}}{C}-O-CH_3 + H_2O$

2. $H-\overset{\overset{O}{\|}}{C}-O-CH_2-CH_3 + H_2O$

3. 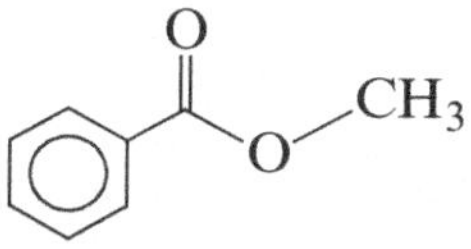 $\text{C}_6\text{H}_5-\overset{\overset{O}{\|}}{C}-O-CH_3 + H_2O$

4. $CH_3-CH_2-\overset{\overset{O}{\|}}{C}-O-CH_2-CH_3 + H_2O$

Chapter 14

	Solution Guide to Naming Esters
STEP 1	Write the name for the carbon chain from the alcohol as an *alkyl* group.
STEP 2	Change the *ic acid* of the acid name to *ate*.

♦ **Learning Exercise 14.3B**

Write the IUPAC and common name, if any, for each of the following:

The odor of grapes is due to an ester.
Credit: Lynn Watson / Shutterstock

1. 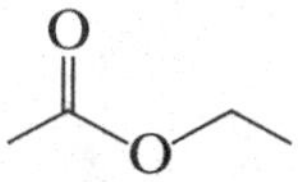 [structure: ethyl ethanoate]

2. $CH_3-CH_2-CH_2-\overset{\overset{O}{\|}}{C}-O-CH_3$

3. $CH_3-CH_2-\overset{\overset{O}{\|}}{C}-O-CH_2-CH_2-CH_3$

4. 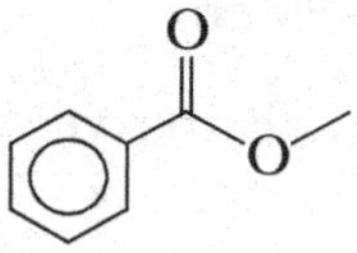 [structure: methyl benzoate]

Answers
1. ethyl ethanoate (ethyl acetate)
2. methyl butanoate (methyl butyrate)
3. propyl propanoate (propyl propionate)
4. methyl benzoate

♦ **Learning Exercise 14.3C**

Draw the condensed structural formulas for **1** and **2** and the line-angle formulas for **3** and **4**.

1. propyl acetate

2. ethyl butyrate

3. ethyl benzoate

4. methyl propanoate

Answers
1. $CH_3-\overset{\overset{O}{\|}}{C}-O-CH_2-CH_2-CH_3$

2. $CH_3-CH_2-CH_2-\overset{\overset{O}{\|}}{C}-O-CH_2-CH_3$

3. 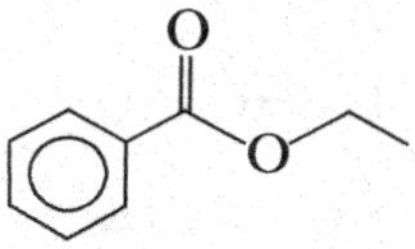 [line-angle structure: ethyl benzoate]

4. 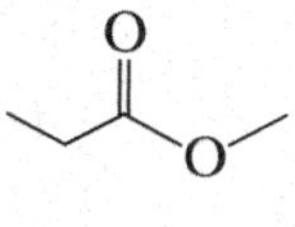 [line-angle structure: methyl propanoate]

14.4 Hydrolysis of Esters

Learning Goal: Draw the condensed structural or line-angle formulas for the products from acid and base hydrolysis of esters.

- In hydrolysis, esters are split apart by a reaction with water. When the catalyst is an acid, the products are a carboxylic acid and an alcohol.

$$CH_3-\underset{\underset{O}{\|}}{C}-O-CH_3 + H_2O \underset{}{\overset{H^+, \text{ heat}}{\rightleftharpoons}} CH_3-\underset{\underset{O}{\|}}{C}-OH + HO-CH_3$$

Methyl ethanoate   Ethanoic acid   Methanol
(methyl acetate)   (acetic acid)   (methyl alcohol)

- Saponification is the hydrolysis of an ester in the presence of a base, which produces a carboxylate salt and an alcohol.

$$CH_3-\underset{\underset{O}{\|}}{C}-O-CH_3 + NaOH \overset{\text{Heat}}{\longrightarrow} CH_3-\underset{\underset{O}{\|}}{C}-O^- Na^+ + HO-CH_3$$

Methyl ethanoate   Sodium ethanoate   Methanol
(methyl acetate)   (sodium acetate)   (methyl alcohol)

♦ Learning Exercise 14.4

CORE CHEMISTRY SKILL
Hydrolyzing Esters

Draw the condensed structural or line-angle formulas for the products of hydrolysis or saponification for each of the following reactions:

1. $CH_3-CH_2-CH_2-\underset{\underset{O}{\|}}{C}-O-CH_3 + H_2O \overset{H^+, \text{ heat}}{\rightleftharpoons}$

2. $CH_3-\underset{\underset{O}{\|}}{C}-O-CH_3 + NaOH \overset{\text{Heat}}{\longrightarrow}$

3. 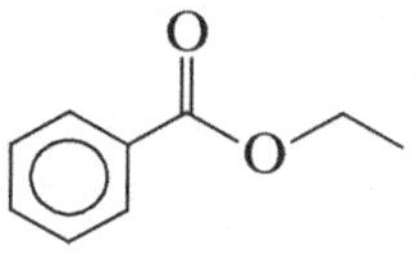 PhC(O)O–CH₂CH₃ + KOH →(Heat)

4. 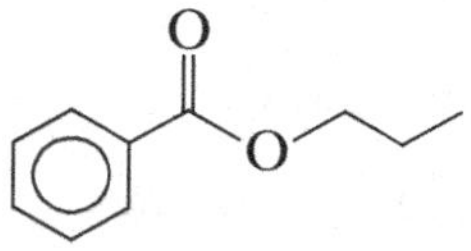 PhC(O)O–CH₂CH₂CH₃ + H₂O ⇌ (H⁺, heat)

Answers

1. $CH_3-CH_2-CH_2-\underset{\underset{O}{\|}}{C}-OH + HO-CH_3$ 2. $CH_3-\underset{\underset{O}{\|}}{C}-O^- Na^+ + HO-CH_3$

3. 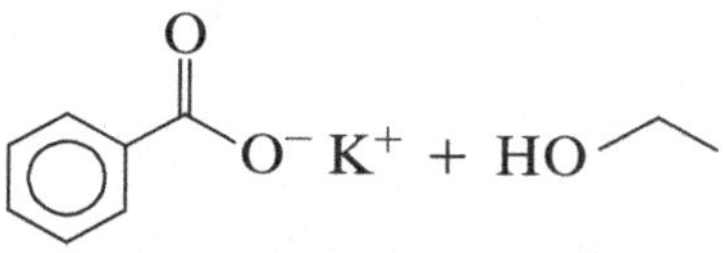 PhC(O)O⁻ K⁺ + HO–CH₂CH₃

4. 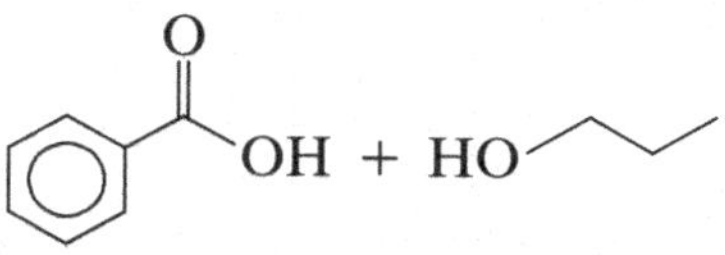 PhC(O)OH + HO–CH₂CH₂CH₃

14.5 Amines

Learning Goal: Write the common names for amines; draw the condensed structural or line-angle formulas when given their names. Describe the solubility, dissociation, and neutralization of amines in water.

Methylamine (1°)

- Amines are derivatives of ammonia (NH_3), in which the alkyl or aromatic groups replace one or more hydrogen atoms.
- In the common names of simple amines, the alkyl groups are listed alphabetically followed by *amine*.
- Amines are classified as primary, secondary, or tertiary when the nitrogen atom is bonded to one, two, or three alkyl or aromatic groups.

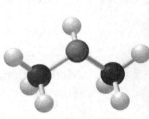
Dimethylamine (2°)

$$\underset{\text{Primary (1°)}}{CH_3-NH_2} \quad \underset{\text{Secondary (2°)}}{\overset{\overset{\displaystyle CH_3}{|}}{CH_3-N-H}} \quad \underset{\text{Tertiary (3°)}}{\overset{\overset{\displaystyle CH_3}{|}}{CH_3-N-CH_3}} \quad \underset{\text{Primary (1°)}}{\overset{NH_2}{\bigcirc}}$$

Trimethylamine (3°)

- The N—H bonds in primary and secondary amines form hydrogen bonds.
- Hydrogen bonding allows amines with up to six carbon atoms to be soluble in water.
- In water, amines act as weak bases by accepting protons from water to produce ammonium and hydroxide ions.

$$\underset{\text{Methylamine}}{CH_3-\ddot{N}H_2} + H_2O \rightleftarrows \underset{\text{Methylammonium ion}}{CH_3-\overset{+}{N}H_3} + \underset{\text{Hydroxide ion}}{OH^-}$$

- Strong acids neutralize amines to yield ammonium salts.

$$\underset{\text{Methylamine}}{CH_3-\ddot{N}H_2} + HCl \longrightarrow \underset{\text{Methylammonium chloride}}{CH_3-\overset{+}{N}H_3\ Cl^-}$$

- Many amines, which are prevalent in synthetic and naturally occurring compounds, have physiological activity.

♦ Learning Exercise 14.5A

Name each of the following amines:

1. $CH_3-\overset{\overset{\displaystyle H}{|}}{N}-CH_2-CH_3$

2. (propyl-N(H)-ethyl line structure)

3. (cyclohexyl-NH_2 line structure)

4. (benzene ring with NH_2 and methyl, meta)

5. $CH_3-CH_2-\overset{\overset{\displaystyle CH_3}{|}}{N}-CH_3$

Answers
1. ethylmethylamine
2. butylethylamine
3. cyclohexylamine
4. 3-methylaniline
5. ethyldimethylamine

Carboxylic Acids, Esters, Amines, and Amides

♦ **Learning Exercise 14.5B**

Classify each of the amines in Learning Exercise 14.5A as primary (1°), secondary (2°), or tertiary (3°).
1. _____   2. _____   3. _____   4. _____   5. _____

Answers   **1.** 2°   **2.** 2°   **3.** 1°   **4.** 1°   **5.** 3°

♦ **Learning Exercise 14.5C**

Draw the condensed structural formulas for **1** and **2** and the line-angle formulas for **3** and **4**.
1. isopropylamine
2. butylethylmethylamine

3. 3-bromoaniline
4. butylamine

Answers

1. 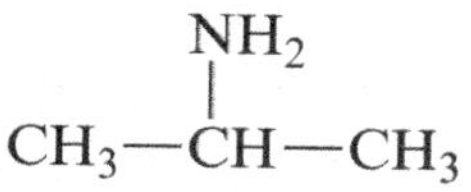 $CH_3-\underset{\underset{NH_2}{|}}{CH}-CH_3$

2. $CH_3-CH_2-\underset{\underset{CH_3}{|}}{N}-CH_2-CH_2-CH_2-CH_3$ (with CH$_3$ on top)

3. 3-aminobromobenzene (NH$_2$ and Br on benzene ring)

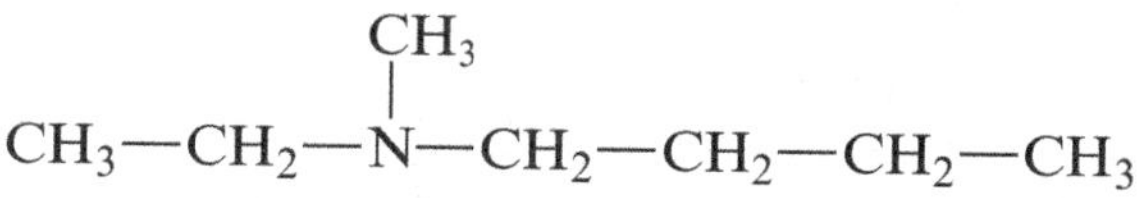

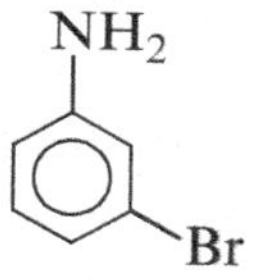

4. 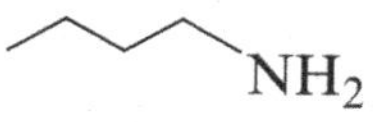 line-angle structure with NH$_2$

♦ **Learning Exercise 14.5D**

Draw the condensed structural or line-angle formula for the products for each of the following reactions:

1. $CH_3-CH_2-NH_2 + H_2O \rightleftharpoons$

2. $CH_3-CH_2-CH_2-NH_2 + HCl \longrightarrow$

3. 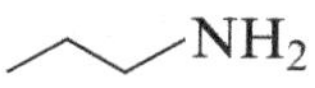 $\sim\!\!\sim\!\!NH_2 + H_2O \rightleftharpoons$

4. 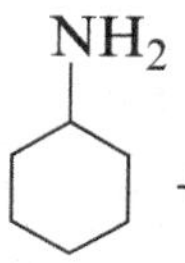 cyclohexylamine + HBr $\longrightarrow$

5. 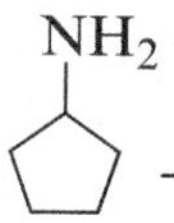 cyclopentylamine + $H_2O \rightleftharpoons$

Answers

1. $CH_3-CH_2-\overset{+}{N}H_3 + OH^-$
2. $CH_3-CH_2-CH_2-\overset{+}{N}H_3\ Cl^-$
3. 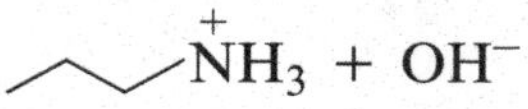 ⌁$\overset{+}{N}H_3 + OH^-$
4. 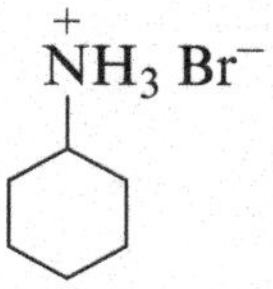 (cyclohexyl)-$\overset{+}{N}H_3\ Br^-$
5. 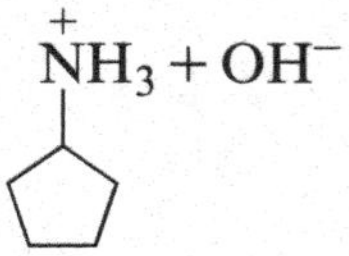 (cyclopentyl)-$\overset{+}{N}H_3 + OH^-$

14.6 Amides

Learning Goal: Write the IUPAC and common names for amides and draw the condensed structural or line-angle formulas for the products of formation and hydrolysis.

- Amides are derivatives of carboxylic acids in which an amine group replaces the —OH group in the carboxylic acid.
- Amides are named by replacing the *ic acid* or *oic acid* ending of the corresponding carboxylic acid with *amide*. When an alkyl group is attached to the N atom, it is listed as *N*-alkyl.

$$CH_3-CH_2-CH_2-\overset{O}{\underset{\|}{C}}-NH_2$$
Butanamide
(butyramide)

$$CH_3-\overset{O}{\underset{\|}{C}}-\overset{H}{\underset{|}{N}}-CH_3$$
N-Methylethanamide
(*N*-methylacetamide)

- When a carboxylic acid reacts with ammonia or an amine, an amide is produced.
- Amides with one to five carbons are soluble in water.
- Amides undergo acid hydrolysis to produce the carboxylic acid and an ammonium salt.

$$CH_3-\overset{O}{\underset{\|}{C}}-NH_2 + H_2O + HCl \xrightarrow{Heat} CH_3-\overset{O}{\underset{\|}{C}}-OH + NH_4^+\ Cl^-$$

- Amides undergo base hydrolysis to produce the carboxylate salt and an amine.

$$CH_3-\overset{O}{\underset{\|}{C}}-NH_2 + NaOH \xrightarrow{Heat} CH_3-\overset{O}{\underset{\|}{C}}-O^-\ Na^+ + NH_3$$

Key Terms for Sections 14.1 to 14.6

Match each of the following key terms with the correct description:

  a. carboxylic acid **b.** saponification **c.** amide **d.** esterification
  e. hydrolysis **f.** ester **g.** amine

1. ____ an organic compound containing the carboxyl group (—COOH)
2. ____ a reaction of a carboxylic acid and an alcohol in the presence of an acid catalyst
3. ____ a derivative of ammonia in which one or more hydrogen atoms is replaced by a carbon
4. ____ a type of organic compound that produces pleasant aromas in flowers and fruits
5. ____ the hydrolysis of an ester with a strong base, producing a salt of the carboxylic acid and an alcohol
6. ____ the splitting of a molecule by the addition of water in the presence of an acid
7. ____ a derivative of a carboxylic acid in which a nitrogen group replaces the hydroxyl group

Answers 1. a 2. d 3. g 4. f 5. b 6. e 7. c

Carboxylic Acids, Esters, Amines, and Amides

♦ **Learning Exercise 14.6A**

CORE CHEMISTRY SKILL
Forming Amides

Draw the condensed structural or line-angle formula for the amide formed in each of the following reactions:

1. $CH_3-CH_2-\overset{\overset{O}{\|}}{C}-OH + NH_3 \xrightarrow{\text{Heat}}$

2. 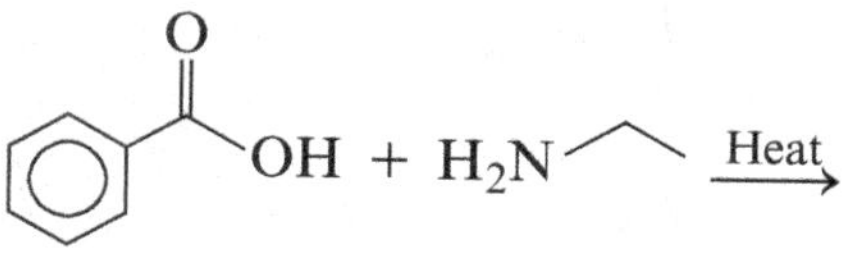 $C_6H_5-\overset{\overset{O}{\|}}{C}-OH + H_2N-CH_2-CH_3 \xrightarrow{\text{Heat}}$

3. $CH_3-\overset{\overset{O}{\|}}{C}-OH + \overset{\overset{CH_3}{|}}{NH}-CH_3 \xrightarrow{\text{Heat}}$

Answers

1. $CH_3-CH_2-\overset{\overset{O}{\|}}{C}-NH_2$

2. 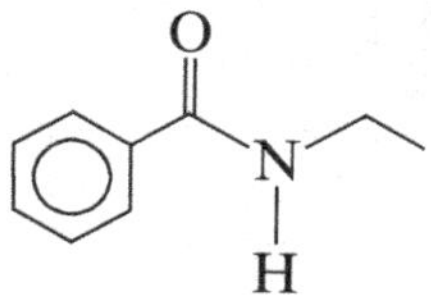 benzamide with $-\overset{\overset{O}{\|}}{C}-\underset{\underset{H}{|}}{N}-CH_2-CH_3$

3. $CH_3-\overset{\overset{O}{\|}}{C}-\overset{\overset{CH_3}{|}}{N}-CH_3$

Solution Guide to Naming Amides	
STEP 1	Replace *oic acid* in the carboxylic acid name with *amide*.
STEP 2	Name each substituent on the N atom using the prefix *N-* and the alkyl name.

♦ **Learning Exercise 14.6B**

Write the IUPAC name (and common name, if any) for each of the following amides:

1. $CH_3-CH_2-\overset{\overset{O}{\|}}{C}-NH_2$

2. 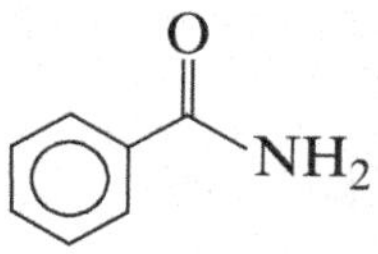 benzamide structure with $-\overset{\overset{O}{\|}}{C}-NH_2$

3. 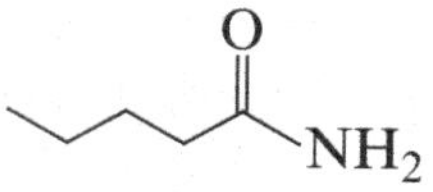 line-angle structure with $-NH_2$

4. $CH_3-\overset{\overset{O}{\|}}{C}-\overset{\overset{H}{|}}{N}-CH_2-CH_3$

Answers
1. propanamide (propionamide)
2. benzamide
3. pentanamide
4. *N*-ethylethanamide (*N*-ethylacetamide)

Chapter 14

♦ **Learning Exercise 14.6C**

Draw the condensed structural formulas for **1** and **2** and the line-angle formulas for **3** and **4**.

1. methanamide
2. butanamide

3. 3-chloropentanamide
4. *N*-ethylbenzamide

Answers

1. $H-\overset{\overset{O}{\|}}{C}-NH_2$

2. $CH_3-CH_2-CH_2-\overset{\overset{O}{\|}}{C}-NH_2$

3. 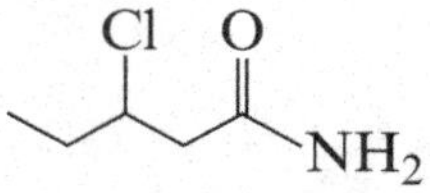 [line-angle structure with Cl on carbon 3 and C(=O)NH₂ group]

4. 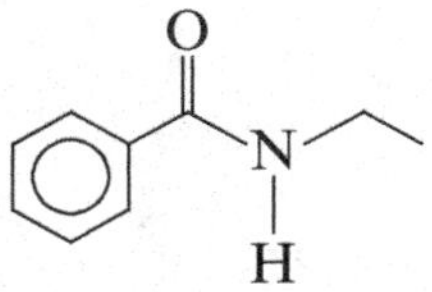 [benzene ring with C(=O)–N(H)–CH₂CH₃ group]

♦ **Learning Exercise 14.6D**

Draw the condensed structural formulas for the products formed by the hydrolysis of each of the following with HCl and with NaOH:

1. $CH_3-CH_2-\overset{\overset{O}{\|}}{C}-NH_2$

2. $CH_3-\overset{\overset{O}{\|}}{C}-\overset{\overset{H}{|}}{N}-CH_2-CH_3$

Answers

1. (HCl) $CH_3-CH_2-\overset{\overset{O}{\|}}{C}-OH + NH_4^+ \; Cl^-$ (NaOH) $CH_3-CH_2-\overset{\overset{O}{\|}}{C}-O^- \; Na^+ + NH_3$

2. (HCl) $CH_3-\overset{\overset{O}{\|}}{C}-OH + CH_3-CH_2-\overset{+}{N}H_3 \; Cl^-$ (NaOH) $CH_3-\overset{\overset{O}{\|}}{C}-O^- \; Na^+ + H_2N-CH_2-$

Checklist for Chapter 14

You are ready to take the Practice Test for Chapter 14. Be sure you have accomplished the following learning goals for this chapter. If not, review the Section listed at the end of the goal. Then apply your new skills and understanding to the Practice Test.

After studying Chapter 14, I can successfully:

____ Write the IUPAC and common names, and draw the condensed structural and line-angle formulas for carboxylic acids. (14.1)

____ Describe the solubility and dissociation of carboxylic acids in water. (14.2)

_____ Describe the behavior of carboxylic acids as weak acids and draw the condensed structural and line-angle formulas for the products of neutralization. (14.2)

_____ Write equations for the preparation of esters. (14.3)

_____ Write the IUPAC or common names and draw the condensed structural and line-angle formulas of esters. (14.3)

_____ Write equations for the hydrolysis and saponification of esters. (14.4)

_____ Write the common names of amines and draw their condensed structural and line-angle formulas. (14.5)

_____ Classify amines as primary, secondary, or tertiary. (14.5)

_____ Write equations for the dissociation and neutralization of amines. (14.5)

_____ Write the IUPAC and common names of amides and draw their condensed structural and line-angle formulas. (14.6)

_____ Write equations for the acid and base hydrolysis of amides. (14.6)

Practice Test for Chapter 14

The chapter Sections to review are shown in parentheses at the end of each question.

For questions 1 through 5, match each condensed structural or line-angle formula to its functional group: (14.1, 14.3)

    A. alcohol   **B.** aldehyde   **C.** carboxylic acid   **D.** ester   **E.** ketone

1. _____ $CH_3-CH(CH_3)-CH_2-OH$

2. _____ $CH_3-CH_2-\underset{\underset{O}{\|}}{C}-OH$

3. _____ $CH_3-CH_2-\underset{\underset{O}{\|}}{C}-H$

4. _____ $CH_3-\underset{\underset{O}{\|}}{C}-O-CH_3$

5. _____ (line-angle structure of a ketone)

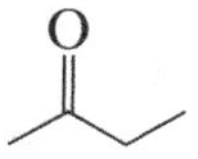

For questions 6 through 10, match the names of the compounds with their condensed structural or line-angle formulas: (14.1, 14.2, 14.3)

A. $CH_3-\underset{\underset{O}{\|}}{C}-O-CH_2-CH_3$ **B.** $CH_3-CH_2-CH_2-\underset{\underset{O}{\|}}{C}-O^-\,Na^+$ **C.** $CH_3-\underset{\underset{O}{\|}}{C}-O^-\,Na^+$

D. 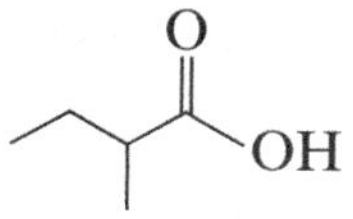 (line-angle structure of 2-methylpropanoic acid with OH)   **E.** $CH_3-CH_2-\underset{\underset{O}{\|}}{C}-O-CH_3$

6. _____ 2-methylbutanoic acid   7. _____ methyl propanoate   8. _____ sodium butanoate

9. _____ ethyl acetate   10. _____ sodium acetate

11. What is the product when a carboxylic acid reacts with sodium hydroxide? (14.2)
    A. carboxylate salt   **B.** alcohol   **C.** ester   **D.** aldehyde   **E.** no reaction

Chapter 14

12. Carboxylic acids are water soluble due to their (14.2)
- **A.** nonpolar nature
- **B.** ionic bonds
- **C.** ability to lower pH
- **D.** ability to hydrogen bond
- **E.** high melting points

13. The products of the reaction are: (14.2)

$$H-\overset{O}{\underset{\|}{C}}-OH + H_2O \rightleftharpoons$$

- **A.** $H-\overset{O}{\underset{\|}{C}}-O^- + H_3O^+$
- **B.** $H-\overset{O}{\underset{\|}{C}}-\overset{+}{O}H_2 + OH^-$
- **C.** $H-\overset{O}{\underset{\|}{C}}-O-CH_3$
- **D.** $CH_3-\overset{O}{\underset{\|}{C}}-OH$
- **E.** $H-\overset{O}{\underset{\|}{C}}-O^- Na^+ + H_2O$

For questions 14 through 17, refer to the following reactions: (14.2, 14.3)

A. $CH_3-\overset{O}{\underset{\|}{C}}-OH + HO-CH_3 \underset{}{\overset{H^+, heat}{\rightleftharpoons}} CH_3-\overset{O}{\underset{\|}{C}}-O-CH_3 + H_2O$

B. $CH_3-\overset{O}{\underset{\|}{C}}-OH + NaOH \overset{Heat}{\longrightarrow} CH_3-\overset{O}{\underset{\|}{C}}-O^- Na^+ + H_2O$

C. $CH_3-\overset{O}{\underset{\|}{C}}-O-CH_3 + H_2O \underset{}{\overset{H^+, heat}{\rightleftharpoons}} CH_3-\overset{O}{\underset{\|}{C}}-OH + HO-CH_3$

D. $CH_3-\overset{O}{\underset{\|}{C}}-O-CH_3 + NaOH \longrightarrow CH_3-\overset{O}{\underset{\|}{C}}-O^- Na^+ + HO-CH_3$

14. _____ is an acid hydrolysis of an ester

15. _____ is a neutralization

16. _____ is a saponification

17. _____ is an esterification

18. What is the name of the organic product of this reaction? (14.3)

$$CH_3-\overset{O}{\underset{\|}{C}}-OH + HO-CH_3 \underset{}{\overset{H^+, heat}{\rightleftharpoons}} CH_3-\overset{O}{\underset{\|}{C}}-O-CH_3 + H_2O$$

- **A.** methyl acetate
- **B.** acetic acid
- **C.** methyl alcohol
- **D.** acetaldehyde
- **E.** ethyl methanoate

19. What is the IUPAC name of the following compound? (14.3)

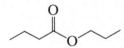

- **A.** butanoic acid
- **B.** butyl propionate
- **C.** butyl propanoate
- **D.** propyl butanoate
- **E.** propyl butyrate

20. The name of $CH_3-CH_2-\overset{O}{\underset{\|}{C}}-O-CH_2-CH_3$ is (14.3)

- **A.** ethyl acetate
- **B.** ethyl ethanoate
- **C.** ethyl propanoate
- **D.** propyl ethanoate
- **E.** ethyl butyrate

Carboxylic Acids, Esters, Amines, and Amides

21. The ester produced from the reaction of 1-butanol and propanoic acid is (14.3)
  A. butyl propanoate **B.** butyl propanone **C.** propyl butyrate
  D. propyl butanone **E.** heptanoate

22. The reaction of methyl acetate with NaOH produces (14.3)
  A. ethanol and formic acid **B.** ethanol and sodium formate
  C. ethanol and sodium ethanoate **D.** methanol and acetic acid
  E. methanol and sodium acetate

23. Esters (14.3)
  A. have pleasant odors
  B. can undergo hydrolysis
  C. are formed from alcohols and carboxylic acids
  D. have IUPAC names ending in *oate*
  E. all of the above

24. Identify the carboxylic acid and alcohol needed to produce the compound shown below (14.3)

$$CH_3-CH_2-CH_2-\overset{\overset{O}{\|}}{C}-O-CH_2-CH_3$$

  A. propanoic acid and ethanol **B.** acetic acid and 1-pentanol
  C. acetic acid and 1-butanol **D.** butanoic acid and ethanol
  E. hexanoic acid and methanol

25. What are the IUPAC names of the products of the acid hydrolysis of: (14.4)

$$H-\overset{\overset{O}{\|}}{C}-O-CH_2-CH_2-CH_3$$

  A. propanoic acid and methanol **B.** methanoic acid and propanol
  C. ethanoic acid and methanol **D.** ethanoic acid and propanol
  E. propanoic acid and ethanol

26. In a hydrolysis reaction, (14.4)
  A. an acid reacts with an alcohol **B.** an ester reacts with NaOH
  C. an ester reacts with H_2O **D.** an acid neutralizes a base
  E. water is added to an alkene

*For questions 27 through 30, classify each amine as **A**, **B**, or **C**. (14.5)*
  A. primary amine **B.** secondary amine **C.** tertiary amine

27. ____ $CH_3-\overset{\overset{CH_3}{|}}{CH}-NH_2$

28. ____ (CH₃)₂N-CH₃ structure with N

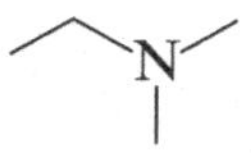

29. ____ $CH_3-CH_2-\overset{\overset{NH_2}{|}}{CH}-CH_2-CH_3$

30. ____ $CH_3-\overset{\overset{H}{|}}{N}-CH_2-CH_3$

31. Amines used in drugs are converted to their amine salt because the salt is (14.5)
  A. a solid at room temperature **B.** soluble in water
  C. odorless **D.** soluble in body fluids
  E. all of these

Chapter 14

For questions 32 through 34, match the structure with the name **A** *to* **E**. (14.5, 14.6)

    A. ethyldimethylamine   **B.** butanamide   **C.** *N*-methylacetamide
    D. benzamide   **E.** *N*-ethylbutyramide

32. $CH_3-CH_2-\underset{\underset{CH_3}{|}}{N}-CH_3$ 33. $CH_3-CH_2-CH_2-\overset{O}{\underset{\|}{C}}-NH_2$ 34. 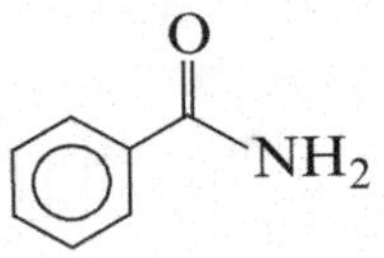 $C_6H_5-\overset{O}{\underset{\|}{C}}-NH_2$

For questions 35 through 38, identify the product as **A** *to* **D**. (14.5, 14.6)

    A. $CH_3-CH_2-\overset{O}{\underset{\|}{C}}-OH + \overset{+}{N}H_4\ Cl^-$ **B.** $CH_3-CH_2-\overset{+}{N}H_3\ Cl^-$

    C. $CH_3-CH_2-CH_2-\overset{+}{N}H_3 + OH^-$ **D.** $CH_3-\overset{O}{\underset{\|}{C}}-NH_2$

35. _____ dissociation of propylamine in water

36. _____ hydrolysis of propanamide with hydrochloric acid

37. _____ reaction of ethylamine and hydrochloric acid

38. _____ amidation of acetic acid

Answers to the Practice Test

1. A	2. C	3. B	4. D	5. E
6. D	7. E	8. B	9. A	10. C
11. A	12. D	13. A	14. C	15. B
16. D	17. A	18. A	19. D	20. C
21. A	22. E	23. E	24. D	25. B
26. C	27. A	28. C	29. A	30. B
31. E	32. A	33. B	34. D	35. C
36. A	37. B	38. D		

Carboxylic Acids, Esters, Amines, and Amides

Selected Answers and Solutions to Text Problems

14.1 Methanoic acid (formic acid) is the carboxylic acid that is responsible for the pain associated with ant stings.

14.3
a. Ethanoic acid (acetic acid) is the carboxylic acid with two carbons.
b. Butanoic acid (butyric acid) is the four-carbon carboxylic acid.
c. 3-Methylhexanoic acid is a six-carbon carboxylic acid with a methyl group on carbon 3 of the chain.
d. 3-Bromobenzoic acid has a carboxylic acid group on the benzene ring and a bromine atom on carbon 3.

14.5
a. 3,4-Dibromobutanoic acid is a carboxylic acid that has a four-carbon chain with bromine atoms on carbon 3 and carbon 4.

$$Br-CH_2-\underset{Br}{CH}-CH_2-\underset{O}{\overset{\|}{C}}-OH$$

b. 2-Chloropropanoic acid is a carboxylic acid that has a three-carbon chain with a chlorine atom on carbon 2.

$$CH_3-\underset{Cl}{CH}-\underset{O}{\overset{\|}{C}}-OH$$

c. Benzoic acid is the carboxylic acid of benzene.

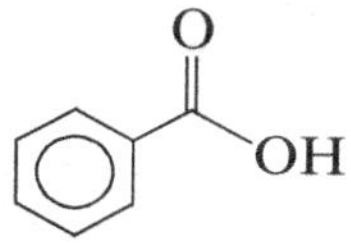

d. Heptanoic acid is a carboxylic acid that has a seven-carbon chain.

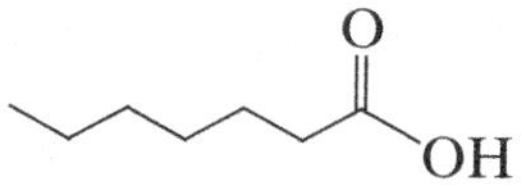

14.7
a. Propanoic acid is the most soluble of the group because it has the fewest number of carbon atoms in its hydrocarbon chain. Solubility of carboxylic acids decreases as the number of carbon atoms in the hydrocarbon chain increases.
b. Propanoic acid is more soluble than 1-hexanol because it has fewer carbon atoms in its hydrocarbon chain. Propanoic acid is also more soluble because the carboxyl group forms more hydrogen bonds with water than does the hydroxyl group of an alcohol. An alkane is not soluble in water.

14.9
a. $CH_3-CH_2-CH_2-CH_2-\overset{O}{\overset{\|}{C}}-OH + H_2O \rightleftharpoons CH_3-CH_2-CH_2-CH_2-\overset{O}{\overset{\|}{C}}-O^- + H_3O^+$

b. $CH_3-\overset{O}{\overset{\|}{C}}-OH + H_2O \rightleftharpoons CH_3-\overset{O}{\overset{\|}{C}}-O^- + H_3O^+$

14.11
a. $H-\overset{O}{\overset{\|}{C}}-OH + NaOH \longrightarrow H-\overset{O}{\overset{\|}{C}}-O^-Na^+ + H_2O$

b. $Cl-CH_2-CH_2-\overset{O}{\overset{\|}{C}}-OH + NaOH \longrightarrow Cl-CH_2-CH_2-\overset{O}{\overset{\|}{C}}-O^-Na^+ + H_2O$

c. (benzoic acid) + NaOH ⟶ (sodium benzoate) + H_2O

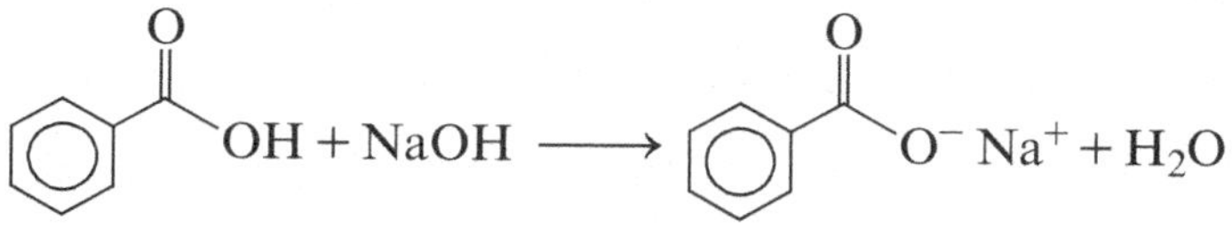

Chapter 14

14.13 a. sodium methanoate (sodium formate)
   b. sodium 3-chloropropanoate
   c. sodium benzoate

14.15 A carboxylic acid reacts with an alcohol to form an ester and water. In an ester, the —H of the carboxylic acid is replaced by an alkyl group.

   a. CH$_3$—C(=O)—O—CH$_3$

   b. CH$_3$—CH$_2$—CH$_2$—CH$_2$—C(=O)—O—CH$_3$

14.17 A carboxylic acid and an alcohol react to give an ester with the elimination of water.

   a. CH$_3$—CH$_2$—CH$_2$—CH$_2$—C(=O)—O—CH(CH$_3$)—CH$_3$

   b. [structure: propanoate propyl ester]

14.19 a. The alcohol part of the ester is from methanol (methyl alcohol) and the carboxylic acid part is from methanoic acid (formic acid). The ester is named methyl methanoate (methyl formate).
   b. The alcohol part of the ester is from ethanol (ethyl alcohol) and the carboxylic acid part is from propanoic acid (propionic acid). The ester is named ethyl propanoate (ethyl propionate).
   c. The alcohol part of the ester is from propanol (propyl alcohol) and the carboxylic acid part is from ethanoic acid (acetic acid). The ester is named propyl ethanoate (propyl acetate).

14.21 a. The alkyl part of the ester comes from the three-carbon 1-propanol and the carboxylate part from the four-carbon butyric acid.

   CH$_3$—CH$_2$—CH$_2$—C(=O)—O—CH$_2$—CH$_2$—CH$_3$

   b. The alkyl part of the ester comes from the four-carbon 1-butanol and the carboxylate part from the one-carbon formic acid.

   H—C(=O)—O—CH$_2$—CH$_2$—CH$_2$—CH$_3$

   c. The alkyl part of the ester comes from the two-carbon ethanol and the carboxylate part from the five-carbon pentanoic acid.

   [structure: ethyl pentanoate]

   d. The alkyl part of the ester comes from the one-carbon methanol and the carboxylate part from the three-carbon propanoic acid.

   [structure: methyl propanoate]

14.23 a. The flavor and odor of bananas is pentyl ethanoate (pentyl acetate).
   b. The flavor and odor of oranges is octyl ethanoate (octyl acetate).
   c. The flavor and odor of apricots is pentyl butanoate (pentyl butyrate).

14.25 Acid hydrolysis of an ester adds water in the presence of acid and gives a carboxylic acid and an alcohol.

14.27 Acid hydrolysis of an ester gives the carboxylic acid and the alcohol which were combined to form the ester; base hydrolysis of an ester gives the salt of carboxylic acid and the alcohol which were combined to form the ester.

a. $CH_3-CH_2-\overset{O}{\underset{\|}{C}}-O^-\ Na^+ + HO-CH_3$

b. $CH_3-\overset{O}{\underset{\|}{C}}-OH + HO-CH_2-CH_2-CH_3$

c. $CH_3-CH_2-CH_2-\overset{O}{\underset{\|}{C}}-OH + HO-CH_2-CH_3$

d. $C_6H_5-\overset{O}{\underset{\|}{C}}-O^-\ Na^+ + HO-CH_2-CH_3$

14.29 The common name of an amine consists of naming the alkyl groups bonding to the nitrogen atom in alphabetical order.
  a. A three-carbon alkyl group attached to $-NH_2$ is propylamine.
  b. A one-carbon and a three-carbon alkyl group attached to nitrogen form methylpropylamine.
  c. A one-carbon and two two-carbon alkyl groups attached to nitrogen form diethylmethylamine.

14.31 a. $CH_3-CH_2-NH_2$

b. $C_6H_5-N(H)-CH_3$

c. $CH_3-CH_2-CH_2-CH_2-\overset{H}{\underset{|}{N}}-CH_2-CH_2-CH_3$

14.33 a. This is a primary (1°) amine; there is only one alkyl group attached to the nitrogen atom.
  b. This is a secondary (2°) amine; there are two alkyl groups attached to the nitrogen atom.
  c. This is a tertiary (3°) amine; three are two alkyl groups and one aromatic group attached to the nitrogen atom.
  d. This is a tertiary (3°) amine; there are three alkyl groups attached to the nitrogen atom.

14.35 a. Yes; amines with fewer than seven carbon atoms hydrogen bond with water molecules and are soluble in water.
  b. Yes; amines with fewer than seven carbon atoms hydrogen bond with water molecules and are soluble in water.
  c. No; an amine with eight carbon atoms has large hydrocarbon sections that make it insoluble in water.
  d. Yes; amines with fewer than seven carbon atoms hydrogen bond with water molecules and are soluble in water.

14.37 Amines, which are weak bases, accept a hydrogen ion from water to give a hydroxide ion and the related ammonium ion. When neutralized with acid, amines form water and the related amine salt.

  a. (1) $CH_3-NH_2 + H_2O \rightleftharpoons CH_3-\overset{+}{N}H_3 + OH^-$

   (2) $CH_3-NH_2 + HCl \longrightarrow CH_3-\overset{+}{N}H_3\ Cl^-$

b. (1) $CH_3-\underset{\underset{H}{|}}{N}-CH_3 + H_2O \rightleftharpoons CH_3-\overset{+}{N}H_2-CH_3 + OH^-$

(2) $CH_3-\underset{\underset{H}{|}}{N}-CH_3 + HCl \longrightarrow CH_3-\overset{+}{N}H_2-CH_3\ Cl^-$

c. (1) $C_6H_5-NH_2 + H_2O \rightleftharpoons C_6H_5-\overset{+}{N}H_3 + OH^-$

(2) $C_6H_5-NH_2 + HCl \longrightarrow C_6H_5-\overset{+}{N}H_3\ Cl^-$

14.39 Carboxylic acids react with amines to form amides with the elimination of water.

a. $CH_3-\underset{\underset{}{\overset{\overset{O}{\|}}{}}}{C}-NH_2$

b. $CH_3-\underset{\underset{}{\overset{\overset{O}{\|}}{}}}{C}-\underset{\underset{H}{|}}{N}-CH_2-CH_3$

c. $C_6H_5-\underset{\underset{}{\overset{\overset{O}{\|}}{}}}{C}-\underset{\underset{H}{|}}{N}-CH_2CH_2CH_3$

14.41 a. Ethanamide (acetamide) has a two-carbon carbonyl portion bonded to an amino group.
b. Butanamide (butyramide) has a four-carbon carbonyl portion bonded to an amino group.
c. *N*-Ethylmethanamide (*N*-ethylformamide); the "*N*-ethyl" means that there is a two-carbon alkyl group attached to the nitrogen. The "methanamide" component tells us that the carbonyl portion has one carbon atom.

14.43 a. Propionamide is the common name of the amide of propanoic acid, which has three carbon atoms.

$CH_3-CH_2-\underset{\underset{}{\overset{\overset{O}{\|}}{}}}{C}-NH_2$

b. 2-Methylpentanamide is an amide of the five-carbon pentanoic acid with a methyl group attached to carbon 2.

$CH_3-CH_2-CH_2-\underset{\underset{}{\overset{\overset{CH_3}{|}}{}}}{CH}-\underset{\underset{}{\overset{\overset{O}{\|}}{}}}{C}-NH_2$

c. Methanamide is the simplest of the amides, with only one carbon atom.

$H-\underset{\underset{}{\overset{\overset{O}{\|}}{}}}{C}-NH_2$

d. *N,N*-Dimethylethanamide is an amide of the two-carbon ethanoic acid, with two one-carbon alkyl groups attached to the nitrogen.

$CH_3-\underset{\underset{}{\overset{\overset{O}{\|}}{}}}{C}-\underset{\underset{}{\overset{\overset{CH_3}{|}}{}}}{N}-CH_3$

14.45 Acid hydrolysis of an amide gives the carboxylic acid and the ammonium salt.

a. $CH_3-\underset{\underset{}{\overset{\overset{O}{\|}}{}}}{C}-OH + \overset{+}{N}H_4\ Cl^-$

b. $CH_3-CH_2-\underset{\underset{}{\overset{\overset{O}{\|}}{}}}{C}-OH + \overset{+}{N}H_4\ Cl^-$

c. $CH_3-CH_2-CH_2-\overset{\overset{O}{\|}}{C}-OH + CH_3-\overset{+}{N}H_3\ Cl^-$

d. 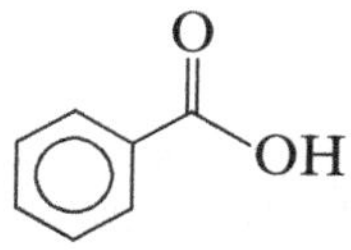 $C_6H_5-\overset{\overset{O}{\|}}{C}-OH + \overset{+}{N}H_4\ Cl^-$

e. $CH_3-CH_2-CH_2-CH_2-\overset{\overset{O}{\|}}{C}-OH + CH_3-CH_2-\overset{+}{N}H_3\ Cl^-$

14.47 a. amine

b. $70.\ \text{kg body mass} \times \dfrac{65\ \text{mg dicyclanil}}{1\ \text{kg body mass}} \times \dfrac{1\ \text{mL spray}}{50.\ \text{mg dicyclanil}}$
= 91 mL of dicyclanil spray (2 SFs)

14.49 $CH_3-CH_2-CH_2-\overset{\overset{O}{\|}}{C}-OH$ Butanoic acid

$CH_3-\overset{\overset{CH_3}{|}}{CH}-\overset{\overset{O}{\|}}{C}-OH$ 2-Methylpropanoic acid

14.51 a. $CH_3-CH_2-CH_2-\overset{\overset{O}{\|}}{C}-O-CH_3$

b. butanoic acid and methanol

c. $CH_3-CH_2-CH_2-\overset{\overset{O}{\|}}{C}-O-CH_3 + H_2O \xrightleftharpoons{H^+,\ \text{heat}}$
$CH_3-CH_2-CH_2-\overset{\overset{O}{\|}}{C}-OH + HO-CH_3$

14.53 phenol, alcohol, amine

14.55 $CH_3-CH_2-CH_2-NH_2$ $CH_3-CH_2-\overset{\overset{H}{|}}{N}-CH_3$
Propylamine (1°)   Ethylmethylamine (2°)

$CH_3-\overset{\overset{CH_3}{|}}{N}-CH_3$ $CH_3-\overset{\overset{CH_3}{|}}{CH}-NH_2$
Trimethylamine (3°)   Isopropylamine (1°)

14.57 a. 3-methylbutanoic acid   **b.** ethyl benzoate
c. ethyl propanoate (ethyl propionate)   **d.** 2-chlorobenzoic acid
e. pentanoic acid

14.59 a. $CH_3-\overset{\overset{O}{\|}}{C}-O-CH_3$

b. $CH_3-CH_2-\overset{\overset{Cl}{|}}{\underset{\underset{Cl}{|}}{C}}-\overset{\overset{O}{\|}}{C}-OH$

c. 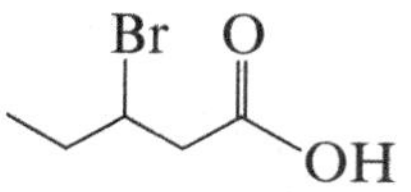 $CH_3-CH_2-\overset{\overset{Br}{|}}{CH}-CH_2-\overset{\overset{O}{\|}}{C}-OH$

d. 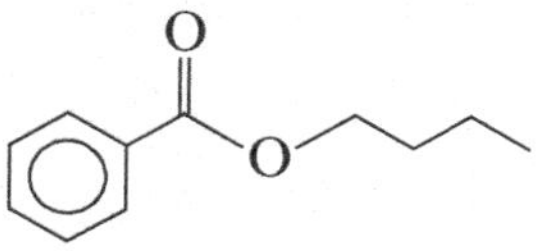 $C_6H_5-\overset{\overset{O}{\|}}{C}-O-CH_2-CH_2-CH_2-CH_3$

14.61 a. $CH_3-CH_2-\overset{\overset{O}{\|}}{C}-O^- K^+ + H_2O$

b. $CH_3-CH_2-\overset{\overset{O}{\|}}{C}-O-CH_3 + H_2O$

c. 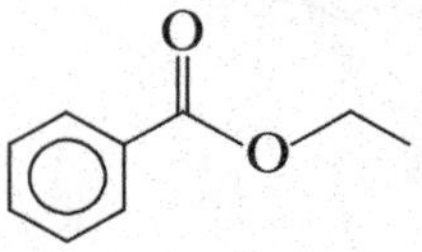 [benzoic acid ethyl ester structure: phenyl–C(=O)–O–CH₂CH₃]

d. $CH_3-CH_2-\overset{\overset{O}{\|}}{C}-\underset{\underset{H}{|}}{N}-CH_3 + H_2O$

14.63 a. $CH_3-CH_2-\overset{\overset{O}{\|}}{C}-OH + HO-\overset{\overset{CH_3}{|}}{CH}-CH_3$

b. 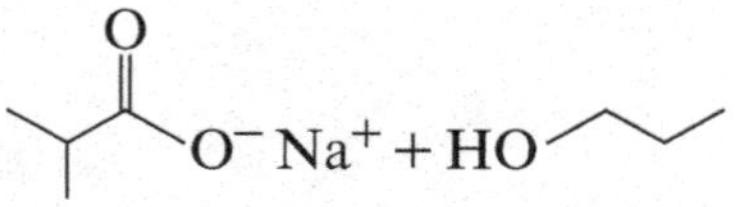 [CH₃CH(–)C(=O)–O⁻ Na⁺ + HO–CH₂CH₂CH₃]

c. [phenyl–C(=O)–O⁻ Na⁺ + H₂N–CH₂CH₂CH₃]

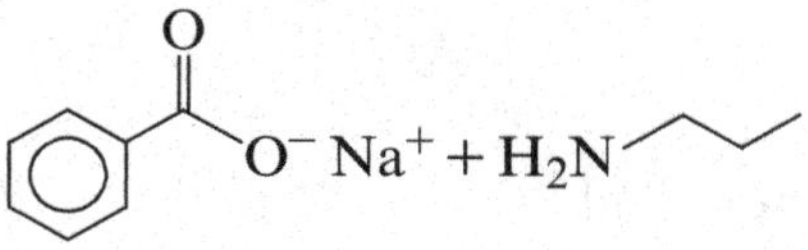

14.65 a. ethylmethylamine (2°)   **b.** *N*-ethylaniline (2°)   **c.** butylamine (1°)

14.67 a. Two methyl groups are bonded to a nitrogen atom.

$CH_3-\overset{\overset{H}{|}}{N}-CH_3$

b. The amino group is attached to a six-carbon cycloalkane ring.

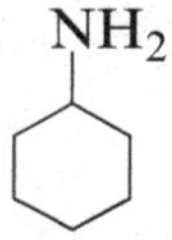

[cyclohexane with NH₂ substituent]

c. This is an ammonium salt with two methyl groups bonded to the nitrogen atom.

$CH_3-\overset{\overset{CH_3}{|}}{\overset{+}{N}H_2}\ Cl^-$

d. A two-carbon ethyl group is bonded to the nitrogen atom of aniline.

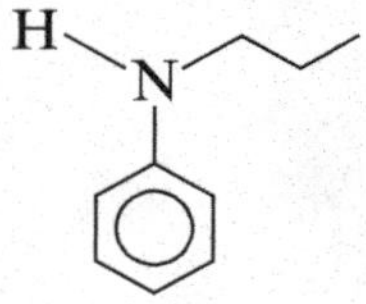

[phenyl–N(H)–CH₂CH₃]

14.69 a. An amine accepts a hydrogen ion from water, which produces an ammonium ion and OH⁻.

$CH_3-CH_2-\overset{+}{N}H_3 + OH^-$

b. The amine accepts a hydrogen ion to give an ammonium salt.

$CH_3-CH_2-\overset{+}{N}H_3\ Cl^-$

14.71 a. *N*-Ethylethanamide; the "*N*-ethyl" means that there is a two-carbon alkyl group attached to the nitrogen. The "ethanamide" component tells us that the carbonyl portion has two carbon atoms.
  b. Propanamide has a three-carbon carbonyl portion bonded to an amino group.
  c. 3-Methylbutanamide has a four-carbon chain with a methyl substituent on carbon 3 as its carbonyl portion bonded to an amino group.

Carboxylic Acids, Esters, Amines, and Amides

14.73 Acid hydrolysis of an amide gives the carboxylic acid and the ammonium salt. Base hydrolysis of an amide yields the carboxylate salt and the amine.

a. $CH_3-\overset{\overset{O}{\|}}{C}-OH + CH_3-\overset{+}{N}H_3\ Cl^-$

b. $H-\overset{\overset{O}{\|}}{C}-O^-\ Na^+ + NH_3$

c. $\overset{\overset{O}{\|}}{\diagup\!\!\!\diagdown}OH + \overset{+}{N}H_4\ Cl^-$

14.75 aromatic, amine, carboxylate salt

14.77 a. $\diagup\!\!\!\diagdown\!\!\!\diagup\!\!\!\diagdown\overset{\overset{O}{\|}}{C}-OH$ Hexanoic acid

b. $\diagup\!\!\!\diagdown\!\!\!\overset{\diagup}{\underset{\diagdown}{C}}\overset{\overset{O}{\|}}{C}-OH$ 2-Ethylbutanoic acid

14.79 a. $CH_3-\overset{\overset{O}{\|}}{C}-O-CH_2-CH_2-CH_3$

b. $CH_3-\overset{\overset{O}{\|}}{C}-OH + HO-CH_2-CH_2-CH_3 \underset{}{\overset{H^+,\ heat}{\rightleftharpoons}} CH_3-\overset{\overset{O}{\|}}{C}-O-CH_2-CH_2-CH_3 + H_2O$

c. $CH_3-\overset{\overset{O}{\|}}{C}-O-CH_2-CH_2-CH_3 + H_2O \underset{}{\overset{H^+,\ heat}{\rightleftharpoons}} CH_3-\overset{\overset{O}{\|}}{C}-OH + HO-CH_2-CH_2-CH_3$

d. $CH_3-\overset{\overset{O}{\|}}{C}-O-CH_2-CH_2-CH_3 + NaOH \overset{Heat}{\longrightarrow} CH_3-\overset{\overset{O}{\|}}{C}-O^-\ Na^+$
$+ HO-CH_2-CH_2-CH_3$

e. Molar mass of propyl acetate ($C_5H_{10}O_2$)
= $5(12.01\ g) + 10(1.008\ g) + 2(16.00\ g)$ = 102.13 g/mole

$1.58\ g\ \cancel{C_5H_{10}O_2} \times \dfrac{1\ mole\ \cancel{C_5H_{10}O_2}}{102.13\ g\ \cancel{C_5H_{10}O_2}} \times \dfrac{1\ mole\ \cancel{NaOH}}{1\ mole\ \cancel{C_5H_{10}O_2}} \times \dfrac{1000\ mL\ solution}{0.208\ \cancel{mole\ NaOH}}$

= 74.4 mL of a 0.208 M NaOH solution (3 SFs)

14.81 a. $H_2N-\!\!\bigcirc\!\!-\overset{\overset{O}{\|}}{C}-O-CH_2-CH_2-\overset{+}{N}\overset{\diagup CH_2-CH_3}{\underset{\diagdown CH_2-CH_3}{-H\ Cl^-}}$

b. The ammonium salt (Novocaine) is more soluble in water and body fluids than the amine procaine.

15 Lipids

Rebecca changed her diet by decreasing the fat content and increasing the amount of vegetables for more dietary fiber. She also drinks six glasses of water every day and is on a regular exercise program and walks every day. Recently, Rebecca had a gallbladder attack. In the body, the gallbladder stores bile, which is used in the intestine for the digestion of fats. If the bile is not moving out of the gallbladder, it can form gallstones, which are mostly cholesterol. If a gallstone blocks the bile duct, gallbladder disease can occur. Rebecca's doctor prescribed a medication that dissolves gallstones and recommended a liver flush to remove small gallstones.

Draw the line-angle formula for cholesterol.

Credit: michaeljung / Fotolia

LOOKING AHEAD

15.1 Lipids
15.2 Fatty Acids
15.3 Waxes and Triacylglycerols
15.4 Chemical Properties of Triacylglycerols
15.5 Phospholipids
15.6 Steroids: Cholesterol, Bile Salts, and Steroid Hormones
15.7 Cell Membranes

 The Health icon indicates a question that is related to health and medicine.

15.1 Lipids

Learning Goal: Describe the classes of lipids.

- Lipids are biomolecules that are not soluble in water but are soluble in organic solvents.
- Classes of lipids include waxes, triacylglycerols, glycerophospholipids, sphingolipids, and steroids.

♦ **Learning Exercise 15.1**

Match the correct class of lipids with the compositions below:
    a. wax   **b.** triacylglycerol   **c.** glycerophospholipid   **d.** steroid   **e.** sphingomyelin

1. _____ a fused structure of four cycloalkanes
2. _____ a long-chain alcohol and a long-chain saturated fatty acid
3. _____ glycerol and three fatty acids
4. _____ sphingosine, fatty acid, phosphate group, and an amino alcohol
5. _____ glycerol, two fatty acids, phosphate group, and an amino alcohol

Answers   **1.** d   **2.** a   **3.** b   **4.** e   **5.** c

Lipids

15.2 Fatty Acids

Learning Goal: Draw the condensed structural or line-angle formula for a fatty acid and identify it as saturated or unsaturated.

- Fatty acids are unbranched carboxylic acids that typically contain an even number (12 to 20) of carbon atoms.
- Fatty acids may be saturated, monounsaturated with one carbon–carbon double bond, or polyunsaturated with two or more double bonds. The double bonds in naturally occurring unsaturated fatty acids are almost always cis.

◆ **Learning Exercise 15.2A**

CORE CHEMISTRY SKILL
Identifying Fatty Acids

Draw the condensed structural and line-angle formulas for each of the following fatty acids:

1. linoleic acid (18:2)

2. stearic acid (18:0)

3. palmitoleic acid (16:1)

Answers

1. $CH_3-(CH_2)_4-CH=CH-CH_2-CH=CH-(CH_2)_7-\overset{O}{\underset{\|}{C}}-OH$

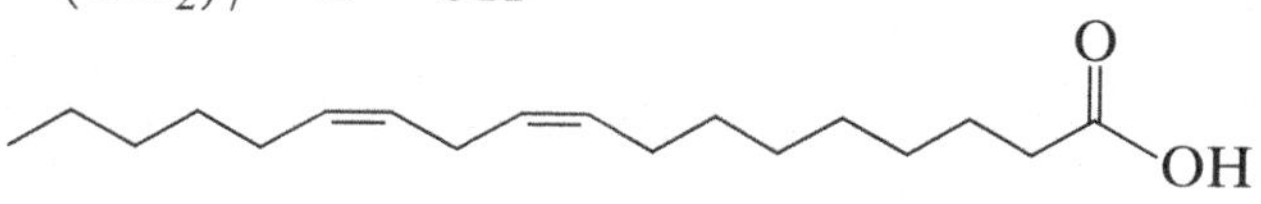

2. $CH_3-(CH_2)_{16}-\overset{O}{\underset{\|}{C}}-OH$

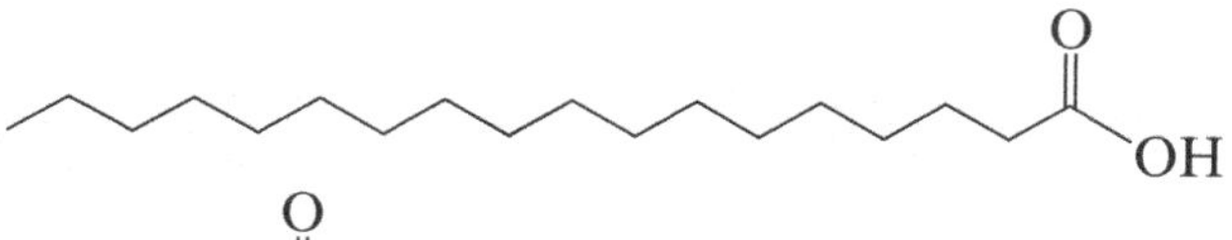

3. $CH_3-(CH_2)_5-CH=CH-(CH_2)_7-\overset{O}{\underset{\|}{C}}-OH$

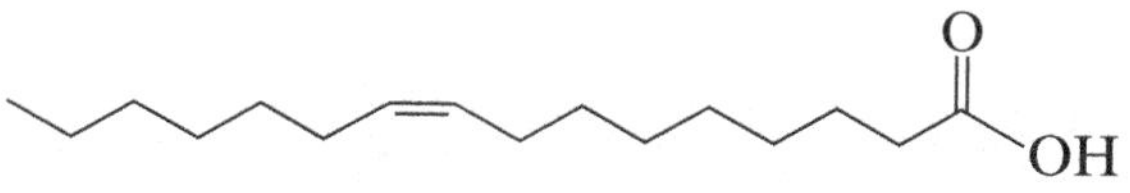

◆ **Learning Exercise 15.2B**

For the fatty acids in Learning Exercise 15.2A, identify which

a. _____ is the most saturated
b. _____ is the most unsaturated
c. _____ has the lowest melting point
d. _____ has the highest melting point
e. _____ is/are found in vegetables
f. _____ is/are from animal sources

Answers   **a.** 2   **b.** 1   **c.** 1   **d.** 2   **e.** 1, 3   **f.** 2

Chapter 15

♦ **Learning Exercise 15.2C**

For the following questions, refer to the line-angle formula for the fatty acid:

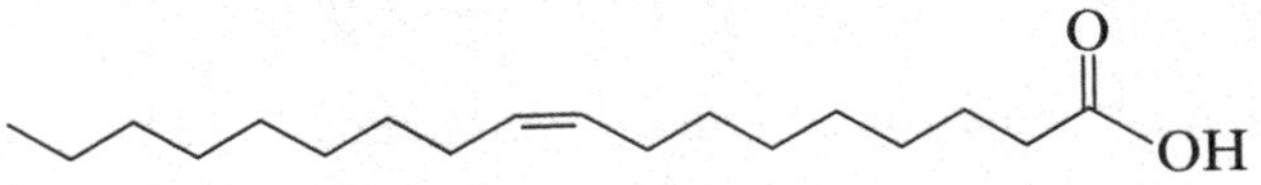

[line-angle structure of oleic acid with cis double bond and COOH group]

1. What is the name of this fatty acid? _____

2. Is it a saturated or an unsaturated compound? Why? _____

3. Is the double bond cis or trans? _____

4. Is it likely to be a solid or a liquid at room temperature? _____

5. Why is it insoluble in water? _____

Answers 1. oleic acid   2. unsaturated; double bond   3. cis
4. liquid   5. It has a long hydrocarbon chain.

15.3 Waxes and Triacylglycerols

Learning Goal: Draw the condensed structural or line-angle formula for a wax or triacylglycerol produced by the reaction of a fatty acid and an alcohol or glycerol.

- A wax is an ester of a long-chain saturated fatty acid and a long-chain alcohol.
- The triacylglycerols in fats and oils are esters of glycerol with three fatty acids.
- Fats from animal sources contain more saturated fatty acids and have higher melting points than most vegetable oils.

♦ **Learning Exercise 15.3A**

Draw the condensed structural formula for the wax formed by the reaction of palmitic acid,

$$CH_3-(CH_2)_{14}-\overset{\overset{O}{\|}}{C}-OH$$, and cetyl alcohol, $CH_3-(CH_2)_{14}-CH_2-OH$.

Answer $CH_3-(CH_2)_{14}-\overset{\overset{O}{\|}}{C}-O-CH_2-(CH_2)_{14}-CH_3$

Lipids

	Solution Guide to Drawing the Structure for a Triacylglycerol
STEP 1	Draw the condensed structural formulas for glycerol and the fatty acids.
STEP 2	Form ester bonds between the hydroxyl groups on glycerol and the carboxyl groups on each fatty acid.

♦ **Learning Exercise 15.3B**

Draw the condensed structural formula for the triacylglycerol formed from glycerol and three molecules of each of the following and give the name:

CORE CHEMISTRY SKILL
Drawing Structures for Triacylglycerols

1. palmitic acid, $CH_3-(CH_2)_{14}-\overset{\overset{O}{\|}}{C}-OH$

2. myristic acid, $CH_3-(CH_2)_{12}-\overset{\overset{O}{\|}}{C}-OH$

Answers

1. 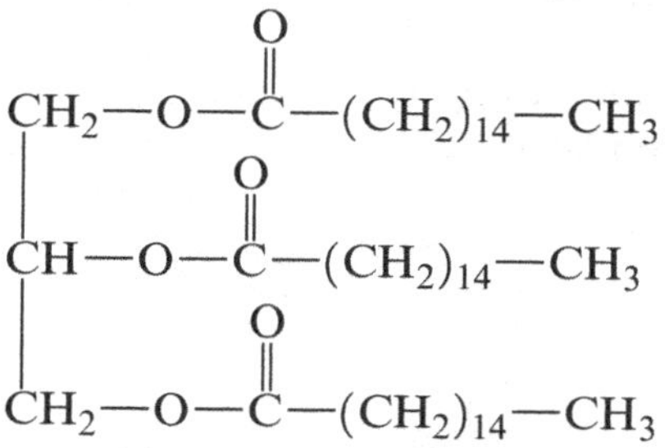
$$CH_2-O-\overset{\overset{O}{\|}}{C}-(CH_2)_{14}-CH_3$$
$$CH-O-\overset{\overset{O}{\|}}{C}-(CH_2)_{14}-CH_3$$
$$CH_2-O-\overset{\overset{O}{\|}}{C}-(CH_2)_{14}-CH_3$$
Glyceryl tripalmitate
(tripalmitin)

2. 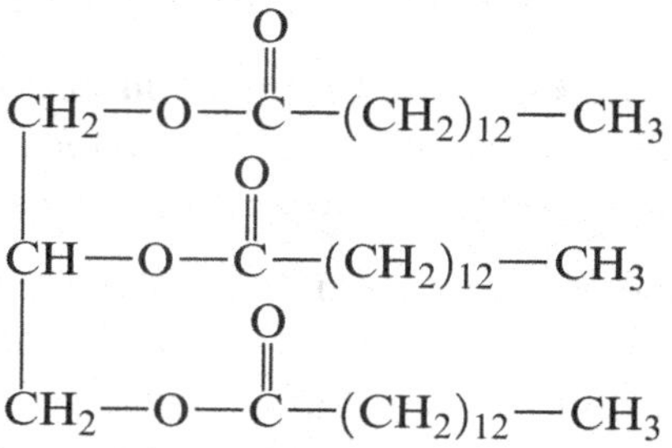
$$CH_2-O-\overset{\overset{O}{\|}}{C}-(CH_2)_{12}-CH_3$$
$$CH-O-\overset{\overset{O}{\|}}{C}-(CH_2)_{12}-CH_3$$
$$CH_2-O-\overset{\overset{O}{\|}}{C}-(CH_2)_{12}-CH_3$$
Glyceryl trimyristate
(trimyristin)

♦ **Learning Exercise 15.3C**

Draw the line-angle formula for each of the following triacylglycerols:

1. glyceryl tristearate (tristearin)

2. glyceryl trioleate (triolein)

Answers

1. Glyceryl tristearate (tristearin)

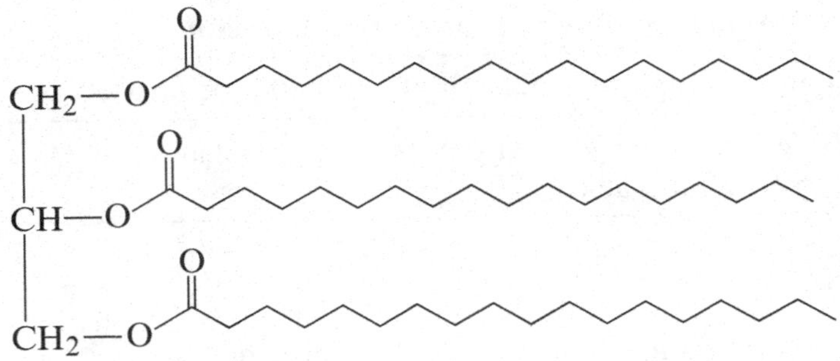

2. Glyceryl trioleate (triolein)

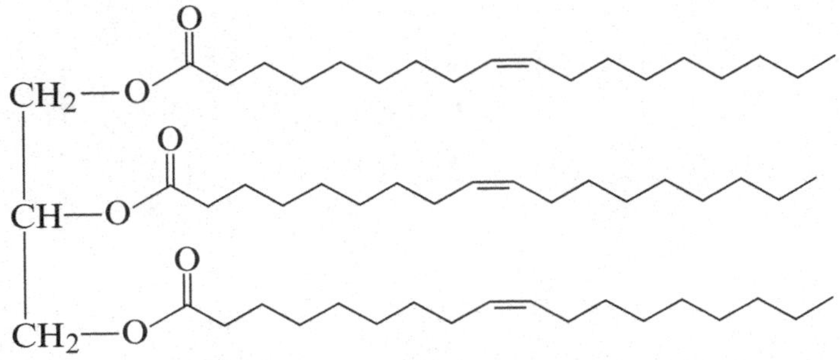

15.4 Chemical Properties of Triacylglycerols

Learning Goal: Draw the condensed structural or line-angle formula for the products of a triacylglycerol that undergoes hydrogenation, hydrolysis, or saponification.

- The hydrogenation of unsaturated fatty acids converts carbon–carbon double bonds to carbon–carbon single bonds.
- The hydrolysis of the ester bonds in fats or oils, in the presence of acids or enzymes, produces glycerol and fatty acids.
- In saponification, a triacylglycerol heated with a strong base produces glycerol and the salts of the fatty acids (soaps).

REVIEW
Writing Equations for Hydrogenation and Hydration (11.7)
Hydrolyzing Esters (14.4)

Vegetable oils (liquids)

Shortening (soft)

Tub (soft) margarine Stick margarine (solid)

Many soft margarines, stick margarines, and solid shortenings are produced by the partial hydrogenation of vegetable oils.

Credit: Pearson Education/Pearson Science

◆ Learning Exercise 15.4

Use condensed structural formulas to write the balanced chemical equation for the following reactions of glyceryl trioleate (triolein):

CORE CHEMISTRY SKILL
Drawing the Products for the Hydrogenation, Hydrolysis, and Saponification of a Triacylglycerol

1. hydrogenation with a nickel catalyst

2. acid hydrolysis

3. saponification with NaOH

Answers

1.
$$CH_2-O-\overset{O}{\underset{\|}{C}}-(CH_2)_7-CH=CH-(CH_2)_7-CH_3$$
$$CH-O-\overset{O}{\underset{\|}{C}}-(CH_2)_7-CH=CH-(CH_2)_7-CH_3 + 3H_2 \xrightarrow{Ni}$$
$$CH_2-O-\overset{O}{\underset{\|}{C}}-(CH_2)_7-CH=CH-(CH_2)_7-CH_3$$

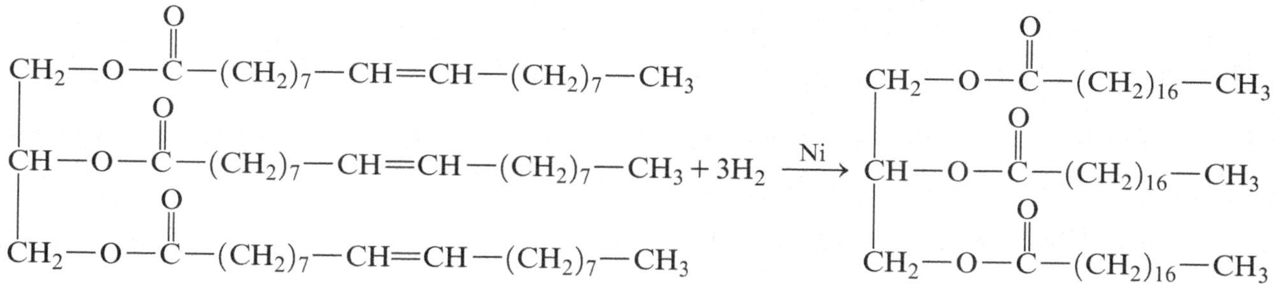

$$CH_2-O-\overset{O}{\underset{\|}{C}}-(CH_2)_{16}-CH_3$$
$$CH-O-\overset{O}{\underset{\|}{C}}-(CH_2)_{16}-CH_3$$
$$CH_2-O-\overset{O}{\underset{\|}{C}}-(CH_2)_{16}-CH_3$$

2.
$$CH_2-O-\overset{O}{\underset{\|}{C}}-(CH_2)_7-CH=CH-(CH_2)_7-CH_3$$
$$CH-O-\overset{O}{\underset{\|}{C}}-(CH_2)_7-CH=CH-(CH_2)_7-CH_3 + 3H_2O \xrightarrow{H^+, heat}$$
$$CH_2-O-\overset{O}{\underset{\|}{C}}-(CH_2)_7-CH=CH-(CH_2)_7-CH_3$$

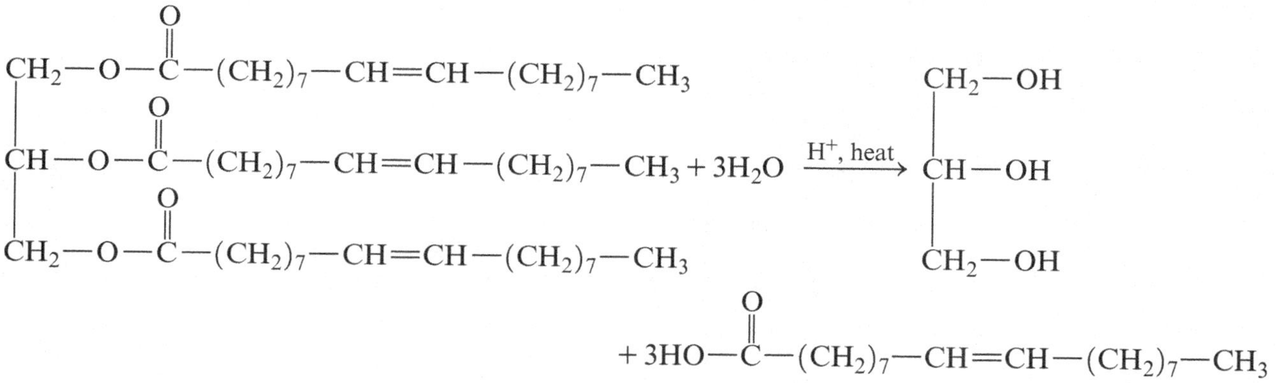

$$CH_2-OH$$
$$CH-OH$$
$$CH_2-OH$$

$$+ 3HO-\overset{O}{\underset{\|}{C}}-(CH_2)_7-CH=CH-(CH_2)_7-CH_3$$

3.

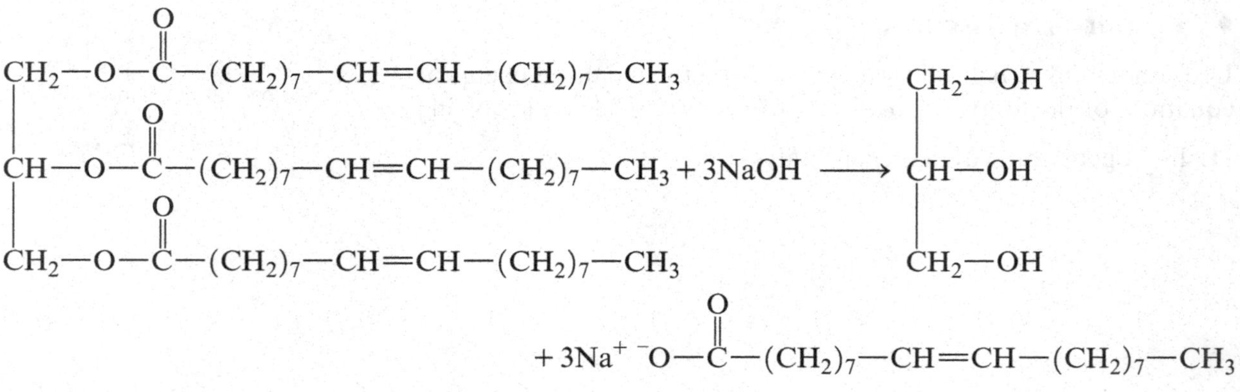

$$\begin{array}{l}CH_2-O-\overset{O}{\underset{\|}{C}}-(CH_2)_7-CH=CH-(CH_2)_7-CH_3\\ CH-O-\overset{O}{\underset{\|}{C}}-(CH_2)_7-CH=CH-(CH_2)_7-CH_3 + 3NaOH\\ CH_2-O-\overset{O}{\underset{\|}{C}}-(CH_2)_7-CH=CH-(CH_2)_7-CH_3\end{array} \longrightarrow \begin{array}{l}CH_2-OH\\ CH-OH\\ CH_2-OH\end{array}$$

$$+ 3Na^+ \; ^-O-\overset{O}{\underset{\|}{C}}-(CH_2)_7-CH=CH-(CH_2)_7-CH_3$$

15.5 Phospholipids

Learning Goal: Describe the structure of a phospholipid containing glycerol or sphingosine.

- The phospholipids are a family of lipids similar in structure to triacylglycerols.
- Glycerophospholipids are esters of glycerol with two fatty acids and a phosphate group attached to an amino alcohol.

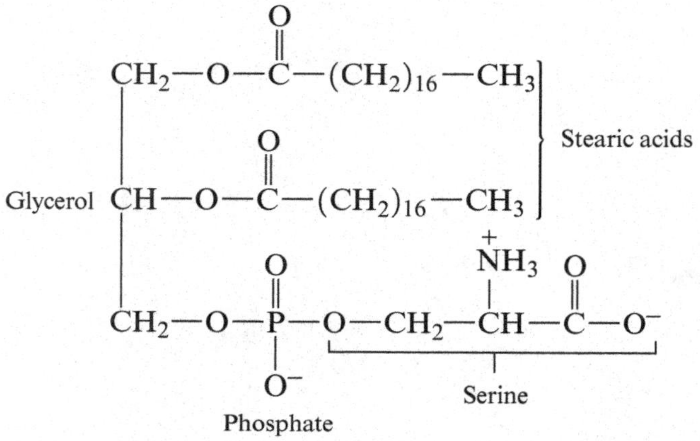

Glycerol
$$\begin{array}{l}CH_2-O-\overset{O}{\underset{\|}{C}}-(CH_2)_{16}-CH_3\\ CH-O-\overset{O}{\underset{\|}{C}}-(CH_2)_{16}-CH_3\\ CH_2-O-\overset{O}{\underset{\|}{P}}-O-CH_2-\overset{\overset{+}{NH_3}}{\underset{\|}{CH}}-\overset{O}{\underset{\|}{C}}-O^-\\ \underset{O^-}{} \end{array}$$ Stearic acids

Phosphate   Serine

- The fatty acids form a nonpolar region, whereas the phosphate group and the ionized amino alcohol make up a polar region.
- In sphingomyelins, the amine group in the long-chain alcohol sphingosine forms an amide bond with a fatty acid and the hydroxyl group forms a phosphoester bond with an amino alcohol.

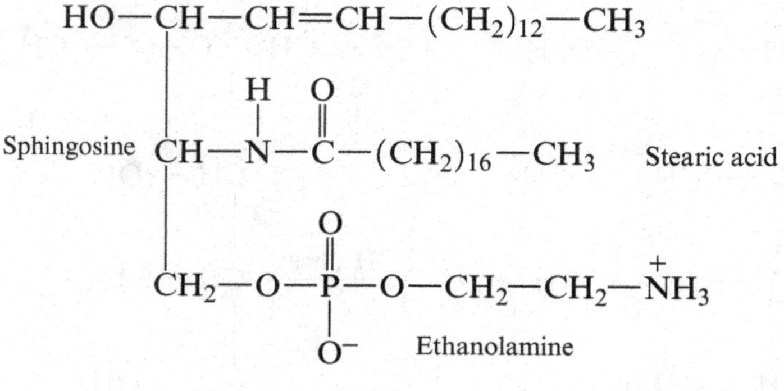

Sphingosine
$$\begin{array}{l}HO-CH-CH=CH-(CH_2)_{12}-CH_3\\ CH-\overset{H}{\underset{|}{N}}-\overset{O}{\underset{\|}{C}}-(CH_2)_{16}-CH_3\\ CH_2-O-\overset{O}{\underset{\underset{O^-}{\|}}{P}}-O-CH_2-CH_2-\overset{+}{NH_3}\end{array}$$ Stearic acid

Ethanolamine

Lipids

♦ **Learning Exercise 15.5A**

Draw the condensed structural formula for a glycerophospholipid that is formed from two molecules of palmitic acid and serine, an amino alcohol:

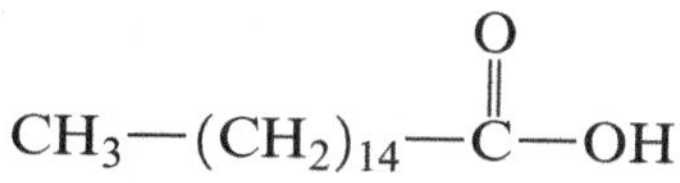

$$CH_3-(CH_2)_{14}-\underset{\underset{\text{Palmitic acid}}{}}{\overset{\overset{O}{\|}}{C}}-OH \qquad HO-CH_2-\underset{\underset{\text{Serine}}{}}{\overset{\overset{+}{NH_3}}{\underset{|}{CH}}}-\overset{\overset{O}{\|}}{C}-O^-$$

Answer

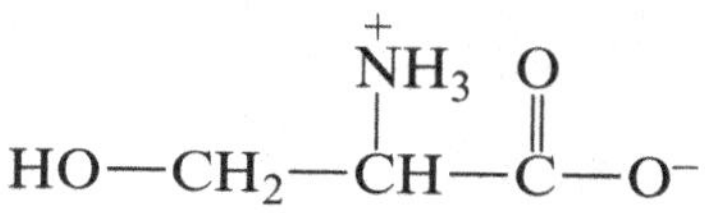

$$\begin{array}{l} CH_2-O-\overset{\overset{O}{\|}}{C}-(CH_2)_{14}-CH_3 \\ | \\ CH-O-\overset{\overset{O}{\|}}{C}-(CH_2)_{14}-CH_3 \\ | \\ CH_2-O-\underset{\underset{O^-}{|}}{\overset{\overset{O}{\|}}{P}}-O-CH_2-\overset{\overset{+}{NH_3}}{\underset{|}{CH}}-\overset{\overset{O}{\|}}{C}-O^- \end{array}$$

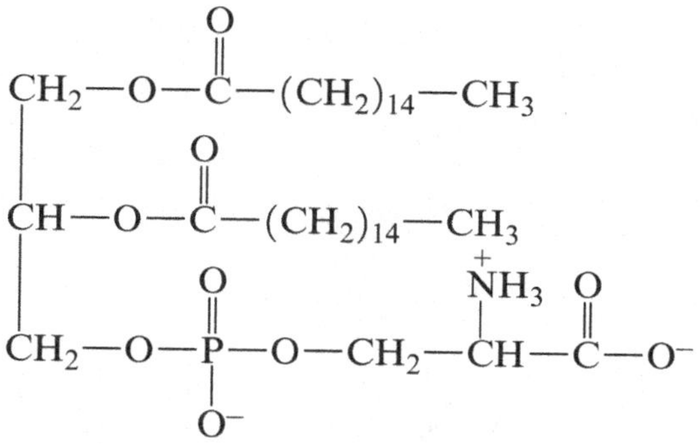

♦ **Learning Exercise 15.5B**

Use the following glycerophospholipid to answer **A** to **F** and **1** to **3**:

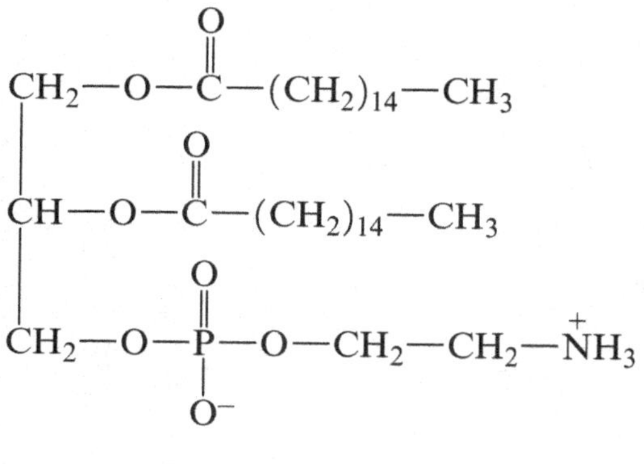

$$\begin{array}{l} CH_2-O-\overset{\overset{O}{\|}}{C}-(CH_2)_{14}-CH_3 \\ | \\ CH-O-\overset{\overset{O}{\|}}{C}-(CH_2)_{14}-CH_3 \\ | \\ CH_2-O-\underset{\underset{O^-}{|}}{\overset{\overset{O}{\|}}{P}}-O-CH_2-CH_2-\overset{+}{N}H_3 \end{array}$$

On the above condensed structural formula, indicate the:

A. two fatty acids B. glycerol
C. phosphate group D. amino alcohol
E. nonpolar region F. polar region

Chapter 15

1. What is the name of the amino alcohol? _____

2. What type of glycerophospholipid is it? _____

3. Why is a glycerophospholipid more soluble in water than most lipids? _____

Answers

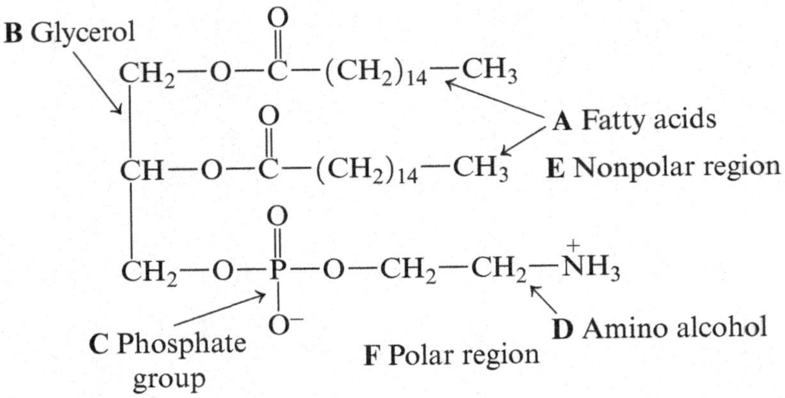

B Glycerol →
$CH_2-O-\overset{O}{\overset{\|}{C}}-(CH_2)_{14}-CH_3$
$CH-O-\overset{O}{\overset{\|}{C}}-(CH_2)_{14}-CH_3$ ← A Fatty acids, E Nonpolar region
$CH_2-O-\overset{O}{\overset{\|}{\underset{O^-}{P}}}-O-CH_2-CH_2-\overset{+}{N}H_3$
C Phosphate group, F Polar region, D Amino alcohol

1. ethanolamine
2. cephalin (ethanolamine glycerophospholipid)
3. The polar portion of the glycerophospholipid is attracted to water, which makes it more soluble in water than other lipids.

♦ **Learning Exercise 15.5C**

Answer the following questions for the phospholipid shown below, which is needed for brain and nerve tissues:

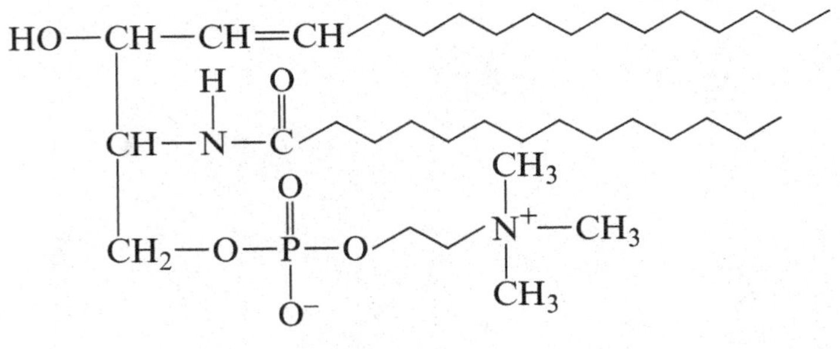

$HO-CH-CH=CH\text{~~~~~~~~~~~~~}$
$\quad\quad|\quad\quad H\;\; O$
$\quad\;CH-N-\overset{\|}{C}\text{~~~~~~~~~~~~~}$
$\quad\quad|\quad\quad\quad\quad\quad\quad CH_3$
$\quad CH_2-O-\overset{O}{\overset{\|}{\underset{O^-}{P}}}-O-CH_2-CH_2-\overset{+}{N}-CH_3$
$\quad\quad\quad\quad\quad\quad\quad\quad\quad\quad\quad\quad\quad CH_3$

1. Is the phospholipid formed from glycerol or sphingosine? _____

2. What is the fatty acid? _____

3. What type of bond connects the fatty acid? _____

4. What is the name of the amino alcohol? _____

Answers   1. sphingosine   2. myristic acid   3. an amide bond   4. choline

Lipids

15.6 Steroids: Cholesterol, Bile Salts, and Steroid Hormones

Learning Goal: Draw the structures of steroids.

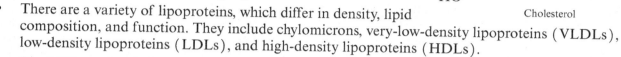

- Steroids are lipids containing the steroid nucleus, which is a fused structure of three cyclohexane rings and one cyclopentane ring.
- Steroids include cholesterol, bile salts, and steroid hormones.
- Bile salts, synthesized from cholesterol, mix with water-insoluble fats and break them apart during digestion.
- There are a variety of lipoproteins, which differ in density, lipid composition, and function. They include chylomicrons, very-low-density lipoproteins (VLDLs), low-density lipoproteins (LDLs), and high-density lipoproteins (HDLs).
- The LDL carries cholesterol to the tissues where it can be used for the synthesis of cell membranes and steroid hormones.
- The HDL picks up cholesterol from the tissues and carries it to the liver, where it can be converted to bile salts, which are eliminated from the body.
- The steroid hormones are closely related in structure to cholesterol and depend on cholesterol for their synthesis.
- The sex hormones such as estrogen and testosterone are responsible for sexual characteristics and reproduction.
- The adrenal corticosteroids include aldosterone, which regulates water balance, and cortisone, which regulates glucose levels in the blood.

CORE CHEMISTRY SKILL
Identifying the Steroid Nucleus

♦ **Learning Exercise 15.6A**

1. Draw the line-angle structure for the steroid nucleus.

2. Draw the line-angle structure for cholesterol.

Answers 1. [structure] 2. [structure]

♦ **Learning Exercise 15.6B**

Match each of the following compounds with the correct phrase below:

    a. estrogen   **b.** testosterone   **c.** cortisone   **d.** aldosterone   **e.** bile salts

1. _____ increases the blood glucose level
2. _____ increases the reabsorption of Na^+ by the kidneys
3. _____ stimulates the development of secondary sex characteristics in females
4. _____ stimulates the retention of water by the kidneys
5. _____ stimulates the development of secondary sex characteristics in males
6. _____ secreted by the gallbladder into the small intestine to emulsify fats in the diet

Answers   **1.** c   **2.** d   **3.** a   **4.** d   **5.** b   **6.** e

Chapter 15

15.7 Cell Membranes

Learning Goal: Describe the composition and function of the lipid bilayer in cell membranes.

- Cell membranes surround all of our cells and separate the cellular contents from the external aqueous environment.
- In the fluid mosaic model (lipid bilayer) of a cell membrane, two layers of phospholipids are arranged with their hydrophilic heads at the outer and inner surfaces of the membrane, and their hydrophobic tails in the center.
- Proteins and cholesterol are embedded in the lipid bilayer, and carbohydrates are attached to its surface.
- Nutrients and waste products move through the cell membrane using diffusion (passive transport), facilitated transport, or active transport.

Key Terms for Sections 15.1 to 15.7

Match each of the following key terms with the correct description:

    a. lipid   **b.** fatty acid   **c.** triacylglycerol   **d.** lipid bilayer
    e. saponification   **f.** glycerophospholipid   **g.** steroid

1. ____ a compound consisting of glycerol bonded to two fatty acids and a phosphate group that is attached to an amino alcohol
2. ____ a type of compound that is not soluble in water, but is soluble in nonpolar solvents
3. ____ the hydrolysis of a triacylglycerol with a strong base, producing salts called soaps and glycerol
4. ____ a compound consisting of glycerol bonded to three fatty acids
5. ____ a compound composed of a multicyclic ring system
6. ____ a model for cell membranes in which phospholipids are arranged in two layers
7. ____ a long-chain carboxylic acid found in triacylglycerols

Answers   **1.** f   **2.** a   **3.** e   **4.** c   **5.** g   **6.** d   **7.** b

♦ **Learning Exercise 15.7A**

1. What is the function of the lipid bilayer in cell membranes?

2. What type of lipid makes up the lipid bilayer?

3. What is the arrangement of the lipids in a lipid bilayer?

Answers

1. The lipid bilayer separates the contents of a cell from the surrounding aqueous environment.
2. The lipid bilayer is primarily composed of glycerophospholipids.
3. The nonpolar hydrocarbon tails are in the center of the bilayer, whereas the polar sections are aligned along the outside of the bilayer.

Learning Exercise 15.7B

Identify the type of transport described by each of the following:

1. K^+ ions move from low to high concentration in a cell. _____

2. Water moves through a cell membrane. _____

3. A molecule moves through a protein channel. _____

Answers   **1.** active transport   **2.** diffusion (passive transport)   **3.** facilitated transport

Checklist for Chapter 15

You are ready to take the Practice Test for Chapter 15. Be sure you have accomplished the following learning goals for this chapter. If not, review the Section listed at the end of the goal. Then apply your new skills and understanding to the Practice Test.

After studying Chapter 15, I can successfully:

_____ Describe the classes of lipids. (15.1)

_____ Identify a fatty acid as saturated or unsaturated. (15.2)

_____ Draw the condensed structural or line-angle formula for a wax or triacylglycerol produced by the reaction of a fatty acid and an alcohol or glycerol. (15.3)

_____ Draw the condensed structural or line-angle formula for the product from the reaction of a triacylglycerol with hydrogen, an acid, or a base. (15.4)

_____ Describe the components of glycerophospholipids and sphingomyelins. (15.5)

_____ Draw the structures of a steroid and cholesterol. (15.6)

_____ Describe the composition and function of the lipid bilayer in cell membranes. (15.7)

Practice Test for Chapter 15

The chapter Sections to review are shown in parentheses at the end of each question.

1. A biomolecule that is soluble in organic solvents but not in water is a (15.1)
   A. carbohydrate   **B.** lipid   **C.** protein   **D.** nucleic acid   **E.** soap

2. A fatty acid that is unsaturated is usually (15.2)
   A. from animal sources and liquid at room temperature
   B. from animal sources and solid at room temperature
   C. from vegetable sources and liquid at room temperature
   D. from vegetable sources and solid at room temperature
   E. from both vegetable and animal sources and solid at room temperature

3. The following condensed structural formula is a(an): (15.2)

$$CH_3-(CH_2)_{16}-\overset{O}{\underset{\|}{C}}-OH$$

   A. unsaturated fatty acid   **B.** saturated fatty acid   **C.** wax
   D. triacylglycerol   **E.** sphingomyelin

Chapter 15

4. The following compound is a (15.2)

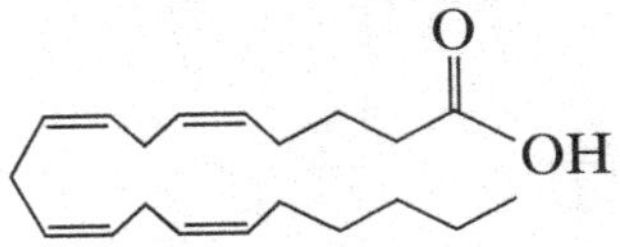

   A. cholesterol
   B. sphingosine
   C. fatty acid
   D. sphingomyelin
   E. steroid

For questions 5 through 8, consider the following compound: (15.3, 15.4)

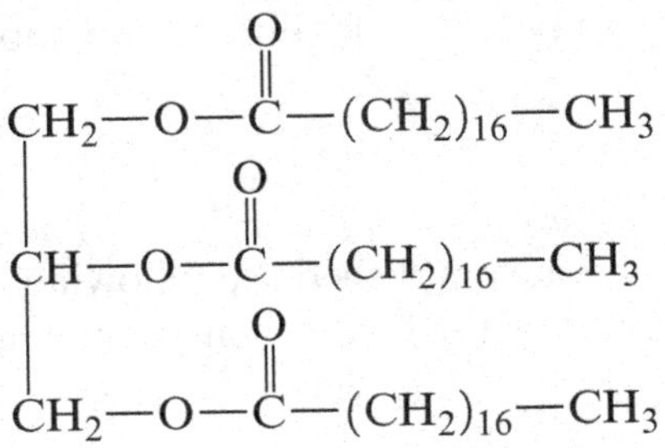

$$CH_2-O-\overset{O}{\underset{\|}{C}}-(CH_2)_{16}-CH_3$$
$$CH-O-\overset{O}{\underset{\|}{C}}-(CH_2)_{16}-CH_3$$
$$CH_2-O-\overset{O}{\underset{\|}{C}}-(CH_2)_{16}-CH_3$$

5. This compound belongs to the family called
   A. waxes
   B. triacylglycerols
   C. glycerophospholipids
   D. sphingomyelins
   E. steroids

6. The compound was formed by
   A. esterification
   B. acid hydrolysis
   C. saponification
   D. emulsification
   E. oxidation

7. The compound would be expected to be
   A. saturated and a solid at room temperature
   B. saturated and a liquid at room temperature
   C. unsaturated and a solid at room temperature
   D. unsaturated and a liquid at room temperature
   E. supersaturated and a liquid at room temperature

8. If this compound reacts with a strong base such as NaOH, the products would be
   A. glycerol and fatty acid
   B. glycerol and water
   C. glycerol and the sodium salt of the fatty acid
   D. an ester and the sodium salt of the fatty acid
   E. an ester and fatty acid

For questions 9 and 10, consider the following reaction: (15.4)

Triacylglycerol + 3NaOH $\xrightarrow{\text{Heat}}$ glycerol + 3 sodium salts of fatty acids

9. The reaction of a triacylglycerol with a strong base such as NaOH is called
   A. esterification   B. lipogenesis   C. hydrolysis   D. saponification   E. β oxidation

10. What is another name for the sodium salts of the fatty acids?
    A. margarines   B. fat substitutes   C. soaps   D. perfumes   E. vitamins

For questions 11 through 16, consider the following phospholipid:

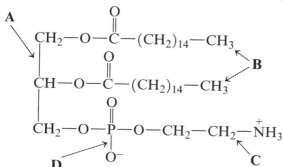

Match the labels, **A** to **D**, with the following: (15.5)

11. _____ glycerol
12. _____ the phosphate group
13. _____ the amino alcohol
14. _____ the polar region
15. _____ the nonpolar region

16. This phospholipid belongs to the family called (15.5)
   A. cholines
   B. cephalins
   C. sphingomyelins
   D. glycolipids
   E. lecithins

17. Which of these are found in glycerophospholipids? (15.5)
   A. fatty acids
   B. glycerol
   C. amino alcohol
   D. phosphate
   E. all of these

For questions 18 through 21, classify each lipid as a (15.1, 15.2, 15.3, 15.5, 15.6)
   A. wax
   B. triacylglycerol
   C. glycerophospholipid
   D. steroid
   E. fatty acid

18. _____ cholesterol

19. _____ $CH_3-(CH_2)_{14}-\overset{O}{\underset{\|}{C}}-OH$

20. _____ $CH_3-(CH_2)_{14}-\overset{O}{\underset{\|}{C}}-O-(CH_2)_{29}-CH_3$

21. _____ an ester of glycerol with three palmitic acid molecules

For questions 22 through 26, select answers from the following: (15.6)

   A. testosterone
   B. estrogen
   C. prednisone
   D. cortisone
   E. aldosterone

22. _____ stimulates development of the female sexual characteristics
23. _____ increases the retention of water by the kidneys
24. _____ stimulates development of the male sexual characteristics
25. _____ increases the blood glucose level
26. _____ used medically to reduce inflammation and treat asthma

Chapter 15

27. The type of lipoprotein that transports cholesterol to the liver for elimination is called a (15.6)
    A. chylomicron
    B. high-density lipoprotein
    C. low-density lipoprotein
    D. very-low-density lipoprotein
    E. all of these

28. The lipid bilayer of a cell consists of (15.7)
    A. cholesterol
    B. phospholipids
    C. proteins
    D. carbohydrates
    E. all of these

29. The type of transport that allows chloride ions to move through the integral proteins in the cell membrane is (15.7)
    A. passive transport
    B. active transport
    C. facilitated transport
    D. all of these
    E. none of these

30. The movement of small molecules through a cell membrane from a higher concentration to a lower concentration is (15.7)
    A. passive transport
    B. active transport
    C. facilitated transport
    D. all of these
    E. none of these

Answers to the Practice Test

1. B	2. C	3. B	4. C	5. B
6. A	7. A	8. C	9. D	10. C
11. A	12. D	13. C	14. C, D	15. B
16. B	17. E	18. D	19. E	20. A
21. B	22. B	23. E	24. A	25. D
26. C	27. B	28. E	29. C	30. A

Selected Answers and Solutions to Text Problems

15.1 Because lipids are not soluble in water, a polar solvent, they are nonpolar molecules.

15.3 Lipids are an important part of cell membranes and steroid hormones.

15.5 All fatty acids contain a long chain of carbon atoms with a carboxylic acid group. Saturated fatty acids contain only carbon–carbon single bonds; unsaturated fatty acids contain one or more double bonds.

15.7 a. ~~~~~~~~~~~~~~COOH

b. ~~~~~~=~~~~~~COOH

15.9 a. Lauric acid (12:0) has only carbon–carbon single bonds; it is saturated.
b. Linolenic acid (18:3) has three carbon–carbon double bonds; it is polyunsaturated.
c. Palmitoleic acid (16:1) has one carbon–carbon double bond; it is monounsaturated.
d. Stearic acid (18:0) has only carbon–carbon single bonds; it is saturated.

15.11 In a cis fatty acid, the hydrogen atoms are on the same side of the double bond, which produces a kink in the carbon chain. In a trans fatty acid, the hydrogen atoms are on opposite sides of the double bond, which gives a carbon chain without any kink.

15.13 Arachidonic acid and PGE_1 are both carboxylic acids with 20 carbon atoms. The differences are that arachidonic acid contains four cis double bonds and no other functional groups, whereas PGE_1 has one trans double bond, one ketone functional group, and two hydroxyl functional groups. In addition, a part of the PGE_1 chain forms a cyclopentane ring.

15.15 Prostaglandins raise or lower blood pressure, stimulate contraction and relaxation of smooth muscle, and may cause inflammation and pain.

15.17 In an omega-3 fatty acid, there is a double bond beginning at carbon 3 counting from the methyl group. In an omega-6 fatty acid, there is a double bond beginning at carbon 6 counting from the methyl group.

15.19 Palmitic acid is the 16-carbon saturated fatty acid (16:0), and myricyl alcohol is a 30-carbon long-chain alcohol.

$$CH_3-(CH_2)_{14}-\overset{O}{\underset{\|}{C}}-O-(CH_2)_{29}-CH_3$$

15.21 Triacylglycerols are composed of fatty acids and glycerol. In this case, the fatty acid is stearic acid, an 18-carbon saturated fatty acid (18:0).

$$\begin{array}{l} CH_2-O-\overset{O}{\underset{\|}{C}}-(CH_2)_{16}-CH_3 \\ CH-O-\overset{O}{\underset{\|}{C}}-(CH_2)_{16}-CH_3 \\ CH_2-O-\overset{O}{\underset{\|}{C}}-(CH_2)_{16}-CH_3 \end{array}$$

15.23 Glyceryl tricaprylate (tricaprylin) has three caprylic acids (an 8-carbon saturated fatty acid, 8:0) forming ester bonds with glycerol.

$$\begin{array}{l}\text{CH}_2-\text{O}-\text{CO}-(\text{CH}_2)_6-\text{CH}_3\\ |\\ \text{CH}-\text{O}-\text{CO}-(\text{CH}_2)_6-\text{CH}_3\\ |\\ \text{CH}_2-\text{O}-\text{CO}-(\text{CH}_2)_6-\text{CH}_3\end{array}$$

15.25 Safflower oil has a lower melting point because it contains mostly polyunsaturated fatty acids, whereas olive oil contains a large amount of monounsaturated oleic acid. A polyunsaturated fatty acid has two or more kinks in its carbon chain, which means it does not have as many dispersion forces compared to the hydrocarbon chains in olive oil.

15.27 Sunflower oil has about 24% monounsaturated fats, while safflower oil has about 18%. Sunflower oil has about 66% polyunsaturated fats, while safflower oil has about 73%.

15.29 a. The reaction of palm oil with KOH is saponification, and the products are glycerol and the potassium salts of the fatty acids, which are soaps.
  b. The reaction of glyceryl trilinoleate from safflower oil with water and HCl is hydrolysis, which splits the ester bonds to produce glycerol and three molecules of linoleic acid.

15.31 Hydrogenation of an unsaturated triacylglycerol adds H_2 to each of the double bonds, producing a saturated triacylglycerol containing only carbon–carbon single bonds.

$$\begin{array}{l}\text{CH}_2-\text{O}-\overset{\text{O}}{\overset{\|}{\text{C}}}-(\text{CH}_2)_7-\text{CH}=\text{CH}-(\text{CH}_2)_5-\text{CH}_3\\ |\\ \text{CH}-\text{O}-\overset{\text{O}}{\overset{\|}{\text{C}}}-(\text{CH}_2)_7-\text{CH}=\text{CH}-(\text{CH}_2)_5-\text{CH}_3 + 3\text{H}_2 \xrightarrow{\text{Ni}}\\ |\\ \text{CH}_2-\text{O}-\overset{\text{O}}{\overset{\|}{\text{C}}}-(\text{CH}_2)_7-\text{CH}=\text{CH}-(\text{CH}_2)_5-\text{CH}_3\end{array}$$

$$\begin{array}{l}\text{CH}_2-\text{O}-\overset{\text{O}}{\overset{\|}{\text{C}}}-(\text{CH}_2)_{14}-\text{C}\\ |\\ \text{CH}-\text{O}-\overset{\text{O}}{\overset{\|}{\text{C}}}-(\text{CH}_2)_{14}-\text{CH}\\ |\\ \text{CH}_2-\text{O}-\overset{\text{O}}{\overset{\|}{\text{C}}}-(\text{CH}_2)_{14}-\text{C}\end{array}$$

15.33 Acid hydrolysis of a fat gives glycerol and the fatty acids.

$$\begin{array}{l}\text{CH}_2-\text{O}-\text{CO}-(\text{CH}_2)_{n}-\text{CH}_3\\ |\\ \text{CH}-\text{O}-\text{CO}-(\text{CH}_2)_{n}-\text{CH}_3 + 3\text{H}_2\text{O} \xrightarrow{\text{H}^+,\text{ heat}}\\ |\\ \text{CH}_2-\text{O}-\text{CO}-(\text{CH}_2)_{n}-\text{CH}_3\end{array}$$

$$\begin{array}{l}\text{CH}_2-\text{OH}\\ |\\ \text{CH}-\text{OH} + 3\text{HO}-\text{CO}-(\text{CH}_2)_{n}-\text{CH}_3\\ |\\ \text{CH}_2-\text{OH}\end{array}$$

Lipids

15.35 Basic hydrolysis (saponification) of a fat gives glycerol and the salts of the fatty acids.

$$\begin{array}{c}
CH_2-O-\overset{O}{\underset{\|}{C}}-(CH_2)_{12}-CH_3 \\
| \\
CH-O-\overset{O}{\underset{\|}{C}}-(CH_2)_{12}-CH_3 + 3NaOH \\
| \\
CH_2-O-\overset{O}{\underset{\|}{C}}-(CH_2)_{12}-CH_3
\end{array} \xrightarrow{\text{Heat}} \begin{array}{c} CH_2-OH \\ | \\ CH-OH \\ | \\ CH_2-OH \end{array} + 3Na^+ \; {}^-O-\overset{O}{\underset{\|}{C}}-(CH_2)_{12}-CH_3$$

15.37
$$\begin{array}{c}
CH_2-O-\overset{O}{\underset{\|}{C}}-(CH_2)_{16}-CH_3 \\
| \\
CH-O-\overset{O}{\underset{\|}{C}}-(CH_2)_{16}-CH_3 \\
| \\
CH_2-O-\overset{O}{\underset{\|}{C}}-(CH_2)_{16}-CH_3
\end{array}$$

15.39 A triacylglycerol consists of glycerol and three fatty acids. A glycerophospholipid also contains glycerol, but has two fatty acids with the hydroxyl group on the third carbon attached by a phosphate bond to phosphoric acid, which forms another phosphoester bond to an amino alcohol.

15.41
$$\begin{array}{c}
CH_2-O-\overset{O}{\underset{\|}{C}}-(CH_2)_{14}-CH_3 \\
| \\
CH-O-\overset{O}{\underset{\|}{C}}-(CH_2)_{14}-CH_3 \\
| \\
CH_2-O-\underset{\underset{O^-}{|}}{\overset{O}{\underset{\|}{P}}}-O-CH_2-CH_2-\overset{+}{N}H_3
\end{array}$$

15.43 This glycerophospholipid is a cephalin. It contains glycerol, oleic acid, stearic acid, a phosphate group, and ethanolamine.

15.45 **a.** This phospholipid is formed from sphingosine.
b. The fatty acid is palmitic acid (16:0).
c. An amide bond connects the fatty acid to sphingosine.
d. The amino alcohol is ethanolamine.

15.47

15.49 Bile salts act to emulsify fat globules, allowing the fat to be more easily digested by lipases.

15.51 Chylomicrons have a lower density than VLDL (very-low-density lipoprotein). They pick up triacylglycerols from the intestine, whereas VLDL transports triacylglycerols synthesized in the liver.

15.53 "Bad" cholesterol is the cholesterol carried by LDL that can form deposits in the arteries called plaque, which narrows the arteries.

15.55 Both estradiol and testosterone contain the steroid nucleus and a hydroxyl group. Testosterone has a ketone group, a double bond, and two methyl groups. Estradiol has an aromatic ring, a hydroxyl group in place of the ketone, and a methyl group.

15.57 Cortisol (b), estradiol (c), and testosterone (d) are steroid hormones.

15.59 The function of the lipid bilayer in the cell membrane is to keep the cell contents separated from the outside environment and to allow the cell to regulate the movement of substances into and out of the cell.

15.61 Because the molecules of cholesterol are large and rigid, they reduce the flexibility of the lipid bilayer and add strength to the cell membrane.

15.63 The peripheral proteins in the membrane emerge on the inner or outer surface only, whereas the integral proteins extend through the membrane to both surfaces.

15.65 a. A molecule moving through a protein channel is an example of facilitated transport.
b. A molecule, like O_2, moving from higher concentration to lower concentration is an example of passive transport (diffusion).

15.67 The functional groups in Pravachol are ester, alcohol, alkene, and carboxylic acid.

15.69 a. $\dfrac{80.\text{ mg Pravachol}}{1 \text{ day}} \times \dfrac{1 \text{ g Pravachol}}{1000 \text{ mg Pravachol}} \times \dfrac{7 \text{ days}}{1 \text{ week}} = 0.56$ g of Pravachol/week (2 SFs)

b. $\dfrac{18 \text{ mg cholesterol}}{5.0 \text{ mL blood}} \times \dfrac{100 \text{ mL blood}}{1 \text{ dL blood}} = 360$ mg/dL (2 SFs)

15.71
$$\begin{array}{l}
CH_2-O-\overset{\overset{O}{\|}}{C}-(CH_2)_{14}-CH_3 \\
\;\;| \\
CH-O-\overset{\overset{O}{\|}}{C}-(CH_2)_{14}-CH_3 \\
\;\;| \\
CH_2-O-\overset{\overset{O}{\|}}{C}-(CH_2)_{14}-CH_3
\end{array}$$

15.73 a. monounsaturated fatty acid **b.** polyunsaturated omega-6 fatty acid
c. polyunsaturated omega-3 fatty acid

15.75 a. Beeswax and carnauba are waxes. Vegetable oil and glyceryl tricaprate (tricaprin) are triacylglycerols.

b.
$$\begin{array}{l}
CH_2-O-\overset{\overset{O}{\|}}{C}-(CH_2)_{8}-CH_3 \\
\;\;| \\
CH-O-\overset{\overset{O}{\|}}{C}-(CH_2)_{8}-CH_3 \\
\;\;| \\
CH_2-O-\overset{\overset{O}{\|}}{C}-(CH_2)_{8}-CH_3
\end{array}$$

15.77 a. $46 \text{ g fat} \times \dfrac{9 \text{ kcal}}{1 \text{ g fat}} = 410$ kcal from fat $\quad \dfrac{410 \text{ kcal}}{830 \text{ kcal}} \times 100\% = 49\%$ fat (2 SFs)

b. $29 \text{ g fat} \times \dfrac{9 \text{ kcal}}{1 \text{ g fat}} = 260$ kcal from fat $\quad \dfrac{260 \text{ kcal}}{520 \text{ kcal}} \times 100\% = 50.\%$ fat (2 SFs)

c. $18 \text{ g fat} \times \dfrac{9 \text{ kcal}}{1 \text{ g fat}} = 160$ kcal from fat $\quad \dfrac{160 \text{ kcal}}{560 \text{ kcal}} \times 100\% = 29\%$ fat (2 SFs)

The fats in these foods are from animal sources and would be mostly saturated fats.

15.79 a. Beeswax is a wax.
  b. Cholesterol is a steroid.
  c. Lecithin is a glycerophospholipid.
  d. Glyceryl tripalmitate (tripalmitin) is a triacylglycerol.
  e. Sodium stearate is a soap.
  f. Safflower oil is a triacylglycerol.

15.81 a. Estrogen contains the steroid nucleus (5).
  b. Cephalin contains glycerol (1), fatty acids (2), phosphate (3), and an amino alcohol (4).
  c. Waxes contain fatty acid (2).
  d. Triacylglycerols contain glycerol (1) and fatty acids (2).

15.83 Cholesterol (a) and carbohydrates (c) are found in cell membranes.

15.85
$$\begin{array}{l} CH_2-O-\overset{\overset{O}{\|}}{C}-(CH_2)_{16}-CH_3 \\ | \\ CH-O-\overset{\overset{O}{\|}}{C}-(CH_2)_{16}-CH_3 \\ | \\ CH_2-O-\overset{\overset{O}{\|}}{\underset{O^-}{P}}-O-CH_2-CH_2-\overset{+}{N}H_3 \end{array}$$

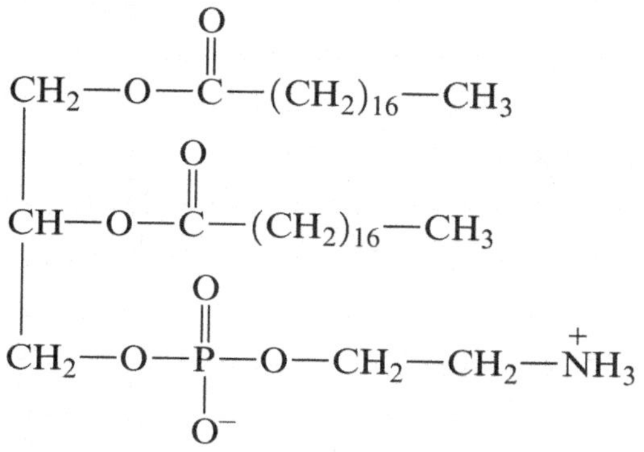

15.87 a. HDL (4) is known as "good" cholesterol.
  b. LDL (3) transports most of the cholesterol to the cells.
  c. Chylomicrons (1) carry triacylglycerols from the intestine to the fat cells.
  d. HDL (4) transports cholesterol to the liver.

15.89 a. Adding NaOH will saponify lipids such as glyceryl tristearate (tristearin), forming glycerol and salts of the fatty acids that are soluble in water and would wash down the drain.

  b.
$$\begin{array}{l} CH_2-O-\overset{\overset{O}{\|}}{C}-(CH_2)_{16}-CH_3 \\ | \\ CH-O-\overset{\overset{O}{\|}}{C}-(CH_2)_{16}-CH_3 \quad +3NaOH \xrightarrow{Heat} \\ | \\ CH_2-O-\overset{\overset{O}{\|}}{C}-(CH_2)_{16}-CH_3 \end{array}$$

$$\begin{array}{l} CH_2-OH \\ | \\ CH-OH \quad +3Na^+ \ {}^-O-\overset{\overset{O}{\|}}{C}-(CH_2)_{16}-CH_3 \\ | \\ CH_2-OH \end{array}$$

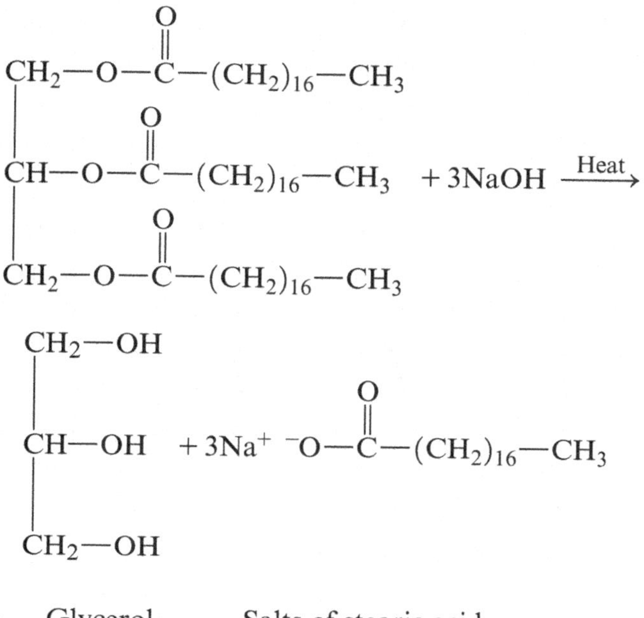

      Glycerol   Salts of stearic acid

Chapter 15

c. Tristearin = $C_{57}H_{110}O_6$

Molar mass of tristearin ($C_{57}H_{110}O_6$)
= 57(12.01 g) + 110(1.008 g) + 6(16.00 g) = 891.5 g/mole

$$10.0 \text{ g tristearin} \times \frac{1 \text{ mole tristearin}}{891.5 \text{ g tristearin}} \times \frac{3 \text{ moles NaOH}}{1 \text{ mole tristearin}} \times \frac{1000 \text{ mL solution}}{0.500 \text{ mole NaOH}}$$

= 67.3 mL of NaOH solution (3 SFs)

15.91 a.

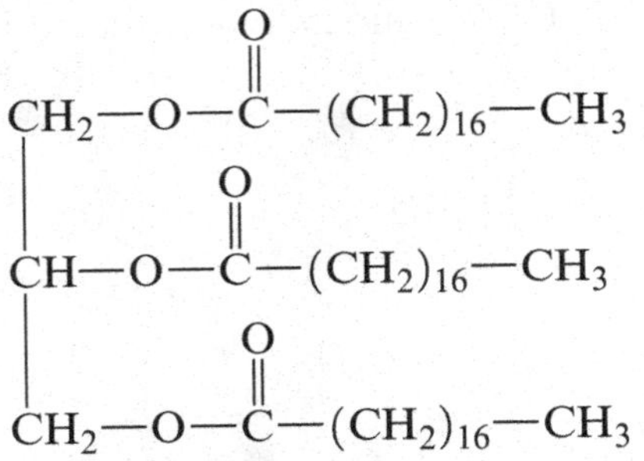

$$\begin{array}{l} CH_2-O-\overset{O}{\underset{\|}{C}}-(CH_2)_{16}-CH_3 \\ | \\ CH-O-\overset{O}{\underset{\|}{C}}-(CH_2)_{16}-CH_3 \\ | \\ CH_2-O-\overset{O}{\underset{\|}{C}}-(CH_2)_{16}-CH_3 \end{array}$$

b. The hydrogenation of the oleic acids converts them into stearic acids, which gives the product tristearin. Each mole of triolein requires three moles of H_2 to completely hydrogenate the fat.

$$1.00 \text{ mole triolein} \times \frac{3 \text{ moles } H_2}{1 \text{ mole triolein}} = 3.00 \text{ moles of } H_2 \text{ (3 SFs)}$$

c. $3.00 \text{ moles } H_2 \times \frac{2.016 \text{ g } H_2}{1 \text{ mole } H_2} = 6.05$ g of H_2 are needed for the reaction.

d. $3.00 \text{ moles } H_2 \times \frac{22.4 \text{ L } H_2 \text{ (STP)}}{1 \text{ mole } H_2} = 67.2$ L of H_2 (at STP) are needed for the reaction.

Selected Answers to Combining Ideas from Chapters 13 to 15

CI.27 a. [structure: HO-C(=O)-benzene(para)-C(=O)-O-CH₂CH₂-OH]

b. [structure: HO-CH₂CH₂-O-C(=O)-benzene(para)-C(=O)-O-CH₂CH₂-OH]

c. $1.5 \times 10^9 \text{ lb PETE} \times \dfrac{1 \text{ kg PETE}}{2.20 \text{ lb PETE}} = 6.8 \times 10^8 \text{ kg of PETE (2 SFs)}$

d. $6.8 \times 10^8 \text{ kg PETE} \times \dfrac{1000 \text{ g PETE}}{1 \text{ kg PETE}} \times \dfrac{1 \text{ mL PETE}}{1.38 \text{ g PETE}} \times \dfrac{1 \text{ L PETE}}{1000 \text{ mL PETE}}$
$= 4.9 \times 10^8 \text{ L of PETE (2 SFs)}$

e. $4.9 \times 10^8 \text{ L PETE} \times \dfrac{1 \text{ landfill}}{2.7 \times 10^7 \text{ L PETE}} = 18 \text{ landfills (2 SFs)}$

CI.29 a. [structure of DEET: m-methylbenzamide with N,N-diethyl]

b. The molecular formula of DEET is $C_{12}H_{17}NO$.

c. Molar mass of DEET ($C_{12}H_{17}NO$)
$= 12(12.01 \text{ g}) + 17(1.008 \text{ g}) + 14.01 \text{ g} + 16.00 \text{ g} = 191.3 \text{ g/mole (4 SFs)}$

d. $6.0 \text{ fl oz} \times \dfrac{1 \text{ qt}}{32 \text{ fl oz}} \times \dfrac{946 \text{ mL}}{1 \text{ qt}} \times \dfrac{25 \text{ g DEET}}{100 \text{ mL solution}} = 44 \text{ g of DEET (2 SFs)}$

e. $6.0 \text{ fl oz} \times \dfrac{1 \text{ qt}}{32 \text{ fl oz}} \times \dfrac{946 \text{ mL}}{1 \text{ qt}} \times \dfrac{25 \text{ g DEET}}{100 \text{ mL solution}} \times \dfrac{1 \text{ mole DEET}}{191.3 \text{ g DEET}}$
$\times \dfrac{6.02 \times 10^{23} \text{ molecules DEET}}{1 \text{ mole DEET}} = 1.4 \times 10^{23} \text{ molecules of DEET (2 SFs)}$

CI.31 a. **A**, **B**, and **C** are all glucose units.
b. An $\alpha(1\rightarrow 6)$-glycosidic bond connects monosaccharides **A** and **B**.
c. An $\alpha(1\rightarrow 4)$-glycosidic bond connects monosaccharides **B** and **C**.
d. The structure drawn is β-panose.
e. Panose is a reducing sugar because it has a free hydroxyl group on carbon 1 of structure **C**, which allows glucose **C** to form an aldehyde.

16
Amino Acids, Proteins, and Enzymes

Credit: Tetra Images/Alamy Stock Photo

Jeremy, a 9-month old boy, has a fever and painful swelling of his hands and feet. Emma, his physician assistant, schedules blood tests to determine if Jeremy has sickle-cell anemia. Hemoglobin is a protein that carries oxygen in the blood. When a mutation occurs, resulting in a change in the amino acid sequence of hemoglobin, the shape of the molecule may be altered, affecting its ability to carry oxygen properly.

Jeremy's blood tests indicate that he has sickle-cell anemia. Emma prescribes hydroxyurea, which is a medication that stimulates the production of fetal hemoglobin. Fetal hemoglobin, which is not produced after birth, is a variation of the hemoglobin molecule that binds more tightly to oxygen molecules than normal adult hemoglobin. How is the function of hemoglobin changed in sickle-cell anemia?

Hydroxyurea

LOOKING AHEAD

16.1 Proteins and Amino Acids
16.2 Proteins: Primary Structure
16.3 Proteins: Secondary, Tertiary, and Quaternary Structures
16.4 Enzymes
16.5 Factors Affecting Enzyme Activity

🮱 The Health icon indicates a question that is related to health and medicine.

16.1 Proteins and Amino Acids

Learning Goal: Classify proteins by their functions. Give the name and abbreviations for an amino acid and draw its structure at physiological pH.

- Some proteins are enzymes or hormones, while others are important in structure, transport, protection, storage, and contraction of muscles.
- A group of 20 amino acids provides the building blocks of proteins.
- In an amino acid, a central (alpha) carbon is attached to an ammonium group ($-NH_3^+$), a carboxylate group ($-COO^-$), a hydrogen atom ($-H$), and a side chain or R group, which is unique for each amino acid.
- Each specific R group determines whether an amino acid is nonpolar, polar neutral, acidic, or basic. Nonpolar amino acids contain hydrogen, alkyl, or aromatic side chains, whereas polar amino acids contain electronegative atoms such as oxygen or sulfur. Acidic side chains contain a carboxylate group ($-COO^-$), and basic side chains contain an ammonium group ($-NH_3^+$).

Amino Acids, Proteins, and Enzymes

♦ **Learning Exercise 16.1A**

Match one of the following functions with each protein below:

    a. structural   **b.** contractile   **c.** storage   **d.** transport
    e. hormone   **f.** enzyme   **g.** protection

1. _____ hemoglobin, carries oxygen in blood
2. _____ trypsin, hydrolyzes proteins
3. _____ casein, a protein in milk
4. _____ vasopressin, regulates blood pressure
5. _____ myosin, found in muscle tissue
6. _____ immunoglobulin, an antibody
7. _____ keratin, a major protein of hair
8. _____ lipoprotein, carries lipids in blood

Answers   **1.** d   **2.** f   **3.** c   **4.** e   **5.** b   **6.** g   **7.** a   **8.** d

♦ **Learning Exercise 16.1B**

Using the appropriate R group, complete the condensed structural formula of each of the following amino acids. Indicate whether the amino acid would be nonpolar, polar neutral, acidic, or basic, and give its three- and one-letter abbreviations:

CORE CHEMISTRY SKILL
Drawing the Structure for an Amino Acid at Physiological pH

1. 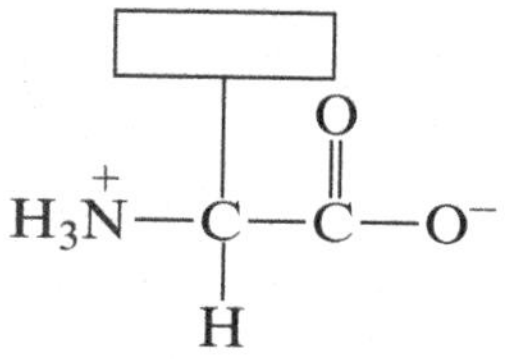 $H_3\overset{+}{N}-\underset{H}{\overset{\square}{C}}-\overset{O}{\underset{}{C}}-O^-$ Glycine _____

2. 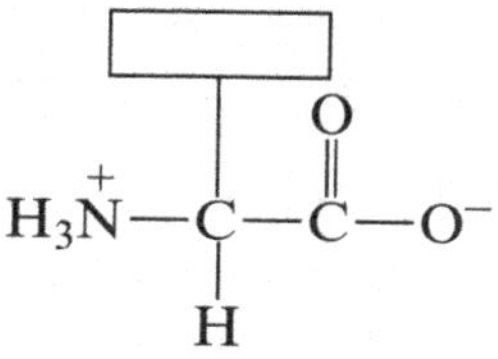 $H_3\overset{+}{N}-\underset{H}{\overset{\square}{C}}-\overset{O}{\underset{}{C}}-O^-$ Alanine _____

3. 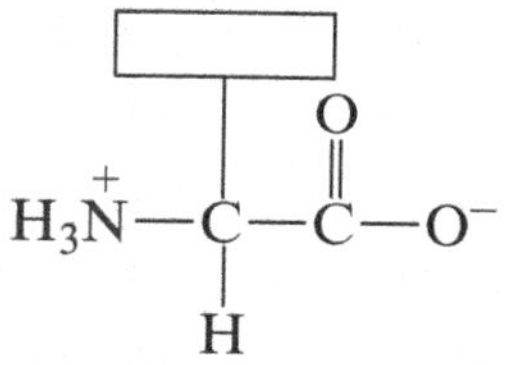 $H_3\overset{+}{N}-\underset{H}{\overset{\square}{C}}-\overset{O}{\underset{}{C}}-O^-$ Serine _____

4. 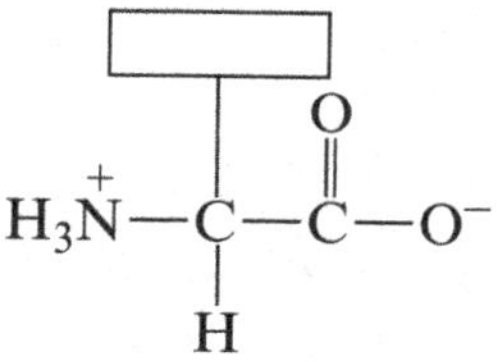 $H_3\overset{+}{N}-\underset{H}{\overset{\square}{C}}-\overset{O}{\underset{}{C}}-O^-$ Aspartate _____

Answers

1. 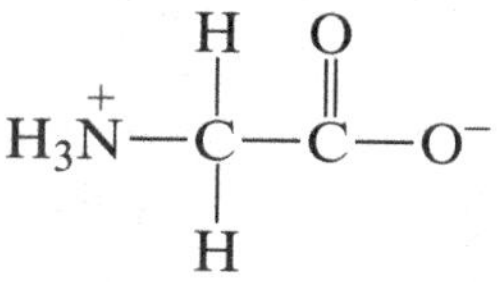 $H_3\overset{+}{N}-\underset{H}{\overset{H}{C}}-\overset{O}{C}-O^-$
   Glycine, nonpolar, Gly, G

2. 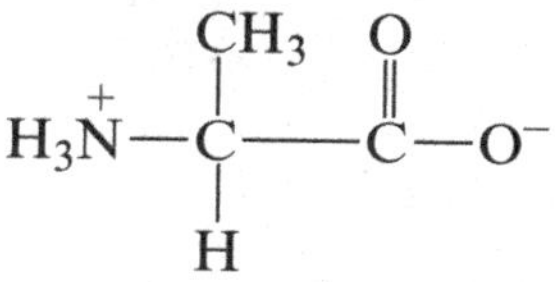 $H_3\overset{+}{N}-\underset{H}{\overset{CH_3}{C}}-\overset{O}{C}-O^-$
   Alanine, nonpolar, Ala, A

3. 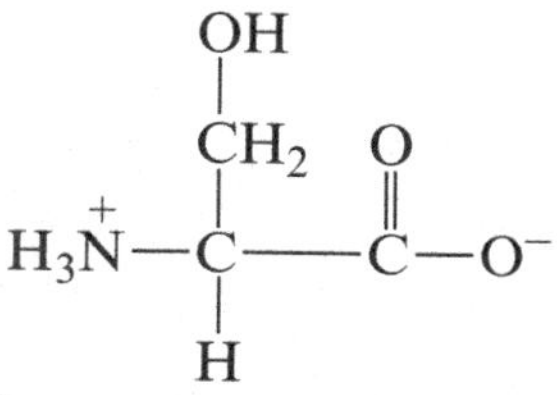 $H_3\overset{+}{N}-\underset{H}{\overset{CH_2-OH}{C}}-\overset{O}{C}-O^-$
   Serine, polar neutral, Ser, S

4. 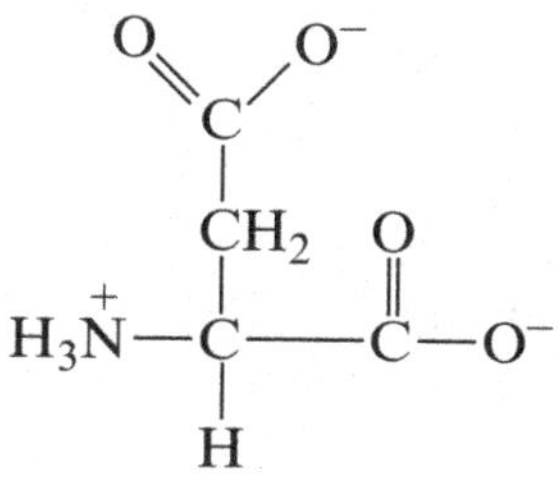 $H_3\overset{+}{N}-\underset{H}{\overset{CH_2-COO^-}{C}}-\overset{O}{C}-O^-$
   Aspartate, acidic, Asp, D

Chapter 16

♦ **Learning Exercise 16.1C**

Draw the structure for each of the following amino acids at physiological pH:

a. Val

b. C

c. threonine

d. F

Answers

a. 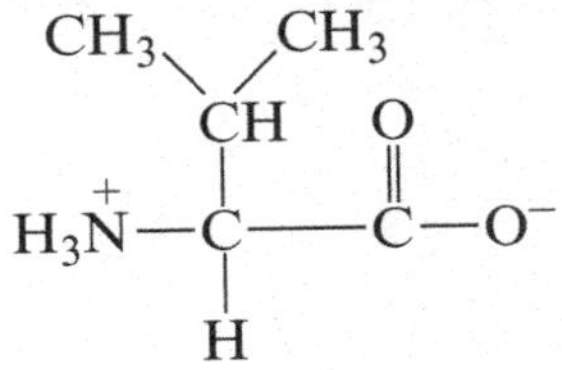
```
    CH₃   CH₃
      \  /
       CH    O
       |     ||
  H₃N⁺—C——C—O⁻
       |
       H
```

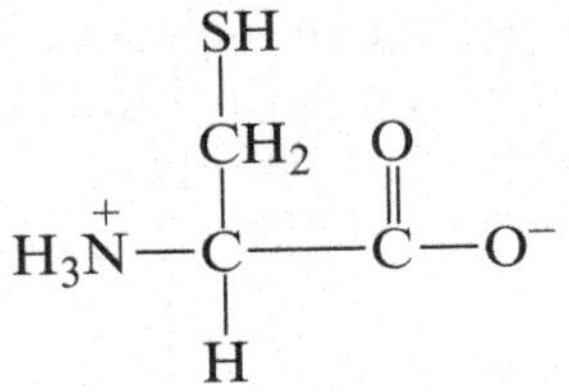

b.
```
        SH
        |
        CH₂   O
        |     ||
  H₃N⁺—C——C—O⁻
        |
        H
```

c.
```
    CH₃   OH
      \  /
       CH    O
       |     ||
  H₃N⁺—C——C—O⁻
       |
       H
```

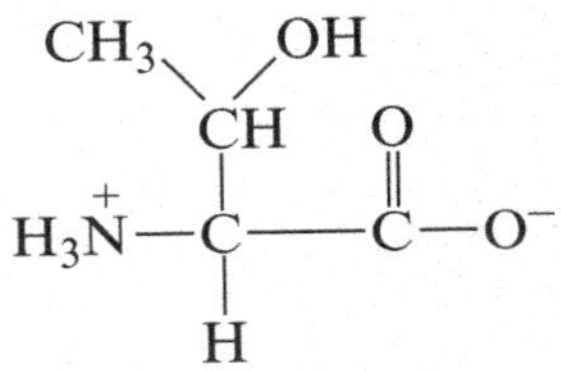

d. 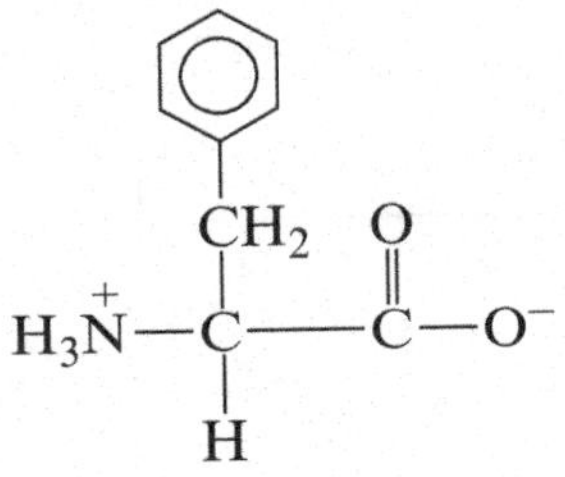
```
        ⌬ (phenyl)
        |
        CH₂   O
        |     ||
  H₃N⁺—C——C—O⁻
        |
        H
```

16.2 Proteins: Primary Structure

Learning Goal: Draw the condensed structural formula for a peptide and give its name. Describe the primary structure for a protein.

- The primary structure of a protein is the sequence of amino acids connected by peptide bonds.
- A peptide bond is an amide bond between the carboxylate group ($-COO^-$) of one amino acid and the ammonium group ($-NH_3^+$) of another amino acid.

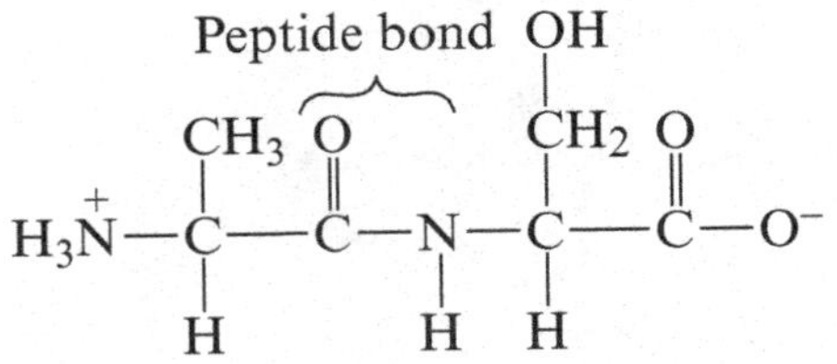

```
          Peptide bond   OH
                         |
         CH₃   O         CH₂   O
         |     ||        |     ||
   H₃N⁺—C——C—N—C——C—O⁻
         |     |   |
         H     H   H
```

> **REVIEW**
> Forming Amides (14.6)

- Short chains of amino acids are called peptides. Long chains of amino acids that are biologically active are called proteins.

- Essential amino acids must be obtained from the proteins in the diet.
- Complete proteins, which are found in most animal products, contain all of the essential amino acids.
- Peptides are named from the N-terminus by replacing the *ine* or *ate* of each amino acid name with *yl* followed by the amino acid name of the C-terminus.

Solution Guide to Drawing a Peptide	
STEP 1	Draw the structure for each amino acid in the peptide, starting with the N-terminus.
STEP 2	Remove the O atom from the carboxylate group of the N-terminus and two H atoms from the ammonium group in the adjacent amino acid.
STEP 3	Use peptide bonds to connect the amino acids.

♦ **Learning Exercise 16.2A**

Draw the condensed structural formulas at physiological pH, and write the name for each of the following:

1. Leu–Phe
2. YV
3. Gly–Ser–Cys

Answers

1. 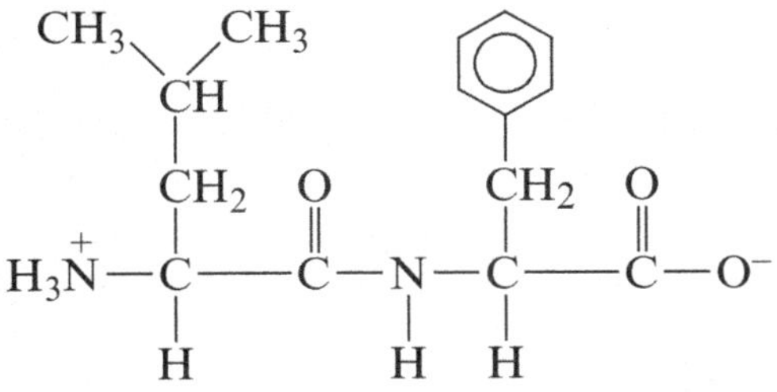 Leucylphenylalanine

2. 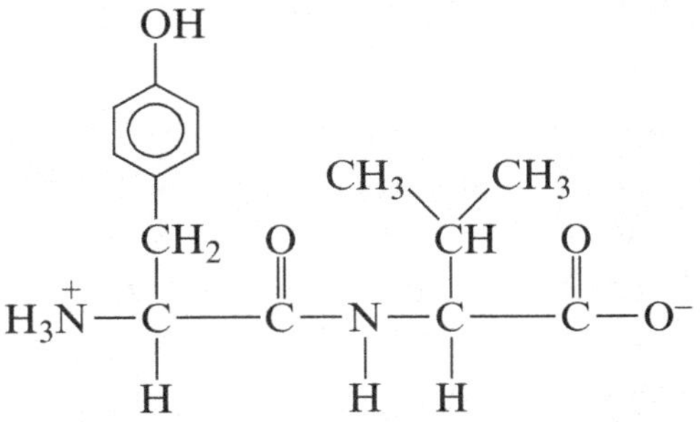 Tyrosylvaline

3. 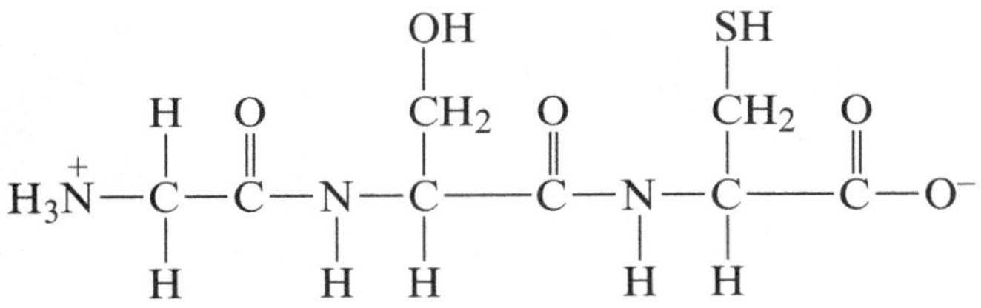 Glycylserylcysteine

Chapter 16

♦ **Learning Exercise 16.2B**

Many seeds, vegetables, and legumes are low in one or more of the essential amino acids tryptophan, isoleucine, and lysine.

	Tryptophan	Isoleucine	Lysine
Sesame seeds	OK	low	low
Sunflower seeds	OK	OK	low
Garbanzo beans	low	OK	OK
Rice	OK	OK	low
Cornmeal	low	OK	low

Indicate whether the following protein combinations are complementary or not:

1. _____ sesame seeds and sunflower seeds
2. _____ sunflower seeds and garbanzo beans
3. _____ sesame seeds and rice
4. _____ sesame seeds and garbanzo beans
5. _____ garbanzo beans and rice
6. _____ cornmeal and garbanzo beans

Answers
1. not complementary; both are low in lysine
2. complementary
3. not complementary; both are low in lysine
4. complementary
5. complementary
6. not complementary; both are low in tryptophan

16.3 Proteins: Secondary, Tertiary, and Quaternary Structures

Learning Goal: Describe the secondary, tertiary, and quaternary structures for a protein; describe the denaturation of a protein.

- In the secondary structure, hydrogen bonds between different sections of the peptide produce a characteristic shape such as an α helix and β-pleated sheet.
- Collagen, which contains a triple helix of peptide chains, makes up as much as one-third of all protein in the body.
- In a tertiary structure, hydrophobic R groups are found on the inside and hydrophilic R groups are found on the surface. The tertiary structure is stabilized by interactions between R groups.
- In a quaternary structure, two or more subunits combine for biological activity. They are held together by the same interactions found in tertiary structures.
- Denaturation of a protein occurs when denaturing agents destroy the secondary, tertiary, and quaternary structures (but not the primary structure) of the protein which causes a loss of biological activity.
- Denaturing agents include heat, acids, bases, organic solvents, agitation, and heavy metal ions.

Amino Acids, Proteins, and Enzymes

♦ Learning Exercise 16.3A

CORE CHEMISTRY SKILL
Identifying the Primary, Secondary, Tertiary, and Quaternary Structures of Proteins

Identify the following descriptions of protein structure as primary or secondary:

1. _____ hydrogen bonding forming an α helix
2. _____ hydrogen bonding occurring between C=O and N—H within a peptide chain
3. _____ the order of amino acids linked by peptide bonds
4. _____ hydrogen bonds between protein chains forming a β-pleated sheet

Answers   1. secondary   2. secondary   3. primary   4. secondary

♦ Learning Exercise 16.3B

Identify the following descriptions of protein structure as tertiary, quaternary, or both:

1. _____ a disulfide bond joining distant parts of a protein chain
2. _____ the combination of two or more protein subunits
3. _____ hydrophilic side groups seeking contact with water
4. _____ a salt bridge forming between two oppositely charged side chains
5. _____ hydrophobic side groups forming a nonpolar center

Answers
1. tertiary
2. quaternary
3. both tertiary and quaternary
4. both tertiary and quaternary
5. both tertiary and quaternary

♦ Learning Exercise 16.3C

Indicate the denaturing agent in the following examples:

   a. heat
   b. pH change
   c. organic solvent
   d. heavy metal ions
   e. agitation

1. _____ placing surgical instruments in a 120 °C autoclave
2. _____ whipping cream to make a dessert
3. _____ applying tannic acid to a burn
4. _____ placing $AgNO_3$ drops in the eyes of newborns
5. _____ using alcohol to disinfect a wound
6. _____ using a *Lactobacillus* culture to produce acid that converts milk to yogurt

Answers   1. a   2. e   3. b   4. d   5. c   6. b

Chapter 16

16.4 Enzymes

Learning Goal: Describe enzymes and their role in enzyme-catalyzed reactions.

- Enzymes are proteins that act as biological catalysts.
- Enzymes accelerate the rate of biological reactions by lowering the activation energy of a reaction.
- Within the structure of an enzyme, there is a small pocket called the active site, which has a specific shape that fits a specific substrate.
- In the induced-fit model, a substrate and a flexible active site adjust their shapes to form an enzyme–substrate complex.
- The names of many enzymes end with *ase*.
- Enzymes are classified by the type of reaction they catalyze: oxidation–reduction, transfer of groups, hydrolysis, the addition or removal of a group, isomerization, or the joining of two smaller molecules.

♦ **Learning Exercise 16.4A**

Indicate whether each of the following is a characteristic of an enzyme: *yes* or *no*

1. _____ is a biological catalyst
2. _____ does not change the equilibrium position of a reaction
3. _____ greatly increases the rate of a cellular reaction
4. _____ is needed for every reaction that takes place in the cell
5. _____ catalyzes at a faster rate at lower temperatures
6. _____ lowers the activation energy of a biological reaction

Answers   1. yes   2. yes   3. yes   4. yes   5. no   6. yes

♦ **Learning Exercise 16.4B**

CORE CHEMISTRY SKILL
Describing Enzyme Action

Match the terms (**a** to **e**) with the following descriptions:

   a. active site
   d. lock-and-key
   b. substrate
   e. induced-fit
   c. enzyme–substrate complex

1. _____ the combination of an enzyme with a substrate
2. _____ a model of enzyme action in which the rigid shape of the active site exactly fits the shape of the substrate
3. _____ has a structure that fits the shape of the active site
4. _____ a model of enzyme action in which the shape of the active site adjusts to fit the shape of a substrate
5. _____ the portion of an enzyme that binds to the substrate and catalyzes the reaction

Answers   1. c   2. d   3. b   4. e   5. a

Amino Acids, Proteins, and Enzymes

♦ Learning Exercise 16.4C

Write an equation to illustrate the following:

1. the formation of an enzyme–substrate complex _____

2. the conversion of an enzyme–substrate complex to product _____

Answers 1. $E + S \rightleftharpoons ES$ 2. $ES \longrightarrow E + P$

♦ Learning Exercise 16.4D

Match the class of enzymes with each of the following types of reactions:

   a. oxidoreductase   b. transferase   c. hydrolase
   d. lyase   e. isomerase   f. ligase

1. _____ combines small molecules using energy from ATP
2. _____ transfers phosphate groups
3. _____ hydrolyzes a disaccharide into two glucose units
4. _____ converts a substrate to an isomer of the substrate
5. _____ adds hydrogen to a substrate
6. _____ removes H_2O from a substrate
7. _____ adds oxygen to a substrate
8. _____ converts a cis structure to a trans structure

Answers 1. f   2. b   3. c   4. e
           5. a   6. d   7. a   8. e

16.5 Factors Affecting Enzyme Activity

Learning Goal: Describe the effect of temperature, pH, and inhibitors on enzyme activity.

- Enzymes are most effective at their optimum temperature and pH. The rate of an enzyme-catalyzed reaction decreases considerably at temperatures and pH above or below the optimum.
- An enzyme can be made inactive by changes in pH or temperature or by chemical compounds called inhibitors.
- A competitive inhibitor has a structure similar to the substrate and competes for the active site. When the active site is occupied by a competitive inhibitor, the enzyme cannot catalyze the reaction of the substrate.
- A noncompetitive inhibitor attaches elsewhere on the enzyme, changing the shape of both the enzyme and the active site. As long as the noncompetitive inhibitor is attached to the enzyme, the altered active site cannot bind or catalyze the substrate.
- A reversible inhibitor causes a reversible loss of enzymatic activity.
- An irreversible inhibitor forms a permanent covalent bond with an amino acid side group in the active site of the enzyme, which prevents enzymatic activity.

Chapter 16

♦ **Learning Exercise 16.5A**

An enzyme has an optimum pH of 7 and an optimum temperature of 37 °C.

a. Draw a graph to represent the activity of the enzyme at the pH values shown. Indicate the optimum pH.

b. Draw a graph to represent the activity of the enzyme at the temperature shown. Indicate the optimum temperature.

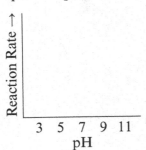
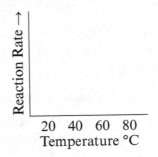

Answers

a.
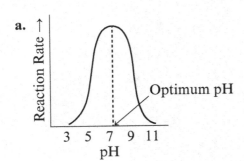

b.
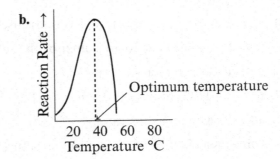

♦ **Learning Exercise 16.5B**

Urease has an optimum pH of 7.0. Indicate the effect of each of the following on the reaction rate of a urease-catalyzed reaction:

    a. increases   b. decreases   c. doesn't change

1. _____ adding an inhibitor
2. _____ running the reaction at pH 9.0
3. _____ lowering the temperature to 0 °C
4. _____ running the reaction at 85 °C
5. _____ adjusting pH to its optimum

Answers   **1.** b   **2.** b   **3.** b   **4.** b   **5.** a

♦ **Learning Exercise 16.5C**

Identify each of the following as characteristic of competitive inhibition (C), noncompetitive inhibition (N), or irreversible inhibition (I):

1. _____ An inhibitor binds to the surface of the enzyme away from the active site.
2. _____ An inhibitor resembling the substrate molecule blocks the active site on the enzyme.
3. _____ An inhibitor causes permanent damage to the enzyme with a total loss of biological activity.

4. ___ The action of this inhibitor can be reversed by adding more substrate.
5. ___ The action of this inhibitor is not reversed by adding more substrate, but can be reversed when chemical reagents remove the inhibitor.
6. ___ Sulfanilamide stops bacterial infections because its structure is similar to 4-aminobenzoic acid, which is essential for bacterial growth.

Answers   1. N   2. C   3. I   4. C   5. N   6. C

Key Terms for Sections 16.1 to 16.5

Match each of the following key terms with the correct description:

a. lock-and-key   **b.** protein   **c.** tertiary structure   **d.** peptide   **e.** inhibitor
f. enzyme   **g.** alpha helix   **h.** denaturation   **i.** active site

1. ___ the portion of an enzyme structure where a substrate undergoes reaction
2. ___ a protein that catalyzes a biological reaction in the cells
3. ___ the folding of a protein into a compact structure
4. ___ a model of enzyme action in which the substrate exactly fits the shape of the active site of an enzyme
5. ___ a substance that makes an enzyme inactive by interfering with its ability to react with a substrate
6. ___ the loss of secondary, tertiary, and/or quaternary structure of a protein
7. ___ the combination of two or more amino acids
8. ___ a polypeptide that has biological activity
9. ___ a secondary level of protein structure

Answers   1. i   2. f   3. c   4. a   5. e   6. h   7. d   8. b   9. g

Checklist for Chapter 16

You are ready to take the Practice Test for Chapter 16. Be sure you have accomplished the following learning goals for this chapter. If not, review the Section listed at the end of the goal. Then apply your new skills and understanding to the Practice Test.

After studying Chapter 16, I can successfully:

___ Classify proteins by their functions in the body. (16.1)

___ Draw the ionized structural formula for an amino acid at physiological pH. (16.1)

___ Describe a peptide bond; draw the structural formulas for a peptide. (16.2)

___ Describe the primary structure of a protein. (16.2)

___ Describe the secondary, tertiary, and quaternary structures of a protein. (16.3)

___ Describe the ways that denaturation affects the structure of a protein. (16.3)

___ Describe the induced-fit model of enzyme action. (16.4)

___ Classify enzymes according to the type of reaction they catalyze. (16.4)

___ Discuss the effects of changes in temperature, pH, and inhibitors on enzyme action. (16.5)

Chapter 16

Practice Test for Chapter 16

The chapter Sections to review are shown in parentheses at the end of each question.

1. Which amino acid is nonpolar? (16.1)
   A. serine
   B. aspartate
   C. valine
   D. cysteine
   E. lysine
2. Which amino acid will form disulfide bonds in a tertiary structure? (16.3)
   A. serine
   B. aspartate
   C. valine
   D. cysteine
   E. lysine
3. Which amino acid has a basic side chain? (16.1)
   A. serine
   B. aspartate
   C. valine
   D. cysteine
   E. lysine
4. All amino acids (16.1)
   A. have the same side chains
   B. have ionized structures
   C. have the same charge at physiological pH
   D. show hydrophobic tendencies
   E. are essential amino acids

For questions 5 through 8, match the function to the correct protein: (16.1)
   A. enzyme
   B. structural
   C. transport
   D. storage

5. myoglobin in muscle
6. α-keratin in skin
7. peptidase for protein hydrolysis
8. ferritin in the liver

9. Essential amino acids (16.2)
   A. are the amino acids that must be supplied by the diet
   B. are not synthesized by the body
   C. are missing in incomplete proteins
   D. are present in proteins from animal sources
   E. all of the above

10. The sequence Tyr–Ala–Gly (16.2)
    A. is a tripeptide
    B. has two peptide bonds
    C. has tyrosine at the N-terminus
    D. has glycine at the C-terminus
    E. all of these

11. The bonding in a tertiary structure expected between lysine and aspartate is a (16.3)
    A. salt bridge
    B. hydrogen bond
    C. disulfide bond
    D. hydrophobic interaction
    E. hydrophilic interaction
12. What type of bond is used to form the α helix structure of a protein? (16.3)
    A. peptide bond
    B. hydrogen bond
    C. salt bridge
    D. disulfide bond
    E. hydrophobic interaction
13. What type of bond places portions of the protein chain in the center of a tertiary structure? (16.3)
    A. peptide bond
    B. salt bridge
    C. disulfide bond
    D. hydrophobic interaction
    E. hydrophilic interaction

For questions 14 through 18, identify the protein structural level that each of the following phrases describe: (16.2, 16.3)
   A. primary
   B. secondary
   C. tertiary
   D. quaternary
   E. pentenary

14. peptide bonds
15. a β-pleated sheet
16. two or more protein subunits
17. an α helix
18. disulfide bonds

19. Denaturation of a protein (16.3)
  A. occurs at a pH of 7
  B. causes a change in protein structure
  C. hydrolyzes a protein
  D. oxidizes the protein
  E. adds amino acids to a protein

20. Which of the following will not cause denaturation? (16.3)
  A. 0 °C   **B.** $AgNO_3$ **C.** 80 °C   **D.** ethanol   **E.** pH 1

21. Enzymes (16.4)
  A. are biological catalysts
  B. are polysaccharides
  C. are insoluble in water
  D. are lipids
  E. are named with an *ose* ending

For questions 22 through 26, select answers from the following (E = enzyme; S = substrate; P = product): (16.4)
  A. S $\longrightarrow$ P   **B.** E + S $\rightleftarrows$ ES   **C.** ES $\longrightarrow$ E + P

22. ____ the enzymatic reaction occurring at the active site

23. ____ the release of product from the enzyme

24. ____ the first step in enzyme action

25. ____ the formation of the enzyme–substrate complex

26. ____ the final step in enzyme action

The enzyme sucrase has an optimum pH of 6.2 and optimum temperature of 37 °C. The changes in questions 27 through 30 would result in which of the following? (16.5)
  A. increases the rate of reaction
  B. decreases the rate of reaction
  C. denatures the enzyme, and no reaction occurs

27. ____ setting the reaction tube in a beaker of water at 100 °C

28. ____ running the reaction at 10 °C

29. ____ adding ethanol to the reaction system

30. ____ adjusting the pH to optimum pH

For questions 31 through 35, identify each description of inhibition as one of the following: (16.5)
  A. competitive   **B.** noncompetitive

31. ____ this alteration affects the overall shape of the enzyme

32. ____ a molecule that closely resembles the substrate interferes with enzymatic activity

33. ____ the inhibition can be reversed by increasing substrate concentration

34. ____ the heavy metal ion, Pb^{2+}, bonds with an —SH side group

35. ____ the inhibition is not affected by increased substrate concentration

Answers to the Practice Test

1. C	**2.** D	**3.** E	**4.** B	**5.** C
6. B	**7.** A	**8.** D	**9.** E	**10.** E
11. A	**12.** B	**13.** D	**14.** A	**15.** B
16. D	**17.** B	**18.** C, D	**19.** B	**20.** A
21. A	**22.** A	**23.** C	**24.** B	**25.** B
26. C	**27.** C	**28.** B	**29.** C	**30.** A
31. B	**32.** A	**33.** A	**34.** B	**35.** B

Chapter 16

Selected Answers and Solutions to Text Problems

16.1
 a. Hemoglobin, which carries oxygen in the blood, is a transport protein.
 b. Collagen, which is a major component of tendons and cartilage, is a structural protein.
 c. Keratin, which is found in hair, is a structural protein.
 d. Amylases, which catalyze the hydrolysis of starch, are enzymes.

16.3 All α-amino acids contain a carboxylate group and an ammonium group on the alpha carbon.

16.5

a. $H_3\overset{+}{N}$—CH(H)—C(=O)—O$^-$ (with H on alpha carbon)

b. $H_3\overset{+}{N}$—C(H)(CH(CH_3)(OH))—C(=O)—O$^-$

c. $H_3\overset{+}{N}$—C(H)(CH_2CH_2COO$^-$)—C(=O)—O$^-$

d. $H_3\overset{+}{N}$—C(H)(CH_2–C_6H_5)—C(=O)—O$^-$

16.7
 a. Glycine has an H as its R group which makes it a nonpolar amino acid. It is hydrophobic.
 b. Threonine has an R group that contains the polar —OH group which makes threonine a polar neutral amino acid. It is hydrophilic.
 c. Glutamate has an R group containing a polar carboxylic acid group. Glutamate is a polar acidic amino acid. It is hydrophilic.
 d. Phenylalanine has an R group containing an aromatic ring; this makes phenylalanine a nonpolar amino acid. It is hydrophobic.

16.9 Amino acids have both three-letter and one-letter abbreviations.
 a. Ala is the three-letter abbreviation for the amino acid alanine.
 b. Q is the one-letter abbreviation for the amino acid glutamine.
 c. K is the one-letter abbreviation for the amino acid lysine.
 d. Cys is the three-letter abbreviation for the amino acid cysteine.

16.11 In a peptide, the amino acids are joined by peptide bonds (amide bonds). The first amino acid has a free —NH$_3^+$ group, and the last one has a free —COO$^-$ group.

a. $H_3\overset{+}{N}$—C(CH_3)(H)—C(=O)—N(H)—C(CH_2SH)(H)—C(=O)—O$^-$
 Ala–Cys, AC

b. $H_3\overset{+}{N}$—C(CH_2OH)(H)—C(=O)—N(H)—C(CH_2–C_6H_5)(H)—C(=O)—O$^-$
 Ser–Phe, SF

c. 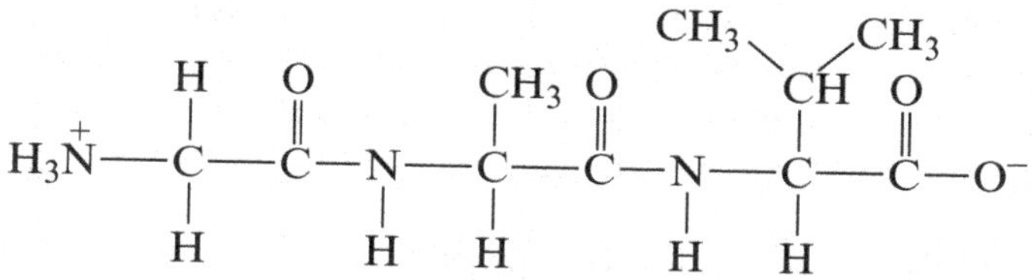 Structure of Gly–Ala–Val, GAV

d. 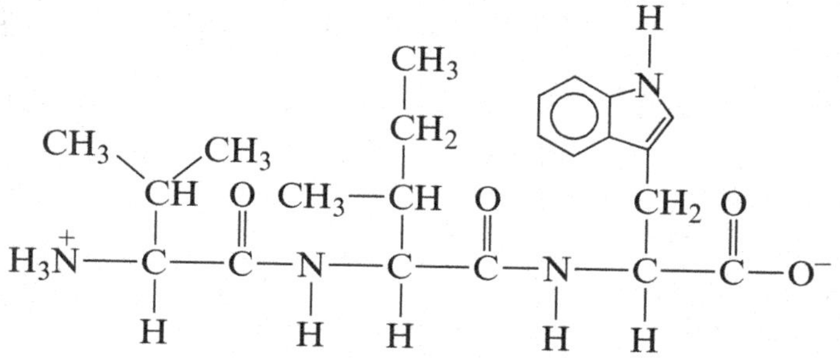 Structure of Val–Ile–Trp, VIW

16.13 a. 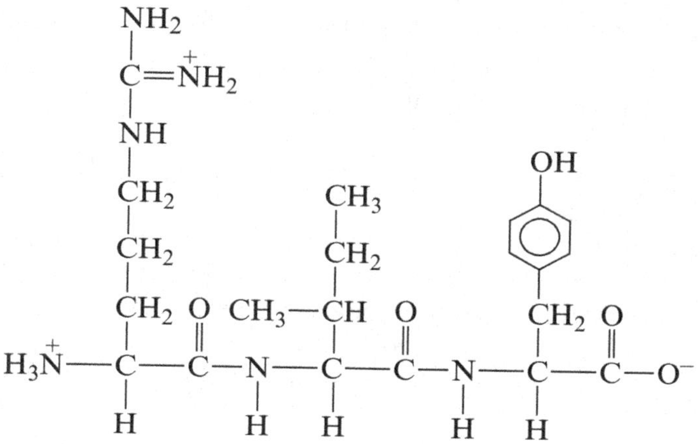 Structure of RIY

b. 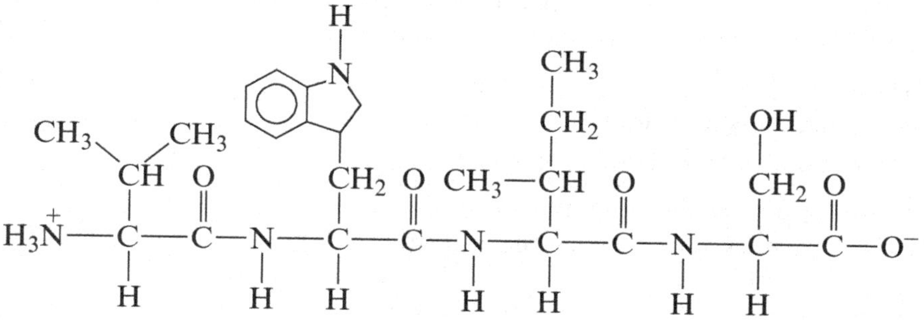 Structure of VWIS

16.15 a. Lysine and tryptophan are lacking in corn but supplied by peas; methionine is lacking in peas but supplied by corn.
  b. Lysine is lacking in rice but supplied by soy; methionine is lacking in soy but supplied by rice.

16.17 The oxygen atoms of the carbonyl groups and the hydrogen atoms attached to the nitrogen atoms form alpha helices or beta-pleated sheets.

16.19 In the α helix, hydrogen bonds form between the carbonyl oxygen atom and the amino hydrogen atom of a peptide bond in the next turn of the helical chain. In the β-pleated sheet, hydrogen bonds occur between parallel sections of a long polypeptide chain.

Chapter 16

16.21 a. The two cysteines have —SH groups, which react to form a disulfide bond.
  b. Aspartate is acidic and lysine is basic; an ionic attraction called a salt bridge, is formed between the —COO$^-$ in the R group of Asp and the —NH$_3^+$ in the R group of Lys.
  c. Serine has a polar —OH group that can form a hydrogen bond with the carboxyl group of aspartate.
  d. Two leucines have R groups that are hydrocarbons and nonpolar. They would have a hydrophobic interaction.

16.23 a. Leucine and valine will be found on the inside of the protein structure because they have nonpolar R groups that are hydrophobic.
  b. Cysteine and aspartate would be on the outside of the protein because they have R groups that are polar.
  c. The order of the amino acids (the primary structure) provides the R groups that interact to determine the tertiary structure of the protein.

16.25 a. Disulfide bonds and salt bridges join different sections of the protein chain to give a three-dimensional shape. Disulfide bonds and salt bridges are important in the tertiary and quaternary structures.
  b. Peptide bonds join the amino acid building blocks in the primary structure of a polypeptide.
  c. Hydrogen bonds that hold adjacent polypeptide chains together are found in the secondary structures of β-pleated sheets.
  d. In the secondary structure of α helices, hydrogen bonding occurs between amino acids in the same polypeptide to give a coiled shape to the protein.

16.27 a. Placing an egg in boiling water coagulates the proteins of the egg because the heat disrupts hydrogen bonds and hydrophobic interactions.
  b. The alcohol on the swab coagulates the proteins of any bacteria present on the surface of the skin by forming hydrogen bonds and disrupting hydrophobic interactions.
  c. The heat from an autoclave will coagulate the proteins of any bacteria on the surgical instruments by disrupting hydrogen bonds and hydrophobic interactions.
  d. Heat will coagulate the surrounding proteins to close the wound by disrupting hydrogen bonds and hydrophobic interactions.

16.29 Chemical reactions in the body can occur without enzymes, but the rates are too slow at the relatively mild conditions of normal body temperature and pH. Catalyzed reactions, which are many times faster, provide the amounts of products needed by the cell at a particular time.

16.31 a. The reactant for the enzyme galactase is the sugar galactose.
  b. Lipase catalyzes the hydrolysis of lipids.
  c. The reactant for the enzyme aspartase is aspartate.

16.33 a. A hydrolase enzyme would catalyze the hydrolysis of sucrose.
  b. An oxidoreductase enzyme would catalyze the addition of oxygen (oxidation).
  c. An isomerase enzyme would catalyze converting glucose to fructose.
  d. A transferase enzyme would catalyze moving an amino group from one molecule to another.

16.35 a. An enzyme (2) has a tertiary structure that recognizes the substrate.
  b. The combination of the enzyme and substrate is the enzyme–substrate complex (1).
  c. The substrate (3) has a structure that fits the active site of the enzyme.

16.37 a. The equation for an enzyme–catalyzed reaction is:
  E + S $\rightleftharpoons$ ES $\longrightarrow$ E + P
  E = enzyme, S = substrate, ES = enzyme–substrate complex, P = products
  b. The active site is a region or pocket within the tertiary structure of an enzyme that accepts the substrate, aligns the substrate for reaction, and catalyzes the reaction.

16.39 Isoenzymes are slightly different forms of an enzyme that catalyze the same reaction in different organs and tissues of the body.

16.41 A doctor might run tests for the enzymes CK, LDH, and AST to determine if the patient had a heart attack.

16.43 a. Running the reaction at a pH below optimum pH will decrease the rate of reaction.
  b. Temperatures above 37 °C (optimum temperature) will denature the enzymes and decrease the rate of reaction.

16.45 From the graph, the optimum pH values for the enzymes are approximately: pepsin, pH 2; sucrase, pH 6; trypsin, pH 8.

16.47 a. If the inhibitor has a structure similar to the substrate, the inhibitor is competitive.
  b. If adding more substrate cannot reverse the effect of the inhibitor, the inhibitor is noncompetitive.
  c. If the inhibitor competes with the substrate for the active site, it is a competitive inhibitor.
  d. If the structure of the inhibitor is not similar to the substrate, the inhibitor is noncompetitive.
  e. If adding more substrate reverses the inhibition, the inhibitor is competitive.

16.49 a. Methanol has the condensed structural formula CH_3-OH, whereas ethanol is CH_3-CH_2-OH.
  b. Ethanol has a structure similar to methanol and could compete for the active site.
  c. Ethanol is a competitive inhibitor of alcohol dehydrogenase.

16.51 $CH_4N_2O_2$

16.53 $21 \text{ lb} \times \dfrac{1 \text{ kg}}{2.20 \text{ lb}} \times \dfrac{20. \text{ mg hydroxyurea}}{1 \text{ kg body weight}} = 190$ mg of hydroxyurea (2 SFs)

16.55 a. The oxidation of a glycol to an aldehyde and carboxylic acid is catalyzed by an oxidoreductase.
  b. At high concentration, ethanol, which acts as a competitive inhibitor of ethylene glycol, would saturate the alcohol dehydrogenase enzyme to allow ethylene glycol to be removed from the body without producing oxalic acid.

16.57 a. The amino acids in aspartame are aspartate and phenylalanine.
  b. The dipeptide in aspartame would be named aspartylphenylalanine.

16.59 a. Asparagine and serine are both polar neutral amino acids; their R groups can interact by hydrogen bonding.
  b. Aspartate is a polar acidic amino acid, and lysine is a polar basic amino acid; their R groups interact by forming a salt bridge.
  c. The polar neutral amino acid cysteine contains the $-SH$ group; two cysteines can form a disulfide bond.
  d. Leucine and alanine are both nonpolar amino acids; their R groups have a hydrophobic interaction.

16.61 a.

(tripeptide structure with side chains: $-CH_2-OH$ (serine), $-(CH_2)_4-NH_3^+$ (lysine), $-CH_2-COO^-$ (aspartate))

  b. This segment contains polar R groups, which would be found on the surface of a protein where they can hydrogen bond with water.

16.63 a. Yes, a combination of rice and garbanzo beans provides all the essential amino acids; garbanzo beans contain the lysine missing in rice.
  b. No, a combination of lima beans and cornmeal does not provide all the essential amino acids; both are deficient in the amino acid tryptophan.

c. No, a combination of garbanzo beans and lima beans does not provide all the essential amino acids; both are deficient in the amino acid tryptophan.

16.65 a. The secondary structure of a protein depends on hydrogen bonds to form a helix or a pleated sheet; the tertiary structure is determined by the interactions of R groups such as disulfide bonds, hydrogen bonds, and salt bridges, and determines the three-dimensional shape of the protein.
  b. Nonessential amino acids can be synthesized by the body; essential amino acids must be supplied by the diet.
  c. Polar amino acids have hydrophilic R groups, whereas nonpolar amino acids have hydrophobic R groups.
  d. Dipeptides contain two amino acids, whereas tripeptides contain three amino acids.

16.67 Glutamate is a polar acidic amino acid, whereas proline is nonpolar. The polar acidic R group of glutamate may form hydrogen bonds or salt bridges with another amino acid R group. It may also move to the outer surface of the protein where it would form hydrophilic interactions with water. However, the nonpolar proline would move to the hydrophobic center of the protein.

16.69 In chemical laboratories, reactions are often run at high temperatures using catalysts that are strong acids or bases. Enzymes are catalysts that are proteins and function only at mild temperatures and pH. Catalysts used in chemical laboratories are usually inorganic materials that can function at high temperatures and in strongly acidic or basic conditions.

16.71 a. The reactants are lactose and water, and the products are glucose and galactose.
  b.

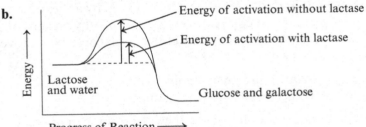

  c. By lowering the energy of activation, the enzyme furnishes a lower energy pathway by which the reaction can take place.

16.73 a. The disaccharide lactose is a substrate (S).
  b. The suffix *ase* in lipase indicates that it is an enzyme (E).
  c. The suffix *ase* in amylase indicates that it is an enzyme (E).
  d. Trypsin is an enzyme which hydrolyzes polypeptides (E).
  e. Pyruvate is a substrate (S).
  f. The suffix *ase* in transaminase indicates that it is an enzyme (E).

16.75 a. Urea is the substrate of urease.
  b. Succinate is the substrate of succinate dehydrogenase.
  c. Aspartate is the substrate of aspartate transaminase.
  d. Tyrosine is the substrate of tyrosinase.

16.77 Sucrose fits the shape of the active site in sucrase, but lactose does not.

16.79 A heart attack may be the cause. Normally the enzymes LDH and CK are present only in low levels in the blood.

16.81 a. Valine has a nonpolar R group which would be found in hydrophobic regions.
  b. Lysine and aspartate have polar R groups which would be found in hydrophilic regions.
  c. Lysine and aspartate have polar R groups containing —COO^- and —NH_3^+ groups which can form hydrogen bonds.
  d. Lysine is a polar basic amino acid, and aspartate is a polar acidic amino acid; they can form salt bridges.

17 Nucleic Acids and Protein Synthesis

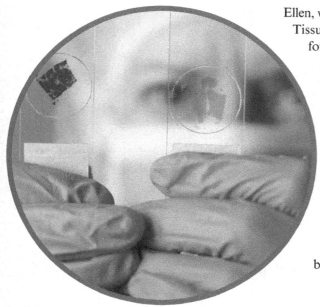

Ellen, who was diagnosed with breast cancer, has had a lumpectomy. Tissue samples from the tumor and lymph nodes were prepared for a pathologist by placing a thin section, 0.001 mm thick, on a microscope slide. The tissues, which tested positive for estrogen receptor, confirmed that Ellen's breast cancer tumors required estrogen for their growth. The bonding of estrogen to the estrogen receptor increases the production of mammary cells and potential mutations that can lead to cancer. The tissues were also tested for altered genes BRCA1 and BRCA2. The normal genes suppress tumor growth, whereas the mutated genes lose their ability to suppress tumor growth, which increases the risk for breast cancer. Ellen's test results for a BRCA mutation were negative. How does tamoxifen reduce the risk of breast cancer?

Credit: Malcom Park/RGB Ventures/SuperStock/Alamy

LOOKING AHEAD

17.1 Components of Nucleic Acids
17.2 Primary Structure of Nucleic Acids
17.3 DNA Double Helix and Replication
17.4 RNA and Transcription
17.5 The Genetic Code and Protein Synthesis
17.6 Genetic Mutations
17.7 Recombinant DNA
17.8 Viruses

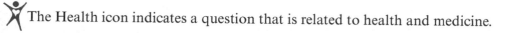 The Health icon indicates a question that is related to health and medicine.

17.1 Components of Nucleic Acids

Learning Goal: Describe the bases and ribose sugars that make up the nucleic acids DNA and RNA.

- Nucleic acids are composed of bases, five-carbon sugars, and phosphate groups.
- In DNA, the bases are adenine, thymine, guanine, and cytosine. In RNA, uracil replaces thymine.
- In DNA, the sugar is deoxyribose; in RNA, the sugar is ribose.
- A nucleoside is composed of a base and a sugar.
- A nucleotide is composed of a base, a sugar, and a phosphate group.
- Deoxyribonucleic acid (DNA) and ribonucleic acid (RNA) are linear chains of nucleotides.

♦ **Learning Exercise 17.1A**

1. Write the names and abbreviations for the bases in each of the following:

   DNA _____   RNA _____

Chapter 17

2. Write the name of the sugar in each of the following:

DNA _____   RNA _____

Answers   1. DNA: adenine (A), thymine (T), guanine (G), cytosine (C)
RNA: adenine (A), uracil (U), guanine (G), cytosine (C)
2. DNA: deoxyribose; RNA: ribose

◆ **Learning Exercise 17.1B**

Name each of the following, and classify it as a purine or a pyrimidine:

a. 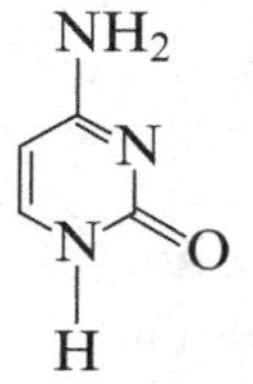 [structure] b. 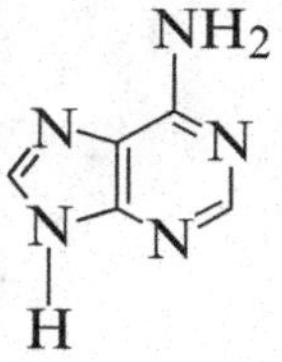 [structure]

c. 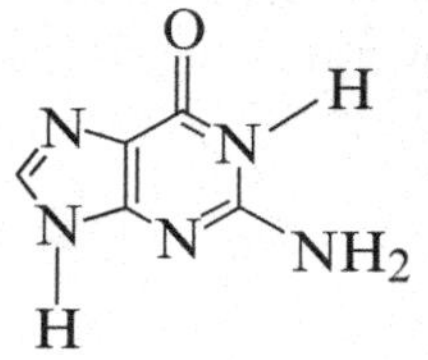 [structure] d. 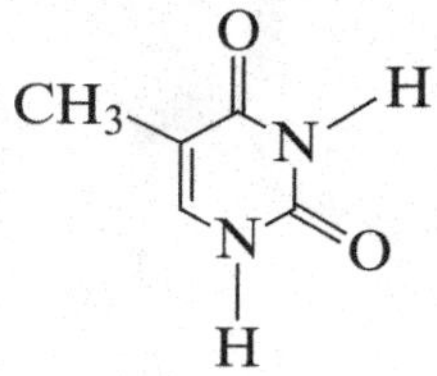 [structure]

Answers   a. cytosine; pyrimidine   b. adenine; purine
c. guanine; purine   d. thymine; pyrimidine

◆ **Learning Exercise 17.1C**

Identify the nucleic acid (DNA or RNA) in which each of the following is found:

1. _____ adenosine monophosphate   2. _____ guanosine monophosphate
3. _____ dCMP   4. _____ cytidine monophosphate
5. _____ deoxythymidine monophosphate 6. _____ UMP
7. _____ dGMP   8. _____ deoxyadenosine monophosphate

Answers   1. RNA   2. RNA   3. DNA   4. RNA
5. DNA   6. RNA   7. DNA   8. DNA

◆ **Learning Exercise 17.1D**

Draw the condensed structural formula for deoxythymidine monophosphate. Indicate the 5′ and the 3′ carbon atoms on the sugar.

Answer

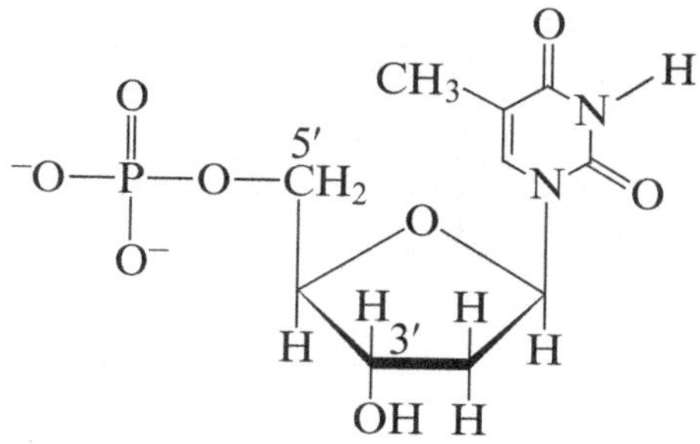

Deoxythymidine monophosphate (dTMP)

17.2 Primary Structure of Nucleic Acids

Learning Goal: Describe the primary structures of RNA and DNA.

- Nucleic acids are linear chains of nucleotides in which the —OH group on the 3′ carbon of a sugar in one nucleotide bonds to the phosphate group attached to the 5′ carbon of a sugar in the adjacent nucleotide.

♦ **Learning Exercise 17.2A**

Draw the structure of a dinucleotide that consists of cytidine monophosphate (free 5′ phosphate) bonded to guanosine monophosphate (free 3′ hydroxyl). Identify each nucleotide, the phosphodiester bond, the free 5′ phosphate group, and the free 3′ hydroxyl group.

Answer

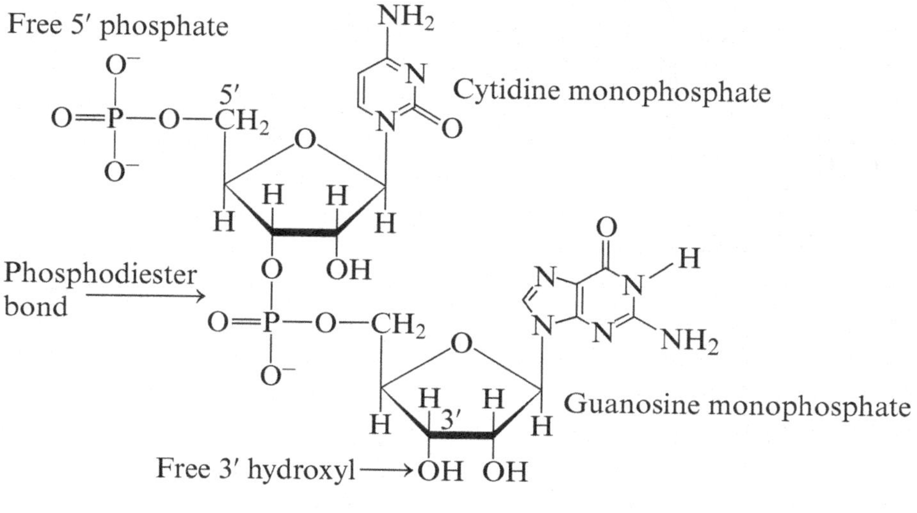

Chapter 17

♦ **Learning Exercise 17.2B**

Consider the following sequence of nucleotides in RNA: A G U C

1. Write the names of the nucleotides in this sequence. _____
   _____

2. Which nucleotide has the free 5' phosphate group? _____

3. Which nucleotide has the free 3' hydroxyl group? _____

Answers   1. adenosine monophosphate, guanosine monophosphate, uridine monophosphate, cytidine monophosphate

2. adenosine monophosphate   3. cytidine monophosphate

17.3 DNA Double Helix and Replication

Learning Goal: Describe the double helix of DNA; describe the process of DNA replication.

- The two strands in DNA are held together by hydrogen bonds between complementary base pairs, A with T and G with C.
- One DNA strand runs in the 5' to 3' direction, and the other strand runs in the 3' to 5' direction.
- During replication, polymerase makes new DNA strands along each of the original DNA strands that serve as templates.
- Complementary base pairs ensures the correct pairing of bases to give identical copies of the original DNA.

♦ **Learning Exercise 17.3A**

Complete the following statements:

1. The structure of the two strands of nucleotides in DNA is called a _____.

2. In one strand of DNA, the sugar–phosphate backbone runs in the 5' to 3' direction, whereas the opposite strand goes in the _____ direction.

3. On the DNA strands, the free phosphate group is at the _____ end and the free hydroxyl group is at the _____ end.

4. The only combinations of base pairs that connect the two DNA strands are _____ and _____.

5. The base pairs along one DNA strand are _____ to the base pairs on the opposite strand.

Answers   1. double helix   2. 3' to 5'   3. 5'; 3'   4. AT; GC   5. complementary

♦ **Learning Exercise 17.3B**

Write the complementary base sequence for the each of following segments of a strand of DNA:

CORE CHEMISTRY SKILL
Writing the Complementary DNA Strand

1. A A A T T T C C C G G G

2. A T G C T T G G C T C C

3. G C G C T C A C A T G C

Answers **1.** TTTAAAGGGCCC   **2.** TACGAACCGAGG
3. CGCGAGTGTACG

♦ **Learning Exercise 17.3C**

Describe the process by which DNA replicates to produce identical copies.

Answer   The DNA replication process begins when an enzyme catalyzes the unwinding of a portion of the DNA helix by breaking the hydrogen bonds between the complementary bases. The bases on the resulting single strands act as templates for the synthesis of new complementary strands of DNA. As the complementary base pairs form, an enzyme catalyzes the formation of phosphodiester linkages between them. Eventually the entire double helix of the parent DNA is copied.

17.4 RNA and Transcription

Learning Goal: Identify the different types of RNA; describe the synthesis of mRNA.

- The three types of RNA differ by their function in the cell: ribosomal RNA makes up most of the structure of the ribosomes and is the site of protein synthesis; messenger RNA carries information from the DNA to the ribosomes; and transfer RNA places the correct amino acids in the protein.
- Transcription is the process by which genetic information for the synthesis of a protein is copied from a gene in DNA to make mRNA.
- The bases in the mRNA are complementary to the DNA, except U in RNA is paired with A in DNA.
- The RNA polymerase enzyme moves along an unwound section of DNA.
- The production of mRNA occurs when certain proteins are needed in the cell.

♦ **Learning Exercise 17.4A**

Match each of the following characteristics with mRNA, tRNA, or rRNA:

CORE CHEMISTRY SKILL
Writing the mRNA Segment for a DNA Template

1. _____ is most abundant in a cell
2. _____ has the shortest chain of nucleotides
3. _____ carries information from DNA to the ribosomes for protein synthesis
4. _____ is the major component of ribosomes
5. _____ carries specific amino acids to the ribosome for protein synthesis
6. _____ consists of a large and a small subunit

Answers   **1.** rRNA   **2.** tRNA   **3.** mRNA   **4.** rRNA   **5.** tRNA   **6.** rRNA

Chapter 17

♦ **Learning Exercise 17.4B**

Fill in the blanks with a word or phrase that answers each of the following questions:

1. Where in the cell does transcription take place? _____
2. How many strands of a DNA molecule are involved in transcription? _____
3. What are the abbreviations for the four nucleotides in mRNA? _____
4. What is the corresponding section of an mRNA produced from each of the following sections of DNA template strand?

    a. C A T T C G G T A   b. G T A C C T A A C G T C C G

Answers   1. nucleus   2. one   3. A, U, G, C
4. a. G U A A G C C A U   b. C A U G G A U U G C A G G C

17.5 The Genetic Code and Protein Synthesis

Learning Goal: Use the genetic code to write the amino acid sequence for a segment of mRNA.

- The genetic code consists of a sequence of three nucleotides (triplets) called codons that specify the amino acids in a protein.
- The 64 codons for 20 amino acids allow several codons for most amino acids.
- The codon AUG signals the start of transcription, and codons UAG, UGA, and UAA signal the end of transcription.
- Proteins are synthesized at the ribosomes in a translation process that includes three steps: initiation, elongation, and termination.
- During translation, the appropriate tRNA brings the correct amino acid to the ribosome, where the amino acid is bonded by a peptide bond to the growing peptide chain.
- When the polypeptide is released, it takes on its secondary, tertiary, or quaternary structures to become a functional protein in the cell.

♦ **Learning Exercise 17.5A**

CORE CHEMISTRY SKILL
Writing the Amino Acid for an mRNA Codon

Give the three-letter and one-letter abbreviations for each of the amino acids coded for by the following mRNA codons:

1. UUU _____   2. AGG _____

3. AGC _____   4. GAC _____

5. GGA _____   6. ACA _____

7. AUG _____   8. CUC _____

9. CAU _____   10. GUU _____

Answers   1. Phe, F   2. Arg, R   3. Ser, S   4. Asp, D   5. Gly, G
6. Thr, T   7. Start/Met, M   8. Leu, L   9. His, H   10. Val, V

♦ Learning Exercise 17.5B

Match the components **a** to **e** of the translation process with the correct description **1** to **5** below:

    a. initiation   **b.** activation   **c.** anticodon   **d.** translocation   **e.** termination

1. _____ the three bases in tRNA that complement a codon on the mRNA
2. _____ the combining of an amino acid with a specific tRNA
3. _____ the placement of methionine on the large ribosomal subunit
4. _____ the shift of the ribosome from one codon on mRNA to the next
5. _____ the process that occurs when the ribosome reaches a UAA, UGA, or UAG codon on mRNA

Answers   **1.** c   **2.** b   **3.** a   **4.** d   **5.** e

♦ Learning Exercise 17.5C

Write the mRNA that forms for each of the following template sections of DNA. For each codon in the mRNA, write the three-letter abbreviation of the amino acid that would be placed in the protein by a tRNA.

1. DNA: ACG   TCA   CCA   CGC

    mRNA: _____ _____ _____ _____

    amino acid: _____ _____ _____ _____

2. DNA: ATG   GCC   TTT   GGC   AAC

    mRNA: _____ _____ _____ _____ _____

    amino acid: _____ _____ _____ _____ _____

Answers
1. mRNA: UGC   AGU   GGU   GCG
    amino acid: Cys   Ser   Gly   Ala

2. mRNA: UAC   CGG   AAA   CCG   UUG
    amino acid: Tyr   Arg   Lys   Pro   Leu

♦ Learning Exercise 17.5D

A segment of DNA that codes for a protein contains 270 nucleotides. How many amino acids would be present in the protein for this DNA segment?

Answer   Assuming that the entire segment codes for a protein, there would be 90 (270 ÷ 3) amino acids in the protein produced.

Chapter 17

17.6 Genetic Mutations

Learning Goal: Identify the type of change in DNA for a point mutation, a deletion mutation, and an insertion mutation.

> **REVIEW**
> Identifying the Primary, Secondary, Tertiary, and Quaternary Structures of Proteins (16.2, 16.3)
> Describing Enzyme Action (16.4)

- A genetic mutation is a change of one or more bases in the DNA sequence that may alter the structure and ability of the resulting protein to function properly.
- In a point mutation, one base is changed, which may result in a different amino acid being placed in the protein.
- In an addition or deletion mutation, the insertion or deletion of one base alters all of the codons following the base change, which affects the amino acid sequence that follows the mutation.
- Mutations result from X-rays, overexposure to UV light, chemicals called mutagens, and some viruses.

♦ Learning Exercise 17.6

Consider the following segment of a DNA template strand: AAT CCC GGG

1. Write the mRNA produced.

   _____ _____ _____

2. Write the amino acid order for the mRNA codons.

   _____ _____ _____

3. A point mutation replaces the thymine in this DNA template strand with guanine. Write the mRNA produced.

   _____ _____ _____

4. Write the new amino acid order.

   _____ _____ _____

5. Why is this mutation called a point mutation?

6. How is a point mutation different from a deletion mutation?

7. What are some possible causes of genetic mutations?

Answers

1. UUA GGG CCC
2. Leu–Gly–Pro
3. UUC GGG CCC
4. Phe–Gly–Pro
5. One base is replaced with another.
6. In a point mutation, only one codon is affected and only one amino acid may be different. In a deletion mutation, all of the codons that follow the mutation are shifted by one base, which causes a different amino acid order in the remaining sequence of the protein.
7. X-rays, UV light, chemicals called mutagens, and some viruses are possible causes of genetic mutations.

17.7 Recombinant DNA

Learning Goal: Describe the preparation and uses of recombinant DNA.

- Recombinant DNA is synthesized by opening a piece of DNA and inserting a DNA section from another source.
- Much of the work with recombinant DNA is done with small circular DNA molecules called plasmids found in *E. coli* bacteria.
- Recombinant DNA is used to produce large numbers of copies of foreign DNA that are useful in genetic engineering techniques.

♦ **Learning Exercise 17.7**

Match each of the following terms with the correct description:

    a. plasmids   b. restriction enzymes   c. polymerase chain reaction
    d. recombinant DNA   e. Human Genome Project

1. _____ a synthetic form of DNA that contains a piece of foreign DNA
2. _____ research that determined the DNA sequence of all the genes in a human cell
3. _____ small, circular, DNA molecules found in *E. coli* bacteria
4. _____ a process that makes multiple copies of DNA in a short amount of time
5. _____ enzymes that cut open the DNA strands in plasmids

Answers   1. d   2. e   3. a   4. c   5. b

17.8 Viruses

Learning Goal: Describe the methods by which a virus infects a cell.

- Viruses are small particles of 3 to 200 genes that cannot replicate unless they invade a host cell.
- A viral infection involves using the host cell's machinery to replicate the viral nucleic acid.
- A retrovirus contains RNA as its genetic material.

Key Terms for Sections 17.1 to 17.8

Match each of the following key terms with the correct description:

    a. DNA   b. RNA   c. double helix   d. mutation
    e. virus   f. recombinant DNA   g. transcription

1. _____ the formation of mRNA to carry genetic information from DNA to allow for protein synthesis
2. _____ the genetic material that contains the bases adenine, cytosine, guanine, and thymine
3. _____ the shape of DNA with a sugar–phosphate backbone and base pairs linked in the center
4. _____ the combination of DNA from different organisms which forms new DNA
5. _____ a small particle containing DNA or RNA in a protein coat that requires a host cell for replication
6. _____ a change in the DNA base sequence that may alter the shape and function of a protein
7. _____ a type of nucleic acid with a single strand containing the nucleotides adenine, cytosine, guanine, and uracil

Answers   1. g   2. a   3. c   4. f   5. e   6. d   7. b

Chapter 17

♦ **Learning Exercise 17.8**

Match each of the following terms with the correct description:

   a. host cell   **b.** retrovirus   **c.** vaccine   **d.** protease   **e.** virus

1. _____ the enzyme inhibited by drugs that prevent the synthesis of viral proteins
2. _____ a small, disease-causing particle that contains either DNA or RNA as its genetic material
3. _____ a type of virus that must use reverse transcriptase to make a viral DNA
4. _____ required by viruses to replicate
5. _____ inactive form of a virus that boosts the immune response by causing the body to produce antibodies

Answers   **1.** d   **2.** e   **3.** b   **4.** a   **5.** c

Checklist for Chapter 17

You are ready to take the Practice Test for Chapter 17. Be sure you have accomplished the following learning goals for this chapter. If not, review the Section listed at the end of each goal. Then apply your new skills and understanding to the Practice Test.

After studying Chapter 17, I can successfully:

_____ Identify the components of the nucleic acids RNA and DNA. (17.1)

_____ Describe the nucleotides contained in DNA and RNA. (17.1)

_____ Describe the primary structure of nucleic acids. (17.2)

_____ Describe the structures of RNA and DNA; show the relationship between the bases in the double helix. (17.3)

_____ Explain the process of DNA replication. (17.3)

_____ Describe the structure and characteristics of the three types of RNA. (17.4)

_____ Describe the synthesis of mRNA (transcription). (17.4)

_____ Describe the function of the codons in the genetic code. (17.5)

_____ Describe the role of translation in protein synthesis. (17.5)

_____ Describe some ways in which DNA is altered to cause mutations. (17.6)

_____ Describe the process used to prepare recombinant DNA. (17.7)

_____ Explain how retroviruses use reverse transcription to synthesize DNA. (17.8)

Practice Test for Chapter 17

The chapter Sections to review are shown in parentheses at the end of each question.

1. A nucleotide contains (17.1)
   A. a base
   B. a base and a sugar
   C. a phosphate group and a sugar
   D. a base and deoxyribose
   E. a base, a sugar, and a phosphate group

Nucleic Acids and Protein Synthesis

2. The double helix in DNA is held together by (17.2)
   A. hydrogen bonds
   B. ester linkages
   C. peptide bonds
   D. salt bridges
   E. disulfide bonds

3. The process of producing DNA in the nucleus is called (17.3)
   A. complementation
   B. replication
   C. translation
   D. transcription
   E. mutation

4. Which occurs in RNA but *not* in DNA? (17.1)
   A. thymine
   B. cytosine
   C. adenine
   D. phosphate
   E. uracil

5. Which type of molecule carries amino acids to the ribosomes? (17.4)
   A. DNA   B. mRNA   C. tRNA   D. rRNA   E. protein

6. Which type of molecule determines protein structure in protein synthesis? (17.5)
   A. DNA   B. mRNA   C. tRNA   D. rRNA   E. ribosomes

For questions 7 through 15, select answers from the following nucleic acids: (17.1, 17.2, 17.3, 17.4, 17.5)
   A. DNA   B. mRNA   C. tRNA   D. rRNA

7. _____ a major component of the ribosomes

8. _____ a double helix consisting of two chains of nucleotides held together by hydrogen bonds between bases

9. _____ a nucleic acid that uses deoxyribose as the sugar

10. _____ a nucleic acid produced in the nucleus that migrates to the ribosomes to direct the formation of a protein

11. _____ places the proper amino acid into the peptide chain

12. _____ contains the bases adenine, cytosine, guanine, and thymine

13. _____ contains the codons for the amino acid order

14. _____ contains a triplet called an anticodon loop

15. _____ replicated during cellular division

For questions 16 through 20, select answers from the following: (17.3, 17.4, 17.5)

A. A G C C T A
   ⋮ ⋮ ⋮ ⋮ ⋮ ⋮
   T C G G A T

B. A U U G C U C

C. A G T U G U
   ⋮ ⋮ ⋮ ⋮ ⋮ ⋮
   T C A A C A

D. G U A

E. A T G T A T

16. _____ a section of an mRNA

17. _____ a strand of DNA that is not possible

18. _____ a codon

19. _____ a section from a DNA molecule

20. _____ a single strand that would not be possible for mRNA

For questions 21 through 25, indicate the correct order of protein synthesis: (17.4, 17.5)
   A. tRNA assembles the amino acids at the ribosomes.
   B. DNA forms a complementary copy of itself called mRNA.
   C. Protein is formed and breaks away.
   D. tRNA picks up specific amino acids.
   E. mRNA goes to the ribosomes.

21. _____ first step
22. _____ second step
23. _____ third step
24. _____ fourth step
25. _____ fifth step

Chapter 17

26. A mutation that occurs by elimination of one base in a DNA sequence is a(n) _____. (17.6)
    A. insertion mutation
    B. viral mutation
    C. double mutation
    D. deletion mutation
    E. point mutation

27. A mutation that occurs by substitution of one base for another in a DNA sequence is a(n) _____. (17.6)
    A. insertion mutation
    B. viral mutation
    C. double mutation
    D. deletion mutation
    E. point mutation

28. Small circular pieces of DNA used in recombinant DNA are called _____. (17.7)
    A. genes   B. plasmids   C. bacteria   D. viruses   E. tumors

29. Small living particles that cannot replicate without a host cell are called _____. (17.8)
    A. genes   B. plasmids   C. bacteria   D. viruses   E. tumors

Answers to the Practice Test

1. E	2. A	3. B	4. E	5. C
6. A	7. D	8. A	9. A	10. B
11. C	12. A	13. B	14. C	15. A
16. B, D	17. C	18. D	19. A	20. E
21. B	22. E	23. D	24. A	25. C
26. D	27. E	28. B	29. D	

Nucleic Acids and Protein Synthesis

Selected Answers and Solutions to Text Problems

17.1 Purine bases (e.g., adenine, guanine) have a double-ring structure; pyrimidines (e.g., cytosine, thymine, uracil) have a single ring.
  a. Thymine is a pyrimidine base.
  b. This double-ring base is the purine adenine.

17.3 DNA contains two purines, adenine (A) and guanine (G), and two pyrimidines, cytosine (C) and thymine (T). RNA contains the same bases, except thymine (T) is replaced by the pyrimidine uracil (U).
  a. Thymine is a base present in DNA only.
  b. Adenine is a base present in both DNA and RNA.

17.5 Nucleotides contain a base, a sugar, and a phosphate group. The nucleotides found in DNA would all contain the sugar deoxyribose. The four nucleotides are deoxyadenosine monophosphate (dAMP), deoxyguanosine monophosphate (dGMP), deoxycytidine monophosphate (dCMP), and deoxythymidine monophosphate (dTMP).

17.7
  a. Adenosine is a nucleoside.
  b. Deoxycytidine is a nucleoside.
  c. Uridine is a nucleoside.
  d. Cytidine monophosphate is a nucleotide.

17.9
  a. Phosphate is present in both DNA and RNA.
  b. Ribose is present in RNA only.
  c. Deoxycytidine monophosphate is present in DNA only.
  d. Adenine is present in both DNA and RNA.

17.11 [structure of deoxyadenosine monophosphate]

17.13 [structure of guanosine monophosphate]

17.15 The nucleotides in nucleic acids are held together by phosphodiester linkages between the 3' OH group of a sugar (ribose or deoxyribose) and the phosphate group on the 5' carbon of another sugar.

17.17 The backbone of a nucleic acid consists of alternating sugar and phosphate groups.
  —sugar—phosphate—sugar—phosphate—

17.19 The free 5' end is determined by the free phosphate group on the 5' carbon of ribose or deoxyribose of a nucleic acid.

17.21 [Structure showing a dinucleotide with Guanosine (G) and Cytidine (C) connected by phosphate groups]

17.23 Structural features of DNA include that it is shaped like a double helix, it contains a sugar–phosphate backbone, the nitrogen-containing bases are hydrogen bonded between strands, and the strands run in opposite directions. A forms two hydrogen bonds to T, and G forms three hydrogen bonds to C.

17.25 The two DNA strands are held together by hydrogen bonds between the complementary bases in each strand.

17.27 a. Since T pairs with A, if one strand of DNA has the sequence A A A A A A, the second strand would be T T T T T T.
b. Since C pairs with G, if one strand of DNA has the sequence G G G G G G, the second strand would be C C C C C C.
c. Since T pairs with A, and C pairs with G, if one strand of DNA has the sequence A G T C C A G G T, the second strand would be T C A G G T C C A.
d. Since T pairs with A, and C pairs with G, if one strand of DNA has the sequence C T G T A T A C G T T A, the second strand would be G A C A T A T G C A A T.

17.29 The enzyme helicase unwinds the DNA helix by breaking the hydrogen bonds between the complementary bases so that the parent DNA strands can be replicated into daughter DNA strands.

17.31 First, the two DNA strands separate in a way that is similar to the unzipping of a zipper. Then the enzyme DNA polymerase begins to copy each strand by allowing the pairing of each of the bases in the strands with its complementary base: A pairs with T and C with G. Finally, a phosphodiester linkage joins the base to the new, growing strand. This process produces two exact copies of the original DNA.

17.33 The three types of RNA are messenger RNA (mRNA), ribosomal RNA (rRNA), and transfer RNA (tRNA).

17.35 A ribosome consists of a small subunit and a large subunit that contain rRNA combined with proteins.

17.37 In transcription, the sequence of nucleotides on a DNA template strand is used to produce the base sequence of a messenger RNA. The DNA unwinds, and one strand is copied as complementary bases are placed in the mRNA molecule. In RNA, U (uracil) is paired with A in DNA.

17.39 To form mRNA, the bases in the DNA template strand are paired with their complementary bases: G with C, C with G, T with A, and A with U. The strand of mRNA would have the following sequence: GGC UUC CAA GUG.

17.41 A codon is a three-base sequence (triplet) in mRNA that codes for a specific amino acid in a protein.

17.43 a. The codon CCA in mRNA codes for the amino acid proline (Pro).
  b. The codon AAC in mRNA codes for the amino acid asparagine (Asn).
  c. The codon GGU in mRNA codes for the amino acid glycine (Gly).
  d. The codon AGG in mRNA codes for the amino acid arginine (Arg).

17.45 At the beginning of an mRNA, the codon AUG signals the start of protein synthesis; thereafter, the AUG codon specifies the amino acid methionine.

17.47 A codon is a base triplet in the mRNA template. An anticodon is the complementary triplet on a tRNA for a specific amino acid.

17.49 The three steps in translation are initiation, chain elongation, and termination.

17.51 a. The codons ACC, ACA, and ACU in mRNA all code for threonine: Thr–Thr–Thr, TTT.
  b. The codons UUU and UUC code for phenylalanine, and CCG and CCA both code for proline: Phe–Pro–Phe–Pro, FPFP.
  c. The codon UAC codes for tyrosine, GGG for glycine, AGA for arginine, and UGU for cysteine: Tyr–Gly–Arg–Cys, YGRC.

17.53 The new amino acid is joined by a peptide bond to the growing peptide chain. The ribosome moves to the next codon, which attaches to a tRNA carrying the next amino acid.

17.55 a. The mRNA sequence would be: CGA UAU GGU UUU.
  b. The tRNA triplet anticodons would be: GCU, AUA, CCA, and AAA.
  c. From the table of mRNA codons, the amino acids would be: Arg–Tyr–Gly–Phe, RYGF.

17.57 a. AAA CAC UUG GUU GUG GAC
  b. Lys–His–Leu–Val–Val–Asp, KHLVVD

17.59 In a point mutation, a base in DNA is replaced by a different base.

17.61 In a mutation caused by the deletion of a base, all the codons from the mutation onward are changed, which changes the order of amino acids in the rest of the polypeptide chain.

17.63 The normal triplet TTT in DNA forms the codon AAA in mRNA. AAA codes for lysine. The mutation TTC in DNA forms the codon AAG in mRNA, which also codes for lysine. Thus, there is no effect on the amino acid sequence.

17.65 a. Thr–Ser–Arg–Val is the amino acid sequence produced by normal DNA.
  b. Thr–Thr–Arg–Val is the amino acid sequence produced by a mutation.
  c. Thr–Ser–Gly–Val is the amino acid sequence produced by a mutation.
  d. Thr–STOP; protein synthesis would terminate early. If this mutation occurs early in the formation of the polypeptide, the resulting protein will probably be nonfunctional.
  e. The new protein will contain the sequence Asp–Ile–Thr–Gly.
  f. The new protein will contain the sequence His–His–Gly.

17.67 a. Both codons GCC and GCA code for alanine.
  b. A vital ionic interaction in the tertiary structure of hemoglobin cannot be formed when the polar acidic glutamate is replaced by valine, which is nonpolar. The resulting hemoglobin is malformed and less capable of carrying oxygen.

17.69 *E. coli* bacterial cells contain several small circular plasmids of DNA that can be isolated easily. After the recombinant DNA is formed, *E. coli* multiply rapidly, producing many copies of the recombinant DNA in a relatively short time.

17.71 The cells of the *E. coli* are soaked in a detergent solution that disrupts the plasma membrane and releases the cell contents including the plasmids, which are then collected.

17.73 When a gene has been obtained using a restriction enzyme, it is mixed with the plasmids that have been cut by the same enzyme. When mixed together in a fresh *E. coli* culture, the sticky ends of the DNA fragments bond with the sticky ends of the plasmid DNA to form a recombinant DNA.

17.75 In DNA fingerprinting, restriction enzymes cut a sample of DNA into fragments, which are sorted by size by gel electrophoresis. A radioactive probe that adheres to specific DNA sequences exposes an X-ray film placed over the gel and creates a pattern of dark and light bands known as a DNA fingerprint.

17.77 A virus contains either DNA or RNA, but not both, inside a protein coat.

17.79 a. An RNA-containing virus must make viral DNA from the RNA to produce proteins for the protein coat, which allows the virus to replicate and leave the cell to infect new cells.
b. A virus that uses reverse transcription is called a retrovirus.

17.81 Nucleoside analogs such as AZT and ddI are similar to the nucleosides required to make viral DNA in reverse transcription. When they are incorporated into viral DNA, the lack of a hydroxyl group on the 3′ carbon in the sugar prevents the formation of the sugar–phosphate bonds and stops the replication of the virus.

17.83 Estrogen receptors bind to DNA and increase the production of mammary cells, which can lead to mutations and cancer.

17.85 Tamoxifen blocks the binding of estrogen to the estrogen receptor, which prevents activation of the gene.

17.87 a. ACC UUA AUA GAC GAG AAG CGC
b. Thr–Leu–Ile–Asp–Glu–Lys–Arg, TLIDEKR
c. There is no change, Asp is still the fourth amino acid.

17.89 a.
Parent strand: A G G T C G C C T
New strand: T C C A G C G G A

b. A G G U C G C C U

c. Arg — Ser — Pro

17.91 DNA contains two purines, adenine (A) and guanine (G), and two pyrimidines, cytosine (C) and thymine (T). RNA contains the same bases, except thymine (T) is replaced by the pyrimidine uracil (U).
  a. Cytosine is a pyrimidine base.
  b. Adenine is a purine base.
  c. Uracil is a pyrimidine base.
  d. Thymine is a pyrimidine base.
  e. Guanine is a purine base.

17.93 a. Deoxythymidine contains the base thymine and the sugar deoxyribose.
b. Adenosine contains the base adenine and the sugar ribose.
c. Cytidine contains the base cytosine and the sugar ribose.
d. Deoxyguanosine contains the base guanine and the sugar deoxyribose.

17.95 Thymine and uracil are both pyrimidines, but thymine has a methyl group on carbon 5.

17.97 [Structure of cytidine 5′-monophosphate: phosphate group linked via $-O-CH_2-$ to a ribose sugar (with OH groups on the 2′ and 3′ carbons) attached to cytosine base (NH_2 on carbon 4, $=O$ on carbon 2).]

17.99 Both RNA and DNA are unbranched chains of nucleotides connected through phosphodiester linkages between alternating sugar and phosphate groups with bases extending out from each sugar.

17.101 Because A bonds with T, DNA containing 28% A will also have 28% T. Thus the sum of A + T is 56%, which leaves 44% divided equally between G and C: 22% G and 22% C.

17.103 There are two hydrogen bonds between A and T in DNA.

17.105 a. C T G A A T C C G
b. A C G T T T G A T C G A
c. T A G C T A G C T A G C

17.107 a. Transfer RNA (tRNA) is the smallest type of RNA.
b. Ribosomal RNA (rRNA) makes up the highest percentage of RNA in the cell.
c. Messenger RNA (mRNA) carries genetic information from the nucleus to the ribosomes.

17.109 a. ACU, ACC, ACA, and ACG are all codons for the amino acid threonine.
b. UCU, UCC, UCA, UCG, AGU, and AGC are all codons for the amino acid serine.
c. UGU and UGC are codons for the amino acid cysteine.

17.111 a. AAG codes for lysine.   **b.** AUU codes for isoleucine.
c. CGA codes for arginine.

17.113 The anticodon on tRNA consists of the three complementary bases to the codon in mRNA.
a. UCG   **b.** AUA   **c.** GGU

17.115 Using the genetic code, the codons indicate the following amino acid sequence:
START–Tyr–Gly–Gly–Phe–Leu–STOP

17.117 The codon for each amino acid contains three nucleotides, plus the start and stop codon consisting of three nucleotides each, which makes a minimum total of (9 × 3) + 3 + 3 = 33 nucleotides.

17.119 A DNA virus attaches to a cell and injects viral DNA that uses the host cell to produce copies of the DNA to make viral RNA. A retrovirus injects viral RNA from which complementary DNA is produced by reverse transcription.

17.121 a. DNA polymerase is involved in replication of DNA (1).
b. mRNA is synthesized from nuclear DNA during transcription (2).
c. Some viruses use reverse transcription (5).
d. Restriction enzymes are used to make recombinant DNA (4).
e. tRNA molecules bond to codons during translation (3).

18
Metabolic Pathways and ATP Production

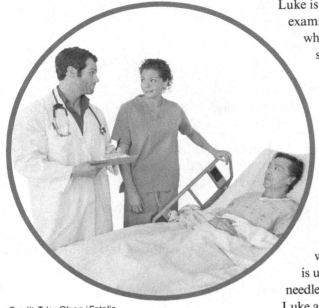

Credit: Tyler Olson/Fotolia

Luke is a paramedic. Recently, bloodwork from a physical examination indicated a plasma cholesterol level of 256 mg/dL, which is elevated. Luke's doctor ordered a liver profile that showed elevated liver enzymes: aspartate transaminase (AST) of 226 Units/L (the normal AST is 5 to 50 Units/L) and alanine transaminase (ALT) of 282 Units/L (the normal ALT is 5 to 35 Units/L). During Luke's career as a paramedic, he was exposed several times to blood and was accidentally stuck by a needle containing infected blood. A hepatitis profile showed that Luke was positive for antibodies to both hepatitis B and C. His doctor diagnosed Luke with chronic hepatitis C virus (HCV) infection. Hepatitis C is an infection caused by a virus that attacks the liver and leads to inflammation. Most people infected with the hepatitis C virus have no symptoms. Hepatitis C is usually passed by contact with contaminated blood or by needles shared during illegal drug use. As part of his treatment, Luke attended a class on living with hepatitis C given by a public health nurse. What are the reactants and products of the reaction catalyzed by alanine transaminase?

LOOKING AHEAD

- **18.1** Metabolism and ATP Energy
- **18.2** Digestion of Foods
- **18.3** Coenzymes in Metabolic Pathways
- **18.4** Glycolysis: Oxidation of Glucose
- **18.5** The Citric Acid Cycle
- **18.6** Electron Transport and Oxidative Phosphorylation
- **18.7** Oxidation of Fatty Acids
- **18.8** Degradation of Amino Acids

 The Health icon indicates a question that is related to health and medicine.

18.1 Metabolism and ATP Energy

Learning Goal: Describe the three stages of catabolism and the role of ATP.

- Metabolism is all of the chemical reactions that provide energy and substances for maintenance of life and for cell growth.
- Catabolic reactions degrade large molecules to produce energy.
- Anabolic reactions utilize energy in the cell to build large molecules for the cell.
- In cells, specialized structures contain the enzymes and coenzymes for the various catabolic and anabolic reactions.
- Energy obtained from catabolic reactions is stored primarily in adenosine triphosphate (ATP), a high-energy compound.
- The hydrolysis of ATP, which releases energy, is linked with many anabolic reactions in the cell.

Metabolic Pathways and ATP Production

♦ **Learning Exercise 18.1A**

Identify the stage of catabolism for each of the following processes:

    a. stage 1   **b.** stage 2   **c.** stage 3

1. _____ involves the oxidation of two-carbon acetyl group
2. _____ polysaccharides undergo digestion to form monosaccharides, such as glucose
3. _____ digestion products such as glucose are broken down to yield to two- or three-carbon compounds

Answers   **1.** c   **2.** a   **3.** b

♦ **Learning Exercise 18.1B**

Match each of the following with its description or function:

    a. ribosome   **b.** mitochondria   **c.** nucleus
    d. cytoplasm   **e.** cell membrane

1. _____ separates the contents of a cell from the external environment
2. _____ contain enzymes that catalyze energy-producing reactions
3. _____ is the cellular material between the cell membrane and the nucleus
4. _____ contains genetic information for the replication of DNA
5. _____ is the site of protein synthesis

Answers   **1.** e   **2.** b   **3.** d   **4.** c   **5.** a

♦ **Learning Exercise 18.1C**

Complete the following statements about ATP:

The ATP molecule is composed of the base (1) _____, the sugar (2) _____, and three (3) _____ groups. ATP undergoes (4) _____, which cleaves a (5) _____ and releases (6) _____. For this reason, ATP is called a (7) _____ compound. The resulting phosphate group, called inorganic phosphate, is abbreviated as (8) _____. The equation for the hydrolysis of ATP is written as (9) _____. The energy from ATP is linked to cellular reactions that are (10) _____.

Answers
1. adenine   2. ribose   3. phosphate   4. hydrolysis
5. phosphate group   6. energy   7. high-energy   8. P_i
9. $ATP + H_2O \longrightarrow ADP + P_i + energy$ 10. energy requiring

18.2 Digestion of Foods

Learning Goal: Identify the sites and products of digestion for carbohydrates, triacylglycerols, and proteins.

> **REVIEW**
> Identifying Fatty Acids (15.2)
> Drawing Structures for Triacylglycerols (15.3)
> Identifying the Primary, Secondary, Tertiary, and Quaternary Structures of Proteins (16.2, 16.3)
> Describing Enzyme Action (16.4)

- Digestion is a series of reactions that break down large food molecules of carbohydrates, lipids, and proteins into smaller molecules that can be absorbed and used by the cells.
- The end products of digestion of polysaccharides are the monosaccharides glucose, fructose, and galactose.
- Dietary fats begin digestion in the small intestine where they are emulsified by bile salts.
- Pancreatic lipases hydrolyze the triacylglycerols to yield monoacylglycerols and free fatty acids.
- The triacylglycerols are reformed in the intestinal lining, where they combine with proteins to form chylomicrons for transport through the lymph system and bloodstream.
- In the cells, triacylglycerols are hydrolyzed to glycerol and fatty acids, which can be used for energy.
- Proteins begin digestion in the stomach, where HCl denatures proteins and activates protease enzymes such as pepsin that begin to hydrolyze peptide bonds.
- In the small intestine, trypsin and chymotrypsin complete the hydrolysis of peptides to amino acids.

♦ **Learning Exercise 18.2A**

Complete the table to describe sites, enzymes, and products for the digestion of carbohydrates.

Food	Digestion site(s)	Enzyme	Products
1. Amylose			
2. Amylopectin			
3. Maltose			
4. Lactose			
5. Sucrose			

Answers

Food	Digestion site(s)	Enzyme	Products
1. Amylose	a. mouth	a. salivary amylase	a. smaller polysaccharides (dextrins), some maltose and glucose
	b. small intestine	b. pancreatic amylase	b. maltose, glucose
2. Amylopectin	a. mouth	a. salivary amylase	a. smaller polysaccharides (dextrins), some maltose and glucose
	b. small intestine	b. pancreatic amylase	b. maltose, glucose
3. Maltose	small intestine	maltase	glucose, glucose
4. Lactose	small intestine	lactase	glucose, galactose
5. Sucrose	small intestine	sucrase	glucose, fructose

♦ **Learning Exercise 18.2B**

Match each statement with the correct location:

   a. stomach   **b.** small intestine

1. _____ HCl activates enzymes that hydrolyze peptide bonds in proteins.
2. _____ Trypsin and chymotrypsin convert peptides to amino acids.

Answers   1. a   2. b

18.3 Coenzymes in Metabolic Pathways

Learning Goal: Describe the components and functions of the coenzymes NAD^+, FAD, and coenzyme A.

- Coenzymes such as FAD and NAD^+ pick up hydrogen and electrons during oxidative processes.
- Coenzyme A carries acetyl (two-carbon) groups produced when glucose, fatty acids, and amino acids are degraded.

♦ **Learning Exercise 18.3**

Select the coenzyme that matches each of the following descriptions:

   a. NAD^+ **b.** NADH   **c.** FAD   **d.** $FADH_2$ **e.** coenzyme A

1. _____ participates in reactions that convert a hydroxyl group to a C=O group
2. _____ contains riboflavin (vitamin B_2)
3. _____ is the reduced form of nicotinamide adenine dinucleotide
4. _____ contains the vitamin niacin
5. _____ is the oxidized form of flavin adenine dinucleotide

Chapter 18

6. _____ contains the vitamin pantothenic acid, ADP, and aminoethanethiol
7. _____ participates in oxidation reactions that produce a carbon–carbon double bond (C=C)
8. _____ transfers acyl groups such as the two-carbon acetyl group
9. _____ is the reduced form of flavin adenine dinucleotide

Answers   1. a   2. c, d   3. b   4. a, b   5. c
           6. e   7. c   8. e   9. d

18.4 Glycolysis: Oxidation of Glucose

Learning Goal: Describe the conversion of glucose to pyruvate in glycolysis and the subsequent conversion of pyruvate to acetyl CoA or lactate.

- Glycolysis is the primary anaerobic pathway for the degradation of glucose to yield pyruvate.
- Glucose is converted to fructose-1,6-bisphosphate, which is then split into two triose phosphate molecules.
- The oxidation of each three-carbon phosphate compound yields 2 NADH and 2 ATP.
- In the absence of oxygen, pyruvate is reduced to lactate and NAD^+ is regenerated for the continuation of glycolysis.
- Under aerobic conditions, pyruvate is oxidized and converted to acetyl CoA, which enters the citric acid cycle.

Key Terms for Sections 18.1 to 18.4

Match each of the following key terms with the correct description:

a. ATP   **b.** anabolic reaction   **c.** glycolysis
d. catabolic reaction   **e.** mitochondria

1. _____ a metabolic reaction that uses energy to build large molecules
2. _____ a metabolic reaction that produces energy for the cell by degrading large molecules
3. _____ a high-energy compound that provides energy for reactions
4. _____ the degradation reactions of glucose that yield two pyruvate molecules
5. _____ the organelles in the cells where energy-producing reactions take place

Answers   1. b   2. d   3. a   4. c   5. e

♦ Learning Exercise 18.4A

Match each of the following terms of glycolysis with the correct description:

a. 2 NADH   **b.** anaerobic   **c.** glucose   **d.** 2 pyruvate
e. energy invested **f.** energy generated **g.** 2 ATP   **h.** 4 ATP

1. _____ the starting material for glycolysis
2. _____ number and type of reduced coenzyme produced

3. _____ steps 1 to 5 of glycolysis
4. _____ steps 6 to 10 of glycolysis
5. _____ operates without oxygen
6. _____ end products of glycolysis
7. _____ net ATP energy produced
8. _____ number of ATP required

Answers   1. c   2. a   3. e   4. f   5. b
            6. a, d, g   7. g   8. g

♦ Learning Exercise 18.4B

CORE CHEMISTRY SKILL
Identifying the Compounds in Glycolysis

Fill in the blanks with the following terms:

lactate NAD^+ acetyl CoA
NADH   aerobic   anaerobic

When oxygen is available during glycolysis, the three-carbon pyruvate may be oxidized to form (1) _____ + CO_2. Under (2) _____ conditions, pyruvate is oxidized to acetyl CoA. The coenzyme (3) _____ is reduced to (4) _____. Under (5) _____ conditions, pyruvate is reduced to (6) _____.

Answers   1. acetyl CoA   2. aerobic   3. NAD^+
            4. NADH   5. anaerobic 6. lactate

♦ Learning Exercise 18.4C

Explain how the formation of lactate from pyruvate during anaerobic conditions allows glycolysis to continue.

Answer   Under anaerobic conditions, the oxidation of pyruvate to acetyl CoA to regenerate NAD^+ cannot take place. Pyruvate is reduced to lactate using NADH in the cytoplasm and regenerating NAD^+.

18.5 The Citric Acid Cycle

Learning Goal: Describe the oxidation of acetyl CoA in the citric acid cycle.

- At the start of a sequence of reactions called the citric acid cycle, acetyl CoA combines with oxaloacetate to yield citrate.
- In one turn of the citric acid cycle, the oxidation of acetyl CoA yields 2 CO_2, 1 GTP, 3 NADH, and 1 $FADH_2$. The phosphorylation of ADP by GTP yields ATP.

Chapter 18

♦ **Learning Exercise 18.5**

CORE CHEMISTRY SKILL
Describing the Reactions in the Citric Acid Cycle

In each of the following steps of the citric acid cycle, indicate if oxidation occurs (yes/no) and any coenzyme or direct phosphorylation product produced (NADH, FADH$_2$, GTP):

Step in Citric Acid Cycle	Oxidation	Coenzyme
1. acetyl CoA + oxaloacetate ⟶ citrate	_____	_____
2. citrate ⟶ isocitrate	_____	_____
3. isocitrate ⟶ α-ketoglutarate	_____	_____
4. α-ketoglutarate ⟶ succinyl CoA	_____	_____
5. succinyl CoA ⟶ succinate	_____	_____
6. succinate ⟶ fumarate	_____	_____
7. fumarate ⟶ malate	_____	_____
8. malate ⟶ oxaloacetate	_____	_____

Answers
1. no   2. no   3. yes, NADH   4. yes, NADH
5. no, GTP   6. yes, FADH$_2$ 7. no   8. yes, NADH

18.6 Electron Transport and Oxidative Phosphorylation

Learning Goal: Describe electron transport and the process of oxidative phosphorylation; calculate the ATP from the complete oxidation of glucose.

- The reduced coenzymes from glycolysis and the citric acid cycle are oxidized to NAD$^+$ and FAD by transferring hydrogen ions and electrons to electron transport.
- In electron transport, electrons are transferred to electron carriers including coenzyme Q and cytochrome *c*.
- The final acceptor, O$_2$, combines with hydrogen ions and electrons to yield H$_2$O.
- The flow of electrons in electron transport moves hydrogen ions across the inner mitochondrial membrane, which produces a high-energy hydrogen ion gradient that provides energy for the synthesis of ATP molecules.
- The process of using the energy of electron transport to synthesize ATP is called oxidative phosphorylation.
- The oxidation of NADH yields 2.5 ATP, whereas the oxidation of FADH$_2$ yields 1.5 ATP.
- The complete oxidation of glucose yields a total of 32 ATP from direct phosphorylation and the oxidation of the reduced coenzymes NADH and FADH$_2$ from electron transport and oxidative phosphorylation.

Metabolic Pathways and ATP Production

Key Terms for Sections 18.5 and 18.6

Match each of the following key terms with the correct description:

   a. citric acid cycle b. oxidative phosphorylation c. coenyzme Q
   d. cytochrome c e. hydrogen ion pump

1. _____ a mobile carrier that passes electrons from NADH and $FADH_2$ to cytochrome c in complex III
2. _____ a mobile carrier that transfers electrons from $CoQH_2$ to complex IV
3. _____ moves hydrogen ions from the matrix into the intermembrane space
4. _____ the synthesis of ATP from ADP and P_i using energy generated from electron transport
5. _____ reactions such as oxidation that convert acetyl CoA to CO_2 and produce reduced coenzymes for energy production via electron transport

Answers   1. c   2. d   3. e   4. b   5. a

♦ **Learning Exercise 18.6A**

Write the oxidized and reduced forms of each of the following electron carriers:
1. flavin adenine dinucleotide   oxidized _____   reduced _____
2. coenzyme Q   oxidized _____   reduced _____

Answers   1. oxidized: FAD   reduced: $FADH_2$
            2. oxidized: CoQ   reduced: $CoQH_2$

♦ **Learning Exercise 18.6B**

1. Write an equation for the transfer of hydrogen from $FADH_2$ to CoQ.

2. What is the function of coenzyme Q in electron transport?

3. What are the end products of electron transport?

Answers   1. $FADH_2 + CoQ \longrightarrow FAD + CoQH_2$
            2. CoQ accepts hydrogen atoms from NADH or $FADH_2$. From $CoQH_2$, the hydrogen atoms are separated into hydrogen ions and electrons, with the electrons being passed on to cytochrome c.
            3. H_2O + ATP

Chapter 18

♦ Learning Exercise 18.6C

Match the following terms with the correct description below:

a. oxidative phosphorylation
b. ATP synthase
c. hydrogen ion pump
d. hydrogen ion gradient

1. _____ the complexes I, III, and IV through which hydrogen ions move out of the matrix into the intermembrane space

2. _____ the protein channel where hydrogen ions flow from the intermembrane space back to the matrix, generating energy for ATP synthesis

3. _____ use of energy from electron transport to form a hydrogen ion gradient that drives ATP synthesis

4. _____ the accumulation of hydrogen ions in the intermembrane space that lowers pH

Answers   1. c   2. b   3. a   4. d

CORE CHEMISTRY SKILL
Calculating the ATP Produced from Glucose

♦ Learning Exercise 18.6D

Complete the following:

Substrate	Reaction	Products	ATP Yield
1. Glucose	glycolysis (aerobic)		
2. Pyruvate	oxidation		
3. Acetyl CoA	citric acid cycle		
4. Glucose	complete oxidation		

Answers

Substrate	Reaction	Products	ATP Yield
1. Glucose	glycolysis (aerobic)	2 pyruvate + $2H_2O$	7 ATP
2. Pyruvate	oxidation	acetyl CoA + CO_2	2.5 ATP
3. Acetyl CoA	citric acid cycle	$2CO_2$	10 ATP
4. Glucose	complete oxidation	$6CO_2 + 6H_2O$	32 ATP

18.7   Oxidation of Fatty Acids

Learning Goal: Describe the metabolic pathway of β oxidation; calculate the ATP from the complete oxidation of a fatty acid.

REVIEW
Writing Equations for Hydrogenation and Hydration (11.7)
Identifying Fatty Acids (15.2)

- When needed for energy, fatty acids are activated by binding to coenzyme A for transport to the mitochondria where they undergo β oxidation.
- In β oxidation, a fatty acyl chain is oxidized to yield a shortened fatty acid, acetyl CoA, and the reduced coenzymes NADH and $FADH_2$.
- The energy obtained from a particular fatty acid depends on the number of carbon atoms.

- Two ATP are required for activation. Then each acetyl CoA produces 10 ATP via the citric acid cycle, and electron transport converts each NADH to 2.5 ATP and each FADH to 1.5 ATP.
- When the oxidation of large amounts of fatty acids causes high levels of acetyl CoA, the acetyl CoA undergoes ketogenesis.
- Two molecules of acetyl CoA form acetoacetyl CoA, which is converted to the ketone bodies acetoacetate, β-hydroxybutyrate, and acetone.

♦ Learning Exercise 18.7A

CORE CHEMISTRY SKILL
Calculating the ATP from Fatty Acid Oxidation (β Oxidation)

Lauric acid is a 12-carbon saturated fatty acid: $CH_3-(CH_2)_{10}-COOH$.

1. How many ATP are needed for activation?

2. How many cycles of β oxidation are required for complete oxidation?

3. How many NADH and $FADH_2$ are produced during β oxidation?

4. How many acetyl CoA units are produced?

5. What is the total ATP produced from complete oxidation?

Answers
1. 2 ATP
2. 5 cycles
3. 5 NADH and 5 $FADH_2$
4. 6 acetyl CoA units
5. activation = -2 ATP
   5 NADH × 2.5 ATP/NADH = 12.5 ATP
   5 $FADH_2$ × 1.5 ATP/$FADH_2$ = 7.5 ATP
   6 acetyl CoA × 10 ATP/acetyl CoA = 60 ATP
   Total = 78 ATP

♦ Learning Exercise 18.7B

Answer each of following questions for myristic acid: $CH_3-(CH_2)_{12}-\overset{\overset{O}{\|}}{C}-OH$

1. the number of β oxidation cycles for the complete oxidation of myristic acid

2. the number of acetyl CoA from the complete oxidation of myristic acid

3. the number of ATP generated from the complete oxidation of myristic acid

Chapter 18

Answers

1. 6 β oxidation cycles
2. 7 acetyl CoA
3.
activation	= −2 ATP
6 NADH × 2.5 ATP/NADH	= 15 ATP
6 FADH$_2$ × 1.5 ATP/FADH$_2$	= 9 ATP
7 acetyl CoA × 10 ATP/acetyl CoA	= 70 ATP
Total =	92 ATP

18.8 Degradation of Amino Acids

Learning Goal: Describe the reactions of transamination, oxidative deamination, and the entry of amino acid carbons into the citric acid cycle.

> **REVIEW**
> Drawing the Structure for an Amino Acid at Physiological pH (16.1)
> Describing Enzyme Action (16.4)

- Amino acids are normally used for protein synthesis.
- When needed for energy, amino acids are degraded by transferring an amino group from an amino acid to an α-keto acid to yield a different amino acid and a new α-keto acid.
- In oxidative deamination, the amino group in glutamate is removed as an ammonium ion, NH_4^+.
- α-Keto acids resulting from transamination can be used as intermediates in the citric acid cycle, in the synthesis of lipids or glucose, or oxidized for energy.

Key Terms for Sections 18.7 and 18.8

Match each of the following key terms with the correct description:

- **a.** β oxidation
- **b.** transamination
- **c.** ketone bodies
- **d.** ketosis (acidosis)
- **e.** oxidative deamination

1. _____ a reaction cycle that oxidizes fatty acids by removing acetyl CoA units
2. _____ the loss of an ammonium ion during the degradation of glutamate
3. _____ a condition in which high levels of ketone bodies lower blood pH
4. _____ the products of ketogenesis that cause ketosis and acidosis
5. _____ the transfer of an amino group from an amino acid to an α-keto acid

Answers 1. a   2. e   3. d   4. c   5. b

♦ Learning Exercise 18.8A

Match each of the following descriptions with transamination (T) or oxidative deamination (D):

1. _____ produces an ammonium ion, NH_4^+
2. _____ transfers an amino group to an α-keto acid
3. _____ usually involves the degradation of glutamate
4. _____ requires NAD^+
5. _____ produces another amino acid and α-keto acid
6. _____ usually produces α-ketoglutarate

Answers 1. D   2. T   3. D   4. D   5. T   6. D

♦ Learning Exercise 18.8B

1. Write an equation for the transamination reaction of serine and oxaloacetate.

2. Write an equation for the oxidative deamination of glutamate.

Answers

1. $\text{HO}-\text{CH}_2-\overset{\overset{+}{\text{NH}_3}}{\underset{|}{\text{CH}}}-\text{COO}^- + {}^-\text{OOC}-\overset{\overset{\text{O}}{\|}}{\text{C}}-\text{CH}_2-\text{COO}^- \xrightarrow{\text{Aminotransferase}}$

   $\text{HO}-\text{CH}_2-\overset{\overset{\text{O}}{\|}}{\text{C}}-\text{COO}^- + {}^-\text{OOC}-\overset{\overset{+}{\text{NH}_3}}{\underset{|}{\text{C}}}-\text{CH}_2-\text{COO}^-$

2. ${}^-\text{OOC}-\overset{\overset{+}{\text{NH}_3}}{\underset{|}{\text{CH}}}-\text{CH}_2-\text{CH}_2-\text{COO}^- + \text{H}_2\text{O} + \text{NAD}^+ \xrightarrow{\text{Glutamate dehydrogenase}}$

   ${}^-\text{OOC}-\overset{\overset{\text{O}}{\|}}{\text{C}}-\text{CH}_2-\text{CH}_2-\text{COO}^- + \text{NH}_4^+ + \text{NADH} + \text{H}^+$

♦ Learning Exercise 18.8C

Match each of the following terms with the correct description:

   a. ketone bodies   b. ketogenesis   c. ketosis
   d. liver   e. acidosis

1. _____ high levels of ketone bodies in the blood
2. _____ a metabolic pathway that produces ketone bodies
3. _____ β-hydroxybutyrate, acetoacetate, and acetone
4. _____ the condition whereby ketone bodies lower the blood pH below 7.4
5. _____ the site where ketone bodies form

Answers   1. c   2. b   3. a   4. e   5. d

Checklist for Chapter 18

You are ready to take the Practice Test for Chapter 18. Be sure you have accomplished the following learning goals for this chapter. If not, review the Section listed at the end of the goal. Then apply your new skills and understanding to the Practice Test.

After studying Chapter 18, I can successfully:

_____ Describe the role of ATP in catabolic and anabolic reactions. (18.1)

_____ Describe the sites, enzymes, and products of digestion for carbohydrates. (18.2)

Chapter 18

_____ Describe the coenzymes NAD$^+$ and FAD, and coenzyme A. (18.3)

_____ Describe the conversion of glucose to pyruvate in glycolysis. (18.4)

_____ Give the conditions for the conversion of pyruvate to acetyl CoA or lactate. (18.4)

_____ Describe the oxidation of acetyl CoA in the citric acid cycle. (18.5)

_____ Identify the electron carriers in electron transport. (18.6)

_____ Describe the process of electron transport. (18.6)

_____ Explain the chemiosmotic theory whereby ATP synthesis is linked to the energy of electron transport and a hydrogen ion gradient. (18.6)

_____ Account for the ATP produced by the complete oxidation of glucose. (18.6)

_____ Describe the oxidation of fatty acids via β oxidation. (18.7)

_____ Calculate the ATP produced by the complete oxidation of a fatty acid. (18.7)

_____ Explain the role of transamination and oxidative deamination in degrading amino acids. (18.8)

Practice Test for Chapter 18

The chapter Sections to review are shown in parentheses at the end of each question.

1. The main function of the mitochondria is (18.1)
   - A. energy production
   - B. protein synthesis
   - C. glycolysis
   - D. genetic instructions
   - E. waste disposal

2. ATP is a(n) (18.1)
   - A. nucleotide unit in RNA and DNA
   - B. end product of glycogenolysis
   - C. end product of transamination
   - D. enzyme
   - E. energy storage molecule

For questions 3 through 6, identify one or more of the following enzymes and end products for the digestion of each: (18.2)
   - A. maltase
   - B. glucose
   - C. fructose
   - D. sucrase
   - E. galactose
   - F. lactase
   - G. pancreatic amylase

3. sucrose _____

4. lactose _____

5. small polysaccharides _____

6. maltose _____

7. The digestion of triacylglycerols takes place in the _____ by enzymes called _____. (18.2)
   - A. small intestine; proteases
   - B. stomach; lipases
   - C. stomach; proteases
   - D. small intestine; lipases
   - E. all of these

8. The final products of the digestion of triacylglycerols are (18.2)
   - A. fatty acids
   - B. monoacylglycerols
   - C. glycerol
   - D. diacylglycerols
   - E. glycerol and fatty acids

9. Chylomicrons formed in the intestinal lining (18.2)
   - A. are lipoproteins
   - B. are triacylglycerols coated with proteins
   - C. transport fats into the lymph system and bloodstream
   - D. carry triacylglycerols to the cells of the heart, muscle, and adipose tissues
   - E. all of these

10. The digestion of proteins takes place in the _____ by enzymes called _____. (18.2)
    A. small intestine; proteases
    B. stomach; lipases
    C. stomach; proteases
    D. small intestine; lipases
    E. stomach and small intestine; proteases

For questions 11 through 15, match the following coenzymes **A** *through* **E** *with the correct description:* (18.3)
    A. NAD^+ B. NADH   C. FAD   D. $FADH_2$ E. coenzyme A

11. _____ converts a hydroxyl group to a C=O group

12. _____ reduced form of nicotinamide adenine dinucleotide

13. _____ participates in oxidation reactions that produce a carbon–carbon double bond (C=C)

14. _____ transfers acyl groups such as the two-carbon acetyl group

15. _____ reduced form of flavin adenine dinucleotide

16. Glycolysis (18.4)
    A. requires oxygen for the catabolism of glucose
    B. represents the aerobic sequence for glucose anabolism and ATP production
    C. represents the splitting off of glucose residues from glycogen
    D. represents the anaerobic catabolism of glucose to pyruvate
    E. produces acetyl units and ATP as end products

For questions 17 through 19, answer for glycolysis: (18.4)

17. _____ number of ATP invested to oxidize one glucose molecule

18. _____ number of ATP (net) produced from one glucose molecule

19. _____ number of NADH produced from one glucose molecule

For questions 20 through 24, match to the compounds **A** *through* **E**: (18.5)
    A. malate   B. fumarate   C. succinate   D. citrate   E. oxaloacetate

20. _____ formed when oxaloacetate combines with acetyl CoA

21. _____ H_2O adds to its double bond to form malate

22. _____ FAD removes hydrogen from this compound to form a double bond

23. _____ formed when the hydroxyl group in malate is oxidized

24. _____ formed by addition of water to a double bond

25. Which is true of the citric acid cycle? (18.5)
    A. Acetyl CoA is converted to CO_2 and H_2O.
    B. Oxaloacetate combines with acetyl units to form citrate.
    C. The coenzymes are NAD^+, FAD, and CoA.
    D. ATP is produced by direct phosphorylation.
    E. all of the above

26. One turn of the citric acid cycle produces (18.5)
    A. 3 NADH
    B. 3 NADH, 1 $FADH_2$
    C. 3 $FADH_2$, 1 NADH, 1 ATP
    D. 3 NADH, 1 $FADH_2$, 1 ATP, 2 CO_2
    E. 1 NADH, 1 $FADH_2$, 1 ATP

Chapter 18

27. How many complexes in electron transport act as hydrogen ion pumps? (18.6)
    A. none   B. 1   C. 2   D. 3   E. 4

28. Electron transport (18.6)
    A. produces most of the body's ATP
    B. carries oxygen to the cells
    C. uses $CO_2 + H_2O$
    D. is only involved in the citric acid cycle
    E. operates during fermentation

For questions 29 through 31, match the components of electron transport with the following activities: (18.6)

    A. NADH   B. $FADH_2$ C. CoQ   D. cytochrome *c*

29. _____ a mobile carrier that transfers electrons to complex III
30. _____ the coenzyme that transfers electrons to complex IV
31. _____ an electron acceptor containing iron

For questions 32 through 36, select the correct answer **A** *through* **F** *for the ATP produced*: (18.5, 18.6)

    A. 2.5 ATP   B. 5 ATP   C. 7 ATP   D. 10 ATP
    E. 20 ATP   F. 32 ATP

32. one turn of the citric acid cycle (acetyl CoA ⟶ $2CO_2$)
33. complete reaction of glucose (glucose + $6O_2$ ⟶ $6CO_2 + 6H_2O$)
34. produced when NADH enters electron transport
35. produced from glycolysis (glucose + O_2 ⟶ 2 pyruvate + $2H_2O$)
36. produced from oxidation of 2 pyruvate (2 pyruvate ⟶ 2 acetyl CoA + $2CO_2$)

For questions 37 through 40, consider the β oxidation of palmitic (C_{16}) *acid*: (18.7)

37. The number of β oxidation cycles required for palmitic acid is
    A. 16   B. 9   C. 8   D. 7   E. 6

38. The number of acetyl CoA groups produced by the β oxidation of palmitic acid is
    A. 16   B. 9   C. 8   D. 7   E. 6

39. The number of NADH produced by the β oxidation of palmitic acid is
    A. 16   B. 9   C. 8   D. 7   E. 6

40. The total ATP produced by the β oxidation of palmitic acid is
    A. 96   B. 106   C. 108   D. 134   E. 136

41. The process of transamination (18.8)
    A. is part of the citric acid cycle
    B. converts α-amino acids to β-keto acids
    C. produces new amino acids
    D. is not used in the metabolism of amino acids
    E. is part of the β oxidation of fats

42. The oxidative deamination of glutamate produces (18.8)
    A. a new amino acid
    B. a new β-keto acid
    C. ammonia, NH_3
    D. ammonium ion, NH_4^+
    E. urea

43. The purpose of the urea cycle in the liver is to (18.8)
    A. remove urea
    B. convert urea to NH_4^+
    C. convert NH_4^+ to urea
    D. synthesize new amino acids
    E. take part in the β oxidation of fats

44. The carbon atoms from various amino acids can be used in several ways such as (18.8)
  A. intermediates of the citric acid cycle
  B. formation of pyruvate
  C. synthesis of glucose
  D. formation of ketone bodies
  E. all of these

45. The carbon skeleton of valine enters the citric acid cycle as (18.8)
  A. pyruvate
  B. oxaloacetate
  C. succinyl CoA
  D. malate
  E. citrate

Answers to the Practice Test

1. A	2. E	3. D, B, C	4. F, B, E	5. G, B
6. A, B	7. D	8. E	9. E	10. E
11. A	12. B	13. C	14. E	15. D
16. D	17. 2	18. 2	19. 2	20. D
21. B	22. C	23. E	24. A	25. E
26. D	27. D	28. A	29. C	30. D
31. D	32. D	33. F	34. A	35. C
36. B	37. D	38. C	39. D	40. B
41. C	42. D	43. C	44. E	45. C

Chapter 18

Selected Answers and Solutions to Text Problems

18.1 The digestion of polysaccharides takes place in stage 1 of catabolism.

18.3 a. The synthesis of large molecules requires energy and is anabolic.
b. The addition of phosphate to glucose requires energy and is anabolic.
c. The breakdown of ATP releases energy and is catabolic.
d. The breakdown of large molecules is catabolic.

18.5 The hydrolysis of the phosphodiester bond in ATP releases energy that is sufficient for energy-requiring processes in the cell.

18.7 Hydrolysis is the main type of reaction involved in the digestion of carbohydrates.

18.9 The bile salts emulsify fat to give small fat globules for lipase hydrolysis.

18.11 The digestion of proteins begins in the stomach and is completed in the small intestine.

18.13 a. Pantothenic acid is a component of coenzyme A.
b. Niacin is the vitamin component of NAD^+.
c. Ribitol is the sugar alcohol that is a component of riboflavin in FAD.

18.15 In biochemical systems, oxidation is usually accompanied by gain of oxygen or loss of hydrogen. Loss of oxygen or gain of hydrogen usually accompanies reduction.
a. The reduced form of NAD^+ is abbreviated NADH.
b. The oxidized form of $FADH_2$ is abbreviated FAD.

18.17 The coenzyme that picks up hydrogen when a carbon–carbon double bond is formed is FAD.

18.19 Glucose is the starting compound of glycolysis.

18.21 In the initial steps of glycolysis, ATP is required to add phosphate groups to glucose (phosphorylation reactions).

18.23 ATP is produced directly in glycolysis in two places. In reaction 7, a phosphate group from 1,3-bisphosphoglycerate is transferred to ADP and yields ATP. In reaction 10, a phosphate group from phosphoenolpyruvate is transferred directly to ADP to yield another ATP.

18.25 When fructose-1,6-bisphosphate splits, the three-carbon intermediates glyceraldehyde-3-phosphate and dihydroxyacetone phosphate are formed.

18.27 a. One ATP is required in the phosphorylation of glucose to glucose-6-phosphate.
b. One NADH is produced in the conversion of glyceraldehyde-3-phosphate to 1,3-bisphosphoglycerate.
c. Two ATP and two NADH are produced when glucose is converted to pyruvate.

18.29 A cell converts pyruvate to acetyl CoA only under aerobic conditions; there must be sufficient oxygen available.

18.31 The oxidation of pyruvate converts NAD^+ to NADH and produces acetyl CoA and CO_2.

$$CH_3-\underset{\text{Pyruvate}}{\overset{\overset{O}{\|}}{C}}-COO^- + NAD^+ + HS-CoA \longrightarrow CH_3-\underset{\text{Acetyl CoA}}{\overset{\overset{O}{\|}}{C}}-S-CoA + CO_2 + NADH$$

18.33 When pyruvate is reduced to lactate, the NAD^+ is used to oxidize glyceraldehyde-3-phosphate, which recycles NADH.

18.35 One turn of the citric acid cycle converts 1 acetyl CoA to 2 CO_2, 3 NADH + $3H^+$, $FADH_2$, GTP (ATP), and HS—CoA.

18.37 a. Two reactions, reactions 3 and 4, involve oxidation and decarboxylation, which reduces the length of the carbon chain by one carbon in each reaction.
  b. Reaction 6 is a dehydrogenation reaction. Two Hs are removed from succinate to produce the C=C bond in fumarate.
  c. NAD^+ is reduced by the oxidation reactions 3, 4, and 8 of the citric acid cycle.

18.39 In reaction 6, a carbon–carbon double bond (C=C) is formed.

18.41 a. The six-carbon compounds in the citric acid cycle are citrate and isocitrate.
  b. In decarboxylation, a carbon atom is lost as CO_2.
  c. The five-carbon compound in the citric acid cycle is α-ketoglutarate.
  d. Reactions 3 (isocitrate dehydrogenase) and 4 (α-ketoglutarate dehydrogenase) are decarboxylation reactions.

18.43 NADH and $FADH_2$ produced in glycolysis, oxidation of pyruvate, and the citric acid cycle provide the electrons for electron transport.

18.45 The mobile carrier coenzyme Q (CoQ) transfers electrons from complex I to complex III.

18.47 When NADH transfers electrons to complex I, NAD^+ is produced (oxidation).

18.49 In oxidative phosphorylation, the energy from the oxidation reactions in electron transport is used to synthesize ATP from ADP and P_i.

18.51 As hydrogen ions return to the lower-energy environment in the matrix, they pass through ATP synthase, where they release energy to drive the synthesis of ATP.

18.53 Glycolysis and the citric acid cycle produce the reduced coenzymes NADH and $FADH_2$, which enter electron transport where they release hydrogen ions and electrons that are used to generate energy for the synthesis of ATP.

18.55 a. 2.5 ATP are produced by the oxidation of NADH in electron transport.
  b. 7 ATP are produced in glycolysis when glucose degrades to 2 pyruvate molecules.
  c. 5 ATP are produced when 2 pyruvate are oxidized to 2 acetyl CoA and 2 CO_2.
  d. 10 ATP are produced when acetyl CoA is broken down to 2 CO_2 in the citric acid cycle.

18.57 Analyze the Problem

Number of Carbon Atoms	Number of β Oxidation Cycles	Number of NADH	Number of $FADH_2$	Number of Acetyl CoA
8	3	3	3	4

  a. A C_8 fatty acid will go through 3 β oxidation cycles.
  b. The β oxidation of a chain of 8 carbon atoms produces 4 acetyl CoA units.
  c. **ATP Production from Caprylic Acid (C_8)**

Activation	−2 ATP
4 ~~acetyl CoA~~ × $\dfrac{10 \text{ ATP}}{\text{acetyl CoA}}$ (citric acid cycle)	40 ATP
3 ~~NADH~~ × $\dfrac{2.5 \text{ ATP}}{\text{NADH}}$ (electron transport)	7.5 ATP
3 ~~$FADH_2$~~ × $\dfrac{1.5 \text{ ATP}}{FADH_2}$ (electron transport)	4.5 ATP
Total	50 ATP

18.59 Analyze the Problem

Number of Carbon Atoms	Number of β Oxidation Cycles	Number of NADH	Number of FADH$_2$	Number of Acetyl CoA
18	8	8	8	9

a. A C$_{18}$ fatty acid will go through 8 β oxidation cycles.
b. The β oxidation of a chain of 18 carbon atoms produces 9 acetyl CoA units.
c. The energy from the oxidation of unsaturated fatty acids is slightly less than the energy from saturated fatty acids. For simplicity, we will assume that the total ATP production is the same for both saturated and unsaturated fatty acids.

ATP Production from Oleic Acid (C$_{18}$)

Activation	−2 ATP
9 ~~acetyl CoA~~ × $\dfrac{10\text{ ATP}}{\text{~~acetyl CoA~~}}$ (citric acid cycle)	90 ATP
8 ~~NADH~~ × $\dfrac{2.5\text{ ATP}}{\text{~~NADH~~}}$ (electron transport)	20 ATP
8 ~~FADH$_2$~~ × $\dfrac{1.5\text{ ATP}}{\text{~~FADH$_2$~~}}$ (electron transport)	12 ATP
Total	**120 ATP**

18.61 Ketogenesis is the synthesis of ketone bodies from excess acetyl CoA from fatty acid oxidation, which occurs when glucose is not available for energy, particularly in severe diabetes, diets high in fat and low in carbohydrates, alcoholism, and starvation.

18.63 High levels of ketone bodies lead to ketosis, a condition characterized by acidosis (a drop in blood pH values), excessive urination, and strong thirst.

18.65 In transamination, an amino group is transferred from an amino acid to an α-keto acid, creating a new amino acid and a new α-keto acid.

a. $\text{H}-\overset{\overset{\displaystyle O}{\|}}{\text{C}}-\text{COO}^-$

b. $\text{CH}_3-\overset{\overset{\displaystyle O}{\|}}{\text{C}}-\text{COO}^-$

18.67 NH$_4^+$ is toxic if allowed to accumulate in the body.

18.69
a. The three-carbon atom structure of alanine is converted to pyruvate.
b. The four-carbon structure of aspartate is converted to oxaloacetate and fumarate.
c. Tyrosine can be converted to the four-carbon compound fumarate.
d. The five-carbon structure from glutamine can be converted to α-ketoglutarate.

18.71 $\underset{\text{Pyruvate}}{\text{CH}_3-\overset{\overset{\displaystyle O}{\|}}{\text{C}}-\text{COO}^-}$ + $\underset{\text{Glutamate}}{{}^-\text{OOC}-\overset{\overset{\displaystyle \overset{+}{\text{NH}_3}}{|}}{\text{CH}}-\text{CH}_2-\text{CH}_2-\text{COO}^-}$

18.73 Analyze the Problem

Number of Carbon Atoms	Number of β Oxidation Cycles	Number of NADH	Number of FADH$_2$	Number of Acetyl CoA
12	5	5	5	6

  a. Five cycles of β oxidation are needed.
  b. Six acetyl CoA units are produced.
  c. ATP Production from Lauric Acid (C$_{12}$)

Activation	−2 ATP
6 acetyl CoA × $\dfrac{10 \text{ ATP}}{\text{acetyl CoA}}$ (citric acid cycle)	60 ATP
5 NADH × $\dfrac{2.5 \text{ ATP}}{\text{NADH}}$ (electron transport)	12.5 ATP
5 FADH$_2$ × $\dfrac{1.5 \text{ ATP}}{\text{FADH}_2}$ (electron transport)	7.5 ATP
Total	**78 ATP**

18.75 a. Glucose is a product of the digestion of carbohydrate.
  b. Fatty acid is a product of the digestion of fat.
  c. Maltose is a product of the digestion of carbohydrate.
  d. Glycerol is a product of the digestion of fat.
  e. Amino acids are the products of the digestion of protein.
  f. Dextrins are produced by the digestion of carbohydrate.

18.77 ATP + H$_2$O ⟶ ADP + P$_i$ + 7.3 kcal/mole (31 kJ/mole)

18.79 Lactose undergoes digestion in the mucosal cells of the small intestine to yield galactose and glucose.

18.81 Glucose is the reactant, and pyruvate is the product of glycolysis.

18.83 Pyruvate is converted to lactate when oxygen is not present in the cell (anaerobic conditions) to regenerate NAD$^+$ for glycolysis.

18.85 The oxidation reactions of the citric acid cycle produce a source of reduced coenzymes for electron transport and ATP synthesis.

18.87 The oxidized coenzymes NAD$^+$ and FAD needed for the citric acid cycle are regenerated by electron transport, which requires oxygen.

18.89 In the chemiosmotic model, energy released as H$^+$ ions flow through ATP synthase and then back to the mitochondrial matrix is utilized for the synthesis of ATP.

18.91 The oxidation of glucose to pyruvate by glycolysis produces 7 ATP. The oxidation of glucose to CO$_2$ and H$_2$O produces 32 ATP.

18.93 a. $24 \text{ h} \times \dfrac{60 \text{ min}}{1 \text{ h}} \times \dfrac{60 \text{ s}}{1 \text{ min}} \times \dfrac{2 \times 10^6 \text{ ATP}}{1 \text{ s cell}} \times 10^{13} \text{ cells} \times \dfrac{1 \text{ mole ATP}}{6.02 \times 10^{23} \text{ ATP}} \times \dfrac{7.3 \text{ kcal}}{1 \text{ mole ATP}}$
= 21 kcal (2 SFs)

  b. $24 \text{ h} \times \dfrac{60 \text{ min}}{1 \text{ h}} \times \dfrac{60 \text{ s}}{1 \text{ min}} \times \dfrac{2 \times 10^6 \text{ ATP}}{1 \text{ s cell}} \times 10^{13} \text{ cells} \times \dfrac{1 \text{ mole ATP}}{6.02 \times 10^{23} \text{ ATP}} \times \dfrac{507 \text{ g ATP}}{1 \text{ mole ATP}}$
= 1500 g of ATP (2 SFs)

18.95 a. Glucose forming two pyruvate in glycolysis yields 7 ATP/glucose.
    b. Pyruvate forming acetyl CoA yields 2.5 ATP/pyruvate.
    c. Glucose forming two acetyl CoA would yield 12 ATP/glucose.
    d. Acetyl CoA going through one turn of the citric acid cycle yields 10 ATP/acetyl CoA.
    e. The complete oxidation of caproic acid (C_6) yields 36 ATP/C_6 acid.
    f. The oxidation of NADH to NAD^+ yields 2.5 ATP/NADH.
    g. The oxidation of $FADH_2$ to FAD yields 1.5 ATP/$FADH_2$.

18.97 a. The disaccharide maltose will produce more ATP per mole than the monosaccharide glucose.
    b. The C_{18} fatty acid stearic acid will produce more ATP per mole than the C_{14} fatty acid myristic acid.
    c. The six-carbon molecule glucose will produce more ATP per mole than two two-carbon molecules of acetyl CoA.
    d. The C_8 fatty acid caprylic acid will produce more ATP per mole than the six-carbon molecule glucose.
    e. The six-carbon compound citrate occurs earlier in the citric acid cycle than the four-carbon compound succinate and will produce more ATP per mole in one turn of the cycle.

Selected Answers to Combining Ideas from Chapters 16 to 18

CI.33 **a.** An alpha-galactosidase is a hydrolase.
b. The substrate of the enzyme is the α-D-galactose unit in polysaccharides.
c. You should not heat or cook beano because high temperatures will denature the hydrolase enzyme so it no longer functions.

CI.35 **a.** Succinate is part of the citric acid cycle.
b. $CoQH_2$ is part of electron transport.
c. FAD is part of both the citric acid cycle and electron transport.
d. cyt *c* is part of electron transport.
e. H_2O is part of both the citric acid cycle and electron transport.
f. Malate is part of the citric acid cycle.
g. NAD^+ is part of both the citric acid cycle and electron transport.

CI.37 **a.** The components of acetyl CoA are aminoethanethiol, pantothenic acid (vitamin B_5), and phosphorylated ADP.
b. Coenzyme A carries an acetyl group to the citric acid cycle for oxidation.
c. The acetyl group links to the sulfur atom (—S—) in the aminoethanethiol part of CoA.
d. Molar mass of acetyl CoA ($C_{23}H_{38}N_7O_{17}P_3S$)
$= 23(12.01\text{ g}) + 38(1.008\text{ g}) + 7(14.01\text{ g}) + 17(16.00\text{ g}) + 3(30.97\text{ g}) + 32.07\text{ g}$
$= 810.\text{ g/mole (3 SFs)}$

CI.39 **a.**

$$\begin{array}{l}
CH_2-O-\overset{O}{\overset{\|}{C}}-(CH_2)_{14}-CH_3 \\
CH-O-\overset{O}{\overset{\|}{C}}-(CH_2)_{14}-CH_3 + 3H_2O \longrightarrow \\
CH_2-O-\overset{O}{\overset{\|}{C}}-(CH_2)_{14}-CH_3
\end{array}
\quad
\begin{array}{l}
CH_2-OH \\
CH-OH \\
CH_2-OH
\end{array}
+ 3HO-\overset{O}{\overset{\|}{C}}-(CH_2)_{14}-CH_3$$

b. Molar mass of glyceryl tripalmitate ($C_{51}H_{98}O_6$)
$= 51(12.01\text{ g}) + 98(1.008\text{ g}) + 6(16.00\text{ g}) = 807\text{ g/mole (3 SFs)}$

c. Analyze the Problem

Number of Carbon Atoms	Number of β Oxidation Cycles	Number of NADH	Number of $FADH_2$	Number of Acetyl CoA
16	7	7	7	8

ATP Production from Palmitic Acid (C_{16})

Activation	−2 ATP
8 acetyl CoA × $\dfrac{10\text{ ATP}}{\text{acetyl CoA}}$ (citric acid cycle)	80 ATP
7 NADH × $\dfrac{2.5\text{ ATP}}{\text{NADH}}$ (electron transport)	17.5 ATP
7 $FADH_2$ × $\dfrac{1.5\text{ ATP}}{FADH_2}$ (electron transport)	10.5 ATP
Total	106 ATP

d. $1 \text{ pat butter} \times \dfrac{0.50 \text{ oz butter}}{1 \text{ pat butter}} \times \dfrac{1 \text{ lb}}{16 \text{ oz}} \times \dfrac{454 \text{ g}}{1 \text{ lb}} \times \dfrac{80. \text{ g glyceryl tripalmitate}}{100. \text{ g butter}}$

$\times \dfrac{1 \text{ mole glyceryl tripalmitate}}{807 \text{ g glyceryl tripalmitate}} \times \dfrac{3 \text{ moles palmitic acid}}{1 \text{ mole glyceryl tripalmitate}} \times \dfrac{106 \text{ moles ATP}}{1 \text{ mole palmitic acid}}$

$\times \dfrac{7.3 \text{ kcal}}{1 \text{ mole ATP}} = 33 \text{ kcal (2 SFs)}$

e. $45 \text{ min} \times \dfrac{1 \text{ h}}{60 \text{ min}} \times \dfrac{750 \text{ kcal}}{1 \text{ h}} \times \dfrac{1 \text{ pat butter}}{33 \text{ kcal}} = 17 \text{ pats of butter (2 SFs)}$